Conserving the Enlightenment

Transformations: Studies in the History of Science and Technology
Jed Buchwald, general editor

Mordechai Feingold, editor, *Jesuit Science and the Republic of Letters*

Sungook Hong, *Wireless: From Marconi's Black-Box to the Audion*

Myles Jackson, *Spectrum of Belief: Joseph von Fraunhofer and the Craft of Precision Optics*

Mi Gyung Kim, *Affinity, That Elusive Dream: A Genealogy of the Chemical Revolution*

Janis Langins, *Conserving the Enlightenment: French Military Engineering from Vauban to the Revolution*

William R. Newman and Anthony Grafton, editors, *Secrets of Nature: Astrology and Alchemy in Early Modern Europe*

Alan J. Rocke, *Nationalizing Science: Adolphe Wurtz and the Battle for French Chemistry*

Conserving the Enlightenment
French Military Engineering from Vauban to the Revolution

Janis Langins

The MIT Press
Cambridge, Massachusetts
London, England

© 2004 Massachusetts Institute of Technology

All rights reserved. No part of this book may be reproduced in any form by any electronic or mechanical means (including photocopying, recording, or information storage and retrieval) without permission in writing from the publisher.

Set in Sabon by The MIT Press. Printed and bound in the United States of America.

Library of Congress Cataloging-in-Publication Data

Langins, Janis.
Conserving the Enlightenment : French military engineering from Vauban to the Revolution / Janis Langins.
  p. cm. — (Transformations)
Includes bibliographical references and index.
ISBN 0-262-12258-8 (hc. : alk. paper)
1. Fortification—France—History—18th century. 2. Military engineering—France—History—18th century. I. Title. II. Transformations (MIT Press)
UG429.F8L36 2004
623'.0944'09033—dc21   2003051355

10 9 8 7 6 5 4 3 2 1

to my family

Contents

Preface   ix

Acknowledgments   xiii

Introduction   1

I   The Setting

1   Early Modern Artillery Fortification   13
2   French Fortification in the Age of Vauban   39
3   Toward a Corps Royal du Génie   85

II   Life And Death in the Corps

4   Siegecraft   107
5   Desk Jobs   137

III   Crisis

6   Engineering and Its Discontents   163
7   Soldiers, Nobles, and Engineers   173
8   Thinking Wars   191
9   Science and Military Engineering   223
10   The Early Career of Marc-René, Marquis de Montalembert   263

11 The Challenge of Montalembert   281
12 The Response to Montalembert   325

IV Engineers and Revolution

13 Montalembert and the Engineers during the Revolution   361
14 The Conservative Art of Military Engineering in Old Regime France   399

Conclusions   429

Notes   435
Bibliography   491
Index   523

Preface

I first encountered the Royal Corps of Military Engineering many years ago when studying the early history of the École Polytechnique, France's most prestigious engineering school from its foundation to the present. Created during the terrible "Year 2 of Liberty," the year of the Terror, it was intended to train engineers at a time when they were in short supply and desperately needed by a republic battling for its life. Although the revolutionary dimensions of the École Polytechnique are easy to see, there were strong connections between it and the old royal school of military engineering at Mézières that perished during the Revolution. Indeed, the École Polytechnique was founded in large part to repair that loss. Mézières was one of the first military engineering schools in Europe and perhaps the finest, training the members of the equally famous and highly respected French military engineering corps. I became fascinated with these royal engineers and the solid roots of modern French engineering in the Old Regime. Here were the men who were perhaps the most modern of the servants of the absolutist monarchy, and they left their stamp on French technical education, on the whole system of grandes écoles, and on the meritocratic French civil service that has served that country so well. Who really were they, and what made them tick?

Military engineers of the eighteenth century were primarily associated with fortifications—building and attacking them. Fortifications occupied an especially prominent place in military thinking in the reign of Louis XIV, and this was exemplified in the career of his great military engineer, Vauban. In my study of the fortification courses taught in the first year of the École Polytechnique, I ran across references to a "graphomaniac" enemy of the engineering corps, who turned out to be the marquis de Montalembert. There had been a bitter polemic lasting for more than a quarter of a century between the engineers and Montalembert, a cavalryman and an amateur fortifier. How could such a polemic have been sustained for so long with the disciples and devotees of Vauban, who was a veritable cult figure not only among engineers but also among an overwhelming majority of Frenchmen? And what was behind it? These were the questions that eventually led to this book.

In the course of writing the book, I believe that I gained a better understanding of eighteenth-century engineers and engineering in France and some insight into the creative tensions generally inherent in engineering. The French Enlightenment spawned many of the creative processes that led to the modern world, and modern engineering was both a significant product and a tool of modernization. I have not been the only one attracted to the history of engineers in this place and time: Ken Alder and Antoine Picon respectively have written seminal books on artillerist-engineers and civil engineers during the Enlightenment. My contribution deals with the fortification engineers, prominent in the army and the state, who were progressive but who nevertheless displayed a stoic attachment to the values the French Revolution was to destroy.

Progress and technology being often mentioned in the same breath, it might be reasonable to assume that engineers, who are at the forefront of technological advance, are naturally inclined to change, even to revolution. The history of the French Revolution—perhaps the defining revolution in modern European history—suggests that there is some validity to this view. Henri de Saint-Simon in the nineteenth century and Thorstein Veblen in the twentieth also believed that engineers were inherently revolutionary and predicted that they would be leaders in movements that would sweep away an old world and replace it with a new and more rational one. Up till now, both have been proved wrong, and it is fair to generalize that engineers have been rather conservative in their attitudes toward society and politics, consistent with their intellectual style of problem solving that favors "conservatism" in the more specific engineering use of the word. Engineering is much concerned with safety and cost, and engineers' solutions have been incremental and specific rather than revolutionary and general.

Engineers increased in number and diversity in eighteenth-century France, but the group that enjoyed the most social prestige, if not always the most bureaucratic prominence, was the corps of military engineers—the Corps du Génie. By the beginning of the eighteenth century the corps had a structure and functions that look modern, prestige that it enjoyed from the reflected aura of Vauban, and an increasing sense of solidarity and confidence. The engineers were conservative in spite of being attuned to the discoveries and ideals of the Scientific Revolution and the Enlightenment. Their conception of the achievements of these movements saw them as compatible with an absolutist monarchical state and with a social order that, on the whole, they saw as worth protecting and consolidating. The majority of these men did not have any nostalgia for traditional ways, nor can they be considered uncritical defenders of the old order simply because it was old. They would have been indignant to be considered as regressive or even stagnant. They saw themselves as true men of the Enlightenment, whose virtues, as they saw them, they appreciated and considered worth conserving. Indeed, they saw some of the Enlightenment's positive features threatened by what they saw as reckless innovation. Being engineers of fortifications, they tried to conserve not only material fortresses but also ways of life and war that they considered to be the great civilizing results of the Enlightenment.

Acknowledgments

This book has been a long time coming, and without the intellectual and emotional support of many people it would never have seen the light of day. I wish to assure those whom it is my duty and pleasure to thank for their indispensable help—and who may think that the appearance of this book with its many flaws is a regrettable outcome of that help—that all the flaws are mine. There would have been many more flaws without their participation. I am deeply thankful to you all. Because of the protracted gestation of this work and the failures of my memory, I will no doubt omit some who should be in the following list, and for some of you my thanks comes too late because you are no longer in this world. I ask you all to accept my apologies.

For encouragement and support in beginning this work, I thank Patrice Bret, Charles Gillispie, Emmanuel Grison, and Trevor Levere. Jed Buchwald deserves my thanks for encouragement in finishing this work, as do the members of my intellectual home, the Institute for the History and Philosophy of Science and Technology at Victoria College in the University of Toronto. Eda Kranakis, Trevor Levere, John Parry, Antoine Picon, Alex Roland, Richard White, and Polly Winsor read all or parts of earlier versions of the manuscript, and their suggestions improved it immeasurably. Numerous archivists and librarians facilitated my work with patience and diligence. Among them are Monsieur Berthon, Archivist at the Academy of Sciences; Claudine Billoux, Archivist at the École Polytechnique, Palaiseau; Benoit Cameron, Librarian of the Royal Military College of Canada, Kingston; M. Lordereau, Conservateur en Chef of the Bibliothèque municipale de Besançon; Francine Masson, Director of the Central Library of the École Polytechnique; Nelly Lacrocq and Nicole Salat, Archivists at the Archives de l'Inspection du Génie at the Service historique de l'armée de terre, Vincennes; Galen R. Wilson, Manuscript Curator, William L. Clements Library, University of Michigan, Ann Arbor; and other archivists at the Archives nationales, Paris; the Bibliothèque de l'École Nationale des Ponts et Chausses, Paris; the Service historique de l'armée de terre; the Musée des Plans-Reliefs at Les Invalides, Paris; National Historic Parks and Sites (Canada); the Bibliothèque Nationale, Paris; the Université de Montréal; and the University of Toronto. For encouragement, ideas, and other services and help I am grateful to Ken Alder, Bruno Belhoste, Evan Bentz, John Bosher, David Buisseret, Scott Campbell, Michel Decker, Jean Dhombres, Per Dahl, Edwin L. Dooley Jr., Nicolas Faucherre, Bruce Fry, Annie Geffroy, Anne Godlewska, Dmitri and Irina Gouzevitch, John and Marie-Thérèse Greenberg, Emmanuel Hublot, David B. Ingram, Monsieur and Madame Maudet, Seymour Mauskopf, the late General Nicolas, James Pritchard, Antoine Prost, Martin Reuss, Aldemaro Romero, Joël Sakharovitch, Todd Shallat, Brett Steele, Mary Thomas, Philippe Truttmann, and Steven Walton.

For financial aid I am indebted to the Social Sciences and Humanities Research Council of Canada and Victoria College. The Department of National Defence of Canada provided photographic services. Permission to reproduce illustrations was graciously granted by the Service historique de l'armée de terre at Vincennes, the University of Michigan Press, Henri Chabrier of Charles-Lavauzelle publishers, Athanasios Migos, editor of *Fort*, Bruno Belhoste, Jean-Pierre de Monza of Jean-Pierre de Monza publishers, the John Robarts Library of the University of Toronto, and the Massey Library of Royal Military College.

I would also like to thank Paul Bethge at The MIT Press.

My family bore a heavy burden in the writing of this book and helped me surmount the many obstacles to its completion. They are an indispensable part of it, and I look upon them almost as co-authors to whom it would be inappropriate to express thanks. Let them accept my dedication instead.

Conserving the Enlightenment

Introduction

In 1793 the French Revolution was at the nadir of its fortunes. Internal revolt and foreign invasion in all parts and on all frontiers of the country menaced the new republic with imminent annihilation. In this supreme moment of crisis, a desperate legislature that had executed a monarch and was pathologically mistrustful of a strong executive power gave virtually all power to a Committee of Public Safety, which instituted a regime of terror and dictatorship that succeeded in the miraculous feat of saving the republic. Robespierre, perhaps the Committee's most famous member, has given it the aura of superhuman energy, sanguinary ruthlessness, and intransigent republicanism. The Committee had all these features, but just as Robespierre was only one of its twelve members, so too the National Convention, which had given the Committee supreme power, expected and got more from it than energy, ruthlessness, and loyalty to the republic. It expected efficiency, and it got it—not from Robespierre, Saint-Just, Couthon, and Barère, the *gens de la haute main* who determined policy and ran the country and its armies, nor from *révolutionnaires*, such as Billaud-Varenne or Collot d'Herbois, who imposed the writ and the faith of the Revolution on Paris and the provinces, but from the *gens d'examen*, such as Lazare Carnot and Claude-Antoine Prieur de la Côte-d'Or.

Carnot and Prieur were the specialists and technicians who were to perform the prodigy of bringing order, discipline, and efficiency—above all efficiency—out of the turmoil of the Revolution. Both were military engineers who had been selected for the Committee of Public Safety primarily because of their technical training. They did their job well and survived both the Committee and Robespierre and his friends. When attacked later in the National Convention as accomplices of the vanquished Robespierre and his fellow Terrorists, Carnot was saved by his reputation as the "Organizer of Victory," a title that has remained in the history textbooks of his country.

Prieur's and Carnot's presence on the Committee is more a tribute to their membership in what used to be the Royal Corps of Military Engineering than to their personal qualities, relatively little known at the time they were elected. Military engineers had an enviable reputation for competence and intellectual superiority that went back to the days of Louis XIV's great engineer Vauban. They also had a reputation for being more loyal to the Revolution than their brother officers in the other arms—allegedly because, in Lafayette's words, they were from a military corps "necessarily composed of enlightened and reflective men," more rational and exempt from prejudice because of their rigorous training in the sciences. Their professional reputation as intelligent and indispensable servants of the state had survived the Revolutionary hostility to officers of the old royal army as well as the attacks of critics during the Old Regime.

The fact that the military engineering corps survived the upheavals of the Revolution not just intact but enhanced might indicate that it was not merely robust but thoroughly modern and solidly established. This was not so, however. Just two decades before Carnot joined the Committee of Public Safety, Louis Le Bègue du Portail, who went on to command the military engineers in the American Revolutionary Army and the siegeworks at Yorktown, complained that military engineers were not accorded the respect they deserved and that they should even abandon the title of "engineer." The Enlightenment and the years before the Revolution were not a period of confident stability and increasing prestige for French military engineering. They were in fact a time of turmoil and intense questioning about their role and their identity. Although connected in some ways with the general crisis of the state and the army on the eve of the Revolution, the difficulties of the military engineers were also caused by conflict over the nature of engineering.

This book will describe the tribulations of the military engineers in France during this period. In doing so, I hope it will improve the understanding of the evolution of modern engineering generally. Anne Blanchard and before her colonel Augoyat have given fine institutional histories of the Corps du Génie, and the innumerable biographies of Vauban have alluded to this history. My aim is to go beyond this work by looking more closely at the internal workings of the corps of military engineering in order to better understand its ideologies, its scientific and technical outlook, and its double role in inserting some military values into engineering in general and some engineering values into modern military establishments.

French military engineers of the Enlightenment left a significant legacy to military engineers and armies everywhere, as well as to modern engineering and engineering education. Indeed, France is a prime locus and the Enlightenment is a prime period in which to study the emergence of engineering as we know it today. The influence of French military engineering was felt from West Point to St. Petersburg. In spite of its characteristically French features, it had a universal aspect that is still recognizable in engineering everywhere today.

It is no revelation that engineering is one of the most important professions of our time. In purely quantitative terms it is the largest profession in the industrialized world, and its impact on the development of technology in a technological world is obvious. Yet it is not a profession like many others, particularly the traditional liberal professions like law and medicine. Both of these conform more closely to the models of professions as traditionally defined by most sociologists and generally conceived by most laymen. The somewhat static and Platonic models of professions, which are perceived to transcend time and space, have had great influence in spite of their acknowledged inadequacies. But they break down when dealing with engineering. More recent work stressing the dynamic nature of professions is more useful in understanding engineering and its history.

That engineering is a heterogeneous profession has been recognized by many scholars. Not only does it have functional divisions, such as the traditional civil, electrical, mechanical, and chemical branches; it also has a host of new ones. Engineers work in a number of different contexts, ranging from bureaucracies to self-employment as independent consulting engineers. They have been less successful than other professionals in preventing interlopers from practicing their profession (one of the more prominent features of traditional professionalism), and many have acceded to managerial posts where their identification with engineering has withered and even disappeared.

This fluidity is perhaps not surprising. The profession of engineering, like the technology it serves, has been continually changing throughout its history. Social, technical, administrative, and political considerations have also molded the configuration of the engineering profession. They have defined and determined the periods of flux and stability that it has undergone in its history.

The story presented here goes beyond engineering to look at technology more generally and consider the problem of innovation in the context of a technical bureaucracy. Here again France is a prime object of study. It was, of course, the country that gave bureaucracy its name and some of its earliest examples. But it was a pioneer in the development of technical and scientific bureaucracies in institutions like the Academy of Sciences, the military engineering corps, the civilian Roads and Bridges (Ponts et Chaussées) administration, and others. The military engineers of the French Enlightenment discussed here provide an early example of a bureaucracy confronting strong external demands for change. Their reaction to these demands provides insight into the intrinsic constraints to which large-scale technologies are subject in the face of novelty. This study will contribute to an extension of Thomas Hughes' concept of technological momentum by concentrating on the related concept of technological inertia.

Not only does this story lead beyond engineers, the military, and technology; it leads to even more general issues affecting French society. Ken Alder has argued that engineers (actually artillerist-engineers) played a major role in shaping the society that emerged from the French Revolution. In his story, artillerist-engineers were remarkably successful, in spite of their invisibility, in profoundly affecting the nature of the post-Revolutionary state in France. Not only did they supply the weapons and the organization for the military success of the Revolutionary and Napoleonic armies, but they imposed their ideology—an ideology of rationality, state service, and action based on deductive thinking and abstract models of artifacts and organization—on the French state, if not on the marketplace. This ideology led to an ultimately unsuccessful attempt to institute the production of uniform interchangeable parts in the arms industry and a more successful attempt to promote a rational, bureaucratic state maintaining a strong mediating presence in industry. The artillery officer Napoleon Bonaparte had many colleagues who worked in less spectacular but just as effective ways in building modern France.

Yet engineers were not an entirely coherent community, as Alder himself admits for his artillerist-engineers. In this book, I will look at a broader and more traditional community of engineers, both in denomination and function: the Royal Corps of Military Engineering of the French army. Alder's artillerist-engineers, with their interests in the mechanical trades and metallurgy, correspond more closely to the mechanical engineers who emerged out of the Industrial Revolution, and they were a minority in Enlightenment France. This was a modernizing yet pre-industrial society, and the more representative Corps due Génie, along with engineers in the civilian Corps des Ponts et Chaussées, were mainly concerned with problems of infrastructure and large-scale construction. There were differences between the military engineers and the civilian engineers, as well as further differences with the artillerist-engineers. In pre-Revolutionary France, the military engineers could be crudely labeled "conservative." But conservative does not mean reactionary. They were men of the Enlightenment with a faith in science and its positive effects on society. They saw their work as a slow and steady construction of a civilized society in which even warfare would be civilized by their art. For them, their society was worth conserving and protecting from the vicissitudes of chance, of war, and from the adventurism of ill-digested novelty.

The engineer at work is an elusive creature. Not given to detailed and public musing on the nature of his work, whether by inclination or (as frequently was the case in the military context) by formal prohibition, the military engineer must be sought in often ephemeral technical papers, memoranda, engineering texts, and syllabi, or in the pronouncements of champions (whose representativeness must always be evaluated carefully). What Antoine Picon has called the "technical rationality" of engineering is equally elusive: in spite of similarities, it is not the same as the habits of mind of scientists, the esprit géométrique of the mathematically inclined, or the traditions of architects and artisans of an earlier age. This technical rationality was evolving rapidly during the Enlightenment.

My point of entry into this complex subject will be the bitter polemic that raged during the last quarter of the eighteenth century between the corps of military engineering and Marc-René, marquis de Montalembert (1714–1800). The narrative of this polemic reveals many things about French military engineering during the Enlightenment and about engineering professionalism more generally. It supports the view that engineering has never been as coherent a profession as the other classical professions—even in France, where professions such as military engineering were created by administrative or political fiat. It also shows that the epistemic space and the professional niche of engineering have contracted and expanded in multiple directions, much like the shifting frontiers of Louis XIV's state during the fortress warfare of the seventeenth and eighteenth centuries. Subsequent episodes in the history of engineering strongly suggest that consolidation of the profession was never permanent and that building and preserving professions, including that of military engineering, is a dynamic and ceaseless process.

Montalembert and the polemic he initiated and sustained give us a prime vantage point from which to examine the complexity of an engineering corps that was often silent and fragmented in more peaceful and less challenging times. Polemics sometimes lead to hardening and sharpening as well as occasionally to simple positions that are defended with vigor. In a fight over essentials, protagonists draw together and submerge finer differences. Essentials become more starkly visible, but the obverse of this advantage is that nuance and contrast can be lost. French military engineers were never a homogeneous group in spite of an undeniable esprit de corps; over time, differences among them and their ideas on professionalism and technology existed and to varying extents and degrees. Although there were those who claimed that the corps of military engineers spoke with one voice, sometimes through a single person, there were contradictions and fissures within it. Being aware of the continuing tension between competing attitudes within the corps is as important for acquiring historical understanding of the corps as is seeing the common front that is usually seen in public and is especially visible in polemics.

Three great debates on technology and tactics agitated the French army during the 1770s and the 1780s. One was over a new kind of artillery proposed by Jean-Baptiste Vacquette de Gribeauval (1715–1789). Another was over the merits of different infantry battle formations in combat. Finally, there was the debate over the virtues of different kinds of fortifications, in which Montalembert was the focal person. He is, however, often neglected by military historians, although mentioned in most histories of fortification as the inventor of a system of fortification that rejected the principles of Vauban.

The fact that Montalembert, although not a military engineer himself, was able to exert such pressure on military engineering says something about that profession. Its occupational domain, its expertise, and its very legitimacy were both more fluid and more fragile than those of the other combat arms. The struggle with their critics therefore throws the features of the corps into relief more strongly. The intensity of Montalembert's attacks required the military engineering corps to articulate a defense against the interloper. As well as counterattacks on those perceived as potential threats and obstacles to the mission of the military engineering corps, this resulted in more than the usual amount of soul searching regarding the nature of that mission. The arguments between military engineers and Montalembert generated not only much heat but also some light, and from this the historian can benefit.

Since much of the business of military engineers was fortification, which involves design and planning, the contrast between Montalembert's systems of fortification and those of the engineers illuminates the science and the technology on which design and planning were based. Montalembert and certain military engineers went further, claiming that fortification was a distinct science. There was an attempt to elaborate the basis for that rather nebulous claim in the course of the polemic.

The polemic with Montalembert reveals the real strains and ambiguities within the community of military engineers. This sharpens our understanding of this exemplary group of technologists at the dawn of an industrial era. Montalembert was no ordinary enemy who could be easily brushed aside or ignored, as most hatchers of new schemes in fortification could be. His influence, the degree of elaboration of his systems, and his energy in fighting for them posed a serious threat to the military engineers, and they responded with exceptional vigor.

Montalembert failed to get his way in fortification. In the artillery, Gribeauval, after a brief eclipse, returned in triumph to institute his reforms. In the line army, the reformer Jacques-Antoine-Hippolyte, comte de Guibert (1743–1790), left a legacy that was passed on to Revolutionary and Imperial generals, including the greatest of them all—Napoleon. The case of Montalembert, however, provides proof that the tide of reform in the army was not unstoppable. He was rejected not only before but also during and after the Revolution, in spite of the respect for his ideas abroad.

What enabled the engineering corps to triumph over its nemesis? One part of the answer is that the same forces that led to the triumph of reformers in the artillery led to the triumph of conservatives in military engineering. The Montalembert episode shows solitary genius, no matter how well placed in networks of patronage, losing in the struggle with the managers of a technological system, no matter how effectively subordinated to administrative fiat and peremptory orders from above. Not just two systems of fortification confronted each other, but two different kinds of technology. In spite of its originality and novelty of conception, one was a technology of the past; in spite of its rather hidebound inertia—and to some extent because of it—the other was a technology of the future. The future, we know, as always in history, won, but it was created in a struggle both with the past and with a concept of a different kind of future. This struggle did not have any single or inevitable outcome; in the event, however, it helped change how wars were imagined and fought and how military engineers saw their role and the roles of other kinds of military officers and other kinds of engineers.

This book is divided into four parts, each with a brief introduction.

Part I presents the background to the story: the birth of classical artillery fortifications in Italy and their transmission to France (chapter 1), the beginnings of a bureaucratic-military profession of military engineering under Vauban (chapter 2), and the evolution of this profession toward coherence and respect up to 1776, when its members officially became part of the Royal Corps of Military Engineering (chapter 3).

Part II looks at the daily life, in wartime (chapter 4) and in peacetime (chapter 5), of the typical eighteenth-century military engineer.

Part III examines the post-1776 crisis of military engineering—a crisis that was provoked in large part by the attacks of Montalembert. Chapter 6 attempts to position engineers in the technological world and argues that there is always tension within engineering because engineers are at the interface of technology and society. Chapter 7 examines the social context in which French military engineers functioned, especially their place in an army with an aristocratic officers corps. Chapter 8 deals with the debates on military doctrine, which were particularly vigorous while military engineering was undergoing a crisis. Chapter 9 looks at the birth of something that many scholars have attributed to the nineteenth century: an engineering science. The three chapters that follow describe the early career of Montalembert (chapter 10), the challenge he posed to military engineers with his new kind of military architecture (chapter 11), and the response of military engineers, especially in the forum of the Royal Academy of Sciences (chapter 12).

The polemic with Montalembert dragged on into the period of the French Revolution. It lasted much longer than the debates over artillery and field tactics. Part IV looks at the end of the final stage of the polemic (chapter 13) and concludes with a chapter arguing for the essentially conservative nature of French military engineers (chapter 14).

At the end of the eighteenth century, the military engineering corps of the French army had essentially assumed the form of a service corps. It increased in size, diminished its expectations, and lost its centrality in strategic thinking, but it had found stability and acceptance in the army. In doing so, it left an important legacy not only to military engineering (engineering service corps are essential to modern armies) but also to engineering in general. It had also achieved a degree of stability. The story of how all this happened reveals some of the protean and dynamic nature of engineering—a nature often obscured by its concrete creations, whose materiality overwhelms our sensitivity to the life behind them.

I
The Setting

The military engineering of the Enlightenment grew out of radical changes in military technology—and occupational changes accompanying them—that occurred during the Renaissance. Gunpowder weapons (already in use, albeit inefficiently, since the fourteenth century) posed new problems for those using them and for those defending against them as they grew more effective. Gunpowder weaponry also facilitated transformations in social and occupational groups. It all contributed to the elaboration of the classical bastioned artillery fortification that came to be known as the *trace italienne*, a label that indicates the geographical locus of many of these changes and whose language indicates the land that domesticated the new artillery fortification most successfully and gave it much of its jargon. The modern military engineer evolved from these changes.

The Renaissance, which saw the social elevation of the architect and the artist as well as the intellectual elevation of the soldier, also witnessed a revival of interest in classical mathematics and an attempt to use mathematics for practical purposes, such as engineering design, cartography, perspective, and castrametation. Mathematics could mean the "practical geometry" or the "constructive geometry" of the Middle Ages, the purified Euclidean geometry and Archimedean mathematics that appealed to Galileo, or simply the common sense and ingenuity of the enlightened engineer or artist. It was this confusion that sometimes made mathematics a valuable weapon in the arsenal of arguments used by military engineers, among others, to exalt their art and give it dignity. From an early period in the history of military architecture, some of its practitioners made what seems to us the naive claim that the use of mathematics in fortification made it possible to create a science of impregnability. The mathematical virtues of demonstrability and certainty were mapped onto the new methods of fortification as security and predictability in war. Never taken seriously by practical fortifiers and soldiers, the idea of security through science nevertheless retained a place in the dreams of armchair fortifiers and sometimes even in the rhetoric of military engineers.

Security through science was not the only idea that Enlightenment military engineering inherited from the Renaissance, and for this reason I will give an overview of fortification in the Italian Renaissance and the implantation of the new methods in France in the sixteenth century. It will necessarily be superficial here. An entire book would be necessary to do justice to either of these two subjects. Another book would be required to untangle the complex genealogy in the Italian Renaissance of eighteenth-century French military engineers. They had many progenitors: artists, architects, soldiers, artillerists, government officials, the traditional master masons of the Middle Ages, builders of ingenious war machines, and what could anachronistically be called project managers.

Chapter 1 will deal with the origins and early growth of bastioned fortification. It will look at some of the motivations for the particular form of this kind of fortification and the occupational groups that were involved with it. By the middle of the sixteenth century there was already an abundant literature on the subject of fortification, and military architecture was perceived as an autonomous branch of architecture. The chapter will also look at the migration of the new methods of fortification across the Alps to France. Leonardo da Vinci was not the only one lured by French kings to ply his trade away from Italy; there were many others, and most of them had the skills of the military engineer. It was not long before local engineers, in France and elsewhere, emulated Italian models and consolidated an indigenous tradition of fortification. At the end of the sixteenth century, Jean Errard de Bar-le-Duc was publishing treatises in French on the new fortification and was a distinct and respected voice on the subject.

Chapter 2 looks at the colossal figure of Vauban, his immediate predecessors, and his political masters. Vauban looms so large in the story of bastioned fortification in France that he throws his successors, contemporaries, and predecessors into the shade. Indeed, historians' fixation on his achievements has been detrimental to the study of other aspects of the history of bastioned fortification. Blaise Pagan and Antoine de Ville were original and competent fortifiers in their own right, and Vauban adopted Pagan's oeuvre as the core of his own military architecture. Moreover, Louis XIV and his Minister of War Louvois were not merely effective and intelligent supporters of Vauban but can be considered to have been true collaborators. After Vauban's death, Louis de Cormontaigne enjoyed almost as great a reputation in the French military engineering corps. Yet in the century that followed him Vauban's ideas resonated, not only in France but elsewhere, and he was a paragon of the ideal fortifier even to his enemies. The eighteenth-century debates on fortification discussed later in this book all ultimately lead back to Vauban. Vauban was more than a fortifier or even a master of siegecraft or a superb organizer. He was all those things, but he also had a global vision of France as a defensible economic and geographical polity that was highly influential not only through the medium of his own writings and activity but also through engineers' devotion to his memory and his methods.

Chapter 3 deals with the organization that Vauban created, which was, perhaps more than anything else, the invisible ingredient in the success of French military engineering during and after his lifetime. It was difficult to impose order on the baroque administrative landscape of French military engineering. Neither Vauban nor his successors ever succeeded entirely, although they succeeded to an immeasurably greater degree than their rivals in other countries. The institutional history in chapter 3 shows that the corps of military engineering in the French army, despite its success under Vauban and after him, was evolving toward being a remarkably modern and bureaucratically effective organization. Nevertheless, at the same time, it was fragile and vulnerable to the crisis that shook it just as French society was entering its pre-Revolutionary phase.

1
Early Modern Artillery Fortification

Duke: ... you said that you have never fired artillery or musket; and he that makes a judgment of a thing whose effect he has not seen or experienced is deceived the most part of the time, because only the eye is that which renders us true testimony to the thing conceived.

Niccolo: It is very true that the exterior sense speaks the truth about things in particular, but not of universals, because universals are subject only to the intellect and not to any sense.

—Niccolò Tartaglia, *Quesiti et Inventioni Diverse,* Venice, 1546[1]

The human tendency to use natural and artificial obstacles for protection against potential threats from animals or other humans is no doubt linked to our most basic instincts. Relatively elaborate fortifications already appear in prehistory: hill forts and the impressive fortifications of Jericho in the eighth millennium B.C. were followed by the more complex fortifications of the first cities in Mesopotamia.[2] In the parlance of the eighteenth-century French fortifiers, fortification was the "conservative art" without which civilization would have been unthinkable in a threatening environment of envious and hardier nomadic neighbors as well as civilized neighbors competing for resources and power. Long before Mycenae and Troy, fortification had become sophisticated in structure and function, providing strongpoints and citadels for dominant elites and protection for urban populations.

In the late Hellenic and Hellenistic world, fortifications became even more sophisticated; so did methods of attack of these fortifications. The effectiveness of Roman siegecraft and the extended period of peace in the Mediterranean area led to a certain decadence in fortification. Yet the basic techniques and methods were not forgotten and the fortifications of the medieval period in Europe made use of this knowledge. The essential principle of fortification was to create a solid physical obstacle to escalade by the attacker. It was also important to prevent attackers from damaging this obstacle or undermining it. In practice this meant high walls, often with a ditch before the wall, and overhanging battlements (machicolations) from which defenders could prevent attackers from getting near the base of the wall. The walls were thick and solid to counter medieval artillery in its various forms, such as trebuchets. At intervals along the walls, round or square towers protruded from them to provide cross-fire and platforms for concentrations of defending soldiers. Within the perimeter of the walls there was often a donjon providing a redoubt for the defenders if the walls were breached.

Medieval fortifications were usually adequate to repel the attackers who besieged them, and more often than not successful sieges were accomplished by blockade and starvation of the garrison rather than systematic siegecraft. This in itself required a tenacity and logistical organization that was often beyond the capacity of the feudal host, so it is fair to generalize that in the perpetual struggle between offense and defense, the latter held the upper hand during the medieval period. Indeed, this superiority, with the advantages it gave to local powers, could be seen as one of the primary supports of the feudal system in Medieval Europe. When Edward III besieged Calais at the beginning of the Hundred Years' War, its burghers succumbed to starvation only after eleven months, and in 1429 the siege of Orleans was lifted after six and a half months when Joan of Arc broke through the besiegers with supplies.[3] But by the end of the Hundred Years' War, English castles in Normandy and Aquitaine were falling like ripe fruit to the French, thanks to the considerably more effective siege artillery operated by the Bureau brothers. This artillery was characterized by relatively light and mobile cannon firing cast-iron balls instead of stone ones and operating in concert in batteries. These cannon breached walls and radically tipped the balance away from defense to offense at the end of the fifteenth century.

Gunpowder Artillery and Fortification

Artillery had been used in sieges from antiquity and throughout the Middle Ages, but it had been relatively ineffective in breaching walls that were solidly built and thick enough. Gunpowder radically changed artillery in Europe. Weapons that hurled projectiles from a distance by means of mechanisms that used the force of gravity (as in the counterweight trebuchet of the Middle Ages) or the stored energy in twisted fibers (such as the catapults of antiquity) were replaced by cannon that used the chemical energy of exploding gunpowder to propel projectiles from metal tubes with greater force than ever before. Appearing at the beginning of the fourteenth century and used both in battle and in sieges, cannon gradually became more effective with improvements in metallurgy and the quality of the gunpowder, which was used both in artillery and eventually in undermining walls. High, straight walls intended to stop escalade were now vulnerable, and the campaigns of the final years of the Hundred Years' War were a potent demonstration of the new state of affairs.

The spectacularly successful combination of the cast-iron cannonball, improved gunpowder (both as a propellant and as an explosive in mines), and lighter but more numerous, more mobile, and more rapidly firing guns, had a devastating and shocking effect when a French army erupted into Italy in 1494. The promenade militaire of the army of Charles VIII through Italy saw fortresses that had held out for seven years fall in eight hours.[4] Both contemporaries and subsequent historians have attached considerable significance to this event for the history of fortification. The wealthy, technically advanced, but politically fragmented states of northern and central Italy were subjected to a period of intense warfare as Hapsburgs and Valois contended for supremacy in the peninsula. A rapid period of development in military architecture led to transformation of the traditional feudal fortifications. By the middle of the sixteenth century, not only had new kinds of fortifications appeared, but there was a proliferation of treatises on the subject that described not only the basic features of what was to be classical artillery fortification for almost three centuries but also many of its variations.

The prime identifying feature of the new artillery fortifications was the pentagonal bastion that protruded from the curtain wall. With a much larger area than the old medieval towers, permitting the mounting of numerous artillery pieces, the angular bastions and walls were now much thicker and noticeably lower than the walls of feudal castles. Ditches became wider. Lines of fire could sweep all faces of the walls and the bastions, leaving no "dead ground" (areas invisible to the defending gunners). Eventually, multiple defensive perimeters and detached outworks appeared. No longer was escalade the main threat to the defender; it was the powerful effects of the new artillery that had to be thwarted and countered. Defense in height gave way to defense in depth.

The view that the appearance of the angular bastion was directly caused by the French invasion of Italy is attractive. The explosive development of military architecture and the writings on this subject in the immediate decades after the invasion tempt one to use the much-overworked word 'revolution' to describe this phenomenon. There is evidence, however, that the angular bastion, or forms approaching it, appeared before the invasion both inside and outside Italy. Contenders for the honor of being the first bastion can be found from Mont Saint Michel to Rhodes. John Rigby Hale has claimed that the earliest examples of the new fortification appeared in Italy in the middle of the fifteenth century, decades before the French invasion (whose importance therefore lay

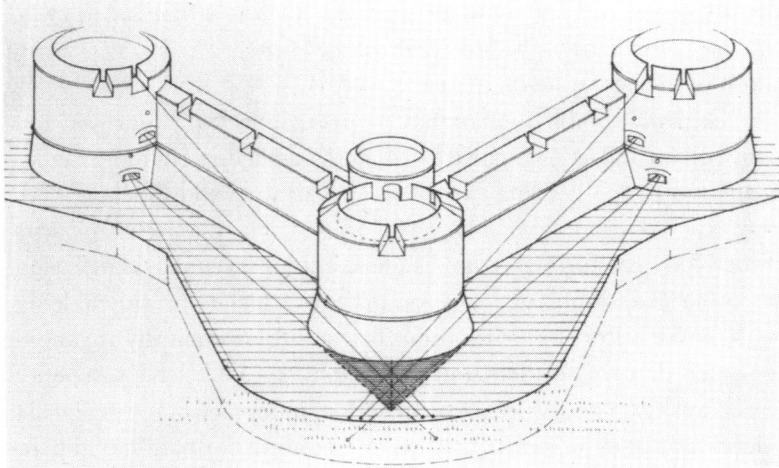

Figure 1.1
Sketch of a medieval curtain wall, with towers and lines of fire to illustrate dead ground. Source: Faucherre 1986, p. 13.

in radically accelerating a trend that was already well established).⁵ This would indicate that the gradual improvement of siege artillery was already being countered by changes in fortification in a process that was more drawn out than hitherto believed.

This view of the history of artillery fortifications suggests a simple scenario of technological challenge and response. A technical challenge is presented and overcome or at least neutralized to some extent by technical innovations. Practitioners such as soldiers and traditional builders are seen as the main vectors of this occurrence. This is undoubtedly true to some extent, but it does not tell the whole story.

There are also strong indications that, besides inadequacy of the old, other considerations played a significant role in the development of the new in fortification. Aesthetics, urban planning, humanist ideals of the duties of architects and builders, anthropomorphic analogies, and changing ideologies of warfare and government all contributed, along with an increasingly manifest tendency to apply mathematics to all spheres of nature and human activity. For example, in 1461 Antonio Filarete proposed a star-shaped city that, though never built exactly, was an archetype of a common motif of urban planning for fortresses on flat terrain such as Francesco Paciotto's citadels at Turin and Antwerp in the 1560s and Vincenzo Scamozzi's fortress of Palmanova in 1593.⁶ It is almost certain that Filarete was not motivated by practical considerations of designing a fortress to withstand gunpowder artillery: mathematical symmetries and

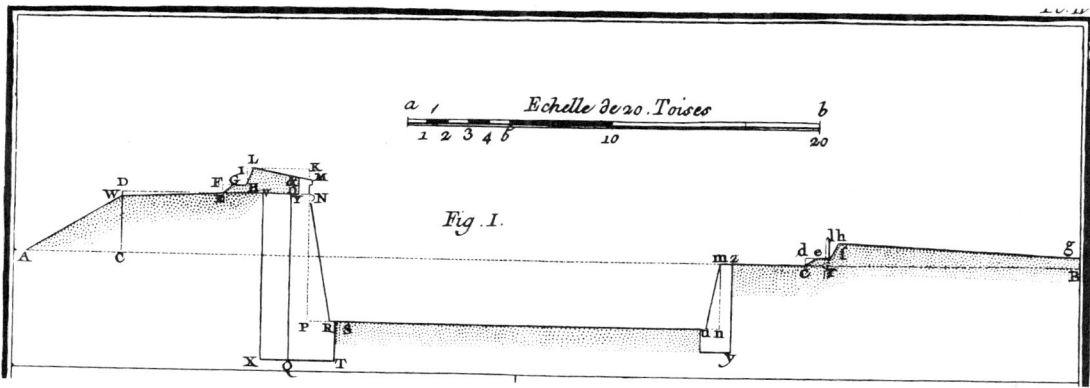

Figure 1.2
Section of a bastion, ditch, and scarp. Ground level: *AB*. Interior talus, or slope, of rampart: *AW*. Terreplein (surface of rampart behind parapet where guns are mounted): *WE*. Talus of banquette (raised way behind rampart whence soldiers fired): *EG*. Banquette: *GH*. Interior side of parapet: *LH*. Top part of slope of parapet: *LM*. Counterfort (or supporting buttress): *VYQX*. Revetment (or masonry facing) of rampart: *MNRSTQY*. Scarp (or inner wall of ditch): *NR*. Cordon (projecting course of stone on wall): *N*. Ditch: *Su*. Counterscarp (or outer wall of ditch): *um*. Revetment of counterscarp: *my*. Covered way: *mc*. Banquette of covered way: *ef*. Interior side of parapet of covered way: *fh*. Glacis: *hg*. Palisade of covered way: *hf*. Source: Diderot and d'Alembert 1751–1765.

aesthetic considerations were undoubtedly stronger.[7] Some scholars have gone so far as to argue that they were even more important than practical concerns of soldiers and builders. The new artillery did not impose a single and exclusive form of a new military architecture. There was in theory a whole spectrum of solutions to the problem of the disrupted balance between offense and defense in fortification that ranged from acceptance and reinforcement of traditional methods with only minor modification to questioning the very necessity of fortification.

It was possible, at least in an initial phase, to try to keep the old methods by increasing the massiveness of the works and improving the quality of workmanship with thicker and better-built walls. The strength and inertia of traditional craft practices are natural obstacles to new technologies. Italian builders had a propensity for stone, even when soldiers pointed out that earth packed behind revetments (retaining walls) withstood artillery better, and they stubbornly resisted using the new material for some time.[8] Hale also mentions a certain marking of time in the development of the bastion between ca. 1440 and 1460.[9] The massive piles of Imola, Pesaro, and Senigallia, with their round bastions and old medieval machicolations, are an indication that there was

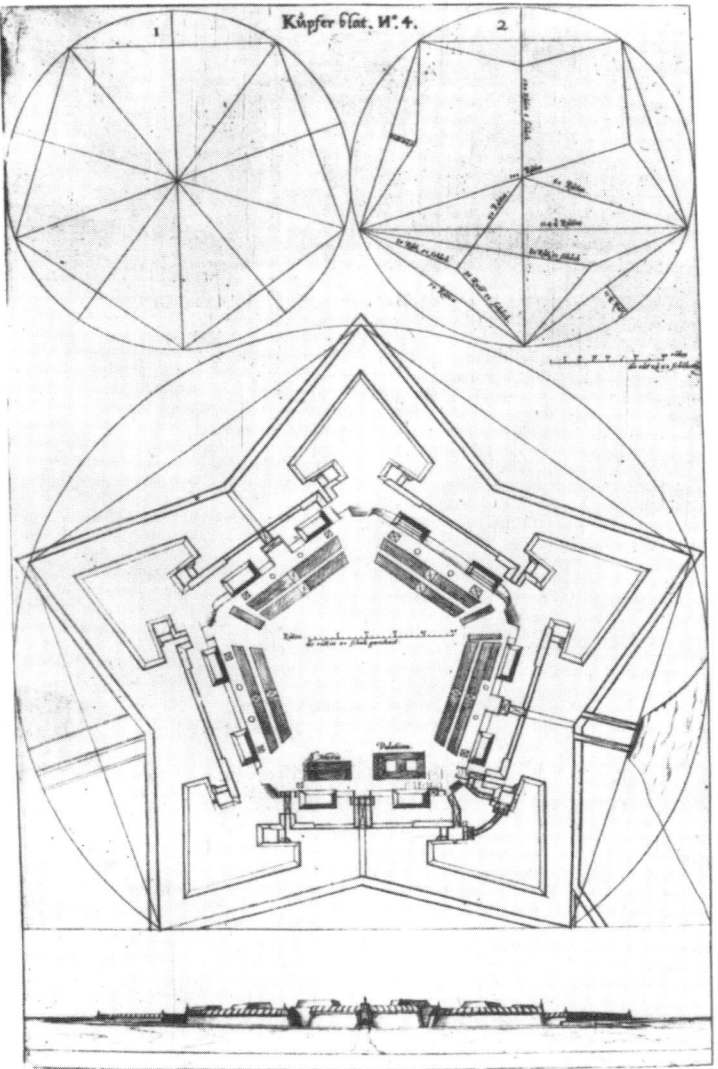

Figure 1.3
The citadel at Antwerp, built between 1567 and 1569 by Francesco Paciotto of Urbino (1504–1576). This plan of this perfect pentagonal fortress was made by an equally famous military architect, the German Daniel Specklin (1536–1589), in 1589. Source: Specklin 1589.

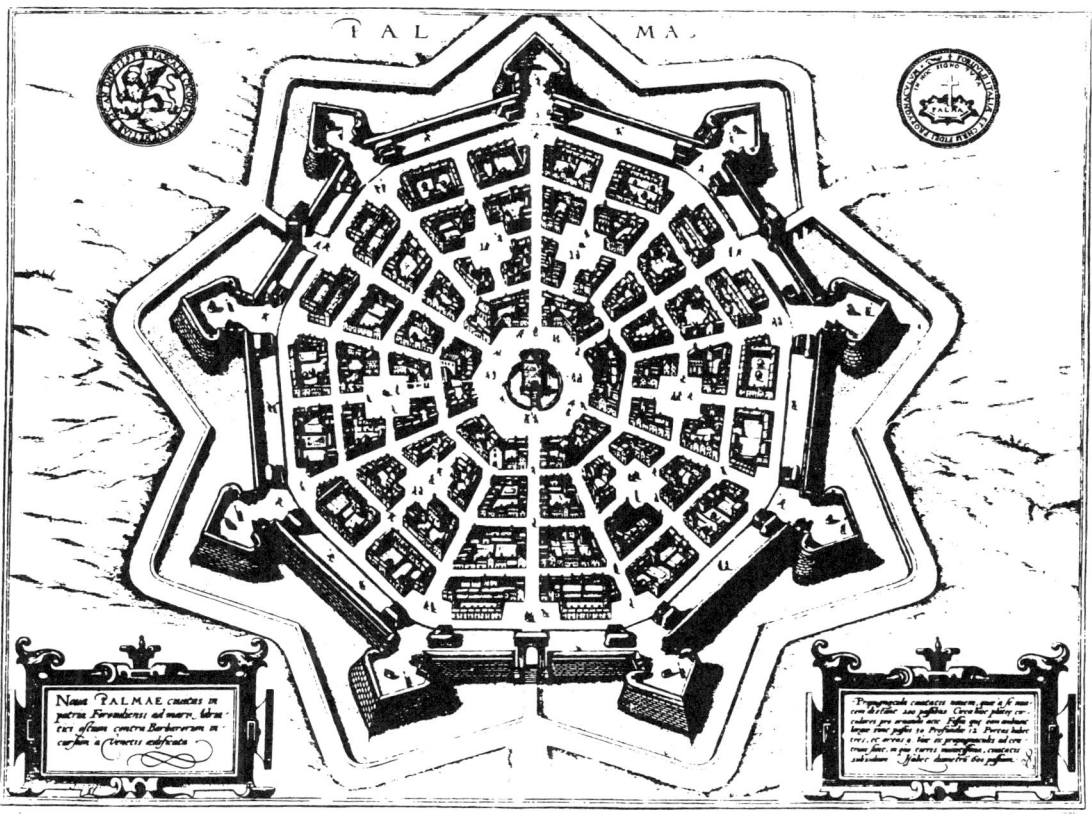

Figure 1.4
The Venetian fortress of Palmanova, planned by the Venetian architect Vincenzo Scamozzi in 1593, is an example, like Vauban's Neuf-Brisach, of a ground plan with a regular geometrical form. Source: Braun and Hogenberg 1572–1618.

an attempt to counter firepower with ever-more-massive fortresses. Although one can detect at the same time attempts to improve angles of fire and to modify the architecture, it is solidity rather than form that impresses in these fortresses. The same can be said for the fortress at Ostia, built between 1482 and 1486, in spite of its rationalized flanking fire and its embrasures for large guns near the bottom of the outer faces of the pentagonal bastion. Although it has a distinctive and hitherto unseen plan with triangular bastions, it is imposing more because of the solidity and massiveness of its construction than because of the new form of its ground plan.

 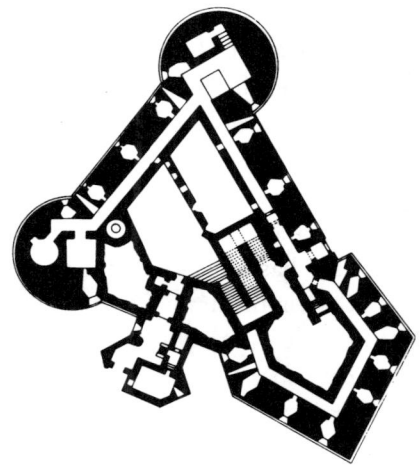

Figure 1.5
The fort or rocca at Ostia, built by Baccio Pontelli between 1483 and 1486, shows an interesting transition from medieval to Renaissance fortification. Two drum towers coexist with a bastion in the same structure. To the right is a plan of the fort showing the towers, bastions, and embrasures for cannon. Source: Hogg 1981, p. 112.

Rather than attempting to hang on to tradition, there was also the radical solution of minimizing or even abandoning fortification as an element in military strategy. For some authors, including Machiavelli, the very idea of fortification was much overrated and even potentially dangerous to the independence of a state. Machiavelli put his faith in new civic militias that in tactics and even weaponry were to resemble the legions of the early Roman republic. No longer would the Italian states rely on mercenaries and foreign allies; they would harness the patriotism of their inhabitants to create powerful field armies that would make fortifications obsolete. Although he did introduce some novel ideas in fortification, such as a ditch *behind* the wall, Machiavelli considered fortresses on the whole to be more dangerous to security and liberty of a country as citadels for potential tyrants than as adequate protection for the state from a well-motivated and well-trained army.[10]

Between these two extremes there was room for variation and experimentation. Italy in the Renaissance, with its wealth, its abundance of technical talent, and its interest in novelty, was ideal soil for this. Distinctive new forms and ground plans designed according to geometrical principles became more frequent at the beginning of the sixteenth century, and by the middle of the century they had become the common and accepted mode of fortification. By then, fairly elaborate fortifications had been built in Verona and Rome according to the new principles of a low-bastioned perimeter that the French called the *trace italienne*.

In the old heroic tradition of historiography of technology, the artist-engineer Francesco di Giorgio (1439–1502) has been credited with the invention of the bastion.[11] His beautiful and brilliant designs, which attracted the admiration and on occasion the imitation of contemporaries and successors, do indicate forms that could be considered bastions. One scholar has seen in the texts accompanying his drawings, rather than in the drawings and designs themselves, a "gestalt switch" from considering the problems of fortifications resisting artillery to abandoning that hope and thinking along new lines of removing fortifications from the sight lines of besieging gunners and providing flanking fire at close quarters during close assault of the fortress.[12] A prime trigger for this switch (no one seems to have taken the extreme position that it was the only one) was the tendency to see fortification more as a theoretical mathematical problem and less as a problem of construction and courage.

The question legitimately arises whether the new military architecture appearing in the sketchbooks of Francesco di Giorgio and Leonardo was merely another of the technical fantasies that would flower with theatres of machines in the following century. These lavishly illustrated works are often more interesting as a symptom of the unleashing of technical imagination and the use of pictorial representation (whose effects were later amplified by the diffusion of images with the printing press) than as immediate and discernible contributions to a practical technology. The community of architects and later military engineers had access to a considerable pool of technical ideas—some undoubtedly impractical or perceived to be so at the time—which served to provide inspiration if not precise models for innovations that were eventually put into practice.[13] Technical fantasy remained an element of military and civil architecture, both to the advantage and to the detriment of those branches of architecture. As the profession of military engineering solidified, fantasy, whether from the genuine innovator or the impractical dreamer, created a breach that was never closed, leading to defensiveness from within and reproaches from without.

Eugene Ferguson sees publications such as Agostino Ramelli's *Various and Ingenious Machines* as "simultaneously disruptive and progressive" for technological design.[14] Some of the machines pictured by Ramelli were not in use at the time he wrote, and some of them were impractical, but his depiction of them is an indication that military engineers were searching for bold new designs. Francesco di Giorgio, working a century before Ramelli, is an example of this in the field of military architecture. The luxuriance of his designs shows a creative spirit of the first rank, untrammeled by conventional designs in fortification, even if his actual execution of these designs was considerably more limited and timid. It also indicates a technological field in rapid flux and imbued with the excitement of genuine novelty. This atmosphere of freedom and change also arose from the freedom that artists and designers had gained from new tools of design, of which perhaps the most important was the concept and tool of *disegno*.[15] New graphical techniques that gave the artist the ability to see potential designs of complete forms on paper in various different projections increased both the pace of innovation and the possibilities of communication between designer and executor, who were beginning to drift apart socially and functionally. The best incarnation of this process was the architect, a title that appears in its modern form during the Renaissance and could well be applied to Francesco di Giorgio himself.

The Renaissance Architect, Mathematics, and Fortification

There never was a guild of architects in the Middle Ages. The birth of a recognizably modern personage of architect occurs over a period of almost 200 years and may be taken to begin with Filippo Brunelleschi (1377–1446), who designed for the church of Santa Maria del Fiore in Florence in the early fifteenth century.[16] Out of more than 100 artists, from Cimabue to Raphael, in Giorgio Vasari's book on the lives of Renaissance architects, painters, and sculptors, there are only seven who are designated solely as architects. Some are referred to as "sculptors and architects" (Brunelleschi himself, for example), "painters and architects," and "painters, sculptors and architects." Among the seven artists designated solely as architects by Vasari are Alberti, two minor figures (Baccio Pontelli and Chimenti Camicia, who "combined simple architectural tasks with military engineering"), Bramante, Giuliano, and Antonio da Sangallo.[17] It is noteworthy that military architecture is already rather important at this early stage for this group.

There is thus a diversity of backgrounds for architects. Brunelleschi apprenticed as a sculptor, Sangallo as a carpenter, and Bramante as a painter. Alberti, a scholar, is better known for his famous book on architecture than for his architectural achievements proper.[18] His book did not contain any figures, and it was devoted in large part to social philosophy and a theory of aesthetics.

One socioeconomic factor favoring the professionalization of the architect in Italy at the time of the Renaissance was the relative weakness of building guilds. The conflicts between Brunelleschi and the stonemasons' guild, which tried unsuccessfully to arrest him for not submitting to their rules, are a sign of this. In Italy, an architect could develop relatively free of traditional medieval technical and social constraints. In France, guilds and master masons were much stronger into the middle of the sixteenth century: the Renaissance castles of Azay-le-Rideau and Chambord, for example, were built mostly by traditional building trades who had an important role in the design.[19] An explosion of construction activity in the wealthy Italian states called for new kinds of specialists who concentrated more on conception and design and tended to separate themselves from the details of actual construction that remained in the hands of traditional builders.

The new architects were by no means ignorant of the details of construction. Although Italian architects on the whole had few ties with building trades (Palladio is a brilliant exception), their technical inexperience should not be exaggerated.[20] Nevertheless, the functions of design and execution, formerly performed by the same individuals, became more separate. Execution at the building site was becoming more distant from design, which more and more took place on paper. This was made possible by the simultaneous development of new graphical techniques, a fluid socioeconomic context, and a spurt in hitherto unseen kinds of building that demanded new approaches and careful planning.[21] That architects often were removed from frequent direct supervision of the work site and needed to communicate clearly and precisely with the executors of their work encouraged development of drafting techniques and the widespread use of plan and elevation views.

According to Wilkinson there were two major unifying characteristics of architects at the end of the fifteenth century. One was "self-apprenticeship to the art of antiquity . . . normally served in Rome."[22] The other was a knowledge of disegno. For Vasari, it was the most important thing and common to the other fine arts, sculpture and painting, which he saw as "sisters" whose "father" was disegno.[23] It meant more than the mechanical skill of drawing, subsuming also design and imaginative depiction of reality both external and subjective. Alberti had not only given his seal of approval for architecture as a fine art; he had also, in *Della Pittura* (1436), set down the details of the art of perspective, pioneered by his friend Brunelleschi, which was being considered as an indispensable part of the new kind of painting.[24] Disegno, which relied so much on the science of perspective, was therefore both highly technical (requiring a knowledge of the liberal arts of arithmetic and geometry) and indispensable in the depiction of new kinds of structures and artifacts. Here was a strong link with mathematics and theory that enabled the painter as well as the architect to claim a higher status than that of a mere artisan.

By the middle of the sixteenth century, there had been a profound transformation in architectural drawings. Besides various projections and detail drawings, modern perspective drawing, pioneered by Brunelleschi and codified by Alberti, had developed during the fifteenth century. It had a great aesthetic impact on architecture as well as on painting. What is worth additional emphasis, however, is the fact that these drawings and the whole tradition of disegno now required not just talent in execution but also more intensive employment of mathematics in conception. Brunelleschi's invention of perspective had made architecture an intellectual enterprise, something beyond mere manual proficiency and work-site management. The association of perspective drawing with the liberal arts of geometry and arithmetic raised the importance of architecture even more in the eyes of contemporaries. The benefits to the status of its practitioners were accompanied by a conception of architecture itself as a liberal art requiring the use of the intellect.

The application of geometry to other fields (dialing, cartography, surveying, navigation, naval construction) during the Renaissance went far beyond practical results. It represented a fusion of two distinct medieval traditions of "practical" and "constructive" geometry.[25] Practical geometry, in spite of its name, was the preserve of scholars who were interested in "practical" problems, most frequently in altimetry, planimetry, and cosmometry—problems using Euclidean geometry to prove rigorously the truth of their propositions. Constructive geometry consisted of the mathematical rules of thumb and numerological tricks that craftsmen had found useful in laying out ground plans and construction without much understanding of their theoretical justification. The fruitful fusion of these approaches during the Renaissance was accompanied by a new interest and in development of mathematics. It is perhaps no accident that the first translation of Euclid into the vernacular was by Niccolò Tartaglia (ca. 1499–1557), well known as a writer on ballistics and fortification as well as mathematics, who also published and popularized works by Archimedes. These translations and publications in 1543 paved the way for a revived interest in rational mechanics that would reach a glorious efflorescence with Galileo.[26]

The "rediscovery" of Vitruvius by the humanist Poggio Bracciolini at the Monastery of Saint Gall in 1415 had done much for the dignity of architecture.[27] One of the first books to be printed, it had focused European minds on the building legacy of antiquity as well as on the profession of architect. Vitruvius had argued that architecture was both a theoretical and a practical art. The mere theoretician was "obviously hunting the shadow, not the substance," while the mere practitioner could not give reasons for what he did—"architects who have aimed at acquiring manual skill without scholarship have never been able to reach a position of authority."²⁸ The impressive list of sciences that an architect was supposed to master made him the equal or even the superior of the scholar. Such talk fitted in well with the social aspirations of many Renaissance architects and artists.

Vitruvius had also considered military architecture, and it was military architecture that some historians consider to be the first to gain independent recognition.²⁹ Certainly by the middle of the sixteenth century it is possible to speak of a specialized military architecture. Increasing interest in the history of military architecture has destroyed the older view that during the Renaissance military architecture was considered to be uninteresting and unworthy of the attentions of superior architects, and that what prestige military architecture did have was reflected from civil architecture. Hale has effectively argued that the reverse was the case.³⁰ Association with the great and powerful in the important work of safeguarding the security of the state added luster to the military architect. Virtually all the great names of Renaissance art, excepting only Palladio, had some association with military architecture; Michelangelo himself is often quoted to the effect that he may not have known much about painting and sculpture but he did know his military architecture.³¹ Moreover, the theoretical study of fortifications overlapped with and touched upon problems of architectural theory in general, most visibly in the domain of urban planning.³²

Mentions of military architecture in Vitruvius had been rather sparse and not applicable to the new kinds of problems raised by powerful gunpowder artillery. It was considerably more difficult for military architects than for others to claim that all technical solutions had been discovered in antiquity. While the classical orders of architecture could be both admired and used in construction by contemporary architects, efforts to claim that the ancients had discovered and used gunpowder did not make much headway among military architects. It is in military architecture, then, that the process of moving beyond Vitruvius, with all the intellectual excitement and importance accorded to novelty it implied, was most visible. The dynamism in military architecture between 1450 and 1550 matched and even exceeded that in civil architecture. By the end of that period, the basic forms had been established and an avalanche of treatises describing the new forms of fortification and arguing for the respectability of a new science of fortification had begun.

What is usually considered the first printed treatise on fortification was in fact written by a great artist and pioneer in the art of perspective who lived north of the Alps. Albrecht Dürer's *Etliche Underricht zu Befestigung der Stett, Schloss und Flecken*, published in Nuremberg in 1527, was more of an architectural fantasy than a practical manual for fortifiers.³³ In spite of some interesting features (such as the caponiers in the ditch of the fortress, which caused some German commentators to claim later that Montalembert had taken his ideas from Dürer), the massive curved walls and roundels in the book were never built and stand outside the dominant Italian tradition of fortification, which eventually spread to the German lands. They are yet another magnificent example of technical fantasy and of the plasticity of artillery fortification design in its early stages.

In 1554, a second edition of Niccolò Tartaglia's *Quesiti et Inventioni Diversi* appeared in Venice.³⁴ This book, which contained discussions of mathematics and mechanics, also considered ballistics and fortifications. The ideas that form was central and that the strength of a fortress depended on its form were enunciated clearly by Tartaglia; the works of some treatise writers a few years later virtually excluded other considerations. Discussing the fortifications of Turin, then considered by all to be among the best in the world, Tartaglia distinguished between fortifications that were strong by nature of the site and those whose strength was attributable to art. Among the latter, he distinguished again between those that were strong because of their massiveness and the quantity of material used (which he did not esteem, because they did not really require a great deal of art from the designer) and those that were strong because of their form. It was form that required science and the attention of a mathematician. Although Tartaglia is often credited with the fundamental invention of the covered way in artillery fortification, it was not his experience with military affairs that he invoked to support his ideas.³⁵ Indeed, there is almost a kind of pride in his contention that he was not actually familiar with fortification. What really counted was an examination of the plan of the fortification. With the help of geometry, one could judge the quality of fortifications without ever having been a soldier or participated in a siege. It was not experience but art that was important. One could fortify on paper, as one could build cathedrals on paper, and one could do so better than soldiers and stonemasons. Tartaglia goes so far as to boast that he does not need practical experience to make his inventions. He achieves this by the use of arithmetic and geometry, both traditional liberal arts.

A few years later, Giacomo Lanteri was concerned almost exclusively with the geometry of the plan. The title of his 1557 book *Delle Fortezze Secondo Euclide* indicates his general views. He insisted that practice in itself was not to be depended on, and that ultimate certainty in fortification, as elsewhere, came from mathematics in general and geometry in particular.³⁶ Although his own mathematics left much to be desired, his attitude was clear: fortification was a mathematical science that could have the demonstrability and certainty of geometry.

Such views and the accompanying implication that those best capable of fortress design were mathematicians and theoreticians by no means went unchallenged. Giovanni Battista Belluzzi, a son-in-law of the architect Girolamo Genga, from whom he had learned his trade, was a military architect much esteemed by Cosimo de' Medici.[37] He felt strongly that "no engineer ignorant of the art of war could design effective fortifications." Nevertheless, a knowledge of mathematics was beginning to be considered a valuable and useful attainment for a military man. Castiglione, in his influential book *The Courtier*, had commended drawing and mathematics to the attention of the gentleman because of its military usefulness.[38] Although by no means universal among military men of officer rank, this was a view shared by many eminent men, including Prince Maurice of Nassau and Francesco Maria della Rovere, duke of Urbino. The figure of the "military intellectual"—another novel feature of the Renaissance—was interested not only in the classics and the military writers of antiquity but also in science and mathematics, which were useful in marshaling arrays of infantry, castrametation, and, not least, fortification.[39]

The writers of treatises on military architecture were therefore a mixed lot. Many were soldiers aiming to better their position; some were scholars who became soldiers, like Girolamo Maggi (ca. 1523–1572), who frequented the universities of Perugia, Pisa, and Bologna, wrote verses on the war in Flanders, fought in the siege of Famagusta, and was strangled as a prisoner in Constantinople. Others were mathematicians and mathematical practitioners, like Tartaglia. Some were university professors, like Galileo. The fact that Galileo offered a series of private and public lectures on fortification at Padua in the 1590s indicates that there was a demand for the subject taught by men with advanced mathematical knowledge.[40]

Galileo's short treatise *Breva Instruzzione all'Architettura Militare* begins with a number of elementary geometrical constructions without proofs, such as a method to construct a pentagon taken from Dürer,[41] and includes a nomenclature of terms used in fortification and the assertion that new kinds of fortifications are necessary because of the form, smallness, and weakness of traditional fortifications against artillery. The main problem is form: there is too much dead ground with round towers protruding from the curtains. The two most important requirements for the bastions (baluardi) are that the flanks be perpendicular to the curtains and that there be a possibility of providing flanking fire for the faces of adjacent bastions. The basic requirements of no dead ground and flanking fire—canonical principles of all future bastioned fortifications—were obviously well established. There is a section on drafting and the proper use of scales. At about the same time, Galileo was offering lessons on the military compass of his own invention that permitted quick computation of a variety of essential parameters in fortification design, such as the length of the sides of a polygon inscribed in a circle of a given radius, and for the use of artillerists, such as calibers of cannon necessary for given weights of cannonballs made from materials of different densities.[42]

In the second edition of his notes on fortification, Galileo indicated the three major defects of bastions that were to be avoided. They were the angle of the bastion being too acute and causing crowding at its tip, flanks that were too long, and, again, bastion flank angles that were not perpendicular to the curtain. There is a fair amount of space devoted to various methods of attack of a fortress and a description of tools and methods of building earthwork fortifications. This was a newer kind of construction technology meriting description; construction of masonry fortifications appears not to be a problem for traditional stonemasons and thus did not need any additional comment.[43] The final pages of the treatise are taken up with a number of different ground plans with remarks on their quality. Triangles and squares were not as good—because of the pinched bastions that this necessarily entailed—as polygons with more sides. This was another of the principles of fortress construction in the following centuries. Like Vesalius' *Fabrica*, almost all fortification texts are illustrated and these illustrations dominate the accompanying text. The ultimate example was the posthumous text of Francesco de' Marchi (1504–ca. 1576), who presented 161 different schemes of fortification—a visual feast that was much admired by his successors.[44] One scholar has made the perceptive remark that in fortification, as in many other areas of human endeavor and creativity, the faculty of vision was becoming more and more important at the expense of words.[45] Vision was important not only at the initial level of the design and the drawing but also for the soldier defending the fortress who must see along lines of fire so that no part of the fortress remains unseen by the defenders and hence vulnerable to enemy attack. The steadily increasing importance of the firearms that replaced hand-to-hand combat concurrently increased the importance of lines of fire that could be visualized and rationalized.

Fortification as a science where science meant drawing in the new way that had been developed and popularized during the Renaissance was changing to fortification as a science where science meant mathematics in the particular aspect of geometry. The importance of mathematics, already evident in the earlier conception of the science of fortification, was now beginning to become dominant.

An analysis of the various treatises on fortification published in the second half of the sixteenth century in Italy indicates that, in spite of a certain diversity, there was a strong common theme of the importance of mathematics in designing fortification.[46]

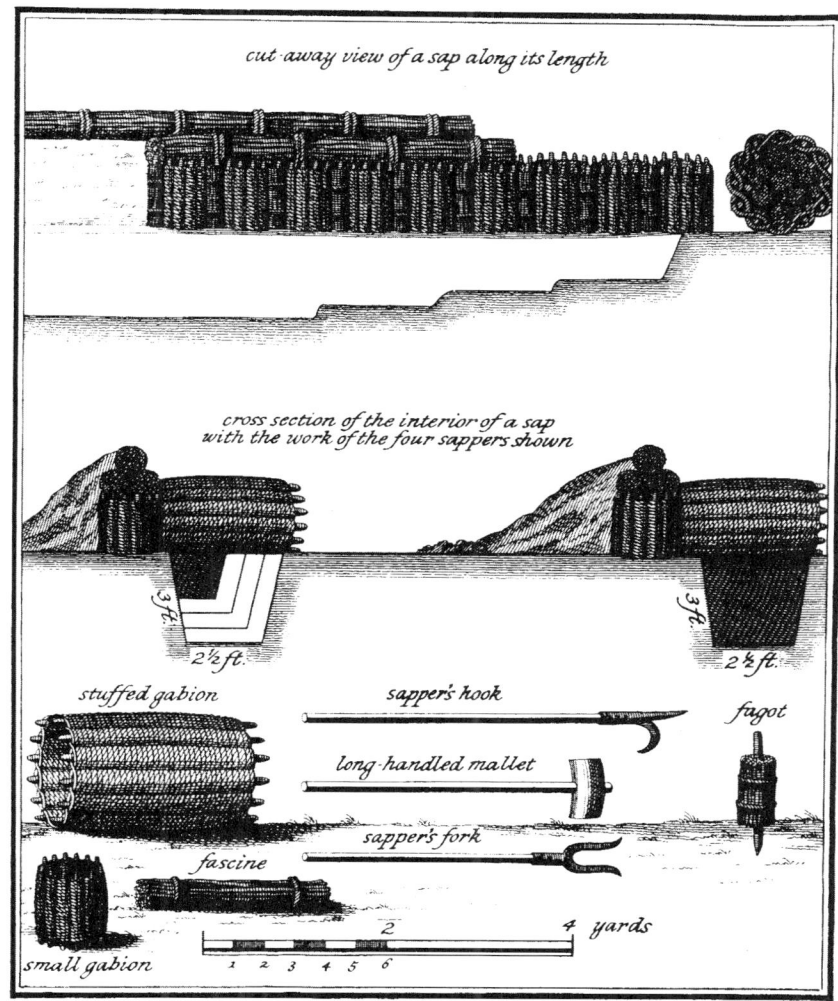

Figure 1.6
Some of the many trenching tools and accessories of siegecraft. The typical siege trench began as a narrow and shallow sap dug at night and protected by gabions and fascines until it was deepened and the protective wall of fascines and wicker gabions was built up and covered with earth. Source: Vauban 1968, plate V.

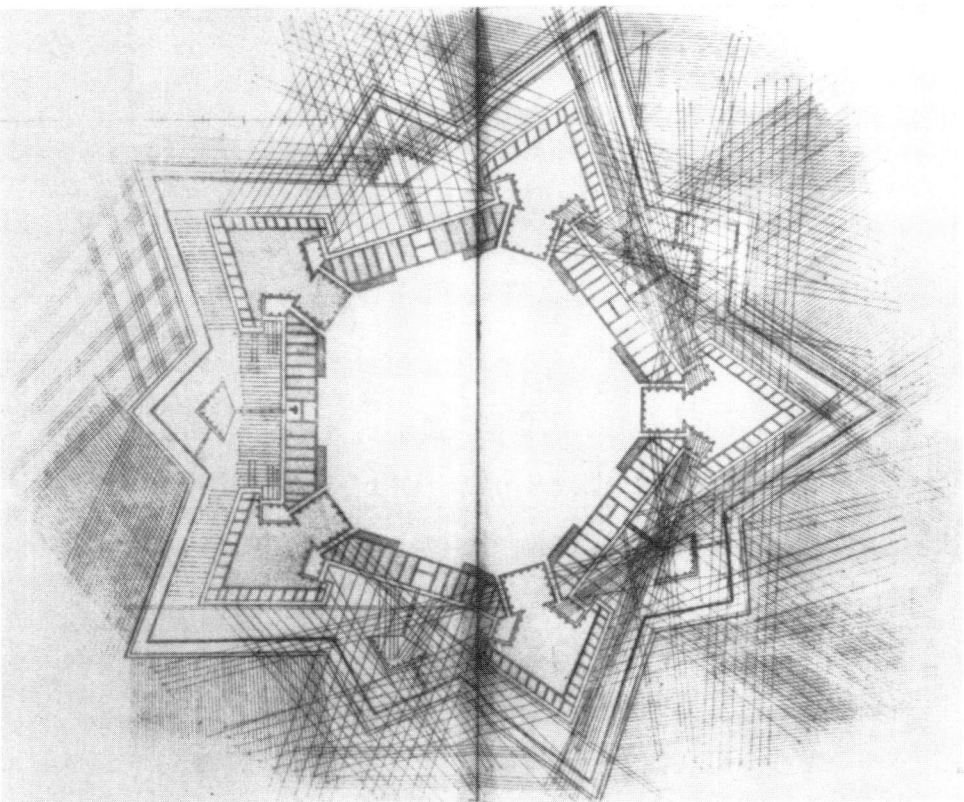

Figure 1.7
One of the many theoretical plans of cities included in Francesco de' Marchi's *Della Architettura militare* (Brescia, 1599). It shows crossing lines of fire in a plan produced by de' Marchi around 1545.

Geometry was perceived to give certitude and demonstrability to fortification. Experienced military men like Jacomo Fusto Castriotto (ca. 1510–1563) had recognized that the dead-ground argument against round bulwarks was a good deal less important in practice than in theory.[47] Yet the basic principle of no dead ground lent itself admirably to mathematical design of fortification. And mathematics was praised as useful. It was pleasant and delightful for those who looked for the higher pleasures of the mind. It was a part of the quadrivium of liberal learning, worthy of the attention of higher social strata of society, and for some it was a key to understanding the underlying order of nature. Above all, it provided true rules and a method in fortification that enabled intelligence (ingegno) to rise above the constraints of experience and to impose order in the confusion of multiple practical possibilities in fortification.

Some mathematicians, including Tartaglia, were claiming that fortification was a rational, Archimedean science of the kind Galileo was attempting to create for mechanics—a science based on mathematics and aspiring to the Euclidean model of formal deduction from axiomatic principles. It is undoubtedly significant that in the frontispiece of the 1550 version of Tartaglia's *Nova Scientia* Euclid is pictured as opening the door for a novice to a walled space wherein the liberal arts surround Tartaglia, who is demonstrating the trajectory of a cannonball. Within this fortified domain is yet another walled area, whose doors are being opened by Aristotle, with Plato standing behind him and pointing to his well-known prohibition that no one ignorant of geometry should enter here. At their most extreme, defenders of fortification as a science seem to go beyond merely coupling its dignity to that of mathematics. There is more than a hint that fortification can be made a science of absolute security. Its certainty, its rationality, and its quantitative nature can provide impregnability, an atavistic dream of the human race and a primary concern of the warring states of early modern Europe.

Along with this view, there was yet another that saw security as always fragile and absolute security as never obtainable. Most soldiers felt that the vicissitudes of war always gave Lady Chance the upper hand, and that fortifications could never be absolutely impregnable. Moreover, their opinion was that experience in the chaos of war was more important in building fortifications than elegant drawings based on the eternal laws of geometry. The hope grew that new sciences might lead to impregnability in fortification and even to increasing predictability in warfare. Not only the technical literature of fortification but also that of the military art in general saw a significant increase from the Renaissance on.[48] Yet the divide between theoretically inclined optimists and practical pessimists on the question of impregnability and predictability persisted.

Most military engineers refused to accept the possibility of impregnability. By the end of the sixteenth century, engineers and the title "engineer" (meaning military engineer) were becoming common, military engineering had established its autonomy from architecture, and the general field of activity of the military engineer was beginning to acquire some coherence. Military architecture, sometimes taken to be synonymous with fortification, was not all there was to fortification. Fortification also required knowledge of men and of machines (both those used in construction and those used in destruction, such as artillery and mines), and it required skill in of soldiering as well as in what today would be called project management. It was a job for the military engineer, a hybrid creature whose ancestors included makers and users of war machines, master builders, and commanders of troops and workmen. In addition to having expertise in fortification, the military engineer had to be expert in the attack and defense of fortifications. These complementary features of his craft were sometimes sources of tension and ambiguity. As Vauban was later to point out, it was difficult to be both builder and destroyer, both attacker and defender, both technician and leader, and rarely were these qualities to be found in the same individual. During the Enlightenment in France, these competing demands on the skills and loyalties of military engineers, along with conditions that were peculiar to the time and place, led to a crisis, which will be discussed later. Yet the divergences within the community of military engineers that became acute in that crisis had been present since the appearance of gunpowder artillery fortifications and even earlier.

There are other significant developments that coincide with the end of the sixteenth century. Galileo's treatises and teaching at that time indicate his interest in fortification. Unknown beyond a relatively restricted circle, Galileo's little texts are nevertheless typical of the attention devoted to fortification by teachers of mathematics, and they mark yet another extension of interest beyond the communities of soldiers, architects, and military engineers. The rich and visually delightful menu of fortress plans in Francesco de' Marchi's posthumous (1599) *Della Archittetura militare libri tre* provides an indication of the variety and degree of development of military architecture since the appearance of the first recognizable bastions a century earlier. Yet another development was the appearance of transalpine practitioners of the science of fortification. Daniel Speckle, or Specklin (1536–1589), whose *Architectura von Vestungen* appeared the year he died, had assimilated the basic concepts of the bastion and wished to promote a German tradition in fortification that could hold its own with the Italian masters. In the Netherlands, another tradition had emerged that was being tested by the bitter warfare between the Spanish and the Dutch. The appearance of a dynamic tradition of military engineering in France with Errard de Bar-le-Duc toward the end of the sixteenth century is the prelude to the flourishing of artillery fortifications that would set the stage for Vauban and French dominance of the field a century later.

Early Artillery Fortification in France

Jean Errard de Bar-le-Duc (1554–1610), a military engineer from Lorraine, had gone to work for Henri IV and had published a number of modest texts—one in 1584 on the fabrication of mathematical instruments and engineering machines in the genre of

Ramelli and Besson and one on elementary geometry in 1594—before he published the text that won him his place as the first great figure in an indigenous tradition of French military engineering that would grow and in the coming centuries come to replace the former Italian domination of European military engineering.[49] Dedicating his *La Fortification reduicte en Art et Demonstrée* (published in Paris in 1600) to King Henri IV, he paid simultaneous homage to his royal master and to practice in the design of fortification: the king had spilled more of his blood and sweat verifying the "rules of the Art" than others could ever use ink in demonstrating them in theory. Yet practice and theory must go hand in hand, for the first is blind without the other and the latter is "armless" without the former. He considered himself an innovator because, he claimed (ignoring the Italian tradition), no other engineer had really written on the science of fortification as a science: "Discourses on mechanical things do not merit this title [of science] at all. There is no question here [in my treatise] . . . about sketches that might work for someone by chance, but about geometrical demonstrations that give infallible assurance to all."[50]

Thus, certainty was possible, and its basis lay in geometry. Unlike many of his successors who wrote treatises in the seventeenth century, Errard refrained from presenting the necessary elementary concepts of geometry and the basic nomenclature of fortification, on the ground that they were too well known to be worth repeating. It is highly doubtful that this belief was justified: many of the treatises in the following century still included introductions to rather basic mathematics, and Errard himself, in his earlier work on geometry, had appealed to the nobility of the country to take up that science not only because it was useful in peace but also because it "produces its most beautiful effects in war."[51] And this geometry was essentially a practical manual of mensuration, surveying, and the use of various mathematical instruments. There was little trace of Euclid. Errard's claims for "geometrical demonstrations" giving "infallible assurance," and the very title of his book, whose proclaimed aim for fortification is for it to be "reduced to an Art and demonstrated," might initially lead the reader to see a rigorous mathematical approach to fortification. Such a view would be erroneous in spite of Errard's contempt for "empirics" in fortification as much as in medicine. His use of the word 'axoims' to describe his working principles is also misleading because these axioms have nothing Euclidean about them; they are merely methods of construction and practical tips. Undoubtedly Errard gives dimensions of the various parts of a standard fortress and indicates how they could be laid out.[52] But besides a discussion of various dimensions, most of the other information is a distillation of practical experience that would be useful to the practical engineer, artillerist, and soldier. There is the information that the Sieur de Linar in Germany in 1572 had discovered that lengthening the barrel of a cannon beyond 12 feet did nothing to increase its projective force and the velocity of the exiting ball, that the amount of powder to be used should be "at least" a third of the weight of the ball, and that the average range of the "canon de France" (firing a ball of slightly less than 20 kilograms) is 600 paces. Penetrations of cannon firing from 200 paces in different kinds of soils (sand, well packed, moderately settled) are given, along with the observation that batteries were usually placed 200–300 paces from the walls to avoid enemy arquebusiers. Errard also presents an interesting analysis of the time needed to repair the damage of one shot from a cannon, and from this he calculates a minimal siege battery to consist of 12 cannon (a cannon can get off one shot every 7½ minutes, and it takes 12 men to repair the damage of an hour's battering by a single cannon). There are tips on materials to be used (stone from Metz and Sedan is best), a recommendation (taken from Dürer) for a batter or talus of 45° for the walls (with the warning that this may be too expensive in all cases), and the dimensions of supporting internal buttresses. The heights of the walls are left to the judgment of the engineer for particular cases.

There are also some thoughts on siegecraft in the form of "principal maxims of the art of assault," from which conclusions can be drawn "necessarily." For example, the attackers should have a bigger front than the defenders, and there should not be a proliferation of attack trenches (which leads to confusion in the attack).

The long list of the requisite qualities of the engineer is interesting. He must primarily know the force of artillery, be experienced in sieges, and be a good commander who avoids confusion in the siegeworks. But "it is also very necessary that the Engineer also speak about all parts of the science so that he can make known that which he will have conceived for the execution of his design by geometrical demonstrations (and not mechanical ones according to the manner of the ignorant)."[53] He must be a "geometer as much for inventing machines, and other instruments serving for the defense of the fortress and the work required, as for demonstrating the utility and profit of his inventions, as much for putting them into practice and apportioning the work to be done for the time and means available, thus avoiding excessive expense, which is usually very inconvenient, for lack of understanding of this beautiful science of geometry. Especially when he is besieged, let him think and find the means of relieving those who work, for there is nothing as insupportable as sleepiness [for both men and beasts]."[54] He also must "know something of ordinary architecture (architecture commune) and masonry."[55] "It is good to haunt the great" to convince them of the necessity and the nature of fortifications, and to be successful in this "that he study to speak briefly and intelligibly" and "that he never consent to a bad design." The final word of advice takes on an elevated tone: "Let him learn rather to teach than to dispute with an ignoramus for it is to be feared, for a man who has science, to meet an ignoramus who has credit, for several reasons that everyone knows."

Here is a portrait of a man of multiple skills: soldier, architect, artisan, diplomat, and skillful propagandist for the usefulness of fortification in general and his solutions in particular. But underlying it all is "geometry," which both discovers and demonstrates. It is the basis of fortification, and the advice mentioned on construction such as the properties of certain building materials is classified among "indifferent things that are not at all the essence of the art of fortification." Yet this "geometry," its Euclidean lineaments easily recognized when it comes to drawing plans and laying out fortifications, is not exactly what the word means to us. "Geometry" for Errard seems to be not only necessary for demonstration of the utility and profit of the engineer's inventions but also necessary for the invention of those machines themselves. In this sense, geometry seems to be similar to the clear-headed and disciplined thought necessary "to avoid excessive expenses" and "to relieve those who work." It is practical skill in computation and mensuration, and it also appears closely related to the more elusive quality of ingenuity and inventive ability that is present in the etymology of 'ingénieur'.

Errard's faith in mathematics is perhaps easier to understand if one remembers that his time was an exciting one for mathematics, with the revival of Greek traditions and new work in the field. Mathematics did not mean the same thing to all people at the time Errard wrote. The distinction between "practical" and "constructive" mathematics during the Middle Ages has already been mentioned. The revival during the Renaissance of Pythagorean and Platonic notions of mathematics as forming the structure of the universe, the increasing interest in Archimedes and his approach in dealing with mechanics, and the new translations of Euclid by humanists all contributed to a lively fluidity in the concept of mathematics.[56]

It is not always clear exactly what engineers meant when they talked of mathematics. For example, Ramelli's curious and somewhat incoherent preface to his great theatre of machines is a eulogy to the importance of mathematics and the mechanic arts.[57] Ramelli blithely blurs the distinction between "pure" mathematics and engineering. For him, Archimedes the mathematician is Archimedes the military engineer, with no sign whatever of Plutarch's implicit condemnation of the mechanic arts and their decided inferiority to mathematics. In his magnificently illustrated plates on various machines there is not a trace of what we would call mathematical or even quantitative considerations, yet they are presented as an example of this "foremost art of mathematics." Here mathematics seems nothing more than a rational approach to nature and problem solving.

For Errard, the aim of certainty and demonstrability in geometry cannot be achieved overnight. The science of fortification is a progressive science, as he points out in his concluding remarks: "I will put an end to this work since, God be thanked, I have not at all reached my goal, which has been to put forward some notable principles and reduce fortification to an art [or] at least to make it honorable and to clarify it more than it has ever been."[58] One cannot expect perfection in the "human sciences," of which fortification is the most important, in a single human lifetime, and Errard hopes that his work will be corrected and improved. Woven in with the strong empirical and practical dimension in Errard's geometry is the optimistic view that geometry can and soon will (in two generations) provide the certainty and demonstrability of things that are "reduced to an Art."

Errard is a true military engineer in that he not only designs and builds fortifications but is also involved in siegecraft and defense of fortifications, where military and managerial skills are required as well. Yet the elements of his brand of military engineering at the end of the sixteenth century sometimes exist in a state of uneasy cohabitation. He has an unabashed confidence that fortification can become a true "art." This certainty itself arises from his strong faith in mathematics. In this he is following the tradition of geometrical fortifiers that can be seen with some of the Italian military architects in the second half of the sixteenth century, a tradition that would remain vital during seventeenth and even the eighteenth century. However, the habits of the practical working military engineer continually remind him that he cannot claim absolute precision in his estimates and that good military engineers need the experience of the battlefield and siege to build fortifications and fight in siege warfare. Errard's successors, while maintaining their faith in mathematics and maintaining its primacy as a tool of the engineer, would be more conscious of its limitations in the practical world of the designer.

One of a growing number of specialists in war-torn Europe, Errard still had to contend with non-specialists of various kinds keen to try their hand at designing fortifications and sometimes getting the opportunity to do so. Just a few years after Errard's death, his compatriot Vignère was complaining that in fortification "there is not a painter, carver, mason, joiner, carpenter, architect, or in fact any class of men whatever, that have not employed their pencil at it; as if the thing lay in representation, and in knowing how to draw a line straight or curved with rule and compass, and inventing the design of a fortress out of one's own head, and were not an art acquired by a large experience of sieges, both attack and defense."[59]

Conflict between the specialist and the non-specialist would remain a constant theme until the time of Montalembert, who can be considered the last great non-specialist in fortification. The parting of ways was already clear by the middle of the sixteenth century. Although many of the great Renaissance architects had been involved in the design of fortifications (sometimes even as their primary activity, as in the case of the Sangallo family), they had also done other things. Increasingly, however, soldiers with practical military experience had begun to practice the trade alongside architects who had acquired such experience. In the case of architectural texts, the rupture appears earlier and more cleanly than it does in the actual practice of military engineering and fortification. In 1554 Pietro Cataneo (d. ca. 1569) included military architecture as part of architecture generally; however, he was one of the last authors do so, and this date marks a definite turning point in the process of specialization of texts on military architecture.[60] Nonetheless, for more than a century after this the increasingly specialized military engineers faced competition from primarily civilian architects, from soldiers, and from a variety of builders who participated in the actual design and building of fortifications. Pepper and Adams, in their well-documented study of the siege and fortifications of Siena in the middle of the sixteenth century, have shown that civilian architects and artists still had a preponderant role in the defensive constructions of the Siennese republic.[61] At the same time, the military engineer Belluzzi, who died in this war fighting for the other side, was attacking the participation of civilian architects in designing and building fortifications on the grounds that without military experience one could not be a good military engineer. The threat to the distinct status of the military engineer came from a number of directions—from architects, builders, and soldiers. Indeed, it was only toward the end of the seventeenth century, with the appearance of Vauban, that military engineering began to cohere as a profession and was defined by more than just the contempt for interlopers from related professions. This coincided with what Duffy sees as a cardinal change in warfare. The great generals of the seventeenth century—Montecuccoli, Turenne, and Condé—could still direct a siege themselves, but Eugene and Marlborough had to rely on experts.[62] However, their cosmopolitan experts were rarely as good or as effective as those of their French enemy. The primary reason for this was Vauban and the professional corps of French military engineering he was forging.

2

French Fortification in the Age of Vauban

Do you want me to teach that a curtain is between bastions, that a bastion consists of an angle and two faces, etc.? That is not my way.
—Vauban to his secretary Thomassin[1]

For most of the sixteenth century and the beginning of the seventeenth, peripatetic Italians dominated European fortification—in construction, in siegecraft, and in publication. Paciotto's fortress at Antwerp in 1567 for the Duke of Alba is perhaps the best-known example of their work, but his confreres were active in France, England, and Germany as well. In 1534, François I hired the former papal architect Girolamo Marini, along with his son and five compatriots; Henri II acquired the services of Jacopo Fusto Castriotto (1504–1576) and may have given him the title of General Superintendent of Fortifications in France, and his successors continued the practice of hiring Italian engineers.[2] By 1543, there were 100 Italians working on the fortifications of Luxembourg.[3] In England, Antonio Melloni worked for Henry VIII in 1545 but was killed as an engineer for the French three years later.[4] At the end of the sixteenth century, Italian military architects could be found in Germany, in Sweden, in Poland, in Russia, and even in the European establishments in Asia, Africa, and the Americas. And of course they were active in the Venetian territories of the Adriatic and the eastern Mediterranean.[5]

As the sixteenth century drew to a close, foreign experts were gradually replaced, and cosmopolitan engineers selling their services to a succession of foreign masters disappeared.[6] In many cases, the locals resented the Italian experts As locals acquired expertise, many of them were able to challenge the foreign specialists. Nevertheless, they usually followed the Italian traditions, accepting their basic principles and, sometimes ineptly, modifying them.[7] The bitter and prolonged war in the Netherlands between the Spanish and the Dutch (1566–1648) led to the further development of indigenous traditions of fortification in Northern Europe, especially in the field of siegecraft.

Warfare against the Hapsburgs on all the frontiers of France led to the continuing development of fortification after the death of Errard's royal master Henri IV in 1610. By that time, a dozen engineers bore the title *ingénieur du roy*, and Henri's great minister Sully had added to his other titles that of Surintendant des fortifications.[8] These are signs of attempts to rationalize and improve the chaotic fiscal and administrative organization of fortification of the late Middle Ages, where responsibilities and costs of fortification were often divided among many authorities. Sully promulgated the first regulation on fortifications in France. It was the germ of a state bureaucracy of military engineering, which was to be fueled by expenditures of almost 8 million livres in the first decade of the sixteenth century.[9] But the struggles of the monarchy against Protestants and the Fronde seem also to have led to a pause in the trend toward centralization of administration of fortifications. By the time of Louis XIV, there were two royal ministers who were in charge of fortification, both sharing the services of a chief technical administrator.

Fortification and siegecraft continued vigorously both in warfare and in books after Henri's death. In the Netherlands, the revolt against the Spanish consisted of a number of famous sieges, which led to a flowering of Dutch and Spanish expertise in both siegecraft and fortification. In France, foreign wars and the struggle with the Huguenots (symbolized by the great siege of La Rochelle in 1627–1628) gave French armies plenty of experience of their own. It is perhaps significant that the last Italian military engineer of any stature who served the French crown participated in the siege of La Rochelle.[10] Writers on the subject were not idle either.

Early French Authors on Fortification after Errard

Some of the better-known writers on fortification were schoolmaster-clerics who taught in various colleges: Georges Fournier (1595–1652), Claude-François Milliet de Chales (1621–1678), and Jean Du Breuil (1602–1670), who wrote under the pseudonym of Silvère de Bitainvieu.[11] Du Breuil's *Art Universel des Fortifications, françoises, hollandoises, espagnoles, italiennes, et composées* appeared in Paris in 1665 and was dedicated to Georges de Guiscart, commander of the important frontier fortress of Sedan, in the hope that it would be "useful for men of war." It is rather amusing to see Du Breuil, hiding behind his pseudonym, passing himself off as a practical soldier and referring his readers to Jesuit schoolmasters for theoretical details of mathematics.[12] Like

Errard, Du Breuil commends practical mathematics to nobles who do not see much use for Euclid and who do not seem to be convinced of the worth of the higher intellectual pleasures that are contained in geometry, but he also affects disdain for those interested in the merely theoretical aspects of geometry. In spite of practical details such as the range of muskets and penetration of cannonballs fired from a distance of 100 toises, however, his book is almost entirely a manual of simple mathematics. It is divided into "treatises" [traités] and subdivided into "practices" [pratiques]. The first "treatise" deals with definitions, the second with a geometrical introduction and the construction of the plans of regular fortresses, the third with irregular fortresses, and the final one with surveying and how to draw plans. But it is clear that the "practices" are the practices of the drafting table, not the field. Along with instructions on how to use a proportional compass there are twenty "General Maxims of a good fortification," which are essentially lists of dimensions and the usual injunctions the necessity for flanking and no dead ground in fortification. But unlike Errard, Du Breuil does not give anything on the attack and defense of fortresses. And when it comes to actual construction, he writes: " . . . as for that which concerns walls of stone and brick, one must deal with a mason or a contractor, who will follow that which you have marked out for him and must do nothing except according to your orders."[13]

Clearly Du Breuil is a practical mathematician. He is not interested so much in rigor or proof as in methods of construction. But this construction is essentially the construction on paper of geometers; his instructions for actual construction on the ground are vague and general. This book and others like it seems to be a kind of vulgarized mathematical evangelism to the unconverted and socially influential class of the military nobility. It was they whose sons increasingly attended colleges and academies to acquire the rudiments of a formal education that would serve their future careers.[14] Du Breuil's book could be seen both as part of an already well-established tradition to ground a science of fortification in mathematics and as an attempt to convince influential people of the usefulness and necessity of mathematics and, not least, mathematics masters.

Another major theme of Du Breuil's work is the attempt to establish some kind of classification of fortifications that would bring order to the plethora of systems of fortification. Such systems were continuing to multiply in the various treatises of the sixteenth century. Du Breuil's title proclaims that fortification is a "universal art" and implies that it is common to the French, Dutch, Spanish, and Italian kinds of fortification. He argues that there are only four methods of fortification (associated with the four nations mentioned) and that any kind of fortification can be seen as some kind of composite of these four elemental systems. Treatises on fortification seem to have been dominated by two contradictory impulses: a rigorous impulse toward simplification and analysis in terms of a limited number of principles, preferably the fewer the better, on the one hand, and an almost undisciplined aesthetic urge to multiply visually pleasing ground plans on the other. De' Marchi's 161 systems of fortification in his magnificently illustrated book exemplify the latter trend.

Du Breuil felt that the proliferation of ground plans had gone far enough, even though it permitted the flexibility every practical fortifier saw as necessary. Imposing some kind of order and limiting the number of types was unavoidable if confusion was to be avoided. National styles of fortification had arisen partly because of the physical environment and available resources and the personal influence of a dominant fortifier of that nationality. For example, the Dutch system was associated with the work of Samuel Marolois, who in the early seventeenth century had written a book based on extensive Dutch practice in wars of independence against Spain.[15] Thirty years later Naudin wrote that some people had referred to the different national styles of fortification as "orders" analogous to the classical orders in civil architecture.[16] But Naudin did not approve of this because it implied that there were a number of intrinsically different styles of fortification that could be multiplied ultimately to the point of confusion. This offended his sense of order and his desire to see fortification in unitary terms. Underneath the multiplicity of styles, Naudin felt, was a unity of method rather than a unity of form. Naudin also made some contemptuous remarks about "fortification curieuse," citing as examples de' Marchi and the royal architect François Blondel (1618–1686).[17] Fortification curieuse was like Gothic architecture: spectacular but not solid.[18]

Although the practice of classifying types of fortification along national lines persisted, the end of the seventeenth century increasingly saw systems of fortification being attributed to individuals, such as Vauban. In Guillaume Leblond's article on fortification in the *Encyclopédie* of Diderot and d'Alembert, there is a good discussion of national styles as well as those of individual fortifiers.[19] The problem of proliferation of systems persisted, and Du Breuil's hope of reducing fortification to a problem of four elements was futile. Nevertheless, the attempt does indicate a taxonomic approach midway between the collection of individual examples and the analytical and methodical approach.

A book on fortification published a few years later by Du Breuil's colleague Milliet de Chales (a Jesuit father) is rather similar to Du Breuil's book.[20] Although it does take up the problems of attack and defense of fortifications in a superficial way, it is ostensibly more Euclidean in format, and the emphasis on mathematics is stronger. It begins with some general principles, then offers definitions and "propositions." Proposition VII is that "bastions without flanks are defective." Proposition XXX says that earth-filled bastions are preferable to empty bastions. Proposition XLI is titled "La methode Françoise moderne." Proposition VII in book II is a geometrical construction (describe any polygon on any given line). Proposition XXXVIII in book II is a construction on the ground (to lay out the ground plan of a fortification on a site). Proposition II in book VI is merely a series of operational tips ("How one must defend oneself against the effort of cannon"). In short, the word 'proposition' is used loosely and often. Milliet bemoans the fact that most people dealing with fortification do not know geometry and his entire second book is essentially a simple treatise on practical geometry.[21] Again, though there is nothing of any real use on actual construction of structures, the author expresses the opinion that casemates should not be built because they are smoky and can be easily crushed by artillery fire.

Along with the Jesuit fathers and schoolmasters, there were a number of practicing military engineers who wrote works or left their methods behind them in treatise form. One of the more interesting of these people is the little-known engineer Lalonde, who is credited with the popularization of decimals in France.[22] In his 1665 book *Elémens de Fortification* he identifies himself as a former engineer who had constructed fortresses and participated in sieges.[23] (He was in fact killed in a siege in 1688.) The book was supposed to be the first part of a larger work on fortification; however, the continuation does not seem to have been written, and the first part was reprinted posthumously a quarter-century later under a different title.[24] This book reveals much about the problems of the engineer in the seventeenth century, as well as the problems of architects, surveyors, merchants, and financiers (to whom the work was addressed along with military engineers). Lalonde's aim is a practical arithmetic, not geometry, and the problems he is trying to solve are problems of administration and accounting.

Lalonde's book is simple in the extreme. For more complex problems he refers his readers to the works of clerical schoolmasters. The calculation of cube roots is omitted. (He says that he used them only once, that the contractor was suspicious of his method, and that he was therefore obliged to use a simpler and more comprehensible technique.) Basic rules of counting, arithmetic, and fractions are presented in the form of specific practical problems worked out step by step. Lalonde applies arithmetical methods developed for abstract situations to practical problems of mensuration and estimating volumes of earthworks and wood. He takes the reader, step by step, through problems such as the following (the sixth problem in chapter 7): Twenty-four soldiers have transported 48¾ cubic toises of earth over a period of 6½ days. There are still 542⅔ cubic toises to transport. How many days will it take for the job to be completed? There are similar problems for calculating short rations for a fortress under siege.

A present-day reader, despite all attempts to keep in mind the spread of numeracy since the Industrial Revolution and despite Lalonde's disclaimer that his "intention was not to write for theoreticians," is rather forcibly struck by the low level of numeracy among professionals (military engineers as well as architects, surveyors, and financiers) one would expect to be versed in these matters.[25] This low level of numeracy certainly raises the possibility that some apparently gross arithmetic errors in official reports by the head of the military engineering corps more than a century later could have been due to incompetence as much as to any attempt to deceive or mislead.

Intended no doubt for beginners in their trade, Lalonde's manual does suggest in the kinds of exercises it presents an aspect of fortification that was quite as important—indeed, probably more so—than the actual design: the management of the job site and the need to control expenses. This was by no means new. Fortresses and castles had always been expensive to construct, and special officials (e.g., the provveditore di fortezze in Medicean Florence, who had been in charge of the finances and administration of fortress construction) had always played a major role in the successful completion of any project.[26] But this role becomes increasingly more complex with the appearance of the modern bureaucratic state in France, the increasing size of fortress building programs under Louis XIV, the growing geographical area subjected to a central administration of fortifications, and the increasing inadequacy of forced labor and requisitions in kind to supply the manpower and materials needed. Success in management was as important as success in design, and the same kind of "practical mathematics" that can be found in fortification design makes its appearance in the management of fortress construction. It is "practical arithmetic," uninterested in any kind of rigorous or theoretical analysis of number theory, presented in the form of worked-out problems that can be assimilated by its practitioners, and not even paying lip service to the ideal of a "universal" science that will lead to certainty and demonstrability that is sometimes present in other books on fortification. Yet it, too, implied a concern for order and rationality that suffused practice and attitudes—a concern that has led some scholars to talk of the application of science to warfare.[27] Mathematics in the form of geometry was applied to design, and mathematics in the form of arithmetic was applied to construction and management.

De Ville and Pagan

Although all these texts were well known and frequently used, the two most respected works on fortification that marked definite changes from the ideas of Errard were those of Antoine de Ville (1596?–1656?) and Blaise-François, Comte de Pagan (1604–1665). Both were soldiers and engineers with extensive military experience in the reign of Louis XIII, and their ideas replaced Errard's as the leading models for fortress construction. They were particularly respected because they not only wrote on fortification; they had practical military experience too.

De Ville, who had worked for Venice and for France as a military engineer, claimed not to have described in his book anything that he or his brother had not seen or practiced. He proposed bastions whose flanks were somewhat longer than Errard's and which made a right angle with the curtain. Although the flanking fire from these flanks was *fichant* (oblique rather than parallel to the face of the adjoining bastion), they were considerably more effective than Errard's puny flanks turned inward toward the curtain walls. De Ville, although a practical soldier and contemptuous of the pedantry of some writers of fortification, argues that one can present a single design that can be useful to demonstrate a *method* of design that is based on reason and experience.[28] His mentioning that knowledge of precise angles and dimensions is not the essence of good fortification design underscores the primacy of a method of fortification design over a collection of set designs to cover different possible cases. This places De Ville among those who look for some kind of underlying unity in the science of fortification at the same time as it distances the experienced soldier from the mathematics master.

"It is doubtful," Lynn White has remarked, "whether the 'Cartesian' mentality, which assumed that mathematics is the key to reality, would have become dominant if Europe had not assiduously bankrupted itself by building new military defenses in which assurance of safety was achieved less by tangible masses of masonry than by abstract geometrical patterns of fire."[29] De Ville, an almost exact contemporary of Descartes and a correspondent of Galileo, is perhaps a good personification of this trend. One of his biographers has also seen him as one of a new breed of technicians who are (to use an anachronistic word, but by no means anachronistic concept) technocratic in their impulses, fully aware that their work goes beyond mere military architecture and requires a global view of the state and its security.[30] Thus, there is what could be called a systems engineering approach that considers human factors along with technical ones in the construction, defense, and attack of fortifications. Considerations of politics, morale, and economics enter into de Ville's plans. And according to the subtitle of his best-known book, *Les Fortifications du Chevalier Antoine de Ville*, "Everything is presented in 55 plates with plan, perspective, and background [i.e., views]. The treatise is proved by demonstrations, experience, common and physical reasons, with narrations of ancient and modern history."

Although de Ville is only a decade older than Pagan, there is a major difference between their ideas on fortification. Pagan, coming from a family of naturalized Neapolitan nobles, lives in an age in which artillery is more important than ever before. Other gunpowder assault techniques, such as the petard and "attaching mines," which up until that time were sometimes even more important than artillery battering, were beginning to disappear, in part because of improved designs and the effectiveness of flanking fire. With Pagan, the cannon has finally become not merely dominant but almost exclusive in the attack and defense of fortresses. This is not to say that subterranean mining, as opposed to mines attached to bastions or walls, ceased to be an important technique. Vauban would create a special group of miners to mount such attacks, and their activity was often decisive in many sieges such as the one at Turin in 1706.

It is difficult to point to any discernible technical innovation that led to the uncontested consolidation of cannon as the chief instrument for attacking fortresses in the late seventeenth century. Barely noticeable incremental improvements probably were more important than any major perceptible design changes. Among them were improvements in casting and in gunpowder quality, better handling of the cannon in larger standing armies, and more experienced and more continuously employed specialists to use them.[31]

Like de Ville, Pagan was a soldier. He had participated in 21 sieges and had lost his eyesight while campaigning in Portugal in 1642.[32] Although there is no fortress that can be easily identified as having been designed by Pagan, he developed simple methods of laying out the ground plan of a fortress that were eventually taken up by Vauban and others. Not only did Pagan make his flanks perpendicular to their line of defense (the line along the face of the adjacent bastion extended in both directions), which enabled better flanking fire; he also greatly increased the size of his bastions and turned away from the relatively cramped bastions of Errard and even de Ville. Unlike Errard, who did not like outworks, Pagan designed capacious demi-lunes and counterguards and began the process of deepening the defensive perimeter that Vauban would continue. Indeed, it is fair to credit virtually all the features of Vauban's "first system" to Pagan. Although Vauban abandoned certain features favored by Pagan (such as multi-tiered flanks) and instituted features of his own (such as the tower bastion), Pagan's influence remained primary. Vauban's "first" system was little more than an insignificant modification of Pagan's system. Moreover, the second and third systems, associated with the fortifications of Belfort and Neuf-Brisach respectively, show features that are continuations and developments of ideas also present in Pagan. Determining the paternity of ideas on fortification is virtually impossible. The same arguments put forward so forcefully by Bertrand Gille regarding Renaissance engineers—arguments that emphasize the existence of a large corpus of technical knowledge that circulated extensively even in manuscript form—are valid here.[33] Vauban's originality, if that is the right word, must be sought elsewhere than in the narrow domain of military architecture.

Sébastien le Prestre de Vauban (1633–1707)

Considerable historiographic difficulties are encountered when attempting to address the question of Vauban's originality. Duffy has pointed out that figures as diverse as Carnot, Maginot, and de Gaulle were able to fall back on Vauban to justify their policies because his "oeuvre is so vast that it offers little resistance to men who search in it for vindication rather than instruction."[34] Vauban was born into the minor nobility of Burgundy and after a rather rudimentary and informal education began a military career at the age of 17 that was to last until his death. His great reputation as an engineering genius, already acquired during his lifetime, only continued to grow after his death, and even Montalembert, usually combative and acerbic, was relatively restrained in dealing with Vauban. His direct attacks on Vauban were few, and he was always careful to distinguish between the great man himself and his self-proclaimed followers. Choderlos de Laclos was one of the few to attack Vauban overtly and violently, and the flood of pamphlets in Vauban's defense on that occasion indicates that this only augmented the reputation for scandal the author of *Les liaisons dangereuses* had acquired.[35]

Vauban was equally famous among laymen and military engineers. Even the sour and snobbish Duc de Saint-Simon (1675–1755) called him "the most gentlemanly and the most virtuous man of his century and with the greatest of reputations as the most knowledgeable man in the art of sieges and fortification, the simplest, the truest, and the most modest."[36] In a court seething with intrigue and ambition, the duke thought that there was "never a sweeter, more compassionate, more obliging man, but respectful with not a trace of polish, the most stingily sparing of the lives of men, with a worth that took everything upon himself and gave everything to others. It is inconceivable that with so much uprightness and frankness, incapable of lending himself to anything false or bad, that he could gain to the degree that he did the friendship of Louvois and the king."[37] The care with which Louvois and the king attempted to secure Vauban's personal safety on numerous occasions is an eloquent testimonial of this.[38]

It was not merely Vauban's activity as an engineer and soldier that attracted the attention and admiration of contemporaries, but also his economic and social ideas and the courage with which he was prepared to present them publicly. Although a devoted servant of the Crown and a firm believer in royal absolutism, he saw many of the vices afflicting French society and the state and wished to take decisive steps to remedy them in order to make absolutism viable.[39] Thus, for example, he called for religious toleration of the Huguenots and a military nobility open to merit.[40] Suggestions that internal customs barriers should be abolished, that the country should have a common system of weights and measures, and that there should be equality before the law complete what looks remarkably like the agenda of the French Revolution. A keen and relentless observer of the human geography and economic condition of France, he was always a fertile source of sound schemes for improving the wealth of the country. But it was his radical proposal for an equitable and rational system of taxation published toward the end of his life that attracted not only the admiration of Saint-Simon and others for his courage and originality but allegedly the ire of Louis XIV.[41] It was in this connection that Saint-Simon wrote: "Patriot that he was, he had been touched all his life by the misery of the people and all the vexations that it endured."[42] Thus, when Fontenelle called Vauban "a Roman whom our century seemed to have snatched from the happiest times of the Republic," he was making an allusion as much to his civic courage as to his military virtue.[43] Books on Vauban continue to be published for a general public in France to this day.[44]

Unlike many other historical figures for whom French streets are named, Vauban's presence is still tangible in a very literal sense in spite of the ravages of modern urbanization. His engineering achievements often still stand and delight the eye: sometimes virtually intact as at Neuf-Brisach in Upper Alsace, sometimes overrun by parking lots and girls' schools as at Besançon, sometimes perceptible only in the plan of the town. Naturally enough they form the core of the reputation within the military engineering corps of the French army from his time to our own that led to a virtual cult of Vauban in the eighteenth century. Nor was it merely French military engineers who fell under the spell of Vauban. John Müller, professor at Woolwich, wrote that all the European powers copied Vauban's estimates for the resources necessary in sieges.[45] Müller was to pay Vauban a compliment that he was not alone in giving, in his plagiarizing of substantial portions of Vauban's works on siege warfare. Well into the nineteenth century,

English and Prussian engineers, even when being critical, considered him the greatest of their kind. For another professor of fortification at Woolwich, writing a century after Müller in 1887, "the general principles which Vauban was the first to grasp, and which his rules embodied, remain as applicable as ever."[46]

Vauban directed the building or renovation of more than 160 military sites, not counting ports and canals, ranging from the great fortresses and citadels of Lille and Neuf-Brisach to small forts and outposts such as the fort at Aix that Montalembert wanted to replace.[47] A military engineer of the eighteenth century would have had a considerably better than even chance of serving in a fortress that had been built or rebuilt by Vauban himself. (In 1777 there were 172 fortified sites under the control of the Royal Corps of Military Engineering.[48]) He also conducted 48 sieges—all successful—during his military career. (Twenty of these sieges were under the personal command of the king and four under that of the Dauphin.) His prodigious physical and mental energy is almost incredible and shows forth in his *Oisivetés* (Idle Trifles), a title both apt and ironical for the no less than 12 manuscript volumes of works on siegecraft, economics, and politics he wrote in his rare moments of leisure.[49] Almost always on missions for the king and Louvois, he literally covered France with his handiwork and gave it much of its present shape. Adding to his reputation after his death was the substantial slowing of construction of fortification in France by an increasingly impecunious government suffering in large measure from the legacy of the costly military campaigns of Louis XIV and his massive construction programs.

Choderlos de Laclos, in his attack on Vauban, made much of the cost of Vauban's fortifications. Though the available figures suggest exaggeration on his part (he claimed that 1.4 billion livres had been spent on fortifications during the reign of Louis XIV), the expenditures were indeed substantial. It takes a foolish temerity to claim anything close to a precise knowledge of government budgets during the Old Regime. Not only are historians reduced to dubious estimates; ministers of the period were not in a much better position to know what was being spent.[50] Fortifications budgets were occasionally split between maritime and land fortifications. Cities and provinces contributed funds in addition to those of the royal treasury, and funds were also obtained from various "extraordinary" or other sources such as separate allocations for military buildings and barracks. There were great fluctuations not only in the fortifications budget but also in the total budgets of the country. In 1672, the year the Dutch War began, the projected budget was about 71 million livres, with about 780,000 livres budgeted for fortification.[51] The year after Louis XIV's death and the end of his disastrous and expensive wars, the annual budget was estimated at 83 million livres. In 1705, at the height of the War of the Spanish Succession, the annual budget soared to almost 219 million livres.[52] Fortification budgets fluctuated during this war from about 2 million to 3 million livres per annum, but spending on fortifications before this period was still substantial. It peaked at 12,678,609 livres in 1689; it reached its minimum of 625,018 livres in 1694.[53] Again, the question arises what these figures include. Forbonnais gives an average annual expense of 4,046,054 livres for fortifications, ports and buildings between 1682 to 1715.[54] Fortifying Neuf Brisach cost 4,048,075 livres, and the amount budgeted for fortifying the military port of Toulon and its arsenal was 8,640,000 livres (not all of which may have been spent).[55] In any case, it does appear that overall spending declined over the century after Louis XIV's death. Augoyat gives a figure of an average of 3,700,000 livres per annum over the period 1777–1787, a period that saw a renewal of activity in building fortifications with the major works at Cherbourg.[56] Moreover, it seems that this may include amounts formerly not included in the royal budget that came from towns and provinces. A major proportion of the expense was on maintenance and repair rather than new works. Thus, although the budgets during another period of relatively intense activity (in the ports of Brest and Toulon between 1764 and 1769) show only 32.6 percent spent on maintenance, the accounts for 1774 show 68.5 percent for maintenance.[57] Judging by other accounts, the latter figure was more typical, and maintenance occupied an increasing proportion of engineers' attention.[58] Therefore, because of a general pacific trend in the early reign of Louis XV after the military adventures of his predecessor, there was simply less work to do for military engineers. As a result, Vauban's achievement, magnificently deployed for all to see, was enhanced by the lack of any comparable program of fortification.

A final reason for Vauban's overshadowing other French engineers—a reason of which he was well aware—was his longevity, something that was remarkable in his profession. His military career, which had begun at the age of 17, spanned 57 years of almost ceaseless activity. Mere survival gave him not only seniority but experience:

> Formerly there was nothing more rare in France than the members of this profession [military engineering] and the few there were lasted such a short time that it was rarer still to have any who had been at five or six sieges, and even more rare were those who had been at that many without having received many wounds, which, putting them out of action from the beginning or the middle of the siege, prevented them from seeing the end of it and consequently becoming more knowledgeable. . . .[59]

Vauban remembered that as a young engineer at the siege of Montmédy in 1657 he saw all three of his fellow military engineers killed by the fifth or the sixth day of the siege and before the end he himself had "received four wounds, mostly light [flesh wounds], and several other cuts in his clothes."[60] His own repugnance to the direct and glorious but bloody and often ineffectual assaults on fortifications without long and methodical preparation not only raised the prestige and fostered the improvement of military engineering but was more economical of the lives of military engineers, which led to additional intangible but definite improvements in the competence of its practitioners and the level of the art. By the end of the century, although casualties were still high among engineers, better recruitment, training, and methods of siegecraft had considerably dropped the mortality rate. His own survival, in the early years of his career before the king took special measures to ensure that his life was protected, owed as much to luck as to his own talents.

Although Vauban is a larger-than-life subject, his very stature leads to great difficulties for the historian, who has to avoid succumbing to a well-established tradition of hagiography or to avoid the revisionist temptation of indiscriminate slaughter of sacred cows. The difficulties are compounded by the publication history of Vauban's works. Most of Vauban's enormous literary production is in the form of letters and memoranda to his superiors and subordinates. Many of these, particularly the plans of fortresses, were kept at the Dépôt du Génie where they were difficult of access even for military engineers. The evidence indicates that he was never eager to publish or even to write treatises dealing with the craft of the engineer and may only have done so at the request of the king. There are three plausible reasons for this: his hectic life that left little time and freedom from interruption that a formal treatise would have required, the need for military secrecy, which Vauban himself always emphasized, and his allergy to any kind of "system" that he felt was a characteristic of the armchair fortifiers who were held in such low esteem by professional soldiers.

In spite of Vauban's systematic bent and his insistence on order in design, execution, and reporting, he was never a builder of *systèmes* in the pejorative sense of that word in the eighteenth century. Rather he was a builder of systems in the sense attached to that word by the modern historian of technology Thomas Hughes. Although Vauban wrote on the art of attacking or defending fortresses, he never did publish or even write a treatise of fortification, the theory of designing and building fortresses, as opposed to the art of attacking or defending them.[61] This reflects fairly well his activity and his accomplishment and provides the first characteristic lineaments of his oeuvre: his genius was dynamic rather than static. He interacted with fortresses more than he planned them, and, in spite of such impressive works as Neuf-Brisach, most of his building activity, too, was improvised and hurried, often rebuilding, renovating, and modifying rather than building from scratch. This accorded well with the anti-scholarly and anti-systematic bent that he displayed to his close colleagues, such as his secretary Thomassin, who pleaded with him in vain to publish a work on fortification. The closest that one can probably come to a formal exposition of principles—Vauban would have called them maxims—is the treatise of Joseph Sauveur (1653–1716), mathematician, member of the Academy of Sciences, and Vauban's own appointed successor as examiner of military engineering candidates. (Vauban's promotion to Marshal, according to Fontenelle, had made it undignified for him personally to examine candidates.) Sauveur took the trouble to go on campaign (to Mons in 1691) and observe the progress of sieges, and there is a strong possibility that Vauban had some input into the work.

Vauban's Military Architecture

Montalembert's later criticisms of Vauban were not in the domain of siegecraft, where he paid tribute to Vauban and even argued that his own new system of fortification had become necessary precisely as a result of Vauban's revolutionary innovations in this domain. His criticisms dealt with the accepted way of building fortifications. Many scholars and foreign military engineers concurred with Montalembert's evaluation:

Vauban's improvements in the mode of attacking fortresses were the most considerable and the most lasting of his service to the art of war, and he put a seal to them by writing his treatises. . . . But he did not leave his mark on the art of fortification [in spite of immense labors in fortresses building] to a corresponding degree, nor has he put on paper with the same completeness his ideas respecting it.[62]

Emulators of Vauban's military architecture and many historians of fortification, in a tradition that seems to have started with Sauveur, talk of his three systems.[63] A system of fortification was the trace and profile of a fortified area in ideal conditions. Each particular fortification could have its own "system" or, as was the more common use of the word, there could be a general "system" or general type of fortification of which there were particular examples or applications. The word 'system' suffered the general opprobrium of the word in the eighteenth century. It was much used by both sides in the polemic over Montalembert's fortification, with both sides accusing each other of creating "systems" of fortification. Thus, the word was used in two senses: as a simple synonym for the particular design of a given fortress or as a general theory of design for fortifications. In the latter sense it almost always had the negative connotation of abstract and inappropriate.

Yet it is misleading to speak of Vauban's systems in the latter sense.[64] Vauban's first system, in the second sense of the word described above, was not really his system. It

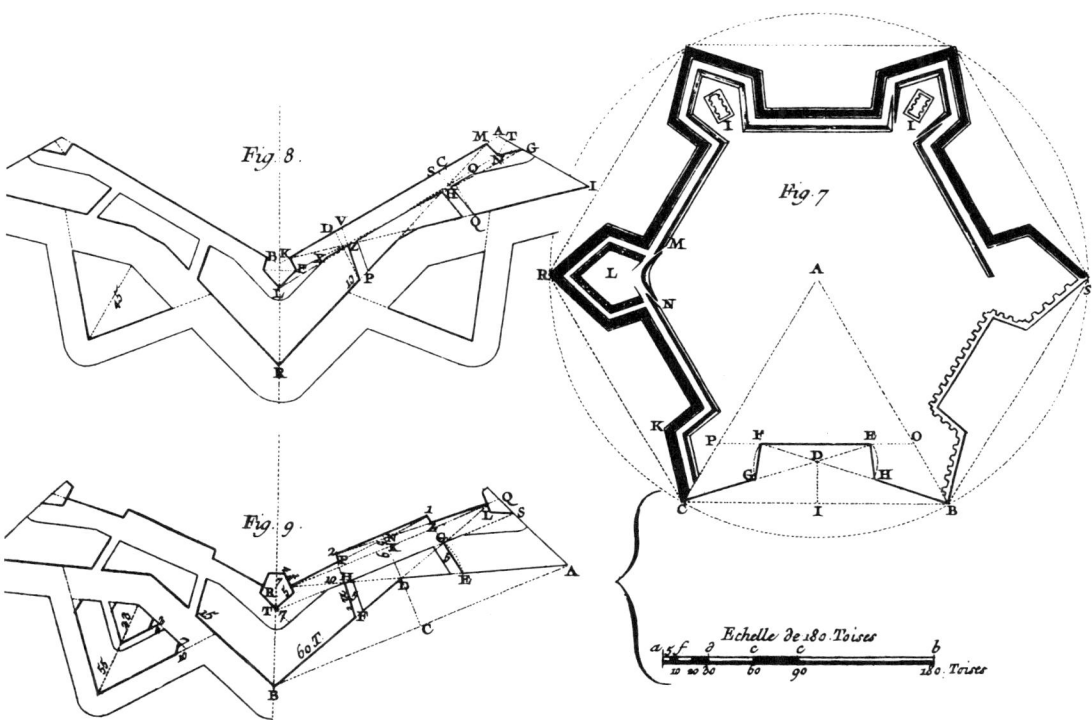

Figure 2.1
The three systems of Vauban as depicted in the *Encyclopédie*. Fig. 7, which also indicates how to lay out the ground plan of the fortress is essentially Vauban's first system; Fig. 8 is the second system; Fig. 9 is the third system, used in the construction of Neuf-Brisach.

was essentially the system of Pagan, who had already laid the groundwork for what was becoming increasingly standard practice in classical artillery fortification as it came to maturity in France in the seventeenth century. The classical system of artillery fortification (sometimes called the trace italienne because of its Italian origins) had three basic elements.

First, it had a characteristic ground plan or trace whose geometry ensured that each part of the perimeter could provide flanking fire in support of other parts and receive such support itself. This gave the trace the overall star-shaped pattern that enabled an effective cross-fire that left no so-called dead ground where the enemy could find a relatively sheltered area unexposed to fire from the fortress. On flat, featureless ground, these plans took the basic shape of regular polygons that were the delight of amateurs and offered ample possibilities for geometrical fantasies in all kinds of aesthetically pleasing variations.

One could say that "earth" was the second feature of the classical system of artillery fortification: they were close to the earth—that is low rather than high—and packed with earth or in some cases even were earthworks. Unlike medieval fortifications, which relied on high and massive masonry walls to provide a barrier to the assailant, artillery fortifications hugged the ground to offer less of a target to artillery. The bastion was a tower that had become angular and had diminished in height and increased in area. The packed earth of the bastion, usually revetted with masonry or (if available) bricks, was more resilient and capable of absorbing artillery fire than hollow masonry, and the increased area provided a spacious gun platform so that attacking fire could be answered with defensive fire.

The final feature was defense in depth. The *enceinte* or perimeter of the fortress was not merely a line enclosing what was to be defended. It was the edge of a zone that stretched farther and farther into the countryside to keep the attackers' cannon away. Beyond the walls of the fortress there was a spacious ditch, sometimes filled with water and sometimes dry, which contained detached defensive structures such as *demi-lunes* and *fleches* (triangular gun platforms protected by a parapet), *tenailles* (similar gun platforms that looked like a wilting bow tie in plan view and were positioned before the curtain walls which joined adjacent bastions of a fortress), and counterguards (earthworks, sometimes revetted, that covered the salient angle of a bastion) in various combinations where the principle again was to ensure flanking fire to and from other parts of the defenses. Each would have to be stormed or neutralized separately by the attackers on their push toward the walls of the enceinte. On the other side of the ditch (the counterscarp), just before the brow of the ditch, was a covered sunken walkway from which defending soldiers could fire under cover at attackers moving up the glacis (a bare artificial pitch sloping away from the fortress). Palisades and sharpened wooden stakes were also placed at various vital positions. This defensive area grew with time. During Vauban's career the distance from the inner walls to the edge of the covered way increased from 290 feet to 702 feet at Neuf-Brisach.[65]

Guillaume Leblond, in his article "Fortification" and the accompanying plates in the *Encyclopédie*, discusses the construction of the trace of Vauban's first system (See fig. 7 in figure 2.1 above). This trace is constructed by taking one side of the basic polygon, which was not to exceed 180 toises, bisecting this distance BC at I, and dropping a perpendicular ID toward the interior of the fortress one-sixth of BC (30 toises in length).⁶⁶ This proportion varied according to the number of sides in the polygon. (For six or more sides it was ⅙, for pentagons it was ⅐, and for squares it was ⅛.) Along the prolonged lines BD and CD (the "lines of defense"), a distance equal to 2/7 of the side of the polygon (in practice this was usually rounded to 50 toises) was marked off from B and C to give the length of the faces (BH and CG) of the bastion. Then, with a compass successively centered on points H and G, with a radius of HG, the lines of defense (BHD extended and CGD extended) were cut at F and E respectively. This defined the lengths of the flanks FG and EH. The curtain was now the line FE. The whole module, consisting of two halves of adjacent bastions and the curtain joining them, was called a "front of fortification." The process was repeated to produce the final trace.⁶⁷

In practice, a builder of a fortress could rarely build a perfect geometrical plan, because of the constraints of the surrounding terrain. The perfect octagonal fortress of Neuf-Brisach was rather exceptional. Vauban always emphasized adaptation to the terrain and avoidance of blind copying of formal models. According to him "the art of fortification does not consist in rules and systems, but only in good sense and experience."⁶⁸ His style is that of a great improviser. Every fortress was different and deserved a different approach. Different approaches were necessary because of different natural settings and, in many cases, existing fortifications, which often required quick and serviceable modifications because time, finances, and (occasionally) local concerns prevented them from being razed to begin afresh. Indeed, this may explain both the generality of some of Vauban's maxims and his reluctance to write a formal treatise of fortification. When pressed by his secretary Thomassin to do so, Vauban is alleged to have replied: "Do you want me to teach that a curtain is between two bastions, that a bastion consists of an angle and two faces, etc.? That is not my way."⁶⁹

By the late 1680s, experience, including his own successes in attacking fortresses, appears to have led Vauban to develop a new type of fortification that came to be more common toward the end of his life, especially where he undertook major new works. The main feature of this new type of fortification, which appeared in both the second system and the third,⁷⁰ was the bastioned tower. Besançon was the first to get bastioned towers in 1687, to be followed by Belfort and Landau. In the 1690s Vauban proposed such towers for other fortresses, and they were included in his magnum opus at Neuf-Brisach, a perfect octagon built on a flat plain.⁷¹ In the second and third systems, the bastion was increased in size and detached from the walls. Where the bastion had stood, there was now a much smaller two-story tower having the familiar pentagonal ground plan of the bastion, with cannon protected from direct enemy fire at both levels. Two important changes were implicit in the bastioned towers. They allowed the general increase in the depth of defense that occurred over Vauban's career. The detached bastion was one more obstacle to be overcome by the attackers and together with the bastion tower functioned as a sort of double bastion. Formerly, the fall or even the breaching of the bastion meant that the enceinte had been penetrated, and according to the accepted traditions of war it enabled the governor of the fortress to surrender with dignity. Further resistance was futile. With the breaching of the bastion, the fortress was now open, and it would be in poor taste to continue fighting and waste the king's men and open the place to the horrors of sack and pillage. Besides providing extra resistance, the detached bastion would also make this attitude, which Vauban despised, psychologically untenable because the fall of the bastion did not mean that the enemy had broken into the fortress. The second major change brought about by the bastion tower was the protection of cannon. Up to that time, cannon were in the open air, subject to the fire of enemy batteries in a generally uneven artillery duel, to which were added the horrors of the ricochet shot extensively used after 1697, as well as to a rain of lethal mortar bombs that would drive the defenders from the ramparts.

The idea of casemated batteries dated back to the Renaissance and to the very beginnings of artillery fortification.[72] But it was not popular in France, where de Ville and others had argued against it. The objections usually raised were the effects of smoke (which could not be easily evacuated from closed spaces, and which hampered the soldiers serving the cannon) and a belief that a national trait of French troops was their unwillingness to be cooped up in dark and smoky caverns.[73] Masonry embrasures for the cannon were also more dangerous than earthen parapets, because flying chips of stone from enemy cannonballs hitting the embrasures could wound and kill the defenders. It was also believed, correctly in most cases, that masonry casemates could be destroyed more easily than the earth-packed bastions once enemy batteries got into range. Finally, masonry, especially vaulted bomb-proof chambers, was more expensive than earthworks.

Vauban's interest in masonry casemates and covered guns in his later years was rather unusual among French military engineers, and after his death it was abandoned.[74] The packed-earth bastion, serving as an open-air gun platform, was preferred in the "modern" system of fortification as it was practiced during the eighteenth century up to Montalembert. In view of the French military engineers' cult of Vauban, it is remarkable that his disciples rejected this important characteristic of his mature work. Some

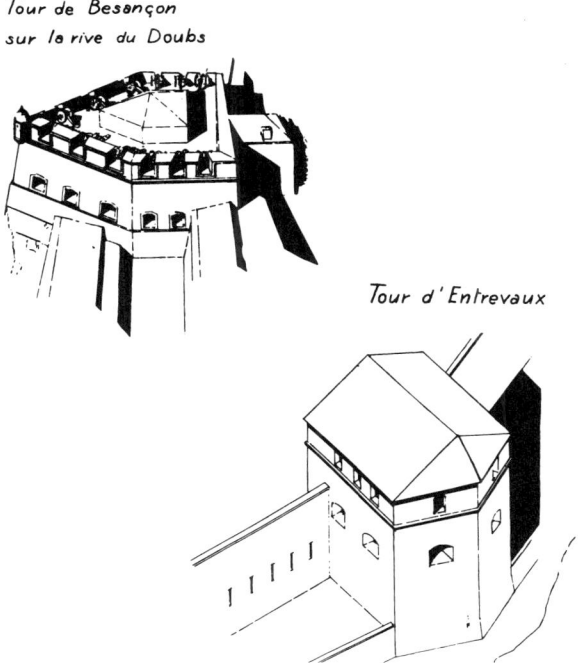

Figure 2.2
Vauban used tower bastions in his second and third systems. Most of the surviving bastions, such as those at Besançon, have roofs. Source: Rocolle 1973, volume 2, p. 130, sketch 141.

historians of fortification, following Montalembert and Carnot, have argued that Vauban's work was denatured and his legacy distorted by the writings and prestige of Louis de Cormontaigne (1696–1752), abetted by Charles-René Fourcroy de Ramecourt (1715–1791), later the unofficial head of the French military engineering corps, who edited and published Cormontaigne's work and used his position to make it the orthodoxy of the corps of military engineering. What eighteenth-century military engineers called Vauban's modern system of fortification was actually that of Cormontaigne. Montalembert and Lazare Carnot even went further and accused Cormontaigne of consciously and surreptitiously distorting Vauban's legacy.

Yet one cannot claim that the mature Vauban had embraced the idea of casemates. The casemates and the tower bastions at Neuf-Brisach are on the last of a number of defensive lines that consist of parapets containing cannon and musketeers in the open. The other examples of casemates, the tour de Camaret and the Château du Taureau,

Figure 2.3
Gun tower at Camaret built by Vauban. Source: *Fort* 23 (1995), p. 64.

are coastal fortifications that could only be attacked by relatively vulnerable warships rather than by siege batteries protected by earthworks. Covered guns were used in certain situations by the ever pragmatic Vauban but were not a general principle of his fortifications.

Vauban's "Method"

If Vauban remained silent about his work as builder, besieger, and defender of fortresses, others did not, and there were many who claimed to offer the public the secrets of his "method" or "system" of fortification. The proliferation of such texts is in itself a tribute to Vauban's prestige and influence. When Hebert, "professeur royal de mathématiques," published a second edition of Pagan's treatise on fortification, he included the "manner of fortifying of M. de Vauban."[75] Other writers included Vauban even more prominently and centrally than Hebert. In 1689, the same year in which Hebert published his work, the chevalier de Cambray published his *Manière de Fortifier de Mr. de Vauban*.[76] It included a separate and succinct treatise on geometry with references to plates and giving detailed constructions in point form.

The "maxims" that follow are familiar to anyone who has read the works of Errard and those following him in France and variants would be repeated a century later by Fourcroy de Ramecourt. The list is worth reproducing as another example of the often incoherent mix of geometrical principles and practical tips that were becoming to be considered the basis of the science of fortification.[77] Vauban would certainly have approved them precisely because they are a mix of geometrical truths, rules of thumb, and plain common sense, leaving much scope for interpretation and individual judgment on the site:

(1) All parts of the fortification must be flanked, i.e., exposed to flanking fire from other parts of the fortification.

(2) Everything must be face, flank, or curtain and capable of resisting the first cannon shots.

(3) Fortification with geometrically regular ground plans is better than irregular fortification.

(4) It is impossible to fortify a triangular ground plan regularly because the gorge angle is less than 90°.[78]

(5) The smaller the central angle of the fortified polygon, the better the defense; i.e., the more sides a polygon has, the better the defense.

(6) The salient angle of the bastion must not be less than 60° and not more than 100°.

(7) The bigger the angles, the easier to fortify.[79]

(8) Curtain angles must be greater than 90° and not less than 110°.

(9) Bigger bastion gorges [the distance between the junctions with the curtain wall of the flanks of the bastion] are better than smaller ones.

(10) Big flanks are better than small flanks.

(11) A bastion's face should never be smaller than half the curtain length.

(12) Bastion faces should be defensible by musketry in adjacent bastions.

(13) Shorter faces are better than long ones.

(14) The curtain should be longer than 60 toises and shorter than 100 toises (unless the fortification is irregular).

(15) The line of defense should be between 120 and 125 toises.

(16) The talus (or batter) must not be excessive.

(17) Wider and deeper ditches are better.

(18) Dry ditches are generally better than wet ditches.

(19) In smaller fortresses where sorties are unnecessary, wet ditches are better than dry ditches.

(20) Outworks should be lower than the main ramparts.

Several years later, without specific acknowledgment, Naudin borrowed large portions of Cambray's work for his own book. He freely admitted, however, that little about his work was original. Naudin's book begins with "La geometrie pratique sur le papier," which is essentially a treatise on the geometrical construction of solids. The second part, "La geometrie pratique sur le terrain," is essentially about laying out the ground plan and estimating earthworks and volumes. The separate treatise on fortification proper that follows gives detailed dimensioned plans and profiles for the first system of Vauban's fortifications as well as the fortifications of Landau and Belfort (Vauban's second system). There is a chapter on irregular fortification (which was to be avoided if possible), and a chapter of tips on site management and construction ("La Fortification effective"), such as how to ensure that contractors don't cheat. The reader is enjoined to calculate dimensions rather than take them off plans, in order to get accurate estimates of earthwork and stonework necessary. There is a chapter devoted to foundations and pilings, palisades, barriers, bridges, and gates. The discussion of how to attack a fortress advises that bribery is by far the cheapest way and should be resorted to whenever possible, and that the conquerors should always be indulgent to the local townsmen so as to gain their loyalty. Naudin gives a case study with a map of the siege of Charleroi by Vauban in 1693.[80] There are chapters on defense and on mining.

Commentaries on Vauban and descriptions of his methods of fortification were common in France and abroad. By the end of the seventeenth century, Vauban's work had already achieved canonical status even though he had published nothing himself. At best the abbé du Fay, another in the schoolmaster tradition of fortification, could boast a curt signed testimonial in his book from Vauban stating that "this little treatise on fortification contains nothing that does not conform to . . . practices in the king's fortresses."[81] The author, whose work was translated into English, made a great point of the authenticity of his work and referred to other unnamed works that falsely claimed to present the systems of Vauban. The title of the work published at the Hague in 1708 by Leonhard Christofle Sturm, professor of mathematics in the Ducal Academy of Wolfenbüttel and later at Frankfurt-on-Oder, *Le Veritable Vauban se montrant au lieu du Faux Vauban* . . . , is symptomatic of the proliferation of books on Vauban attempting to extract from Vauban's work his method, which was considered to be the summit of achievement in fortification.

How is one to interpret the abundant literature purporting to present the "true" Vauban? This literature implies a traditional view that fortification is an activity that is doubly secret owing to its military aspect and its craft aspect, and that its recipes might be stolen and unveiled by Huguenot renegades or by public-spirited authors. True, there is more often talk of "method" than of recipes; nevertheless, there is still a plurality of methods. Deidier talks of the "methods of the most skilled authors of Europe," and there are, of course, the "national" methods already mentioned. And this would seem to imply a diversity of approaches rather than simply a diversity of results, possibly all valid, that undermines Errard's dream of a universal science and the "reduction of fortification to a (true) art."

There is also an underpinning of mathematics in the various methods whose aim is not only to allow easier comprehension and diffusion but also some demonstrability, not to mention association with a prestigious and legitimate science unconditionally accepted by all. Furthermore, not only in France is the method of Vauban more and more considered to be the best. This growing dominance of one method of fortification that overshadowed the others must have strengthened the claims to legitimacy of fortification as a true science both demonstrable and certain. If the presumptuous dream of the late Renaissance that fortification was a mathematical science "according to Euclid" was tempered and even undermined in the actual contents of textbooks, it remains in titles and lurks within prefaces. Vauban himself clearly rejects a view that fortification is a set of mathematical rules, that there is a science of fortification not only grounded in mathematics but mathematical in essence. For Vauban, mathematics is undoubtedly indispensable as a tool, but the science, more properly the art, of fortification is not mathematical. Henry Guerlac's claim that Vauban applied "science" to warfare resides essentially in the view that Vauban was indeed quantitative, methodical, and orderly in his approach to administration, design, and construction.[82] It does not mean that Vauban ever attempted to discover or even believed in the existence of general principles that were mathematically expressible and could be subjected to formal analysis and from which one could derive a limited number of unequivocal design and construction procedures. Oddly, the prestige of his career and his oeuvre may have contributed to keeping alive a belief that had never gone uncontested and was usually rejected, especially by Vauban himself: that there was some kind of "scientific" key to fortification based on mathematics.

No doubt there is some conflation in some minds, particularly those of mathematics' masters, between Vauban's putative "method" and a universal science. But for most working military engineers, Vauban's methods—in the form of earthy advice and working tips—seem to have been accepted for the prestige and success of their illustrious author as well as their down-to-earth practicality. Most thought that Vauban's fortresses were better than that those of others. There was little suggestion—until the embattled Fourcroy de Ramecourt, fighting off the attacks of Montalembert on the corps of military engineering, asserted this toward the end of the century—that Vauban possessed the key to some universal science. This would come later as a defensive reaction to attacks on the methods of military engineering corps for whom Vauban had become creator, model, and guardian spirit.

Vauban and His Masters

By 1651, when Vauban began his career, the number of French engineers had more than quintupled from the time of Henry IV.[83] This was due in large part to the massive expansion of the French army that was accelerated during the reign of Louis XIV.[84] During his reign, France became the first European state where the numbers of military effectives exceeded the numbers of the clergy.[85] It obviously had some effect on the nature of engineers as a social group because the days of the individual, cosmopolitan engineer of fortune were gone. Now engineers formed part of a national organization of specialists administered by a royal bureaucracy. Indeed, they themselves were becoming bureaucrats. It was this bureaucracy that gave French military engineering its particular nature and its unmatched effectiveness. The towering personality and attractive figure of Vauban has tended to obscure this important fact. Without denying Vauban's personal achievement as an engineer and his role in molding the nature of the French military engineering corps, it must be conceded that some of his admirers tended to minimize and even ignore the vital support he received from Louvois, the Minister of War, and King Louis XIV himself. Following Gaston Zeller and others, Philippe Truttmann argues that "that which a little too hastily is called the system of Vauban was the work, if not of the king, then at least of the team Louis XIV—Louvois—Vauban."[86] Without either the unflinching support of political figures committed to fortifications as a central feature of national policy and military strategy or the reliable backing provided by a relatively efficient and powerful administration in the hands of determined masters, many of Vauban's achievements would have been impossible. Vauban never enjoyed a perfectly free hand and was often overruled, not merely on financial matters but also on specific matters dealing with fortification. This was not always fortunate for the development of fortification. One of his forward-looking ideas—that cast iron rather than bronze cannon be tried in fortresses, an idea later taken up by Montalembert—was rejected by the Grand Master of Artillery, the Duc de Maine. Another prescient suggestion by Vauban—a new kind of gun carriage for fortress guns, similar to naval gun carriages—was overruled by Louis XIV, who wanted regular gun carriages such as those used by field guns so that fortress cannon could be used on campaign in case of necessity.[87]

Louvois too participated actively in fortification policy and even in fortification design. The role of Louvois in French history has recently been subjected to re-evaluation, notably by André Corvisier. Earlier historians were inclined to condemn him as the executor or even the inspirer of Louis XIV's disastrous militaristic policies.[88] He was contrasted with Colbert, seen as the pacific architect of lasting French economic and industrial achievements. However, like Colbert, Louvois was a laborious and energetic administrator whose attention to detail and capacity for work were legendary. Vauban was not the only talented engineer in Louvois' service and Louvois corresponded personally with all of them.[89] There is no doubt that Louvois was one of the most eminent representatives of an administrative apparatus that had developed in France from the end of the Middle Ages.

Louvois' unfailing support for Vauban is well known.[90] It is in large part as a result of this support, along with Louvois' painstaking and intelligent attention to logistical and administrative detail, that Vauban could accomplish his own work and leave an organization of military engineers which could dispense with a Vauban altogether—a true corps of military engineers. Vauban's great Dutch rival Menno van Coehoorn (1641–1704) in some respects had a reputation that matched Vauban's as a master of siegecraft and builder of fortifications. Yet the organization he left behind him could not match Vauban's, because of the effectiveness of French governmental bureaucracy. For Duffy, "where the essential difference between the methods of Coehoorn and Vauban resided was not so much in the ability of the two engineers, or even the resources at their command, as in the almost unlimited authority which the Frenchman possessed to shape the progress of the attack according to his wishes."[91] Along with Vauban's authority on the battlefield, he had the support of the organization of military engineers, which he and his masters created. Frederick the Great's French military engineer Lefèbvre complained that "It is only in the French service that you can carry out [siege] operations with real proficiency, for in that army the technical officer is treated with respect and given everything he needs. It is not the individual who carries the burden, but the corps as a whole."[92]

The features of the governmental bureaucracy associated with the corps were not those of today's bureaucracies. In spite of the similarities of Louis XIV's bureaucracy to modern bureaucracies, their common features should not be exaggerated. Those who see the reign of Louis XIV as an age of austere classicism and rationality, comparing his government to a Cartesian machine, sometimes confuse unrealistic ideal models of bureaucracy with the actual daily functioning of his administration.[93] The webs of patronage, intrigue, and favoritism are so difficult for the historian to unravel that they are often ignored. It has been remarked that the initial growth of the modern French state was characterized by successive overlays of more modern features on existing forms that continued to a greater or lesser degree to retain their vitality in spite of their growing anachronism.[94] Traditional forms, clienteles, and power structures continued to coexist and interact with some elements of a model of a rational bureaucracy that some historians have attempted to project backward onto a rich and fluid reality. If the rational bureaucratic ideal existed in the rhetoric and the aims of those on the top of the bureaucratic pyramid, its incarnation in practice was a much slower and more muddled affair. Yet the French military engineering corps, and perhaps to a greater extent the Corps des ponts et chaussées, was one institution where the ideal was made real to a greater extent than elsewhere.

When all caveats have been entered and all qualification have been made, however, Vauban still looms as a colossal personality both in character and achievement. Louis XIV was both greatly interested and competent in fortification. Racine, who accompanied Louis to the siege of Namur in 1692, indicates quite clearly the importance of sieges for his warlike master. In his "Relation of that which took place at the siege of Namur," Racine mentions Vauban only once.[95] Yet, in a personal letter to Boileau from the same siege, where the courtier could forget his duties as the king's official historiographer, Racine writes that "M. de Vauban, with his cannon and his bombs, carried out the whole expedition himself."[96] It is true that Vauban was a member of a team, but what a team it was! To be a distinct and respected member of a trio that included the Sun King and his dictatorial War Minister was no mean achievement. It is true that, in the final analysis, Vauban's role was consultative: he never even attained the executive leadership of his corps. However, his masters always took his advice seriously, even when they sometimes rejected or modified it, and Vauban always spoke his mind forcefully.

Engineering France

According to Michelet, "France made France." In a sense, all countries make themselves, but more than most countries, perhaps, France is a constructed country. It was constructed in the sense that it was expanded from a center to a periphery, constructed because earlier than in most other countries a centralizing and rationalizing administration sought to impose a pattern—social, administrative, productive, and even visual—on it, and constructed as a geometrical territory. Engineers, both military and civil, played a large part in constructing France's infrastructure and its boundaries. The building regulations for civilian houses at Neuf-Brisach, Vauban's chef d'oeuvre on the upper Rhine, are a reflection of the centralizing project of engineers and their superiors that was extended to the country as a whole:

Article 1. All those who build must make the facades of their buildings conform to those that have already been constructed on the main square both for the decoration of these facades and the height of the cornices and for the size of the shops, doors, and windows, which must all be the same, as well as the height of the roofs.[97]

During Louis XIV's reign there was an energetic and on the whole successful push toward the so-called natural frontiers of France and a major attempt to create an artificial frontier—a *ceinture de fer* (belt of iron)—where no natural geographical frontier existed. If the concept of France as a geographical space that could be consolidated, improved, and managed was still in its infancy, it was nevertheless present. There were a number of good reasons for the growing consciousness of France as a territory to be managed. Beyond ideological reasons, the most important ones came from disciplines cultivated by mathematical practitioners and engineers. They were the progress of cartography, the continuing development of hydraulic works in both the military and the civilian sphere, the growing interest in statistics, and the importance of gardens as a special space for the nobility.

Richard Westfall has documented the explosion of "map consciousness"—the realization of maps' capacity to summarize and convey information—in Europe.[98] From fewer than three dozen known maps in England before 1500, the number rises to 200 in the next 50 years and is virtually impossible to count after 1550. The situation in the rest of Europe and France is undoubtedly similar. The engineer, secret agent, and mapmaker Nicolas de Nolay, who held the title of Geographer to the King, was working on drawing maps of all the French provinces for the Regent Catherine de Medici in the 1560s. The purpose of these maps was clearly to improve governance by getting a better idea of the complex administrative divisions of the country. Already cartography sponsored by the state was closely linked to the gathering of information other than geographical. Colbert, with his enormous appetite for facts and information, ordered officials to provide memoranda on every province giving information on the divisions and subdivisions of various ecclesiastical, fiscal, military, and judicial districts. Two of the main tasks assigned to the Royal Academy of Sciences, which functioned under Colbert's aegis from 1666, were making maps and measuring the meridian. Sébastien Leclerc's famous painting of Louis XIV and Colbert visiting the Academy of Sciences (a scene that never in fact occurred) has in the background a formal garden and an observatory under construction and shows Louis perusing a ground plan of a fortification; a map lies on the floor. In very significant ways, all these objects were related.

Jean Picard (1620–1682), one of the first members of the new academy, made geodesic triangulation measurements from Paris to Amiens, an operation that would eventually be extended from Dunkerque to Perpignan in order to obtain precise positions of a triangulated grid that was to serve as a basis for a map of the country. A corrected map of France, made by Picard and La Hire, using more precise determinations of longitude proposed by Jean-Dominique Cassini, was sponsored by the Academy and finished in 1682; it was published in 1693. By making more precise the outlines of the country (and moving back what had been thought to be the position of the coastline), it caused Louis XIV to joke that his academicians had robbed him of more territory than his enemies. The prolonged wars with those same enemies hampered the process of mapping the entire country on a scale of 1:86,400 (1 ligne to 100 toises), and the project would only be resumed in 1747 with the great enterprise run by the Cassini family. After many difficulties, a series of properly triangulated maps that covered the country was almost ready in 1790.[99]

Military engineers were also involved in mapmaking, either as military engineers proper or as ingénieurs géographes, whose duties were the production of maps for the army. Toward the end of the eighteenth century, there was considerable friction between the military engineers who were mapping the frontiers of the country and the Cassini enterprise. The military engineers argued that the Cassini maps sold on the open market were dangerous for national security (if good) or useless (if bad, as they claimed was often the case).[100] Questions of quality aside, it is clear that maps were giving new tools not only to soldiers but also to administrators, diplomats, and rulers. The mapping of the engineers was mostly limited to more local military tasks, but the idea of a country that could be mapped for various purposes was becoming a commonplace.

Along with mapping went the gathering of statistics. One need only think of William Petty (1623–1687) across the Channel to see that French governments were not the only ones to display a keen interest in the resources of their country and the productivity of their inhabitants. Royal intendants were barraged with requests from Colbert for information on their généralités, as were military engineers by Louvois. Vauban himself provides a good example of the thirst for figures and statistics in his "Lettre sur la manière de faire les statistiques," written to the engineer Hüe de Caligny, Director of Fortifications in Flanders, who had sent Vauban a memorandum on the Flemish-speaking parts of Flanders.[101] It was one of a series of memoranda from all parts of France intended to serve as instruction to the heir to the throne, the Duc de Bourgogne. After complimenting de Caligny on his work, Vauban lists other things that should be included. Significantly, the first addition Vauban demands is a map of the area with administrative divisions indicated and marked by dotted lines and distinguished by color. A scale of conversion from local units to those of Paris is also requested. Continuing, Vauban wants to know the productivity per unit of area of different kinds of crops, the size of garrisons, the number of ecclesiastics (broken down by order), the population (broken down by age, gender, and marital status), the number of sailors, the number of people in legal professions, and the extent of wooded and cultivated areas. Vauban is pleased to see the detailed enumeration of the people in Dunkerque by street, along with an indication of the types of crafts practiced in the city. He promises to reimburse de Caligny for hiring someone as an assistant for his further work and asks him not to forget anything of what he tells him and "not to fear adding on your own anything that comes to mind."[102]

Vauban's own *Description géographique de l'Élection de Vezelay* is a model of this kind of statistical report. Fontenelle, in his eulogy, even attributed to Vauban the creation of "modern statistics," crediting him with having taught the intendants of the provinces how to write this kind of report and "to [collect] in their généralités all the documents and notions relating to commerce and agriculture."[103] Vauban used special printed forms to gather the information he required. Along with the detailed economic and demographic inventory of the area, he commented on the quality of the wine, the poverty and demoralization of the inhabitants, and the large-scale indebtedness of the poor to the rich. Vauban the engineer had the eye of an economist before the term was born. His descriptions of the territories he visited and inspected, his appetite for information and data, and his perceptive remarks on the resources and inhabitants of various French provinces still make interesting reading today.[104] The idea of a tax on production to spread to all sections of the population presented in his *Projet d'une Dixme royale* (1707), which in the year of his death allegedly earned him the wrath of the king and the admiration of contemporaries and descendants, has gained him a place among the precursors of economic thought in France.[105] Along with triangulations and more detailed topographical maps, enumerations of the economic and human resources of the country were providing a tableau that was increasingly colorful and rich.

Improving fluvial navigation, draining marshes, constructing hydraulic machinery, and building canals were activities that were by no means unknown in the Middle Ages, but they had accelerated during the demographic and industrial revival after the depression of the late Middle Ages. In the sixteenth and seventeenth centuries, land reclamation in the Netherlands saw a vast expansion and increase in effectiveness, with about twice as much land reclaimed as during the entire Middle Ages.[106] Henri IV's great minister Sully had already planned to develop a system of canals in France to connect the Seine to the Loire, the Loire to the Saône, and the Saône to the Meuse to allow navigation from the Atlantic to the Mediterranean. Work had begun on the Briare canal connecting the Seine to the Loire; it was completed almost 40 years later (1642). Not surprisingly, Colbert worked to improve the navigability of a number of rivers from Languedoc to Flanders, but his greatest impact came with the support he gave to Pierre-Paul Riquet, who between 1661 and 1681 built the magnificent Canal des Deux Mers connecting the Atlantic to the Mediterranean. (It was also called the Canal du Midi or, as Vauban called it, the Canal du Languedoc.) Vauban was ordered to inspect and design parts of the canal and was one of the many who admired it as an outstanding achievement of modern times. Here was a canal truly national in scale that tied together two of the natural frontiers of the country, had mobilized large numbers of workers and financial resources, and in Picon's words, was truly a "pharaonic work," clearly establishing the moderns as superior to the ancients.[107]

Gardens had become an increasingly important part of princely culture since the Renaissance. They provided a locus where technology, ostentation, and art met and where engineers played a role. Gardens being linked not only to botany but also to hydraulics, because of the needs for providing water in ever more elaborate and grandiose fountains) and to architecture (because of the complex designs of the ground plans), it is not surprising that one meets the engineer here as well. Although André Le Nôtre (1613–1700), the great landscape architect of Louis XIV, was responsible for his gardens at Versailles, Vauban was consulted by Louvois (and his excellent advice ignored) on the complex problems of the water supply. The king took enormous pride in his gardens and had drafted a guide for visiting them in which every detail of the carefully chosen route was minutely regulated.[108] The king himself often accompanied distinguished visitors to view the impressive expanses of the Versailles gardens with their beautifully arranged groves, orangerie, fountains, ponds, and menagerie. The garden was a demonstration of wealth, artistic taste, technological virtuosity, and power. It was the domination, but not the destruction, of nature by art and the play of reason. It was a sort of "voluntaristic geography."[109]

Lest the foregoing lead us to think that the French classical garden was an entirely geometrical and brutal imposition on nature, Thierry Mariage reminds us that the simplistic and stark distinction between the English "natural" garden and the French rational garden is blurred in practice. Both gardens are planned and both take into account the terrain that is ignored at one's peril. In spite of its apparent rigidity, the French garden is an attempt to enhance and not to smother nature. Moreover, the best gardeners of all, such as Le Nôtre, always thought beyond the mere frame of the garden and paid attention to inserting it into a larger context in which it had to be integrated harmoniously.[110] According to Hélène Vérin, the design of gardens for princes required the same skills as those of military engineer: choosing terrain, transforming and using it for a human purpose, and managing it.[111] Making terraces and counterscarps, planting trees or palisades, or sculpting a glacis or a grove had much in common in execution and purpose.

We should not forget the military engineers and their creations on the frontiers as facilitators and contributors of the notion of a space that could be conceived as a whole, managed, and improved. They had been encompassing the kingdom with fortifications since the sixteenth century. Although this may have given them a distorted external view of the national space of the country, it was also a view that saw that space as becoming more coherent, more rational, and more defensible. To give some focus and nuance to these generalizations it is useful to look at Vauban's own ideas on the territory he did so much to create.

It may be best to begin with Vauban's conceptions of the navigable waterways that were the basic communication grid of the country, more important in most cases than the road network that thickened so dramatically in the century after his death. Vauban was intimately involved with water: building canals and moats, making waterways more navigable, improving harbors with special sluices (such as those that served to flush the silt out of Dunkerque's harbor), and even commenting on naval policy (against Mahan, he advocated privateers rather than battle fleets). It has been argued, however, that Vauban did not show enough interest in what could be called national canals—those that linked up major river basins and watersheds. Indeed, he felt, like some highhanded feudal lord using forced labor to build roads, that the populations living along the canals should pay for their upkeep, rather than the beneficiaries of the increased trade brought about by the canals. In this regard, Vauban lacked a truly global vision of hydraulic France.[112] This lack of interest in long canals is attributed to the fact that Vauban was Louvois' protégé and did the bulk of his work for the Minister of War on the frontiers. Consequently, he was clearly in the ministerial clan: that of Louvois, which was never friendly with the other great ministerial clan of Colbert. Since most ports, navigation in the interior, and commerce came under Colbert's jurisdiction, Vauban would have had relatively little opportunity to deal with these matters. Yet a rather different picture emerges in his memorandum on the Canal des Deux Mers, which shows a keen sense of the canal's economic benefits to the country, its effect on international trade, and its ability to connect with different climatic and agricultural regions whose productions could complement or, in times of poor harvests in one region, replace the production of another.[113] Moreover, in the newly conquered area of Flanders, Vauban he took an intense and intelligent interest in water and its effects. This was partly because he had to. The water table in the area was high and many of the fortresses were protected by inundations in case of attack. Canals, locks, and weirs were important. Sometimes canals were dug to make the transport of materials for construction easier and cheaper. At other times they were dug to provide a protective barrier with a clear military function or to provide water for flooding areas around individual fortresses. No larger vision than obvious necessity was the impulse for all these activities.

More interesting perhaps is Vauban's clear understanding of modifying traditional communications in order to favor integration into the new polity. For example, he wanted to disrupt the age-old links between Flanders and what were then the Spanish Netherlands by building water communications parallel to the new border. The following passage is one of many that portray Vauban as a soldier and a man of frontiers—frontiers to be made and frontiers to be broken:

The king's conquests having procured him one of the most beautiful, the most lucrative, and the most fertile dominions in the world, it seems necessary to separate it totally from the links and interests that could lead its inhabitants to wish to return to their former masters and compatriots . . . by facilitating their commerce in such a manner that it would pull them completely from dependence on Spain and Holland.[114]

Perhaps the most famous phrase that Vauban left the French language was "pré carré," which encapsulates both his vision of space and that of modern French aménagement (the global organization of space by creating necessary infrastructures and developing resources). A literal translation would be "square meadow." The word 'pré' is also part of an idiom that gives it the meaning of dueling ground, and the element of violence and defense that is implicit is also appropriate for Vauban's term. In a letter to Louvois in 1673 Vauban wrote:

> . . . seriously, Monseigneur, the king should think a bit about squaring his field [à faire son pré carré]. This pell-mell mix of friendly and enemy fortresses pleases me not at all. You are obliged to maintain three for one; your peoples are tormented, your expenses are greatly augmented and your forces much diminished. And I will add that it is almost impossible that you could keep them all in good condition and supplied. Moreover, I say that, if in all the contentions that we so often have with our neighbors, we would come to have a little bad luck or (God forbid) end up on the losing side, most of them would go as they came. That is why, either by a treaty or a good war, if you believe me on this, Monseigneur, always preach the squaring, not of the circle, but of the field [pré]. It is a good and beautiful thing to be able to hang on to what one has with both hands.[115]

This was written during the first year of the Dutch War, when that campaign promised to be more difficult than during the successful and rapid War of Devolution (1667–68) against Spain, in which Louis XIV had easily gained twelve fortified towns, including Lille, Tournai, and Oudenarde on the border of the Spanish Netherlands. Gaining fortresses was not enough for Vauban; the disposition of the fortresses was important too. Indeed, a larger number of fortresses could turn out to be a disadvantage if they strained resources, gobbled up troops for garrisons, or were positioned so as to be vulnerable or useless in war. There had to be a rational pattern that permitted mutual support. Vauban even suggested that some fortresses be decommissioned and razed. At the end of the Dutch War, Vauban would again be writing to Louvois that "the frontier would be very well fortified if it were reduced to two lines of fortifications, just like lines of soldiers in battle."[116] The king had seen Vauban's siege parallels at Maastricht in the same war as a kind of battle array, and Vauban saw the lines of fortresses in a similar light.

The same month he wrote to Louvois, Vauban proposed two lines of fortresses. The first line, consisting of 13 "*grandes places*" and two forts, included Dunkerque, Bergues, Furnes, Fort de la Quenock, Ypres, Ménin, Lille, Tournai, fort de la Mortagne, Condé, Valenciennes, Maubeuge, Philippeville, and Dinant. The second line, consisting of 13 "*places*," included Gravelines, Saint-Omer, Aire, Béthune, Arras, Douai, Bouchain, Cambrai, Landrecies, Avesnes, Marienbourg, Rocroy, and Charleville.[117] These lines stretched from the sea to the Ardennes, and they proved their usefulness when Marlborough and Eugene tried to smash through them in the War of the Spanish Succession. It took the Allied armies 4 years and 47,000 soldiers to barely breach the double line of fortresses and ultimately saved Louis XIV from disastrous defeat.[118] Two features of these fortresses should be noted: First, water was an important element in the whole network. Wet moats, sluices, and connecting canals made them formidable defensive positions. Second, there was what an archaeological survey of the remaining fortresses describes as the "Russian doll" aspect to the fortresses—usually a Vauban fortress was built on top of or around a Spanish fortress, which in turn had a medieval core.[119] Fortress building was expensive, and even a king as willing as Louis XIV had financial limits to his ambitions. Indeed, most of Vauban's fortress building consisted of patching up, renovating, adding to, and otherwise modifying existing fortresses, and the skill with which he did this is a part of his genius. It shows that the military engineer not only had to take into account the constraints of nature (terrain, position, and ground) but also the constraints of artificial surroundings, when building fortresses.

Strictly speaking, Vauban is only talking of the fortresses in Flanders when he writes about his pré carré; however, this concept underpins his conception of France as a whole. In the same memorandum where he outlines the double lines of fortresses, he proposes that in a future war, besides getting some additional fortresses in the Spanish Netherlands in order to disrupt any defensive line against France, Luxembourg and Strasbourg should be captured for their positions. He states that Strasbourg is an easier target than it appears to be and that it would be very useful for "carrying war over the Rhine."[120] Corvisier has implicated Vauban in the aggressive politics of "reunions" pursued by Louis XIV and has argued that there is certain but indirect evidence that Vauban inspired a policy of "aggressive defense" in the period after the treaty of Nijmwegen (which ended the Dutch War in 1678). Vauban was frequently called upon to give advice (and he did not hesitate to give unsolicited advice) on the shape of the frontier. This advice—and consequently French foreign policy—was constantly informed by topographical and engineering preoccupations, such as the ease of fortifying frontiers.[121] It was not only "by treaty" but also by a "good war" that Vauban hoped to tidy up the jumble of intertwined territories created by a long history that offended the mind of the military engineer and the soldier. He was willing to raze some fortresses that he considered superfluous to free more troops for offensive action when he felt that those fortresses would have no value in his defensive chain. Quality and disposition was as important as quantity of fortified works.

The concept of France as a hexagon is rather recent. Not established until the beginning of the twentieth century, it became political and geographical orthodoxy by the end of World War II. The hexagon comparison first appeared in 1795 (in a German work), and a scholar studying the matter has found only four geometrical comparisons in the seventeenth century.[122] One of them is Vauban's pré carré. Vauban was a "man of the frontier," and frontiers absorbed the bulk of his attention.[123] The great pré carré of France was to be enveloped by defensive lines, in no place more obviously than on the northeastern frontier, where the low, almost featureless plain that extends to the Urals was a natural invasion route to France since time immemorial. The other fences around the pré carré, although they needed to be reinforced to a greater or lesser degree with fortifications, provided imposing natural barriers to potential invaders.

Vauban did look inward on occasion, as when he argued that Paris should be fortified because of its importance to France. Louis XIV was compared to the head of the body politic; Paris, however, was compared to its heart, and losing it would indeed have the effect of driving a stake through the heart of the nation.[124] Vauban also looked outward at times, as when he contemplated the colonization of Canada. Generally, he appears to have had little interest in this direction, feeling that colonies should be few and that Canada and Santo Domingo should be defended vigorously and steps taken to populate them. Yet describing Vauban's thinking as linear, static, local, and concerned exclusively with the periphery would be a gross simplification. He always thought of his frontiers as zones with depth and felt the need to understand and manage the economies and populations of these areas. On occasion, this led him to attempt things that military historians feel make no sense either in his time or in ours, such as the attempt to provide a civilian population for a newly constructed fortress in a bleak and isolated area of the Alps.[125]

Being a "man of the frontier," Vauban did, however, look outward toward France's immediate neighbors. The so-called natural boundaries of France were attained over centuries, and there was initially no sense of such natural boundaries. Lille and Strasbourg only became French in the seventeenth century, Nancy in the eighteenth, and Nice in the nineteenth century. André Zeller argued that the Rhine as a "natural boundary" of France is a concept that was only established during the French Revolution, just about that time that Germans began to see the Rhine as a German river.[126] During the Old Regime, there were only a few isolated appearances of this concept in France, and it was again only the Germans who took the concept seriously. However, Vauban appears to have had a very clear idea of an "engineered" boundary that is often conflated with a natural geographic boundary. As proclaimed by all texts on the subject, fortification was the enhancement of natural sites to permit effective defense by outnumbered defenders. On the scale of an entire country like France, its frontiers were its defenses, sometimes requiring substantial artificial construction, as on the northeastern frontier, sometime less, thanks to natural geographic advantages. Even in the latter case, the military engineer needed to exercise his art—to build coastal fortifications along the Atlantic and Mediterranean to protect ports and possible landing sites or to bar mountain passes in the Alps or the Pyrenees. Louis XIV reached the Rhine with the conquest of Strasbourg in 1681, and Vauban was one of those who felt that the Rhine was an indispensable frontier, as he wrote in a letter to Racine when he was worried that Strasbourg might be returned to the Empire.[127] But he also thought of limits to which France should withdraw. In a letter to Le Peletier in 1693, which also displays his earthy and irreverent style, he argues that with only minimal construction, the Alps could be made a very good frontier: "If instead of flying like a butterfly across the Alps we had stuck to arranging the frontier decently, the king would have had his mind at rest. Instead, you are still in for two or three years of uneasiness."[128]

In 1700, with France at the height of its power, having attained most of the boundaries Vauban wanted, he wrote a memorandum titled Present interests of the States of Christendom.[129] If Louis XIV succumbed to the temptation of having his grandson on the throne of Spain with the undivided Spanish dominions under his rule, Vauban did not. The basic outlines of French policy should consist of three fundamental refusals: refusal to found a universal monarchy (the temptations of Empire had ruined Spain), refusal to go beyond what are clearly its natural boundaries, and refusal to go beyond the seas and create a colonial empire (with the possible exception of Canada and Santo Domingo). The document speaks of renunciation of ambition and the tone is modest:

All the ambitions of France should be contained between the summit[s] of the Alps and the Pyrenees, [and] from Switzerland to the two seas. It is there that at opportune times it should propose to establish limits by legitimate means. . . .[130]

Closer reading of the details, however, indicates the almost breath-taking aggressiveness of his proposals. To give France its pré carré, Vauban wanted the eventual annexation of the territory of Nice, Lorraine, Zweibrücken, Montbéliard, the Palatinate, the archbishopric of Trier and those parts of the archbishoprics of Cologne and Mainz that were west of the Rhine, the German Duchies of Cleves and Jülich, the Spanish Netherlands, the dominions of Liège, and the Dutch Brabant. There can be no doubt that Vauban had a clear idea of natural and compact boundaries and that a greater part of the Rhine was one of them.

The year before his death, Vauban was suggesting peace proposals to extricate France from the War of the Spanish Succession, which was bleeding it white, and he seems to have become a little less ambitious. He asserted that the country had "natural limits beyond which it seems that good sense forbids us to turn our thoughts. All that has been undertaken beyond the two seas, the Rhine, the Alps, and the Pyrenees has always turned out badly [for France]."[131] For Vauban, the adjective "natural" now had political, as well as engineering and geographical connotations. Skeptical about French interests being served by a Bourbon on the throne of Spain because blood was not as important as national interests, he advised that Louis XIV give back the poisoned chalice of the Spanish throne for his grandson. Yet he continued to urge the king to demand Montbéliard, Lorraine, Luxembourg, the whole of Alsace, and forts on the Rhine. The natural boundaries continued to be central to his thinking and he had not relented in their pursuit. For France, "it is very important that our frontier should be unified, well arranged, and without enclaves, which are normally very inconvenient and good for causing quarrels and attracting disputes that are only decided with the sword...."[132] By now, however, the tired old warrior believed that all nations, not only France, had "limits that Nature had prescribed," that all countries should keep within them and that none of them should be in a position to trouble the others.[133]

Vauban's ideas on frontiers and the pré carré differ from the modern French idea of aménagement in many ways. Ultimately the frontier should be static, it is linear, and it delineates the territory of the nation without channeling its economic life or defining its networks of exchanges and communication. In a history of the efforts of the French state to develop policies of regional planning, one can read that "Perhaps one of the traits which has been imprinted on the planning [aménagement] of French territory is the ambition of making space reasonable [raisonner l'espace]."[134] It is here that Vauban resembles and inspires his successors who engineered France. If modern concepts of planning only surfaced in France after the Second World War, there is a long history of previous development and Vauban occupies a central place in it because of his particular perception of France's unity. Roland Mousnier has argued that military needs were the dominant influence on the institutions of France. This is no less true for the institutions planning the organization of infrastructures and communications in the country. Military engineers had a direct and substantial influence on the frontiers and many of the ports of the country but they also had an indirect influence in other parts of the country.

Vauban's successors in giving form to the interior of the country were in fact the civil engineers—the ingénieurs des ponts et chaussées. Even they, however, groped for a century to acquire the concept of a network of communications, in the sense of an ensemble of means of transport with a hierarchy and spatial articulation that forms a system, and whose functional qualities depend on the modalities of this articulation and hierarchy.[135] Lepetit dismisses the hierarchical classification of roads by the Minister of Finance Orry in 1738, which would appear to demonstrate the idea of network, as misleading: it is more descriptive than prescriptive in its aims. All through the eighteenth century, French civil engineers continued to think in local terms, they were subjected to powerful local constraints, and their road maps reflect this. Guillerme agrees that the concept of network only appeared in the first third of the nineteenth century, attributing it to concepts of systems of fortification, distribution networks for municipal water supplies, and the circulatory system in medicine. Concepts informing systems of fortification go back to the eighteenth century, most notably to Cormontaigne, who looked at France as a circle with concentric lines of defense on its frontiers with fortresses on each line having uniform characteristics of positioning and defensive capabilities that are distinct from the other lines.[136] The classification of fortresses is based not only on military utility but also on commercial importance. The germ of these concepts undoubtedly comes from Vauban's ideas, which Cormontaigne, as usual, adopts and presents more abstractly and more mathematically. He uses an abstract geometrical model leaving no room for natural features such as rivers or hills. It is therefore fair to say that Vauban not only modified the actual physical territory of France with his fortresses and his frontier towns girdled and constrained with curtains and bastions, but he left an intellectual legacy that was developed by military and civil engineers that inspired later regional planning.

The aspect of economic exchanges and dynamism of Vauban's are effaced, but not entirely absent in his proto-networks.[137] If there is a dynamic element to Vauban's conception systems of fortresses, it is limited to the military sphere and even there he thinks in terms of the relatively static operations of siege warfare. He was a soldier of the Old Regime, but it was an Old Regime that was actively seeking modern methods of ensuring stability and its own survival. After all, the ultimate aim of warfare—that human activity which is most dynamic and disruptive—was to bring "repose" to the king and the country. There is an inherently conservative and static element in his thought. It is on this point that modern aménagement differs profoundly from Vauban, even though there is a fundamental similarity in the conception of France that can be engineered for the good of its inhabitants.

Vauban's Engineers

That Louis XIV and Louvois and his successors in the Ministry of War appreciated Vauban is clear from the honors that he received. In 1688 he was promoted to Lieutenant-General, the first engineer to receive that honor. Later he would become the first engineer to become a Marshal of France. He was awarded one of the first eight Grand Crosses of the Order of Saint Louis, the most prestigious military decoration of the Ancien Regime that the king created in 1693. Vauban's reputation continued to grow after his death, both among supporters and enemies of the royal power. It achieved cult status among military engineers, and this is understandable for a number of reasons. One of the primary reasons is that, in spite of the existence of organized bodies of military engineers in France for centuries before him, it was Vauban who created the essential framework for a corps of military engineering in France.

When Vauban began his career as an engineer in the 1650s, the two major ministers in Louis XIV's government were Colbert and Louvois. The latter had inherited his ministry from his father, and among his responsibilities were the Ministry of War and the civil administration of frontier areas and territories that had been newly conquered by France. His great rival Colbert was in charge of what could be called the Ministry of Finance, the Navy and the Colonies, the King's Household, the King's Buildings (an important responsibility for a regime that built Versailles), and the civil administration of the older interior and maritime provinces of France. This twofold division of the country had important consequences for the administration of fortifications. Colbert not only had the responsibility of fortifying harbors but he also had to build and maintain some fortresses. Louvois not only had to build and maintain fortresses, but he had to organize the attacks of enemy fortresses during the king's frequent wars.

Although at the beginning of Louis XIV's reign all fortifications were in theory in the jurisdiction of the Minister of War, Louis had divided responsibilities between his two ministers, and they had separate organizations of engineers, which each administered in his own fashion.[138] Colbert had further divided responsibility among three intendants de fortifications who were at the top of a well-defined hierarchy of engineers. Louvois' organization was initially less well developed; a clear hierarchy did not exist (except in siege operations). Although engineers reported to the intendants of police, justice, and finance of the areas where the engineers were stationed, the energetic Louvois also kept in touch with many of them personally. Vauban was in Louvois' department and soon acquired an unofficial position (that later was formalized) as chief adviser to Louvois on engineering matters.

The administrative fission of fortifications was, however, accompanied by some unity in the domain of expertise in the form of the General Commissioner of Fortifications, who was called upon to provide technical advice to both departments. Before Vauban officially acquired this position in 1677, it was held by the Chevalier de Clerville, a distinguished veteran of the Thirty Years' War. Clerville's technical competence was apparently noticeably inferior to that of Vauban, and Louvois began to use Vauban as a de facto commissioner of fortification even before Clerville's death. In his typically brutal fashion, Louvois had written to an engineer ordering him to listen politely to all that Clerville said but not to pay any attention to it, and to come and get his orders from Vauban.[139] Although Colbert was more polite and still continued to use Clerville, he too was impressed by Vauban and used his services after 1677.

The administrative division of government engineers reflects more fundamental social and professional cleavages between the engineers of Colbert and those of Louvois. Whereas about two-thirds of Colbert's engineers were commoners, the rest being of noble origin, these proportions were almost exactly reversed in Louvois' department.[140] In general it can be said that Colbert's engineers came from the wealthier, more populated, more urbanized parts of France. Although there is a wide variety in the professions of their families, the number of those coming from families of engineers, architects, and building contractors makes up more than a quarter of the sample. Lawyers, army officers, merchants, and civic officials are also well represented.[141] It is likely, as Blanchard has pointed out, that Colbert, in his capacities as Minister of the Navy and Director of the King's Buildings, was able to draw on a well-established milieu of building professions that already existed in Paris and the major ports of the kingdom as a result of royal construction of public buildings and arsenals.[142]

The origins of Louvois' engineers are rather different. There are considerably fewer sons of engineers, architects, and merchants—the calculating professions that supplied about 40 percent of Colbert's engineers. They come more from the poorer, rural areas and small towns of France. There seems to be a greater diversity of professional origins—the most common being that of army officer. But even this provided only one out of eight engineers. One could almost say that the two departments of engineers reflect a split between a more modern and a more traditional society.

There is also a clear gap in the respective competence of what one could legitimately call two kinds of engineers. The engineers of Colbert were better capable of undertaking complex construction projects. In today's terms, they were better at civil engineering and building. The reasons for this lie in the importance of offensive military operations for the engineers of Louvois. They were, after all, takers of fortresses more often than builders of fortresses—more often soldiers than engineers. The nature of military engineering contained within itself two different kinds of activities that required two different kinds of talents, training, and personal attributes. One can distinguish, then, between what Blanchard calls the ingénieurs des tranchées (trench engineers) and the ingénieurs des places (fortress engineers). As Vauban himself said, in an often cited letter, the latter kind of engineer was more difficult to train and, in his opinion, more valuable than the former:

Since I'm on this subject, I must make you see the difference between those who know how to build and those who only know how to attack fortresses. There is no officer possessing a bit of good sense whom I cannot render capable of directing a trench, a lodging on the counterscarp, a descent into the ditch, the attachment of a mine, etc., in three more or less reasonable sieges. But a good builder is only made after fifteen years of application, [and] even then he must have been employed at different tasks and he must be a man of great application. We presently have a fairly good number of those who are good for sieges, but very few who understand building well and even fewer of those who understand both. This leads me to spare them as much as I can, for if you knew the trouble I have every day to correct and hide the defects of one and the other [kind of engineer] you would pity me. When one can succeed in making oneself intelligent in the works and in sieges and be a good infantry officer, [then] that makes the perfection of the craft. . . .[143]

Vauban himself provides a good example of the how one became an ingénieur des tranchées. He had begun his military education as a cadet in a cavalry regiment after an irregular education by the local parish priest that gave him "a fairly good tincture of mathematics and fortifications," supplemented by a talent for drawing.[144] Volunteering for siege work and—more importantly—surviving, he went on to develop his expertise on the job. The story is a typical one for engineers in Louvois' department before Vauban himself began to set entrance examinations and organize a formal "novitiate" of several years of practical experience before a final examination led to the full status of military engineer.

Even then, it took a lifetime to make a good military engineer:

Military engineering [le Génie] is a craft that is beyond our strength. It contains too many things for it to be mastered by a man to a sovereign degree of perfection. I have a fairly good opinion of myself to believe that I am one of the best of the troop and capable of giving lessons to the most expert, and with all that, when I examine myself, I only find myself a half-engineer, after forty years of very diligent application and of the greatest experience there ever was. The glory for that is to Him who saved me and let me live until now. . . .[145]

It seems, then, that under the ministries of Louvois and Colbert, there was no administrative and little technical unity in their departments which dealt with fortifications. One could say that Colbert was at the head of a civil engineering department and the military engineers—more military than engineers—were under Louvois. There is no homogeneity of origins nor undoubtedly of attitudes in a group of men who are not only administratively divided but who, by the nature of their functions, were usually geographically scattered.

The first major step toward unity came after the death of Louvois in 1691 (Colbert had died in 1683) when the two different departments of fortifications were taken away from their respective ministries, united, and put under the control of a senior administrator who dealt directly with the king himself. There are several reasons that can be suggested for this administrative reshuffle. The role of personal rivalries and simple bureaucratic confusion is not an anomalous condition of bureaucracy, but is born with it. The initial division of responsibility for fortifications at the beginning of Louis XIV's reign may have simply been an example of this. Perhaps Louis merely saw an opportunity of correcting an error after the disappearance of his two powerful and strong-willed ministers. But two other possibilities suggest themselves as explanations of the unification of 1691. Vauban, as Commissioner general of fortifications since the death of Clerville, had already brought the two departments together to a great extent, because of his uncontested competence and credit with both Louvois and Colbert. Furthermore, in an age of fortress warfare, fortifications were always a matter of importance and were becoming even more important during the War of the League of Augsburg (sometimes called the Nine Years' War) that had begun in 1688. It found Louis fighting with virtually the whole of Europe joined in the Grand Alliance (the Dutch and the English under William of Orange, the Empire, Spain, Sweden, and the Electors of Bavaria, Saxony, and the Palatinate, as well as Savoy). The doctrine of the pré carré enunciated by Vauban reflected the thinking of Louis and all his ministers. Fortifications were thrown into the foreground by the desire to rationalize frontiers that doctrine entailed. Siege warfare was perceived also to minimize chance in military operations—a feature dear to a monarch who wanted to exercise personal control in all domains of his government.

Moreover, Louis XIV seems to have had a distinct personal interest in siegecraft and fortifications. He was present at 20 of Vauban's sieges and Christopher Duffy has suggested that the siege—a kind of rational and orchestrated baroque military ballet with a definite and positive outcome thanks to Vauban's expertise—was something that appealed at a fundamental level to the warlike king who had a taste for order and rationality. The new Directeur général des fortifications des places de terre et de mer, Michel le Peletier de Souzy (1640–1725), just like any minister, had his regularly scheduled weekly meeting with the king to inform him on the activities of his department, which Louis followed with genuine and knowledgeable interest.[146] Vauban, too, was often directly consulted by the king. If the functional and technical unity of the Department of fortifications was only incipient and fragile, then administrative unity had been restored after 1691.

Le Peletier was one of the breed of hard-working and devoted administrators of the Sun King who came from the noblesse de robe. At a very early age of 27, after passing through the posts of councilor at the parlement of Paris and Councillor of State, he became the intendant of recently occupied Flanders. There he acquitted himself well and picked up some knowledge of fortifications in this heavily fortified frontier province. Vauban remained at his post of Commissioner General of Fortifications (in 1703 he would be promoted to Marshal of France) and served as the chief technical consultant to Le Peletier, who now became a kind of Minister of Fortifications.

Colbert's intendants de fortifications were abolished in 1692 and a new hierarchy of command was instituted.[147] The country was divided into 23 territorial units called *directions*, further subdivided into *chefferies*, headed by a Directeur and an ingénieur en chef respectively. In wartime, brigades of engineers were drawn from the various directions irrespective of location on an ad hoc basis. Thus, engineers who had formerly been under Colbert who still served mainly in fortresses began going on military campaigns.

After many years of agitating for this, Vauban finally was able to introduce a formal procedure of entrance into the engineering corps that was based on a preliminary examination in mathematics and drawing, followed by a "novitiate" of a year or two of practical work at a fortress, to be completed with a final examination on the theory and practice of fortifications. Success in this examination would bring a *brevet* (commission) of a royal engineer; failure meant transfer to an infantry regiment. According to Vauban,

One must examine publicly and on a number of different occasions the young people who wish to introduce themselves into [the administration] of fortifications to be employed there as engineers, not only in matters of geometry and mensuration, but also on the other more necessary parts of mathematics such as trigonometry, mechanics, arithmetic, geography, civil architecture, and even drafting [dessein]....[148] No one must be accepted in [the administration of] fortifications by favor or recommendation; merit alone and people's capability must bring them their employment....[149]

The system of examinations instituted by Vauban is important not only for an understanding of the later history of the French military engineering corps but also for an understanding of the development of the *concours*, the competitive examination that is still such an important element in the French educational system today.

Vauban himself served as the examiner, but because of his numerous commitments and frequent absences from Paris, he was assisted by Sauveur, member of the Academy of Sciences. The tradition of having as examiner an academician was continued well into the nineteenth century. It had a powerful impact not only in homogenizing the recruitment of the engineering corps but in imposing high standards.

There was a great need for military engineers in the years after 1691. The War of the League of Augsburg (1688–1697) and the War of the Spanish Succession (1701–1714) kept enormous French armies in the field and were also the occasion for many famous sieges in all parts of Europe. Not only French diplomacy but also French fortifications played a major part in deterring the armies of Eugene of Savoy and Marlborough from marching into France after the disastrous French defeats in the Low Countries. Between 1692 and 1717 more engineers were recruited than had been in the corps in 1691: 363 vs. 276.[150]

Most of these new arrivals had been subjected to the process of examination introduced by Vauban. (After 1697, Le Peletier did not want to accept new engineers who had not been examined by Vauban.[151]) The social origins of those recruited indicate a slight preponderance of commoners over nobles (58.1 percent vs. 41.9 percent). These proportions are closer to those in Colbert's old department than that of Louvois. The need for manpower in Louis XIV's armies led throughout his reign to an increase in the number of commoners among his officers. It is likely that Vauban, in view of his demonstrated respect for the better qualified builders, who were commoners in proportionately larger numbers as the figures for Colbert's old department seem to imply, wanted to increase the recruitment of this type of engineer. More than a fifth of the new recruits were sons of former engineers.[152]

Greatly increased opportunities for military experience also may have contributed to instilling a greater esprit de corps and breaking down the divisions between the engineers of the old departments of Colbert and Louvois. Almost eighty of them died during the War of the Spanish Succession alone, thus causing a constant turnover in the ranks, opportunities for combat, and closer relations.[153] Although not all of Colbert's engineers were drafted for military campaigns, for the reasons Vauban had so clearly expressed, more were drafted than previously.

The period of Le Peletier's stewardship of a united Fortifications Department (Vauban had died in 1707) had mixed effects in forging an esprit de corps and a sense of technical and administrative unity. There was quite clearly a swamping of the old personnel by new recruits who were more rigorously selected according to uniform criteria and who possibly (if one can speculate on this from the data on the social and professional origins of their fathers) had backgrounds and educations conducive to greater technical competence. Then, too, the wars must have had a positive effect in cementing ties of comradeship in situations of danger and sacrifice. But the wars must have had some contrary effect. Casualties tore the tissue of human relations, and many of the building projects undertaken by Vauban must have been hurried and remained at an incomplete stage.[154] Furthermore, many (more than a hundred) of the new engineers engaged during the wars were retired after peace came in 1714. Yet on balance, this was period of consolidation of a new profession, whose unity was far from perfect to be sure, but which now had an existence as an administrative unit. Military engineers had certainly been around before 1691, they had even been organized into larger units, but from now on they were on their way to forming a corps both in the sense of a service arm of the military and as part of the body politic of the Old Regime.

The role of Vauban in this process was critical. He served as exemplar and talisman. Having attained the highest rank in the army, admired by friend and foe, respected both by the king and by his opponents, with a matchless record of courage and competence, Vauban loomed large in the Department of Fortifications. Certainly he outshone its nominal head Le Peletier as he had Clerville, his own predecessor as General Commissioner of Fortifications. He had systematized siegecraft and created a consistent strategic view of fortifications for the security of France. He was also primarily responsible for the actual recruitment and organization of the Department of Fortifications. With the full support of the royal power for his methodical and rational methods of recruitment, examination, and promotion, Vauban had succeeded in making future Vaubans unnecessary. That was perhaps his greatest achievement for French military engineering: the laying of the basis for a bureaucratized, rationally organized and selected corps of engineers who had a standard body of practice and procedures for their work. Yet this did not come about at once, in spite of Vauban's clear ideas of his aims and his efforts to attain those aims. The everyday working organization of the Department of Fortifications was, as was the case for all French administrative bodies of the Old Regime, more chaotic than the rational facade of the Sun King's court and his ambition suggested. Moreover, many of the military engineers were not soldiers and many of those who were only held commissions as reserve officers [officiers reformés]. Not accepted as full members of the army, they lacked the prestige that this entailed and which they felt they deserved. The memory of Vauban, who had done so much to create and solidify a true corps of military engineers, to train it, and to exalt it, remained a primary force for cohesion and development as it continued to struggle for what it felt was its rightful place as a true corps of the Old Regime and its army.

3
Toward a Corps Royal du Génie

It was . . . to follow the example of such a celebrated leader as Vauban that the French nobility embraced the art of engineering, and freed itself of the old prejudice to the effect that it was disgraceful to engage in warfare except as an officer of the field arms. . . . Just as the nobility had once abandoned the lance for the pike, so now they readily laid aside the pike to take up the measuring rod.

—Chevalier de Guignard[1]

Le Peletier resigned his post after the death of the old king and was succeeded by Claude-François Bidal, marquis d'Asfeld (1667–1743). D'Asfeld was descended from a wealthy family of silk merchants and bankers, and his father had obtained a Swedish title of nobility from Queen Christina for financial services rendered. He had also served Louis XIV on diplomatic missions in Germany. These unprepossessing origins, from the point of view of the traditional nobility, were overcome by the son in a very distinguished military career that eventually brought him the title of Marshal of France. The ministries having been abolished and replaced by councils in the system of Polysynodie during the first years of the Regency that succeeded Louis XIV, d'Asfeld sat as a member of the Councils of War and the Navy in his position as Director General of Fortifications. He also continued his military career under Louis XV and commanded an army in the War of the Polish Succession that eventually participated in the siege of Philippsburg in 1734, where the young Montalembert got his first experience of a siege.

Louis XIV's most trusted administrators were civilians. Even the French army was administered mainly by civilians, some of whom, including the intendants des armées, accompanied the army on campaign. Indeed, one of the most striking features of Louis XIV's army was the extent to which civilian bureaucrats controlled it.[2] But d'Asfeld, unlike Le Peletier, was a soldier, and one aspect of his administration was a further militarization of engineers that had been advanced and accelerated by the War of the Spanish Succession.

The Marquis d'Asfeld's Department

At the beginning of d'Asfeld's administration, only 45 percent of the engineers in the department of fortifications held a military rank; most of those who did not were from Colbert's old department. By 1726 this figure had climbed to 55 percent, and by the time of d'Asfeld's death (1743) only one-third of the engineers did not have a military rank.[3] In 1732, engineers also received a uniform for the first time.[4]

Most of the engineers in Louvois' old department had been officers who had volunteered for siege work and had gradually built up enough expertise to obtain the brevet of King's Engineer. Gradually, as these officers essentially became specialist engineers, they were given the status of *officer réformé* (reserve officers with the rank but not the duties nor the attachment to a specific unit of a regular officer). Those who had started their careers as civilians and had acquired siege experience were sometimes, on a strictly individual basis, also given the title of officier réformé as a reward for military service and as a mark of status to deal with officers in the course of their duties. Others in Louvois' department, and virtually all those in Colbert's, were not officers; they were simply engineers with more or less seniority. An ingénieur des places could in fact attain a very senior position without ever being a commissioned officer. Situations thus arose where an ordinary engineer could have a higher *military* rank than his superior in technical matters. In 1746, for example, there were thirteen directors of fortifications who had the relatively humble rank of captain while there were two chief engineers who held the rank of lieutenant general. Nineteen chief engineers were colonels or lieutenant colonels, and 94 were captains.[5]

All this illustrates the ongoing problem of the status of engineers in the French army of the Old Regime. For a long time, engineers as well as artillerists, members of the other corps à talents, were considered interlopers and not true soldiers. Indeed, the many civilians on campaign with most armies of the seventeenth and eighteenth centuries—the sutlers, womenfolk, domestics, contractors, and purchasers of war booty—contrasted strongly with the actual fighting troops and at the same time were a very real and substantial presence. For many officers proud of their blue blood and martial qualities, engineers were merely a superior kind of camp follower.

But the ambiguity of military status for the engineers led to more than the inevitable problem of friction with officers of the traditional fighting arms of the army. It caused friction within the corps of engineers itself. Divided along lines of occupational expertise, background, and instincts, separated from one another during 30 years of administrative divorce, engineers had been gradually fusing into a homogeneous group thanks to the work of Vauban(who was supported by the king and aided by the king's competent bureaucrats). The lack of military rank for many engineers and the disparity of military and technical rank for others continued as a source of discord and tension within the corps. Only many years of working together would attenuate and ultimately dissolve this source of internal conflict. Even after the definite disappearance of the conflict after 1776, one may well wonder whether its avatars did not return in other forms. Do not the attacks of a Carnot on the ossified leadership of his corps on the eve of the Revolution express the impatience of scientific and technical talent subjected to the constraints of military discipline?

D'Asfeld contributed to the slow but steady process of unification of the corps on the one hand by exerting himself to obtain military rank for his engineers. He had participated in sieges, and he appears to have appreciated the professional and martial qualities of his subordinates. In another, more organic fashion, he helped to fuse the corps together. During the period of his leadership, one can detect a systematic bias toward hiring the sons or the relatives of engineers or former engineers. There are a number of reasons that could be attributed to this phenomenon that could very well have little to do with d'Asfeld's personal action. Almost a century of continuous warfare had built up a distinct social unit, an unofficial sort of guild of military engineering. There was in place a substantial community of engineers and former engineers, whose offspring would naturally tend to gravitate to their father's occupation. The well-established belief (widespread in the case of military careers) that genuine vocation and even competence were absorbed in some ill-defined, almost biological manner from parents and parental milieu must also have played some role.

After the long wars of Louis XIV there was a drop in the level of recruitment in the army as well as in the engineering corps. Because places were fewer, d'Asfeld had more opportunities for choice and this choice seems to have been exercised preferentially in favor of the sons of engineers. Of the 309 engineers entering the corps between 1716 and 1747, of those whose fathers' professions were known, 47 percent were the sons of engineers and 67 percent were close relatives.[6] The next largest category of paternal profession was army officer, with 15 percent. This shows a steady rise in recruits coming from army backgrounds (only one-tenth of engineers' fathers with known professions were army officers in 1691) and most likely is an indication of the growing prestige of the corps among military officers.

Potential recruits to the engineering corps were still subjected to an entrance examination by François Chevallier (1699–1748), Sauveur's nephew by marriage, who had succeeded him in 1720. This example of apparent nepotism nicely illustrates the case for the corps as a whole. Chevallier was not just the nephew of Sauveur but also a member of the Academy of Sciences. So too, recruits who were related to engineers could not count merely on family connections—they were also subjected to the examination, and there is no reason to believe that they were the beneficiaries of egregiously preferential treatment. But all other factors—especially the all-important one of competence determined by examination—being equal, the sons and relatives of engineers were chosen. The fact that these enfants du corps did better on their examinations than other candidates cannot be necessarily construed as a sign of favoritism.[7] The sons of engineers often had the inestimable benefit of working with their fathers and being trained by them in the tricks of the trade and the examination before becoming official candidates. But connections within the corps seem to have become more helpful and to have undermined the competitive examination to some extent. Military engineering was by no means as closed a milieu as some of the more exalted strata of the legal profession—such as members of the parlement of Paris.[8] But it was beginning to show, in an immature but noticeable way, the signs of exclusiveness and desirability that were a hallmark of both respected professional corporations and entrenched social castes of the final decades of the Old Regime.

D'Asfeld's Director Generalship differs in some important ways from that of Le Peletier aided by Vauban under the watchful eye of the Sun King himself. Although d'Asfeld had some experience as an ingénieur des tranchées himself, he by no means had the technical competence of a Vauban. Thus, the Directors had more freedom of initiative and less centralized control in technical matters than they did in the preceding period. At the same time, Louis XV never displayed the interest or the competence in fortifications of his predecessor, and d'Asfeld had a greater freedom of action than any of his predecessors. But d'Asfeld in turn shared this freedom with his subordinates. Without an imposing technical presence, he was content merely to strengthen the social and personal ties of the members of his department internally and to defend its position externally to the best of his limited ability. As a result of this, the corps of engineers was both more loosely controlled and less centrally controlled than in Vauban's time. This occurred as the growing complexity of fortifications militated for the increasing importance of specialists.

The period of d'Asfeld's personal rule over his "family," as Anne Blanchard has called it, was in some ways a period of stagnation when little work was done on major engineering projects. Engineers seem to have spent at least as much time on consolidating their cozy personal niches within the corps than in carrying out any grandiose feats of construction à la Vauban. Cormontaigne's works around Metz in the 1720s and the 1730s, along with the completion of other fortresses that had not been finished by Vauban at Metz, Thionville, and Briançon, were the only major construction in the period of d'Asfeld's general directorship. It was on the whole a relatively peaceful time, and the government was perpetually in financial straits. Truttmann is even stronger in his verdict: the whole reign of Louis XV (1715–1774) was a period of stagnation, the second of three phases of the military engineering corps in the century before the Revolution. (The first was the most fecund period, the period of constitution of Vauban's pré carré, essentially between 1679 and 1697; the third was the reign of Louis XVI, when construction was essentially limited to work on the great ports, especially Brest and Cherbourg.[9]) Only at the end of d'Asfeld's rule did the War of the Austrian Succession (1740–1748) require a major combat role for engineers. The enormity of their casualties, especially among the young and inexperienced, led to doubts about their training and competence, and was one of the reasons for the creation of the Royal Military Engineering school at Mézières. During the period 1691–1791, 209 French military engineers (14 percent) died in combat. Of these the majority were in junior positions (70.8 percent), but the slaughter among engineers during the War of the Austrian Succession was exceptional: in the last 5 years of the war, 48 were killed in action, died of their wounds, or died of "fatigue" on campaign (of the 13 engineers who died at the great siege of Berg-op-Zoom, four died of "fatigue").[10] This represented one-sixth of the corps, and on an annual basis it was even greater than the number killed during the heavy fighting of the War of the Spanish Succession.[11]

D'Argenson's Age of Reason

After d'Asfeld's death, there occurred another change in the status of the military engineering corps. Rather than continue as a kind of autonomous ministry, it was once again attached to the Minister of War, as was the case almost 70 years before. The post of Director General was abolished, and the various Directors of the corps dealt directly with the Minister of War. This had happened because Louis XV's ambitious and powerful Minister of War, Marc-René de Voyer, Comte d'Argenson (1696–1764), wanted to swallow up the Corps of Engineers, which he considered to be poorly directed by d'Asfeld, a man he did not like and whom he considered weak and ineffectual.[12] (Marshal Belle-Isle, military hero and future successor of d'Argenson, had also wanted to step into d'Asfeld's shoes, but he did not yet have enough credit at court to attain his aim.) This new change had mixed results. On the one hand, the engineers undoubtedly had a more powerful protector at Court, not only in the particular case of d'Argenson, but in the person of future Ministers of War as well. Next to the Contrôleur général des finances, the Minister of War was usually the most important personage among royal administrators. But, on the other hand, the Directors of fortifications lost some of their autonomy, and the minister, responsible for many other functions besides fortifications, was more removed from the day-to-day functioning of the corps. The bureau of fortification in the Ministry of War was one of several specialized departments run by a senior clerk. D'Argenson had replaced the military engineer Charles-Antoine de Rault de Ramsault (1687–1774), who had performed this function in the final years of d'Asfeld's life, with Noël de Regemorte, an engineer of the Corps des ponts et chaussées.

The Regemorte (sometimes called Regemortes) family not only provides an example of an illustrious dynasty of engineers, a not uncommon phenomenon among engineers both civil and military in the eighteenth century, but also illustrates the persisting lateral entrance of outsiders from other corps enjoying ministerial favor or dispensation into what is ostensibly a clearly structured and rigid bureaucracy. This provides a valuable antidote to seeing too much consistency in the fledgling technical bureaucracies of the eighteenth century. Jean-Baptiste de Regemorte (?–1724) was the general contractor at Neuf-Brisach and eventually became an ingénieur géographe (military topographer), serving with distinction as a military engineer in the War of the Spanish Succession in Alsace. His son Antoine (1703–1745) began, like his father, as an ingénieur géographe, but eventually transferred to the Corps of Military Engineers; he died in the siege of Tournai. Another son, Louis (1710–1774), was an engineer in the levées et turcies de la Loire (the administration of dikes and flood control of Loire), itself assimilated into the Corps des ponts et chaussées, who was renowned for a bridge he built at Moulins. Noël (1710–1801) had begun his career as a civil engineer in the dike and flood control administration of the Loire and had also worked in mapping Alsace but was called by d'Argenson to head the bureau of fortification in the Ministry of War.[13] Noël de Regemorte was a competent engineer who maintained an extensive correspondence with Louis de Cormontaigne (1696–1752), the most respected engineer in the corps after the death of Vauban. After his departure in 1755, however, it appears that there was no technician in overall charge of the corps, but an administrative official in the War Office. Regemorte's functions seem to have been taken over by d'Argenson's chief administrative officer of the War Ministry (the premier commis), Dubois, who in turn was replaced by Piètre in 1767.[14] In 1776 the premier commis of the bureau of artillery and fortifications was Jacques-Julien Le Sancquer (1718–1799?), who stayed at this post until the Revolution.[15] These men were not engineers themselves and were undoubtedly more comfortable with the administrative rather than the technical aspects of their functions. To complicate an already confusing bureaucratic terrain, Louis-Hyacinthe Boyer de Cremilles (1700–1768), a former cavalry officer who had superintended some canal building with army troops, was given Le Peletier's and d'Asfeld's old title of Director General of fortifications for a short time in the early 1760s but seems to have provided merely another level of officialdom between the bureau of fortifications and the Minister of War.[16] It was only in 1776, with the arrival of Fourcroy de Ramecourt as the effective head of the military engineering corps as the result of the Saint-Germain ordinance, that an engineer once again had a visible and effective position in the administration of the bureau of fortifications in the Ministry of War. Moreover, the Ministry, in spite of its importance, did not have the administrative clout or the relative financial freedom enjoyed by the Contrôle générale des finances. This was a major reason for the growth and eventual expansion of the Corps des ponts et chaussées, which was a part of this more powerful ministry directly under the control of the de facto prime minister of the country. Picon has argued convincingly that the growing interest in economic and commercial policy—a natural domain of the Contrôle générale—further strengthened the hand of the Corps des ponts et chaussées.[17]

The emphasis of d'Argenson's administration of the corps can nevertheless be described as bureaucratic rationalization, increasing militarization, and formal training. No exceptions were henceforth permitted for entrance to the corps as had been the case under d'Asfeld: from now on it was by competitive examination alone that recruits entered the corps. This examination was under the sole control of the academician examiner, and henceforth serving engineers were deprived of any input into deciding who could enter the corps.[18] D'Argenson was the first to give the corps of engineers a regular charter, officially fixing their conditions of service and ranks in the form of the long and detailed royal ordinance of 7 February 1744.[19] One result of this ordinance was a further order, which was a procedure for evaluating and weighting the military experience of engineers for promotion. It is worth looking at it for it illustrates both the importance of military service in attaining rank in the corps and the ways in which such an elusive quality as engineering achievement in a military context was forced into a straitjacket of apparent rationality whose basis was military performance.

According to the rules, which were actually made up by senior officers in the corps and approved by the minister,

> An expedition such as a battle, a forced [defensive] line, breaking into a fortified camp and other posts under the fire of the enemy will be counted as one trench dug.
>
> Sieges will be compared by the number of trenches dug and not by the number of sieges.
>
> The first ten years of seniority of an engineer under [the command] of another will be counted as one siege, the next five after the first ten as being worth another siege.
>
> The character of a Chief Engineer or a Director of Fortifications counts as nothing in comparing war service, which alone determines rank in wartime.
>
> Finally . . . among commissions of the same kind but of unequal seniority, he who has the most testimonials of military service. . . has priority.[20]

Quantifying a career by trenches was in some ways typical of an attitude, both inside and outside the engineering corps. A kind of perverted and simplistic Cartesianism, more congenial to an absolutist Procrustes than Voltaire, is at work here. The delirium of numbers, more absurd than rational, manifested itself in other ways, such as the moment of resistance of a fortress (which played a large part in the debate between Montalembert and the corps).[21]

D'Argenson's first acts were to introduce more outside inspection and control of the military engineering corps and to continue the process of militarization. All engineers admitted to the corps were automatically awarded lieutenant réformé status (without pay). There was also an increase in the number of senior military commissions for those with greater technical experience to bring their military and technical ranks into closer agreement. Whereas it had not been uncommon for directors to retire with the rank of capitaine réformé, the situation was now improved in this regard. Yet this did not end the old division among the ingénieurs des places and the ingénieurs des tranchées. Indeed, it was maintained and even exacerbated by strict insistence on the priority of military services over technical achievements for promotion. At the same time, the distinct subservience of the corps to other arms of the army was reiterated by the provisions that, aside from a few special cases, engineering officers could never command other units, no matter what their rank.

There was much discontent among engineers as a whole and among the ingénieurs des places in particular. They were also not happy with d'Argenson's idea of subjecting them to external control and supervision. Transferring d'Asfeld's chief engineering aide, Rault de Ramsault, the director of the Bureau of Fortification on the rue Barbette in Paris, and replacing him with his own man, the civil engineer Noël de Regemorte as one of d'Argenson's first acts, was a clear signal of this. If Regemorte was not a member of d'Asfeld's "family," he was at least technically competent. Even this was not necessarily the case with the senior military officers who were ordered to inspect fortifications and report to the Minister of War on their quality and the competence of their personnel. Relentless subterranean opposition by engineers to having generals of the line army inspect them and the apparent lack of interest by generals to do so seem to have resulted in the de facto return to the older situation. Yet d'Argenson's aim seems clear: engineers were to be brought unequivocally within the army but under the control of regular line officers who would also set the criteria for promotion.

Another example of both rationalization and successful resistance to it was the episode of amalgamation of the engineers with the artillery. The artillery, along with other certain other corps of the army, had been in theory subjected to a Grand Master and Captain General, who was usually a Prince of the Blood or some other noble of very exalted rank. In 1755, the Comte d'Eu, who held this post, was prevailed upon to resign, and the king abolished it. This was done so that d'Argenson could unify both engineers and artillerists under a single command. The idea was hatched by Dubois, d'Argenson's premier commis, who happened to be in charge of the administration of both the artillery and engineers, and who was attracted by the merits of bureaucratic tidiness in the proposed arrangement. Neither the engineers nor the artillerists were consulted.

A new Royal Corps of Artillery and Military Engineering was founded. All its members were to be trained identically, and all were supposed to be capable of performing both the functions of the engineer and the artillerist. From the point of view of formal neatness and rationality this may have appeared an excellent idea.[22] Yet the organization led to utter confusion that was compounded by the outbreak of the Seven Years' War the following year. Things got to be so bad that the forced marriage between the artillerists and the engineers was dissolved less than 18 months later.

The approach to the recruitment of engineers into the corps by successive Ministers of War after d'Asfeld's paternalistic preference for the "enfants du corps" was also in accord with d'Argenson's principles—rationalization and establishment of more rigid control by the Minister of War. The percentage of sons of engineers being accepted into the corps dropped from almost 50 in the time of d'Asfeld to about 13.1 between 1748 and 1777; between 1778 and 1791 it dropped to about 5.6.[23] Another significant change in the composition of the corps was the increasing number of nobles: in the same periods of comparison, the percentage of nobles jumped from 57 to 75. If Guignard's opinion in the epigraph above is somewhat exaggerated, it is nevertheless true that for some members of the nobility—most particularly for the rural and military nobility—engineering was becoming an attractive career.

Beyond a changing and a more pronounced social homogeneity, there was another factor that increased the cohesion of the corps and similarity of its members. This was the improvement and more formal organization of the instruction for entrants to the corps. Military engineers had to be not only noble but also smart; along with blue blood they were required to have gray matter. The foundation of the Royal Engineering School at Mézières was also a clear signal that the new distinctive feature of the corps was going to be not family background but, along with noble birth, competence, proven first in an even more rigorous entrance examination and then in a superb formal course of study.[24]

Blanchard claims that a prime motivation of d'Argenson in setting up the school was a desire to remedy the perceived lack of competence that had shown itself in the dreadful casualty figures (especially among the younger engineers) in the War of the Austrian Succession. Along with an order for engineers to wear helmets and body armor in the trenches and to change the color of the uniform from scarlet to gray, d'Argenson seems to have felt that better instruction would also save lives. Yet, along with concerns for safety, the predilection of Enlightenment thinkers for new schemes of formal education in all areas, especially in the technical domain, undoubtedly had an effect. The latter part of the eighteenth century sees the appearance of the École des ponts et chaussées, the École des of Mines, and veterinary, drawing, and trade schools.[25]

Initially d'Argenson appears to have thought not of a single school but of a number of schools located in the chief fortresses under the command of the more experienced engineers, who would teach younger engineers and recruits the niceties of their trade in a more systematic manner than had been the case before. D'Argenson may have been thinking of the regimental artillery schools that had functioned since 1720.[26] It is possible, too, that the foundation of the École des Ponts et Chaussées the year before (in 1747) excited some emulative reflexes. Only a few of the Directors responded to his call for setting up schools. The first to do so, Nicolas-François-Antoine de Chastillon (1699–1765), who was stationed in northeastern France at the frontier town of Mézières, produced a remarkably complete plan for a school and began immediately to put it into effect.[27] The engineer and military theoretician Pierre Bourcet (1700–1780) set up a school for military surveyors at Grenoble.[28] One of the most prominent, Cormontaigne, called for two kinds of schools: a preparatory school dealing with theoretical subjects and a practical school.[29] But only Bourcet's and Chastillon's schools were properly set up, and very quickly Chastillon's became obligatory for all new recruits. Little is known about Bourcet's short-lived school, but the school at Mézières has rightly attracted the attention of historians of science and education.

The School of Mézières

The teaching staff of Chastillon and a few engineers was supplemented by civilian teachers of mathematics, physics, and drawing, as well as skilled craftsmen who taught stonecutting and carpentry. The standard of the civilian teachers was high. Among them would be Charles Bossut (1730–1814) and Gaspard Monge (1746–1818), both future academicians and eminent mathematicians.[30] Another academician, the abbé Jean-Antoine Nollet (1700–1770), whose lectures on electricity had captivated Parisian society and whose theories on the same subject had captivated most European researchers before Franklin, also taught physics in the "belle saison."

The engineering officers were in firm command of the school and this sometimes caused friction with the civilian teachers. But this did not prevent young recruits from receiving an excellent theoretical course of study over their two-year stay at the school. When Bossut eventually left in 1768, he became the examiner of the school, and entrance standards, if anything, became more severe than they had been. Cadets also received an excellent practical education. The content of their studies in military architecture seems to have been primarily influenced by Cormontaigne.

Mézières thus acted as a new focus for the engineering corps in the second half of the eighteenth century. All engineers were required to pass through it, and they received demanding and standardized training there. But in spite of appearances and the emphasis placed on the school by many historians of science, it did not become, as is so often the case, a prime defining characteristic of a profession.[31] The senior positions in the corps were held for a long time by an older generation that had more military experience and came almost exclusively from the old "family" of d'Asfeld. Ironically, too, the superior training that younger engineers received at Mézières tended to become more and more of an ornament. After the Seven Years' War, engineers had fewer opportunities for military action—the sole exception was the American Revolution. The same was true for construction, with the important exception of some new work around military ports. More and more of their old functions were taken over by the civilian Corps des Ponts et Chaussées, the engineers of the Navy, and even the ingénieurs géographes and topographers of the Ministry of External Affairs.

Mézières never played the same role as a focus for the corps of military engineering that the École des ponts et chaussées (whose origins can be dated to 1747) played. Although the latter never had as illustrious a faculty as Mézières (it relied to a great extent on mutual instruction and courses taught in other institutions), or anywhere near the same level of formal theoretical instruction, or the competitive examination administered by a member of the Academy of Sciences, it succeeded in giving the Corps des ponts et chaussées, along with genuine technical competence, a more homogeneous and solid corporate identity than the military engineers.[32]

There are a number of reasons for this. Perhaps the most important is the outstanding personality of Jean-Rodolphe Perronet (1708–1794), who not only served as the director of the school but also was the effective head of all students and junior employees of the Corps des ponts et chaussées before they were promoted to positions of responsibility. He was located in Paris, the intellectual and commercial center of the country and close to the center of power as well. His superior was the energetic Daniel Trudaine (1703–1769), whose official position—intendant de finances, chargé du détail des Ponts et Chaussées in the Contrôle générale—positioned him well to promote his policies successfully. Trudaine also instituted frequent meetings of an assembly of the senior engineers of the Corps des ponts et chaussées who happened to be in Paris from the provinces to deliberate on technical issues connected with their work. There was thus an ideal mixture of unified and vigorous leadership under Perronet and Trudaine along with consultation and exchange of ideas of the members of the corps. Students at the school generally spent more than the two years that was the norm at Mézières under the attentive direction and observation of Perronet who sent them out on summer campagnes (work periods on construction projects under the guidance of practicing engineers). There was thus an ongoing and vital interplay between work in the provinces and administration and teaching in Paris that welded the members of the corps together much more effectively than was the case for the military engineers. Perronet's dominance over the school and the corps only increased under Trudaine's less forceful successors and was continuous from the beginning of the school right up into the Revolution. This was all the easier because there were fewer engineers in the administration of the Corps des ponts et chaussées: in 1790 there were 218 of all ranks, whereas in the same year the number of military engineers was reduced from 329 to 310.[33]

The situation for the military engineers was noticeably different. Chastillon was a good and experienced engineer who had original ideas about pedagogy and about graphical techniques. But he was only one among two dozen Directors of fortification. Indeed, at the time of the foundation of Mézières he only held the rank of chief engineer in that fortress. No matter what his personal credit with the minister, he never played as central a role in the military engineering corps at Mézières (where he was also in charge of the daily administration of the precinct of the Meuse) as Perronet did at Paris. While Perronet ruled supreme and undisturbed over the École des Ponts et

Chaussées, Mézières had three different directors. Correspondence with Versailles went through administrators who were not always technically competent. Although all military engineers passed through Mézières after the 1750s, they normally stayed there only two years, and for all but a very few there was no contact with the school thereafter. While they were there, their practical experience was limited to a mock siege of the town of Mézières; there was none of the circulation between the school and a variety of ongoing projects all over the country enjoyed by the civil engineers. The school and the bureau of fortifications, with its maps, archives, and administrative memory, were physically separate, unlike the case for the administration of the Corps des ponts et chaussées. Moreover, the ethos of Mézières never penetrated the highest echelons of the corps. On the eve of the Revolution, without a single exception, none of its elderly though robust leaders were graduates of the school. They were oriented much more toward a military command structure than toward a school. To be sure, this military command structure did provide one cohesive force that their civilian counterparts lacked, and this was increasingly reinforced toward the end of the century: there was increasing class homogeneity and caste consciousness as the number of genuine nobles among military engineers increased. Even here, however, the ingénieurs des ponts et chaussées did not lag far behind their military counterparts. If virtually all the students at the École des Ponts et Chaussées were technically commoners, they were as adamant as any noble in not wanting students of what they considered disreputable origins as colleagues. Military engineering cadets at Mézières harassed fellow students who were of "low extraction," but so did students at the École des Ponts et Chaussées.[34] On a number of occasions, they demanded that fellow students be expelled on the grounds of parental origins or occupations.[35] These actions received the approval of Perronet, who himself was ennobled for his services.[36] As future functionaries of an absolutist state, the difference between their world view and that of their military counterparts was not as great as one might think.

In short, the corps of military engineering remained dispersed and atomized in spite of the undeniable homogenizing effects of Mézières, the high quality of the engineering education that it dispensed, and the social cachet of nobility of its members. D'Argenson, when approving the foundation of Mézières, had demanded that engineers have their mathematics "at their fingertips," and Mézières had certainly seen to that.[37] It did instill a common language and to a certain extent a common approach to the problems of fortification design—that of Cormontaigne—but it played no other role in instilling cohesiveness into the corps. Up until 1776 there was no central consultative body on technical matters that met regularly to discuss projects. There was a certain amount of informal consultation over new technical proposals, and it was sometimes quite thorough. But after the death of d'Asfeld there was never any continuity of direction.

The bureaucratic cohesiveness of the corps, realized temporarily to some degree by the personal energy and prestige of Vauban, had unraveled somewhat during d'Asfeld's leadership and, in spite of the efforts of d'Argenson and his successors, always remained fragile. It was an ideal toward which ministers and the senior leadership of the military engineers aspired and which they felt was indispensable for the effective functioning of the corps. It was something that the Corps des ponts et chaussées attained but the military engineers did not. It is probable that this is why the senior leadership of the military engineering corps valued this elusive cohesiveness to the extent they did, often at the expense of technical novelty.

Louis de Cormontaigne (1696–1752)

The dominant technical influence on French military engineering for most of the century after Vauban's death was Louis de Cormontaigne. Like Vauban, he began his military career at the age of 17 and participated in the last sieges of the War of the Spanish Succession. Unlike Vauban, he never held high military rank (only achieving the equivalent of brigadier-general 4 years before his death) and spent most of his career in subordinate positions, only becoming a Director of fortifications for the part of Lorraine that included Metz and Thionville in 1745. Nor did he ever lead a siege. His influence is attributable to his prodigious industry and his many writings on all aspects of fortification (which circulated mostly in manuscript form in the Military Engineering Corps, and some of which were published only after his death). Cormontaigne seems to have been one of the few engineers who mastered to a high degree the art of writing memoranda, which brought him to the attention of the Minister of War d'Argenson, who consulted him on many matters, including the foundation of the school at Mézières.[38] The capacity to write articulate memoranda accompanied by numerous and attractive drawings in an increasingly bureaucratic environment must not be underestimated in the making of reputations at this time.[39] His "style was fairly literary, something which is not often met with in this mass of documents."[40]

Like Vauban, Cormontaigne published almost nothing in his lifetime, and if the publicity brought about by the publication in 1741 of his *Architecture militaire ou l'Art de Fortifier* in the Hague brought him to the attention of d'Asfeld, who had previously ignored this same work submitted in manuscript, his further writing remained in ministerial offices and the archives of the military engineering corps.[41] His testy temperament and urge for polemics, as displayed in his bitter controversy with Bélidor, must have also had some impact on officials. The strong bent of d'Asfeld's successor, the minister of War d'Argenson, to rationalize the operations of his ministry and Cormontaigne's ability to present his ideas appears to have led to consultation between the minister in Paris and the officer of nominally undistinguished rank in Metz.

The nineteenth-century military engineering officer and historian of fortification Cosseron de Villenoisy has evaluated Cormontaigne's ability as a fortifier as reasonably good but as having dried up as he became a writer of bureaucratic works on fortification. The struggle of Vauban and Louvois to impose order and method on the administration of engineers was not nearly as complete as they desired or as subsequent historians have claimed. Much had been done, but the relapse under the easygoing regime of d'Asfeld left much to do. Even after the ordonnance of 31 December 1776, the corps was still struggling to achieve the cohesion and administrative regularity that Vauban envisioned and wanted. It is in his attempt to impose procedures and order, much more than in the area of contributions to the art of siegecraft and fortification, that Cormontaigne deserves the title of Vauban's successor awarded to him by Fourcroy. Even here there is a certain dryness and rigidity that would have been alien to Vauban but was becoming ever more the rule for the bureaucratic machine inherited from the regime of the Sun King.

Cormontaigne's style is displayed in a memo on the perennial theme of placing siege batteries, and he predictably argues that they should be placed as Vauban (and all engineers who followed him) wanted them placed: along the prolongations of the faces of the enemy bastions, and behind rather than in front of the siege parallels. He begins by arguing that nobody has matched or surpassed Vauban in siegecraft and "therefore the fact is decided in general, and both one and the other [the engineers and artillerists] must conform to [Vauban's method] in general, without any dispute on the part of one or the other."[42] The proper way of "effacing this would-be discord and remove the secret jealousy that causes it" is to have very clear and detailed regulations signed by the king himself. Thus, the regulatory authority of the monarch, together with the exemplary authority of Vauban, is to decide and fix procedures.

Another view of Cormontaigne in action is provided by a lengthy reply to d'Argenson on the question of tenders for construction work. D'Argenson had wanted more uniformity in contracts and the suspicious minister had wanted to know why it was always the same contractors who were engaged to build fortifications. As for the contracts, he "did not think that it was suitable for the good of the king's service that these details [of building tenders and contracts] should become arbitrary according to the system that each individual engineer adopts."[43] The first part of Cormontaigne's reply was an annotated copy of Vauban's *Directeur Général des Fortifications* bringing the work up to date by commenting on changes in nomenclature of positions (the conducteurs and chasse-avants were now called ingénieurs volontaires) and procedures (contractors were now in charge of paying their labor, and the Corps of Military Engineers no longer directly remunerated workmen hired by the day), and by suggesting additional rules (such as one forbidding people to serve in departments where their relatives were in senior positions).[44] Cormontaigne felt that there had been a decline in the competence of engineers in the time since Vauban until the present "happy time" when the minister was setting things right again. The next section of his memorandum, specifically dealing with abuses in the practices of tendering and how to remedy them, is perceived by the author as a continuation of Vauban's work. The various tricks fraudulent contractors employ are outlined in the form of an imaginary example using "Jean" and "Pierre" as protagonists. The main thing Cormontaigne wants is the institution of fixed procedures. These fixed procedures are suggested in the last section of the memorandum by giving a detailed description of operations at Metz, Cormontaigne's own post. Almost 200 pages of prices for various materials and services used at Metz are given, and this is clearly the place to look if one wants to know how much it cost to plant a tree at Metz in 1749 (19 sols). One gets the strong impression that Cormontaigne is suggesting that the various items listed should serve as a standard list for other *directions* of the corps.

The urge of d'Argenson and Cormontaigne to impose unity on the corps of military engineering and to rationalize its operations merely reflected a trend already clearly visible with Louvois and Vauban. This was true even though the effort to achieve this was sometimes resisted by local engineers in the provinces and occasionally faltered or even regressed, as it seems to have done during d'Asfeld's leadership. Centralization, bureaucratization, and standardization of the corps also promoted the transmission and diffusion of technology, with the attendant advantages and disadvantages that these features implied. One can distinguish two periods in this process that are separated by the watershed of the foundation of the school at Mézières. In the previous period one can speak of a "portfolio culture" in which knowledge and new ideas about fortification were contained in the personal papers of individual engineers—in their portfolios of drafts, plans, estimates, and memoranda for ongoing work and in their more theoretical musings. Seized by the government immediately upon the death of an engineer, these made their way into the archives of the corps in Paris.[45] They rarely found their way into print and reputations were built on those ideas that reached fellow engineers isolated in the provinces or made their way into the offices of the minister.

But although Louvois and Vauban had attempted—often brutally and tactlessly—to impose common standards on subordinates, their complaints and the violence of their attempts indicate the degree of autonomy enjoyed by local engineers and the diverse educational backgrounds from which they came. After the death of the energetic Vauban, who crisscrossed the country and whose physical presence was hard to avoid, the calmer times of d'Asfeld led to a reassertion of older patterns of autonomy and variety. The foundation of Mézières, in undermining the old semi-apprenticeship system, often associated with parental instruction and initiation into the corps, performed a number of important functions for d'Argenson. At the same time that it discouraged dynastic fiefdoms and consequently autonomy and disorder in the domain of administration and personalities, it brought uniformity into the technical domain. While esprit de corps grew at the expense of esprit de famille, local technical particularities faded in the face of a common corpus of knowledge presented in a systematic and uniform way to every new member of the corps. The circulation of ideas was enhanced and channeled. The older individualistic portfolio culture still coexisted with the school culture, but the new projects that came across ministerial desks or were passed on to colleagues were now judged with the yardstick of an official orthodoxy.

Cormontaigne and his reputation benefited from the structural changes in the training of the military engineering corps as well as from contingent factors. His assiduity had given him a reputation in the portfolio culture that was enhanced more by his ordinariness of talent than by any exceptional genius. The ideas that Cormontaigne presented were not unsound, and generally they were a continuation of the basic ideas of classical artillery fortification that had existed even before Vauban. It has been cogently argued that Cormontaigne's popularity within the corps and his identification as the true successor of Vauban were due to his quality of being a "Mr. Ordinary" whose work found perfect resonance with ordinary members of the corps.[46] His above-average capacities for work, for polemics, and for attracting the attention of superiors amplified his message within the portfolio culture. The contingent factor was the adoption of Cormontaigne's ideas as quasi-official doctrine by Fourcroy, the unofficial head of the military engineering corps after 1776, who was less talented but bureaucratically more adept than Cormontaigne and who used his eminent position to spread the gospel of Cormontaigne's "modern" system of fortification. Fourcroy had obtained Cormontaigne's papers, which he and a few colleagues edited, often unintelligently and even dishonestly, in order to put across their own pet ideas, making Cormontaigne look even less original than he was. These manuscript editions were circulated unofficially through the corps and were eventually published after the Revolution. Henri-Jean-Baptiste de Bousmard (1749–1807), who was the first to publish on the fortification instruction at Mézières while an émigré in Prussia during the Revolution, owned a copy of Cormontaigne and published some of his work as well.[47] Mézières incorporated many of the ideas of Cormontaigne into its teaching, thus giving them additional exposure.

Besides his notorious "moments of fortification" and his theoretical siege journals for estimating the power of resistance of a fortress, Cormontaigne's reputation with Montalembert and many later historians of fortification was based on his alleged attitude toward artillery. It was claimed that Cormontaigne downplayed the effectiveness of defensive artillery and called for a reduction in the number of guns used by the defenders even from that of Vauban. Furthermore, he rejected casemates and the idea of protecting guns from enemy fire, and he abandoned the idea of bastioned towers that Vauban had introduced in his "second and third system." He has therefore been portrayed as a regressive thinker out of touch with his time, when it was clear for all to see that the effects of artillery were becoming more effective against fortifications.

It is true that, through the prism of Fourcroy's simplistic vulgarization of Cormontaigne, one could arrive at this opinion. But in fact, Cormontaigne's attitudes toward artillery were much more subtle and much less disrespectful of artillery than appears to be the case from his edited works. His arguments for the diminution of artillery are in fact put forward in the case of smaller fortresses, which he considered to be qualitatively inferior to larger fortresses and even useless. The reason was that more cannon required larger amounts of munitions and soldiers to service them, which put a strain on the capacity of a small fortress. Crowding large amounts of artillery into inadequately sized fortifications would only make the defense more inefficient, encourage excessive and needless gunfire and wastage of ammunition, and provide more valuable booty for the inevitably triumphant besiegers. In fact, Cormontaigne's suggestions for the use of defensive artillery are quite imaginative. Instead of keeping it in the same position at all times, it was to be continually moved around (something that requires a large number of men, and is another reason for diminishing the amount of artillery or increasing the size of the fortress) to different positions as the attack progresses. At different stages of the attack, cannon of different calibers should be at different places in the fortifications and moved elsewhere when they are under fire. On no account should they indulge in artillery duels with the attackers, for they will surely lose, at the cost of lives, materiel, and uselessly expended ammunition.[48]

Even the apparently suicidal idea of uncovered cannon on open bastions and the rejection of covered flanks and casemates had foundations in the experience of the military engineering corps: Vauban's bastioned towers were proving troublesome because of the large amounts of smoke generated in confined spaces, and shortly after his death attempts were being made to obviate the severe interference this appears to have caused to the gunners.[49] Cormontaigne simply shared a belief that was virtually universal in the engineering corps: that defensive artillery could never match offensive artillery. Intelligently used, it was, of course, necessary in slowing the advance of the besiegers, but it could not be pitted against them in a straight contest. Like many other engineers, Cormontaigne felt that countermines built in advance were more effective than artillery in stemming an attack.[50]

In the 70 years after Vauban's death, French military engineers had acquired a more cohesive organization and a school to train rigorously selected entrants into their corps, which still had not been entirely integrated into the army. The techniques of fortification had not changed much. In spite of Cormontaigne's modifications, they were easily identifiable as Vauban's so-called first system, which he had inherited from Pagan. At the same time, the quality and prestige of the Mézières school had increased even more its graduates' reputation for being a corps savant. But it was also, as Coulomb pointed out, a corps militaire, and it had continued Vauban's tradition of successful siegecraft—so much so, perhaps, that it had undermined their own importance in the eyes of other soldiers.

II
Life and Death in the Corps

The life of soldiers in modern armies is full of contradictions. Perhaps the major one is the alternation between periods of excruciating boredom and the extreme and egregiously abnormal intensity of the struggle for survival and victory on the battlefield. In his classic book *The Face of Battle*, John Keegan stresses the difficulties of grasping the mental universe of the soldier in battle—a situation that is virtually unimaginable for civilians unacquainted with war. Even the best art falls short in helping us enter this world of combat and death. Military engineers, as soldiers, also partook of boredom and terror, since they spent time both in dull garrison towns and fortresses and were in the forefront of battle in siege warfare. For them, there were additional contradictions that were not always common to other soldiers: they were involved in both building and tearing down, in thinking about latrines and strategy, and in being bureaucrats and warriors. As warriors, they occupied a central place in the wars of Louis XIV; although this position was gradually undermined throughout the eighteenth century, they continued to shed their blood as much or more so than their fellow officers in the other combat arms. At the same time, they were intimately involved in the planning and design of fortresses—something that was also declining after the huge building programs of Vauban's time—and in managing their maintenance and upkeep. This was by no means an easy or uniform task in a France that was still far from an ideal Cartesian machine responsive to the king and his officials at Versailles and elsewhere. They knew from experience that localisms had to respected and manipulated when they could not be ignored or flattened—something that stubbornly continued to be the case more often than not.

The following two chapters will look at the engineers' techniques on the battlefield—siegecraft—and at their work behind their desks.

Chapter 4 will describe the siegecraft developed by Vauban, arguably his greatest achievement as a military engineer. No fortress ever resisted him successfully, and the record of the French military engineers who succeeded him was almost as good. Vauban introduced or systematized a number of techniques in siegecraft that minimized casualties and virtually banished the concept of impregnability from military thinking. Another feature that appeared in Vauban's time was the attempt to create a collective memory for the military engineering corps in the form of siege journals. Fourcroy de Ramecourt was to attach great importance to this tradition of siege journals and was to claim (at least in public) that it was the basis of a science of fortification that made French military engineers the unassailable authority in matters of fortification. Yet the quality of these journals was uneven and they often were larded with details regarding costs and service credits for engineers participating in the sieges. Nonetheless, they did provide a factual basis for calculation of logistics and materiel requirements. For others, they suggested the possibility of learning from mistakes and being able to predict the course of a siege and to show how to prolong it. Toward the end of the eighteenth century, for a number of reasons, sieges diminished in importance if fortresses did not. This was to pose a number of threats to military engineers. Once again, this coincided with the period of the general crisis I will describe in part III.

Chapter 5 will look at the paper world inhabited by engineers. Keeping records, drafting tenders, managing sites, reporting to their superiors, and dealing with a complex array of subordinates and superiors were skills that slowly became easier to acquire and exercise as the corps became more bureaucratic, its procedures became more routine, and its institutional framework became more solidly established. But paperasserie and dull routine, a mortal sin of a progressive Enlightenment and an indispensable crutch of a bureaucracy, led to more than mere boredom and frustration for overeducated engineers. It stimulated debate about their role, suggested new directions for their activity, and permitted the evolution of military engineering from a combat arm to a service arm of the army.

4
Siegecraft

For 'tis the sport to have the enginer
Hoist with his own petar. . . .
—William Shakespeare, *Hamlet,* act III, scene iv[1]

Vauban, the great fortifier whose methods were copied all over Europe, never wrote a treatise on military architecture. The books that Vauban did write were treatises on siegecraft and fortress defense. Never meant to be published, they circulated among French engineering officers until pirated versions appeared abroad many years later. His first major work on the subject, *Mémoire pour servir d'instruction dans la conduite des sièges et dans la défense des places*, was not published until 1740 but was written around 1670, when he was a little less than halfway through his long career.[2] He had already had 20 years of experience in siege warfare, and the War of Devolution with Spain (1667–68), which saw the capture of Lille, was behind him. The *Mémoire* was, however, written before the capture of Maastricht in the Dutch War (1673). At Maastricht, Vauban had first used the "siege parallel," an innovation in siegecraft that Christopher Duffy considers "the greatest advance in the siege attack since mobile siege artillery was introduced in the 1490s."[3] It was also written before the equally revolutionary innovation of the ricochet shot, which Vauban perfected at the siege of Ath in 1697—an innovation that Montalembert claimed made most artillery fortifications built up to that date obsolete.

Vauban's Siegecraft

The aforementioned innovations in siegecraft point to a feature of Vauban's career and his achievement that have sometimes been overlooked by historians and admirers in the eighteenth century, overwhelmed by the magnificent and massive material accomplishments scattered over the countryside. Vauban's reputation as a maker of fortresses overshadowed, in a misleading way, his reputation as a taker of fortresses. This comes out in the debate between Montalembert and the military engineers: the latter insisted on Vauban's military architecture whereas it was Montalembert who praised his siegecraft. Ephemeral as it was by necessity, in contrast to the almost indestructible remains of his fortresses, his siegecraft was nevertheless the most original aspect of his activity as well as that which influenced most profoundly the success of French arms in the short run and the nature of fortification and warfare itself in the long run. Siege parallels and the ricochet shot were crucial in the first place for accelerating the progress of sieges and making some parts of the fortifications much more vulnerable than they had been previously. They are also indispensable for understanding the nature of Vauban's contributions.

Before Vauban, besiegers had dug under fire a siege trench that snaked toward a vulnerable part of the besieged fortifications. Throwing up gabions and fascines before them and building up earthworks as soon as they had dug a trench, sappers made a zig-zag pattern on their way to the defensive perimeter. The sharp angles of the zig-zag minimized the degree to which defenders of the fortress could shoot along the length [enfilade] of the trench. As the angles got sharper this threat was diminished, but the amount of digging required increased. Just as dangerous was the vulnerability of the trench to sorties from the fortress. Defenders could attack the head of the trench where the number of sappers and troops guarding them was limited by its cramped dimensions and not well defended by supporting fire from the rest of the trench which essentially was perpendicular to the fortress.

The siege parallel solved these problems in a highly satisfactory manner. The original zig-zag trenches now became merely communications trenches between a series (usually three) of long trenches parallel to each other and concentric with the besieged fortress. These trenches could now be filled with soldiers who could provide strong support to the zig-zag communication trenches moving toward the fortress and bring more fire to bear on the fortress in the final assault. Batteries of artillery could also be placed in a number of places in a wide arc to batter the defenses, thus multiplying the potential points of attack. As Louis XIV himself expressed it, "we advanced toward the fortress [Maastricht] in broad and spacious trench lines, almost as if we were drawn up for a field battle."[4] Parallels were a kind of counter-fortress advancing on a fortress.

Vauban's use of parallels was not the first recorded. The plates and suggestions in Vauban's own treatise on the attack of fortresses indicate that his own ideas were evolv-

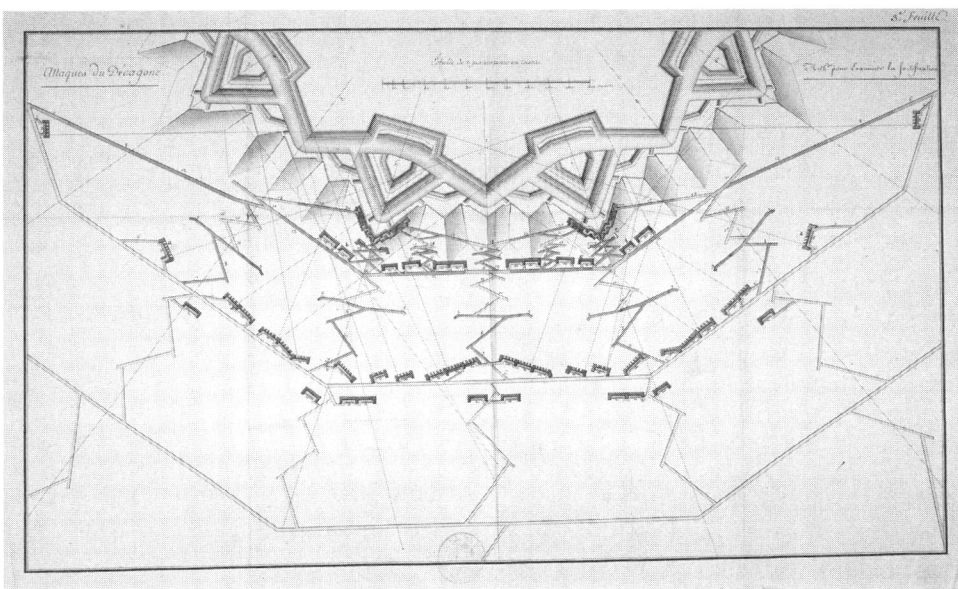

Figure 4.1
A plan view, drawn by Fourcroy de Ramecourt, of the initial phases of a Vauban siege. Note the parallels, the communications trenches, and the emplacements of siege batteries. Source: AIG, ms. article 21, section 1, §1, carton 6, piece 25, feuille 3.

ing in this direction, yet they were by no means developed before the siege of Maastricht when they appear in full flower. There were certainly signs of precursors to the siege parallel in Holland and France in the seventeenth century and even as early as the sixteenth. But the first time they were used in a clearly recognizable form was by the Turks at their prolonged but successful siege of Candia (Iraklion) in Crete (1648–1669).[5] Vauban was certainly aware of this siege, which attracted the attention of the whole of Europe and even saw the participation of French volunteers in the defense, and it is fair to suspect that Candia was his inspiration.[6] Thus, his accomplishment lies not so much in the originality of his idea as in its rapid adoption, development, systematization, and application. Indeed, this was true to such a degree that siege parallels, which immediately became routine after Maastricht, have long been considered, and not just by his hagiographers, as Vauban's invention.

A similar situation exists for the bayonet.[7] The bayonet was certainly around before Vauban, but it was in the form of the plug bayonet, essentially a knife shoved into the muzzle of the musket. This enabled it to be used as a short pike, but at the same time disabled it from use as a firearm. Vauban's innovation was simple and elegant: the baionette à douille (ring bayonet). By attaching the knife to the edge of a ring that would fit around the muzzle, the bayonet could now be used without hindering the musket from firing. This almost insignificant transformation led to the hitherto casual use of the bayonet to become generalized and changed the nature of infantry warfare by accelerating the decline of the pike and the final triumph of the hand-held firearm.

The ricochet battery provides clearer evidence of Vauban's originality, but even here Vauban himself and his predecessors had realized that underpowered cannon shots, with only about half the regular charge were more devastating to the defenders than powerful straight trajectory shots. In a ricochet shot, an underpowered cannonball was lobbed into a bastion, where, instead of embedding itself in the earth or masonry, it bounced around, killing soldiers and shattering gun carriages. Again, Vauban's genius lay in his identification of the essential features of a wide existing body of practice whose lessons were confusing and sometimes contradictory, extracting the lessons, and systematically developing them into new and qualitatively superior procedures.

Thus, Vauban's first known work on siegecraft, *Mémoire pour servir d'instruction dans la conduite des sièges et dans la défense des places*, does not contain his best-known innovations. Yet it is worth more than a passing glance. It does deal with the primary aspect of Vauban's contribution, and it gives us an idea of Vauban's style and his approach to the problems of siegecraft. Its down-to-earth manner, spiced with flashes of dry humor and indignation against silliness in high places, gives it a vivacity and readability far superior to most books on fortification of his time. Based on Vauban's accumulated experience and a merciless dissection of recent French mistakes in siegecraft, it consists of three parts. The main part concerns the attack of fortifications; this is followed by a shorter part on the defense of fortifications; and it ends with a part entitled "Definitions, Maxims, and Examples," which includes a case study of mistakes made by the French in the attack on Lille in 1667. The clear illustrations (29 plates), the names for all the necessary trenching and siege tools, the titles of the short sections (such as "How to build communication bridges," "How to mount the artillery," and "How to establish positions in the breach"), and the pithy concluding remarks clearly indicate that it is intended as a manual for the serving officer. Yet it contains the techniques of siegecraft and defense that would be under the direct supervision of military engineers, whom Vauban defends with passion and eloquence:

> . . . their science demands a great deal of courage and spirit, a solid genius, perpetual study and consummate experience in all the principal parts of war. And if nature rarely brings together in one man the first three qualities, it is even more extraordinary to see him escape the violence of our sieges and live long enough to acquire the other two. With that understood, it is not surprising that among the many who believe themselves to be engineers, or who claim to be, there are so few who are able and deserving of the title. The profession is great and very noble, but it demands a genius especially made for it and continuous application throughout many years, something that nature and the vigor of our sieges accord but rarely.[8]

And Vauban bluntly regrets that engineers must "follow the dictates of some half-learned lieutenant general in a position of authority"—precisely the kind of people for whom he was writing this work.

There are lists of supplies and munitions that should be stocked for a typical fortress, and they appear to be informed by long experience and shrewd judgment of human nature. For example, by the twentieth day of a siege, the governor of a fortress should expect one-tenth of his garrison to be wounded, sick, or "hiding," and Vauban makes the appropriate corrections for his estimates.[9] Commanders should guard the pyres used to illuminate night attacks to prevent pilferage of wood, and a paymaster should be in the trenches, ready to pay sappers on the spot.[10] This "will assure that you will never lack for men; they will even volunteer for the work once they have had a go at it." The banality of some of Vauban's "axioms" ("High positions command lower positions") and "maxims" ("Take care always to attack a fortress at its weakest point and never at its strongest") remind one of the mathematics masters writing on fortification who were of the object of Vauban's raillery.[11] But other "axioms" are greatly useful. For example, "Enfilades from a distance are more dangerous than those close up: projectiles from a distance, being slow and nearly spent, drop and deviate from their flat trajectory so that interposed traverses, even high ones, cannot stop the projectiles from falling into the trench [but] it is not difficult to cover against a close enfilade, since the projectile travels with such force that it maintains a flat trajectory; thus, if it crosses the top of one traverse it will be stopped by the next, causing damage only between the two."[12] This is very likely the germ of Vauban's brilliant idea of the underpowered ricochet shot.

Vauban gives handy tables for the construction of siege batteries with from two to sixteen cannon.[13] The columns in the tables give the length of the battery, soldiers needed for construction, firing the guns, servicing the guns, preparing fascines, diverse tools, timber, and the powder and ball necessary for a day's firing. After the first two lines of the table, the subsequent lines could have easily been constructed by the most simple of linear extrapolations, but Vauban prefers to write out the table in full. There is another table for giving the quantity of powder necessary for charging mines at different depths underground.[14] Again, Vauban indicates that the charge may have to be increased by a fifth or a sixth if there is masonry rather than just earth in the blast zone.

There are also detailed estimates of munitions and supplies necessary to withstand a standard siege, down to such apparently insignificant details as the special dietary requirements of wounded officers and men (mutton for the former and veal for the latter) and plums for the sick.[15] Bélidor's admiring comment that Vauban "could descend to the examination of an infinity of minute subjects" accurately describes Vauban's greatness and limitations as well as that of the military engineering corps throughout the eighteenth century.[16] Keeping one's attention fixed on details was necessary for success in peace and war in the profession of the military engineer in the eighteenth century, and it was by no means easy. In a society and an army where leadership disdained details and left these for their social inferiors, the engineer found himself having to pay attention to these details in order to supervise his subordinates effectively and to execute his projects.

Vauban affirms that he has "tested the effective range of grenadiers," and that "the definitive distance [he has given] for the placing of the palisade [in front of a ditch] is based upon these tests," and there are also empirical formulas for the amount of earth displaced by the explosions of mines.[17] Vauban does look upon siegecraft as one of the "sciences . . . founded upon certain clearly established principles," but it is by no means a mathematical or deterministic science.[18] His style is clearly pragmatic and not doctrinaire. After describing the geometry of trenches of circumvallation for the besieging army and giving a number of tips on how they should be built, he adds this: "Do not be overmuch concerned about the symmetry of the lines, but push them out and pull them in—sometimes considerably and sometimes only slightly—to take every advantage of the terrain."[19] When giving his figures for estimating the duration of a siege, he takes care to add at the beginning that "the following outline is intended as an instructive calculation and not a hard, fast rule."[20]

Toward the end of his career, 30 years after writing this work, Vauban wrote a more finished and corrected version for Louis XIV's grandson the Duc de Bourgogne as a kind of manual of the art of siegecraft and defense.[21] He was nearing the end of a glorious career, and he had had ample opportunity to augment his experience even more in the continuing wars of Louis XIV. Although he considers it disorganized [tout confus qu'il est], it is a more polished work than the previous one. Although much of the earlier material is repeated, there are expansions and additions that display even greater experience and increasing sophistication of his thinking on attacks. For example, he adds the proviso that soldiers (and even officers) should be well supplied with tobacco. To his rather obvious maxim that one should attack the weakest part of a fortress, he adds this qualification: "unless one is obliged to do so by overriding reasons which, compared to the appropriate ones, result in what is the strongest in ordinary cases being the weakest in extraordinary ones, which depends on the places, the times, [and] the seasons in which fortresses are attacked."[22] He goes on to give a detailed example (his siege of Valenciennes commanded by Louis XIV in person in 1677) where the usual rule had to be broken, and he explains why.

Siege Journals of the Battlefield

If one wants to get details of the actual events of the sieges, the account of the famous siege of Namur in 1692 by the great tragedian Racine, who earned his keep as Historiographer of the King, is not much better than the account by Laurence Sterne's Uncle Toby of the second siege of Namur.[23] True, literary scholars have not overlooked the similarity between the classical Vauban siege and the classical theatre of Racine, with its tight structure and its three unities of time, place, and action.[24] But the formalized and elaborate ballet of siegecraft seen from a king's tent was somewhat different from the work in the trenches, where Vauban insisted that the king and other high nobles should never venture because there is no place that is safe in a trench.[25] The best account of a theoretical siege can be found in the writings of Vauban discussed here as well as in the analyses of sieges to determine the theoretical strength of fortifications later in the eighteenth century.[26] But there are also numerous accounts of individual sieges in the French military engineering archives. It was the custom—instituted by the bureaucracy of Louis XIV, probably at the demand of Louvois or Vauban himself—that a journal be kept of every siege and be deposited for future reference in the military engineering archives. Beyond any functions of fiscal control or historiography, these journals were undoubtedly intended to be retained and to provide past experience of siege warfare to educate military engineers for either specific or general duties in the future. They were supposed to provide a trove of empirical knowledge of siege warfare that was often invoked in the debates with Montalembert.

But a look at siege journals in the archives reveals a rather wide range of formats and differing degrees of completeness in the descriptions of events. These variations seem to depend on the competence of the officer drawing up the journal and to some extent on the period at which they were drawn up. A cursory look at some siege journals suggests that some earlier ones, prepared under Vauban, were better than those prepared under the more easygoing directorship of d'Asfeld. At the beginning of d'Argenson's directorship there are signs of concern with documenting military services in terms of trenches as the new Director General required. Yet, in spite of their lack of standardization and their spotty quality, siege journals acquired increasing weight with those military engineers who thought they were creating a Newtonian science.

One of the early siege journals was for the siege of Mons in 1691, famous for the extent and stealth of the preparations (which marshaled 100,000 soldiers, 20,000 additional laborers, and large quantities of supplies and munitions), for the early season of the year in which it was attempted, for the presence of Louis XIV himself, and for a rapid victory. A large part of the army served as an army of observation to keep William of Orange from an attempt at relief. The military engineer Senneton de Chermont drafted a "memorandum on the most considerable events that occurred in the attack on Mons during the whole time of the siege."[27] The town was invested on 14 March 1691, and conscripted peasants built the lines of circumvallation while the army waited the arrival of the king a few days later. Vauban presented the king with a plan of attack, which was apparently followed; it involved a diversionary attack and a main attack on the Bertemont gate, which was protected by a huge hornwork. The town being surrounded by marshes and flooded areas, a river was diverted to drain the main approaches and to facilitate the attack. On the 24th, the siege began in earnest when "3 to 400 workers dug the first parallel." Because of protection available from a raised roadway, this was untypically done in the daytime. Work continued during the night. By morning, 1,500 paces of the trench were ready, and batteries were installed, with 10 men killed and wounded. At 10 o'clock on the morning of the 26th, a battery of 25 mortars (some firing 500-pound bombs) and 38 large cannon of unspecified caliber opened fire "in order to intimidate the bourgeois." In the evening the town was bombarded with red-hot cannonballs by 20 pieces of 12- and 14-pounder cannon in an effort to burn it.

A detailed and labeled map accompanied the account so that the advance of the siege trenches could be followed. This advance was fairly rapid because the garrison was not particularly energetic in attempting to disrupt the work by sorties, as noted by Senneton. This work was difficult:

... one did not dig half a foot before striking water, and with it a mud so soft that one stuck in it without being able to extricate oneself. I speak about this through experience having conducted the main part of the lodging [of the attack batteries, presumably on the third parallel].

To get enough earth to make the parapet of the trench it was necessary to widen [the trench] considerably and consequently to dig it with the spade in two shifts and so that [the parapet] would be solid to add a prodigious number of fascines.[28]

At 3 P.M. on March 31, when the French were at the outworks of the fortress and had descended into the ditch, Marshall Boufflers ordered an attack on the hornwork, which was initially cleared. But a vigorous counter-attack by the Spanish threw the French into confusion and forced them to retreat in panic. Losses were heavy: 130 men, among them "12 officers, 5 engineers, 30 musketeers of the king." The next day, after Louis XIV had browbeat his own royal guards who had ignominiously fled and replaced them with other troops, the hornwork was finally taken. The siege drew to its inevitable end and the fortress asked for terms at six in the evening on 8 April. The terms were generous: "They were accorded almost everything they wanted."[29] Of the 7,000-man garrison, there were 2,500 casualties, attributed by Senneton de Chermont to its patiently submitting to the almost ceaseless barrage of artillery fire. He felt that they would have prolonged the siege by sustaining the same casualties in sorties against the besiegers. French casualties were listed as 300 dead and an equal number wounded. The report ends with a list of the artillery used in the siege, this time along with the calibers. The capture of Mons had taken only 15 days.

The casualties for the French were considered to be exceptionally light. The capture of Mons several years later in 1709, in a siege that lasted a month, was considered to be economical in lives, with 2,300 casualties.[30] At that time, during the War of the Spanish Succession, it was the Allies who were attacking a French fortress fortified by Vauban. The assault on Vauban's fortress of Lille in the same year by Eugene and Marlborough had resulted in almost 14,000 casualties during a siege that had lasted nearly 6 months. Mons provided one of the many examples to justify Vauban's techniques of siegecraft.

More than half a century later Mons was again attacked and taken by the French during the War of the Austrian Succession. Here too there was a siege journal (*Journal du siège de Mons*). The French again attacked the Bertamont gate, but this time their diversionary attack from the heights of Nimy was pushed with such energy that it went nearly as fast as the main attack. The siege lasted about as long as the 1691 siege, and its progress was documented in very much the same style as Vauban's attack. This time, there was no drawing attached to the journal and no reference to such a drawing in describing the events of the siege. The attack was pushed even more vigorously than 50 years before: the trench was opened on the night of 24–25 June (it was becoming customary to make siege journal entries by "nights") by 3,000 workers supported by four

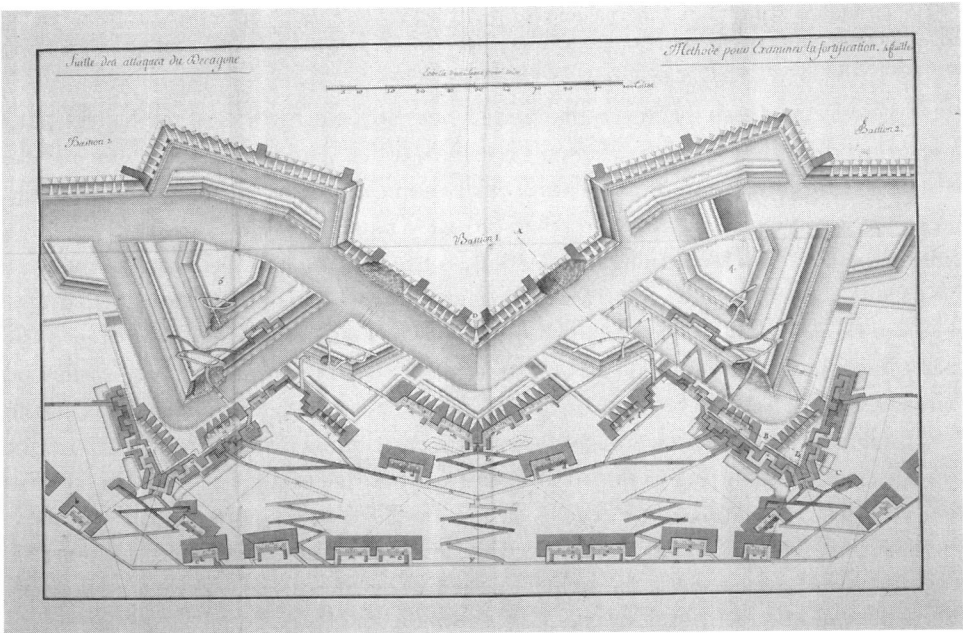

Figure 4.2
The final days of the siege in Fourcroy de Ramecourt's model siege journal. Note the saps in the ditch before the demi-lune and its redoubt, and the collapsed ramparts in the bastion caused by artillery battering. Source: AIG, ms. article 21, section 1, §1, carton 6, piece 25, feuille 4.

battalions of regular troops, six companies of grenadiers, and two pickets of dragoons. In all, 2,900 toises of trench were dug, and the attackers had pushed to within 200 toises of the covered way (compared to about 300 toises for Vauban's siege). There were no casualties. It seems that here is an indication that over time the corps of military engineering had grown more efficient at organizing the movement of men and materials and conducting siege works.

Casualties mounted as the siege progressed, but they were even fewer than those at the earlier siege of Mons by Louis XIV. In all there were 127 soldiers and 12 officers killed, 460 soldiers and 20 officers wounded. Engineers, however, had a disproportionate share of the casualties. Thus, of the 12 officers killed, seven were engineers, and of the 20 officers wounded, four were engineers.[31] During the night of 28–29 June a mortar bomb not only killed an engineer and an officer but left "Mr. Fournier, engineer, with his face smashed by the same bomb and in danger of death." Another engineer was also wounded that same night. Not all people were heroes in the siege, however. The journal recounts how an enemy sergeant manning a redoubt with twelve men successfully negotiated a surrender on condition that the French would fire six cannon shots and he would answer with six musket shots so that honor would be satisfied.

A look at some of the casualties not only suggests rich human drama but also illustrates the existence of engineering dynasties in the corps. There were two brothers Senneton de Chermont, relatives of the Senneton de Chermont who wrote the siege journal of 1691, also at Mons, and wounded, in 1746. There were in fact three branches of this family who supplied the corps with engineers from generation to generation from 1675 to 1768.[32] Pierre de Filley, who would examine Montalembert's prospectus in 1761, was there with both of his sons serving in his brigade, one of whom was killed at the siege. A third engineer son had died almost ten years earlier.

In some ways, this account of the siege is less perfect than that of the earlier one. The author is unknown and it trails off inexplicably on the night before the surrender with the author expressing hope that the fortress would fall the next day.

The year following the siege of Mons, the siege of Bergen-op-Zoom, which lasted more than two months, was eventually brought to a close by a very risky and unusual assault without the usual engineering preparations. It was the greatest siege of the War of the Austrian Succession in Flanders, and the capture of a fortress constructed by Coehoorn, garrisoned by good troops and supplied from the sea by the Dutch and British navies, was a great triumph for French arms, if not for French engineers. The siege had been proceeding very slowly and finally the exasperated commander had ordered an assault in defiance of all the expected rules. This no doubt contributed to its success against the surprised Dutch garrison. Montalembert used the siege as one of his examples of the redundancy of the engineer's art as it was practiced and the ineffectiveness of classical artillery fortifications. The siege journal for Bergen-op-Zoom is similar to others in listing the daily—and more often the nightly—events of the siege, giving the number killed, the distances dug, the artillery installed, and the names of the engineers on duty every day.[33] There is also a strict accounting of costs. For example, 86,056 gabions and 7,504 clayes were placed by the regiments besieging the town, at a cost of 10 sous per gabion and 5 sous per claye.[34] Sappers were paid by the length of trench dug, and the rates varied with the danger. Work in the second parallel on the night of 18–19 July was remunerated at 1 livre and 10 sous per toise dug, while the riskier work in the third parallel near the covered way of the enemy fortress received 3 livres per toise. In a particularly dangerous place, repairing saps ruined by cannon and bombs, the price went up to 4 livres. In all, the wages paid to sappers amounted to 93,938 livres and 10 sous. In addition, a sum of 2,862 livres, 14 sous, 11 deniers was paid to the military engineers for petty expenses, undoubtedly for the tips that Vauban had recommended all engineers be ready to dispense on the spot in the trench when some particularly hazardous or difficult piece of work had to be done. Special sergeants attached to the engineers as aides were given a daily allowance of 50 sous.

There is in fact very little additional technical information aside from the length of trench dug. In the Bergen-op-Zoom journal there is a single profile of the Coehoorn bastion. There were also translations of the siege journals of the Dutch defenders. But even here the technical information is scanty. There does not appear to have been any kind of detailed post-siege analysis or a formal process of saving and diffusing important lessons learned in the siege. In this case, the surviving documentary evidence undermines the supposition that systematic documentation of sieges was useful in improving the efficiency of French military engineers.

In the debates with Montalembert, military engineers made much of their siege journals as a kind of collective memory that gave them valuable cumulative experience, but the modern reader is left perplexed at what this could have been. Accounting and service records seem more important than technical matters in these records. Thus, unlike the siege journal of Mons in 1691, the 1746 siege journal of Mons has a table of all the engineers participating in the siege on its first page. The engineers were divided into 6 brigades with a senior engineer commanding ten subalterns in each brigade. The commander-in-chief of the engineers was assisted by a staff of four engineering aides. There is a great deal of variability in the quality of the journals and some of the later ones are definitely inferior to the ones written during Vauban's time.[35] Yet if practice was far from perfect and often variable in the redaction of siege journals, the intent that they should have an instructive value is quite explicit.

A document that was circulated around this time concerning the duties of engineers on campaign reinforces the impression that documenting questions of precedence and service records was more important to engineers than the performance of any technical duties.[36] It was written by a young engineering officer and was probably submitted to d'Argenson shortly after the death of d'Asfeld and it is very likely that it was an ideal toward which military engineers tended in practice. The preamble indicates that the commanding engineer, besides being someone who inspires confidence and capable of the best and most elevated views must never leave without the best maps of the territory in which the campaign is taking place and he should have along with him books and memoranda on the history of previous military operations in that area. His first task must be to make maps and to assemble his subordinates in a "sort of school" to study the operations of the campaign. He must also see to it that the engineers train the sappers, miners (normally not commanded by the engineers in peacetime), and support troops so that they can effectively carry out the orders of the engineers. But he will also "inspire all the engineers, in particular the ones in charge, the desire to see as often as they can the [commanding] general, the general officers of the army, and all other persons in place." He himself must "dine often" with the commanding general and try to see him every day.[37]

Behind the numbered articles and dry regulations there is an easily discernible concern for fairness regarding the possibilities of accumulating the service record that was so important for promotion. Thus, article 8 specifies that if the general wants to have an engineer particularly attached to him as an adviser, then there should be a rigid order of rotation for this post: regular engineers (ingénieurs ordinaires and not the ingénieurs volontaires who were on probation and of whom there still was a certain number in mid-century) from each brigade starting from the last engineer in the last brigade. Here one can hear the junior engineer who wants to get a chance to attract the attention and the favors of the great. Article 15 mentions among the functions of the engineer quartermaster that of procuring and furnishing for all according to their ranks bread and fodder, "and to take enough pains concerning all this so that no one suffers." Obviously in the days before well-organized commissariats and supply trains, raw ingenuity and "taking pains" were necessary even for engineering officers to get fed.[38]

Siege Journals of the Imagination

For some very influential military engineers, the siege journal, originally an instrument of financial control, military reporting, and record of services, provided the basis for a scientific tool that was central to the profession by the middle of the eighteenth century. Its purpose was to provide a reproducible and objective method of evaluating the "goodness" of fortifications, quantified in terms of expense of construction and days of resistance to a hypothetical attack. In 1743, the engineering general Jean Gourdon de l'Eglizière had compared two proposed designs of fortification by subjecting them to a fictitious attack and calculating the number of days each would resist.[39] The analysis involved a day-by-day description of the attack, such as the one that was supposed to be given in a regular siege journal, complete with detailed maps and drawings of the site and fortress with indications of the progress of the hypothetical siege. Fourcroy de Ramecourt used the methodology to contrast the designs of Cormontaigne, whom he admired, with other possible designs, to come to the decision that Cormontaigne was indeed superior to those of his competitors, including Gourdon de l'Eglizière. In an accompanying letter to Choiseul, the Minister of War, to whom he had sent his analysis, he suggested that this "method . . . is not yet practiced enough even though [it is] the only one known [for] reasoning on fortification in a fashion that is intelligible to men of war" and that it should be developed, perfected, and diffused.

Already in 1761, the year he had first examined and rejected Montalembert's new proposals of fortification, he had sent a résumé of a formal methodology to reconnoiter and inspect fortresses.[40] Once again, the modus operandi is typically Fourcroy de Ramecourt. It is in the form of an elaboration of earlier work considered to be best practice by an eminent engineer, in this case Jean-Baptiste Favart, who had written on the subject in 1715. And it proposed to institute a standard procedure that follows a certain mechanical pattern. It is fair to suggest that it is no doubt because of these bureaucratic skills, devoid of any originality but methodical and always clear, that Fourcroy de Ramecourt preferentially reached the ear of ministers at Versailles and would eventually rise to be the head of the military engineering corps.

In evaluating the "utility of a fortress on the frontier" Fourcroy de Ramecourt begins with general considerations (such as accessibility, terrain, and the quality of the soil besiegers would be forced to dig) that must be kept in mind and always clearly written up in memoranda and again proceeds to argue that there must be a "certain and clear method" to evaluate the actual strength of a fortress, not by "summary views [aperçus sommaires] . . . composed of purely arbitrary ideas without any proof but by memoranda. . . ." Proof "by memoranda" was certainly Fourcroy de Ramecourt's forte.

These memoranda would indicate why certain fronts of the fortifications would never be attacked and were to contain

first a [description of a] regular attack on each of the accessible fronts with a detailed journal of the work by the enemy and a plan where each day of the attack is tinted with a different color. By comparing these well-mounted attacks one could know positively and very surely both the real strength of the fortress and defects of equilibrium in its different fronts as well as the basis on which one must evaluate its needs of all kinds, relative to its capacity. For if the fortress requires 3 to 4000 men for the garrison by its extent and can only defend itself for 15 days, it is useless to provision it for 40 days of trenches.

second a descriptive itemization of the number and use of troops and artillery necessary for its defense, from which would be deduced on the basis of full knowledge of the facts the list of munitions and victuals indispensable for the fortress and consequently the space necessary both for lodging healthy, sick or wounded men and for storing supplies of all kinds.

third accounts of the condition of the fortress with respect to these different objects and the completions and additions necessary both for the fortifications and for the interior.[41]

Once such a system was established, the funds for fortification would be used in the most efficient way possible. There would be an additional, even greater and incalculable advantage in keeping engineers busy drafting standard documents:

> ... work thus directed would keep the officers of the corps occupied with the essential details and foundations of our service and they would banish forever from our letters and memoranda a quantity of systematic propositions that have sometimes occasioned useless expense in our fortresses.[42]

Up to the present, Directors had had a tendency to think in very narrow terms of their particular fortresses and to spend more attention to pursuing pet projects and getting a share of the meager fortification budget at Versailles. Every Director was

> zealous in seeing his ideas on fortification executed and most of them saw nothing brilliant in the idea of a storehouse. I have known some who insistently demanded the construction of a counterguard and to get it they proposed neither that it would be filled with earth or revetted at the gorge. When the necessity for these features was pointed out to them, they said that this was so expensive that the Court would not hear of it; while once the counterguard was constructed, the Court would have to reconsider and finish it. A successor wanted something else and left the first [counterguard] imperfect. Thus, from one [Director] to another much was undertaken and nothing completed.[43]

Even d'Asfeld himself, whom Fourcroy de Ramecourt seems to admire, was accused of not being immune from the temptation of building works for "show."

Preceding the "analysis" proper of a fortress was a procedure to determine the "utility of a fortress on the frontier." Engineers were to indicate the situation of the fortress with respect to the frontier, the reason why it was built (which may have changed in the interim), the usefulness of the fortress in present circumstances and the obstacle that it presents to an enemy attack into the kingdom as a whole. Here is a conception of fortresses as units that are part of a system of national defense that was already quite explicit with Louis XIV, Louvois, and Vauban. Roads the enemy would take to attack and the best way a relieving army could harass should also be consigned to writing.

A major difficulty that led to a host of subsidiary evils was the tendency of each engineer to think that his fortress and his point on the frontier was the most important if not the only one to take into consideration.

> But the king's service essentially requires us to strip ourselves of all personal and badly conceived interests, or that a clear-sighted chief responsible for everything knows how to unravel and to thrust aside our prejudices.[44]

It is more than likely that Fourcroy de Ramecourt already harbored ambitions to be the "clear-sighted chief" who would overlook a military engineering corps that was unified institutionally and methodologically. This was no easy target to achieve.

Louvois and Vauban had struggled to do so; d'Argenson, Saint-Germain, and Fourcroy de Ramecourt continued to follow in their footsteps. Louis XIV's ambition was to control warfare from the center and to eliminate the fickleness of chance on the battlefield. He wanted to achieve a state where a bureaucratic system made individuals interchangeable and it would be possible for command to be delegated to men of only average ability. But it was no doubt this bureaucratization of war in the domain of siegecraft that led to French successes in the War of the Austrian Succession as much as the ineptitude of their opponents. It is also likely that it was also both cause and effect of the domination of the corps by men like the hardworking but mediocre Fourcroy de Ramecourt whose bureaucratic talents were greater than their military ones. For some people, both military engineers and ministers of state, it was no doubt a price worth paying, for it must have had other advantages both technical and political. It would have made the corps run more smoothly in peacetime with less squabbling and strutting, and it would also ensure more effective execution, if not design, of construction projects.

Bureaucracy had not, of course, grown up on the battlefield or at the siege. As its name implies it had been born in the offices of the commis or clerks of royal ministers whence it had spread everywhere. This might suggest that the bureaucratization of warfare as evidenced particularly in the military engineering corps was part of a more general and widespread phenomenon, whose wellspring lay outside it. Yet, as the next chapter will attempt to show, there was an indigenous focus to bureaucracy within the military engineering corps itself, where it had achieved a level of sophistication that was no whit inferior nor posterior to the development of bureaucracy anywhere else in France.

Siege Warfare

In Europe during the seventeenth and eighteenth centuries, fortifications assumed an almost overwhelming importance in military strategy and the concerns of states. In France, during the reign of Louis XIV (1660–1715) in particular, fortresses and fortress warfare assumed a centrality that was only gradually eclipsed during the rest of the eighteenth century. When the Sun King went to war in person, it was almost invariably to a siege. He presided over twenty and he was never present at a major field battle, preferring the parallels of siege warfare to the maneuvering lines of soldiers. Why were fortresses and fortress warfare so important, and why did fortress warfare become less important toward the end of the century?

The traditional view that gunpowder destroyed feudalism by destroying the baronial castles, which were the basis of its power, has a considerable grain of truth in it. It must, of course, be remembered that this was not a simple case of technological determinism because it was the expense of acquiring and maintaining modern artillery that was often beyond the power of individual feudal lords. It was increasingly the appanage of kings who could use the power of organization and taxation of early modern states to get more and better guns than their vassals. A pendant to the view is that these same characteristics of the early modern state enabled it to build and maintain the expensive artillery fortifications that could face modern artillery. Another attribute of the increasingly modern state was the tendency to control the ground of its territory as never before, and this required the royal presence to be respected and to be ubiquitous.

In the debate over Geoffrey Parker's assertion in *The Military Revolution*, where Parker argues that the trace italienne artillery fortifications led to larger armies besieging them and consequently led to much larger army size than the medieval state could support and thus led in turn to the creation of modern fiscal and administrative methods that brought into being a whole gamut of changes that led to modernity, Nicholas Adams has countered that the growth of armies was in fact a result of the defensive strategy of the "Spanish school."[45] The Spanish Hapsburgs, and the powers that had the resources or tried to acquire the resources to imitate them, relied on garrisoning fortresses in peacetime. The empire of Charles V, the first European world power with the means, the dream, and the will of dominating a Catholic Europe, set an example to which other states would aspire. This required control on the ground that was tighter than the loose bonds of vassalage, this required garrisons and hence large armies, something which only the Hapsburgs could initially afford, and it was this, not the *kind* of fortification, that led to the fundamental changes that Parker claims made the modern European world. Yet the fortresses the Hapsburgs used for control were fortresses of the modern kind and there were more of them. Increasingly they became the primary focus of warfare. The bigger armies that were their cause rather than effect also contributed to change the nature of warfare.

As armies grew in size, problems of logistics and movement over the poor roads and bridges of Europe made fortresses more important because they could hinder and disrupt an advance unless they were first reduced, and this slowed down military operations even more and made war even more expensive. This is also a traditional historical view and such eighteenth-century French military engineers as d'Arçon subscribed to it as well. Van Creveld, while he does not believe that the existence of fortresses was an inevitable and logical cause of siege warfare nor that army mobility was impeded by them to the extent that the traditional view suggests, does nevertheless agree that the fortress was a prime target in warfare.[46] Fortresses were concrete objectives in wars that no longer had moral-ideological aims but politico-economic ones. Moreover, fortress warfare increased in importance because generals were unwilling to risk field battles where they could easily lose both battles and soldiers, who were becoming increasingly more expensive and valuable as they became better trained. Duffy also believes that eighteenth-century European armies had less difficulties in moving over bad terrain than he previously thought, which would undermine one of the main explanations for the importance of fortress warfare.[47] Yet siege warfare is still a central element of eighteenth-century warfare. If control of communications is no longer the only reason nor a reason as important as it once may have been for explaining the dominant role fortresses played, it nevertheless remains important.

Along with these reasons, there were others that help explain the importance of fortifications and siege warfare in France. One of the earliest states in Europe to acquire the rudiments of its modern identity, it had begun to do so at the end of the Hundred Years' War in battles against what hindsight identifies as foreign invaders and internal rebels. Capturing fortresses with the more effective artillery of the Bureau brothers was a significant feature of this process. From the time of the Capetians on, French monarchs were more successful than others in moving toward a centralized state, with the help of a judicial-administrative elite and the leadership of wealthy towns that they had managed to enlist or coerce. Internal and external threats to French unity and the central power, often conflated, as French factions appealed for external support, persisted during the religious wars and the wars of the Fronde, and were extremely grave. One does not need to be a practitioner of psycho-history to agree that the experience of fleeing at night from a Paris threatened by revolt and the Fronde left at least some impact on the mind of the prince who later became Louis XIV, who worked with such single-minded determination to crush noble opposition and external threats, and to concentrate power at Versailles. A literary critic has suggested that the impenetrable walls of etiquette Louis created around his royal person, and which much resemble elaborate fortifications in depth, were another form of allaying the craving for security and order that the king shared with many of his subjects.[48]

Fortification policy in France in the seventeenth century proceeded along two planes—destruction in the interior and construction on the periphery to ward off the threat of the Hapsburgs who surrounded the country on all sides. This policy was followed from Francis I through the great cardinals Richelieu and Mazarin to Louis XIV and Vauban. Throughout the seventeenth century the frontier solidified and moved outward to what would come to be called the "natural" boundaries of the country. As wealthy urban areas with traditions of municipal independence came under the control of the monarchy, particularly on the northeastern frontier, control and loyalty was established with the help of fortifications. Nicolas Faucherre has examined Louis XIV's policy of establishing citadels in these towns and argues that this crushed municipal liberties and laid the groundwork for the absolutist state of the eighteenth century.[49] Frontier towns thus secured and fortified were no longer seen as individual towns but as links in a frontier that at once protected the country and made it safe for absolutism. They were part of the ceinture de fer that girded Vauban's pré carré. At the same time, they became more valuable as prizes for other powers thinking along the same lines.

This ever more tightly controlled space, which was becoming a territory, rubbed up against other powers and Louis XIV started on an aggressive policy of war to extend, secure, and rationalize it. His predecessors had used diplomacy as much as military means to gain allies and isolate enemies, but the Sun King adopted an increasingly militaristic approach. This may perhaps be attributed to a bellicose temperament, but no matter what the actual cause, he was able to do this on a larger scale because he possessed certain advantages not available to them. He had an increasingly peaceful interior—the nobility had been brought to heel, its capabilities for armed revolt stunted, its martial energy channeled outward for the king's benefit, and its administrative functions taken over by new categories of people beholden to the royal power rather than to feudal lords. The Protestants had made their peace with a Catholic monarch, and the greater towns were cowed or loyal. Frugality and the remorseless fiscal efficiency of his predecessors had given the crown better resources in a country that was increasing in wealth and population as the disruptions of civil conflict receded and productive capacity improved. Those resources were augmented by the king's early conquests, as French territory expanded into prosperous new areas. The administration and army were the most efficient in Europe, and the king had capable and loyal officials doing his bidding. Since the sixteenth century, France had become a considerably more tightly administered country: it has been estimated that in 1515 there was one royal functionary for every 115 square kilometers and for every 950 inhabitants whereas in 1665 there was one for every 10 square kilometers and every 76 inhabitants.[50] Louis' own instincts for order, hierarchy, and control, along with his energetic and continuous oversight of his administration led to even greater power of the French state and its army, which was used to enhance further the elements of strength enumerated here.

It is perhaps no accident that we can date the true beginning of the French military engineering corps to Louis XIV's reign, and Vauban being the first engineer to be promoted to general officer rank (lieutenant general) by Louvois in 1688 is symbolic of the ever increasing importance of siege warfare.[51] The same qualities of organization and efficiency that were apparent in other military and administrative policies were visible in the Département des fortifications des places de terre et de mer. Thanks to Louis, Louvois, and Vauban, the military engineering corps received a powerful initial impulsion that permitted it to develop to its mature state a century later, and serve a model for other parts of his army and administration.[52] Of his armed forces, this was the arm that was the most important in siege warfare.

The siege was perhaps the essence of the Ludovician military style. Duffy is right to assert that "Of all the operations of war, a grand siege was in fact Louis' favorite. Not only did it follow a predictably successful course (thanks to Vauban), but it provided a magnificent spectacle in the baroque style, at once vigorous and technical."[53] But "classical" rather than Duffy's "baroque" may be a better adjective for Louis' taste in war and siegecraft. Rules, order, and coherence, rather than bizarrerie, extravagance, and emotion, were the desirable features of well-conducted sieges, such as those of Vauban. In 1675, the king established the Ordre du Tableau, a table of seniority to regulate promotions of his officers by seniority, and Louvois tried to adhere to it strictly to minimize favoritism and the attendant squabbles it would bring. Appointing the eponymous Lieutenant-Colonel Martinet as Inspector General of the Infantry by Louvois in 1667 brought the army under tighter and more orderly control.[54] Louis went to the trouble of drawing up elaborate instructions ("Manière de montrer les jardins de Versailles") for the viewing of Le Notre's gardens. The royal tour guide decreed that there be no deviation in the order in which the gardens were to be shown to distinguished visitors.[55] It is entirely conceivable that the man who loved to make nature Euclidean in his gardens at Versailles and who created a table of ranks for his officers had the hubris to want impose regularity and predictability on that most unpredictable of human activities—war. Siege warfare was the best way of doing it.

Siege warfare contributed admirably to what were perceived by contemporaries as a distinctly French military strategy. French armies would take to the field at what was considered to be very early in the season, when their enemies were usually not prepared or unwilling to do so. After descending on a fortress on the frontier, they relied on their excellent siegecraft to snatch a fortress, quickly strengthened it, and settled in to await the counterattack. Relying again on their engineers, as builders and defenders this time, they would hold off the attack until the end of the campaigning season, and attempt to repeat the procedure in the next year. If not spectacular, it was a systematic and to some extent predictable way of waging war and it increased control over territory gradually but surely.⁵⁶ An added benefit was that it occurred on foreign territory that would then be converted into an additional resource for the future. (While widening its security zone and smoothing out the boundaries of its pré carré, it also led to increasingly aggrieved and worried neighbors, who allied themselves to regain what the French had taken from them, attempting to use the same techniques that the French used so successfully.)

Siegecraft made the concept of "war machine" appropriate and the mechanistic idea was easily extended in the country of Descartes to the rest of the army and even society. Military engineering and fortifications were not merely valuable for themselves as controllable aspects of war. Louis XIV's emphasis on fortification clearly shows a desire for control both administrative and political over his army as a whole and the ultimate hope that Mistress Chance, always present on the battlefield could be subdued and bound. According to Jules-Louis de Bolay, marquis de Chamlay,

> by the great conquests that have been made and by the advantageous situation of the towns that have been fortified, the king finds himself in the situation of being able to have his army commanded by whomever he pleases, without having to apprehend anything about the mediocre capacity of him to whom he will entrust it.⁵⁷

In citing this remark, Léonard added that neither did the king have to be apprehensive of *superior* capacities, such as those of Condé during the Fronde. Louis did not trust his soldiers, both because of the potential for disobedience by the better ones that he had witnessed himself and because they lacked the qualities he felt were important for the management and leadership of his kind of army. With the disappearance of Louis' great captains Turenne and Condé from the scene, military operations were increasingly controlled from Versailles. The marquis de Chamlay, just quoted, was a civilian and a *robin*, a man with legal training and administrative experience, and it was he who served as Louvois' effective chief of staff.⁵⁸ He was only one of a horde of civilian administrators much resented by the king's marshals. The army was indeed controlled by civilians and it was only in the middle of the eighteenth century (1758), with the appointment of Marshal Belle-Isle to the Ministry of War, that a soldier held that position. (The predominance of soldiers in the post after that date was a sign of the accelerating professionalization of the army that will be discussed in later chapters.)

Fortresses and siege warfare helped make generals interchangeable and their efforts predictable, something that Louis was also aiming to accomplish in his administrative machine. Along with civilian administrators and military engineers it would help to make war more "industrial"—a concept that, for later French military engineers like d'Arçon, meant systematic and workmanlike in almost the sense used by Thorstein Veblen, relying more, though never entirely, on technology, construction, and patience, than on blood, guts, and dash.[59] It would also allow for the distant but close management of warfare as he was attempting to do for the country as a whole. More than being merely at the center of a Panopticon, Louis was an industrious spider at the center of a web at Versailles. Some historians have argued that this siegemaster's approach to conducting warfare and managing field armies had disastrous effects:

> Following the thinking of Mazarin, Louis XIV and Louvois regarded it as a royal obligation to push everywhere, to its ultimate limits, the spirit of centralization, the unity of the government, [and] to direct the armies like the administration, generals like intendants, to dictate to them not only their campaign plans, but even the smallest details, the marches, the encampments, the daily movements. [This was] a pretension that would not have been justified even by their continual presence at the theatre of war and that their habitual absence made even more unreasonable and more dangerous.[60]

Proper provisioning of a fortress required enormous amounts of supplies and weaponry, and this implied financial and organizational resources that were not routinely available to most countries but France. A more detailed look at the particular example of Landau, an important fortress on the borders of France and the German states that was often attacked and changed hands during Louis' wars, is instructive in giving an idea of the resources necessary and the mentality of the engineer who organized the provisioning. Hüe de Caligny was one of Vauban's best engineers and followed his procedures when writing about Landau.[61] He begins with the necessary personnel, the most important of whom is, of course, the "chief engineer," who requires 9–10 assistants, such as inspectors, works contractors, clerks, craftsmen, and others. A table with three columns, of the kind that Vauban liked so much, gives the desirable amount, actual amount, and the deficiency or surplus of supplies. There is attention to detail of the sort that is so often evident in Vauban as well: 1,000 sewing needles will be needed for a siege. The detail extends to the actual description of the architectural resources available to counter the enemy. Thus, Caligny describes how to use the thorny hedge in Mme. Villeman's garden to mount rallies and harass attackers. Smaller amounts of wine will be needed than might normally be necessary because the good burghers of Landau are known to have large quantities of it in their cellars, it only gets better with age, and can be requisitioned during the siege.

The military provisions and equipment are impressive. For the octagonal fortress, 128 guns are required (20 24-pounders, 24 16-pounders, 26 12-pounders, 28 8-pounders, and 30 4-pounders). In addition to this, 44 pierriers and mortars are necessary. (Only 53 of the 128 guns were actually on site, probably a normal state of affairs considering the cost and scarcity of artillery, even for a king as wealthy and warlike as Louis XIV.) These guns require 1.2 million pounds of gunpowder, 102,400 rounds of shot, 11,000 bombs, 6,000 pierrierloads of pebbles, and 1,430,965 pounds of lead. The 8,800 men of the garrison also need 9,338 200-pound sacks of flour for themselves and 205,700 rations of fodder for their horses. One's powder would be used carefully, it was assumed that for the first nine days of the siege, the fortress artillery would not shoot because the first works the enemy would be out of range. When the fortress did open fire, its guns would be deliberately underpowered to give the enemy a false sense of their real range.

Landau was a powerful fortress with an uncharacteristically large armament, which, as de Caligny indicates, was not up to its desirable level in practice. Yet with de Caligny's ideal complement of weaponry it had slightly fewer guns proportionally than Vauban's specification for a hexagonal fortress—90 guns. (Fourcroy de Ramecourt and Cormontaigne, probably more realistic as well as more pessimistic about the power of defensive artillery specified even fewer guns—68 for a hexagonal fortress.)[62] Usually, the attackers' artillery was less or, at best, at the same strength of the defenders. The difficulties of marshaling and delivering the siege artillery were considerable. Indeed, because of these difficulties, Pepper and Adams have cast serious doubts on the effectiveness of siege artillery (and consequently the necessity for the new kind of Renaissance artillery fortifications) well up into the sixteenth century: "... very real limitations on the availability and mobility of guns meant that fortifications of only moderate strength would rarely have to face heavy artillery."[63] Toward the end of the sixteenth century the Duc de Guise estimated that the siege train to attack a strong fortress would require 24 heavy guns in two batteries along with 6–7 culverins. Along with collecting and transporting these guns were the additional difficulties of transporting the shot (8,000–10,000 pieces), powder (200,000 pounds), and associated vehicles.[64] The size of the support personnel and animals to move the siege-train were enormous and it was here that French organizational abilities gave them an advantage. Twenty thousand peasants were required to provide support and dig trenches at Mons in 1691 and Marlborough mentions 16,000 horses required for the Allied siege-train at the siege of Lille in 1708.[65] In Maurice de Saxe's lightning campaign in the Netherlands in 1747, the French astounded Europe with a siege-train of 90 cannon and 70 mortars. The more typical eighteenth-century siege-train had 30–60 heavy pieces of which something less than a third were mortars and howitzers.[66] For the army of Charles V, 33 horses were required for a 96-pound double cannon (with a dismounted barrel) as well as 6 more horses for the empty gun-carriage, extra horses for the actual powder and shot (156 for 8 days at 30 rounds per day), and still more for tools and baggage.[67] By the time of the Revolution, it was estimated that for the heaviest guns of the lighter field artillery only 22 horses were required per gun and associated caissons and wagons.[68] This was a decline, partly due to lighter guns and better carriages, for the traction required for the guns, but animals required for munitions and powder must have stayed very much the same, considering that rates of fire and consequent expenditure of ammunition increased over the eighteenth century. Yet in the three centuries between the French irruption into Italy that demonstrated the power of artillery on medieval fortifications and their irruption into the Netherlands during the Revolution, siegecraft remained a slow and complicated business whose logistics strained the resources and organizational talents of all European states. Even with the French, who were masters of the art, siege warfare was cumbersome, expensive, and rarely allowed for spectacular general movements that were characteristic of Frederick the Great and Napoleon. Toward the end of the century, military men began to question the usefulness of fortification, and even the usefulness of military engineers and started thinking more seriously about alternatives.

In 1781, the Austrian Emperor Joseph II dismantled most of the so-called Barrier Fortresses in the Austrian Netherlands on the French border, which had been garrisoned by Dutch troops since the treaty of Utrecht. This action seemed to imply the uselessness of permanent fortifications in a most dramatic manner. These were, after all, the fortresses that had been hotly contested a century ago, had been a major theatre of war less than 40 years earlier, and had been considered vital for the security of the Netherlands. Joseph's action was in fact bizarre and by no means was it proof that fortresses had outlived their usefulness. It was motivated more by political than military or technical considerations for it occurred at a time when Austria had made peace with France and did not see a threat from a country that seemed to have renounced any expansionist ambitions and had become its ally. When the war with Revolutionary France broke out a few years later, the Austrians were to rue Joseph's decision. While Vauban's old chain of fortresses once again shielded Paris from the attacks of the Coalition during the initial reverses of the Revolutionary armies, the lack of the old Barrier fortresses enabled them to sweep through the Netherlands with lightning speed when the French were able to go on the offensive.

The French Revolution, the far-flung campaigns of the Revolutionary armies, and Napoleon's fast and mobile warfare did not mean the end of fortresses or sieges. Sieges continued to be important in the Revolutionary wars: Charleroi, Maubeuge,

Valenciennes, and Dunkerque on Vauban's old frontier with Low Countries, Mantua and Genoa in Italy, and even Acre in Palestine were at times the center of attention of generals. The Peninsular War saw a number of hard-fought sieges and the siege of Zaragoza, like the guerrilla war in the field, was a portent of new kinds of wars and sieges. There is a symbolism of sorts in the fact that, Lazare Carnot, military engineer and ardent defender of the importance of military engineering and the usefulness of fortresses, then the "Organizer of Victory" of the Revolutionary armies who preached the offensive, received the news of Napoleon's abdication as the commander of the (still resisting) fortress of Antwerp in 1814.

Napoleon himself, admired as the supreme representative of a new kind of warfare that involved bold and rapid movement and involving vigorous field actions, was a firm believer in the necessity of fortifications. The great offensive warrior believed that without fortifications offensive war was impossible. Much like Guibert, he believed that fortifications were to be support bases [points d'appui] from which offensive operations could be launched and which could serve as places of refuge in the event of defeat. His ideas on the roles of fortresses differed from the earlier view of necklaces of fortresses as barriers; rather he wanted fewer and larger fortresses, not necessarily on the frontier, that would also serve as depots. On campaign, he called for small, unsophisticated, but reasonably robust forts that could provide safety from cavalry raids. A student of Napoleon's ideas on fortification and its uses in war has identified five different kinds of fortification that Napoleon proposed.[69] The military engineers of the Old Regime had also proposed all of them on occasion. New permanent fortifications were built during Napoleon's reign (notably in Alessandria in Italy and at Antwerp).[70] Napoleon allowed his engineers latitude for some innovation in the military architecture of these fortifications, although his own ideas on military architecture were rather conservative.[71]

The period after the Napoleonic wars saw a flurry of fortress building in Russia, Germany, and the Austrian dominions, using some of the new ideas of Montalembert. The sieges of Sevastopol and the Franco-Prussian War, the building of the fortifications of Paris during the Orleanist monarchy, the defensive lines and fortresses of Séré de Rivières and Brialmont at the end of the nineteenth century, all indicate that fortresses survived and functioned into the twentieth century. Brest Litovsk, Sevastopol, Leningrad, Stalingrad, Dien Bien Phu, and Khe San remind us that fortresses and sieges did not simply disappear.

Yet it is undeniable that the fortress was no longer central in warfare and the thinking of generals in the way it had been under Louis XIV. The stronghold and the wars that revolved around it had been displaced as a focus of military action and thought, although it had by no means become irrelevant, as some eighteenth-century military thinkers and later historians claimed. Why was this so?

The argument that a better and more dense network of communications made it easier to bypass fortresses is no longer as compelling to historians of fortification as it once was.[72] It is undeniable that roads, particularly in France, improved during the eighteenth century, in large part because of the excellent work of engineers. When the Revolution came, France had 25,000 kilometers of roads.[73] More and better roads, however, does not necessarily mean that it is easier to avoid enemy fortresses or less risky to leave them in the rear of an advance. There are other more plausible reasons for the disrepute that fortresses had acquired in certain quarters. Some of them have to do with the activities of military engineers themselves.

Fortresses were being captured more quickly and this was due both to improved artillery and to the effectiveness of the siegecraft of military engineers. Taking the length of average sieges in the eighteenth century as a measure of the effectiveness of siegecraft is not entirely reliable, but it is one of the few available. There was in fact a decline in the time taken to capture fortresses and in Maurice de Saxe's campaign in the Netherlands the Barrier fortresses fell considerably more rapidly than 50 years earlier during Vauban's time. In part this was a tribute to the effectiveness of the French military engineering establishment. Even without a charismatic leader like Vauban, enervated to some extent under d'Asfeld's leadership, and with less control over the artillery than in Vauban's time, the corps of military engineering performed extremely well. This performance was used by its opponents as proof that fortresses—at least of the kind military engineers constructed—were useless. Another reason for the declining resistance of fortresses was the increased power of artillery. Duffy's remark that the power and mobility of ordnance had been enhanced "by about 50 per cent," as a result of a host of minor improvements in the design and manufacture of cannon, gun-carriages, gunpowder, and harness, is not particularly helpful or illuminating.[74] However, it is generally correct. There is a consistent and universally shared view by all soldiers in the eighteenth century, including artillerists and military engineers, that artillery had reached an unprecedented and radical increase in its power. Both military engineers and their opponent agreed on this point. It is difficult to determine the precise nature of the improvement of artillery and the contribution of the various factors to it, as well as a little presumptuous to attempt quantification of that improvement, but all of Duffy's suggested reasons for it are undoubtedly valid. The Swedish military engineering general Virgin, accompanying Maurice de Saxe in his sieges in Flanders, refers to the unprecedented power of the artillery as a fact and it inspired him to develop new kinds of mortar casemates to take account of it. Montalembert unceasingly reiterates his argument that artillery of hitherto unseen power was making the classical bastioned fortifications obsolete, and the unprecedented number of guns he prescribes for his new fortifications are another sign of the importance of artillery in his thinking. The military engineers, on the other hand, base much of their argument for the impossibility of impregnability as a realistic objective for fortifications precisely on the power of siege artillery. They are less respectful of the powers of defensive artillery because of its more limited ability to move. For some, the best way to counter the invincible siege artillery was the use of countermines, rather than to confront it directly in an impossible duel.

More generally, increased army size had an important effect on pushing siege warfare out of the center of military operations and military thinking. Louis XIV's army was about seven times greater than that of Henri IV. The French army retained high peacetime levels of about 150,000 men during the eighteenth century and had peaked at 400,000 in wartime.[75] Army size had grown considerably since the beginning of the seventeenth century and the size of siege armies had fallen, again, partly as a result of the superior methods of siegecraft. These large armies were the result, not of a special kind of fortification but more generally because of richer and more populous states that were at the same time more stable and administered more efficiently, nowhere more so than in France. States could more easily afford large armies. They could also use them more effectively than previously. This increased effectiveness manifested itself in two forms—in a managerial and an operational form. The bureaux of the Ministry of War continued the traditions of tight control and administration of the army—the example set by Louvois was there to maintain and enhance. By modern standards, it might be an exaggeration to speak of an efficient administration in Old Regime France; there were only 80 clerks in the offices of the Ministry of War at the beginning of the Revolution.[76] However, the very fact of a permanent staff of officials running the army enabled better management of finances and logistics, care and housing of troops, and regularity of recruitment than was imaginable in earlier ages. All these features had incalculable benefits in wartime.

Operational efficiency, or what might be called tactical efficiency, was the other feature that almost became an inevitable requirement for large armies, which had reached the limits of operation as a single group under a single chief. It was during the debacle of the Seven Years' War that the concept of division was developed by Marshal de Broglie. Breaking up the army in divisions, each with its own units of artillery, engineers, and the other arms, large enough to fend off an entire army until other divisions could come to its rescue, enabled armies to proceed along wider fronts and to be more adventurous than before. This was now possible, even necessary, partly because the numbers of soldiers were available. Another reason, however, was a better trained officer corps in whom the supreme commander could have confidence. The increasing numbers and quality of such officers were another feature of the late eighteenth century, particularly in France. The divisional structure also called for more coordination and staff work. Although army staffs were seen, particularly by military engineers, as nests of favoritism and incompetence—and to some degree this was justified—there was increasing and irresistible pressure to endow them with the kinds of skills, topographical and organizational, that engineers possessed to a superlative degree. Officers of the other arms increasingly sought and used an education that emphasized mathematics and the sciences. The results of this were more initiative in operations, more freedom of movement, and an increasing ability to bypass fortresses. The large armies could also capture fortresses more quickly. Endowed with good engineers, a smaller siege army could attack a fortress, while an observation army could spare it the worry and work of protecting itself from an enemy relief army—a pattern that had already been demonstrated with excellent results by Maurice de Saxe.

Permanent fortifications thus moved to the side of the line of vision of general officers. A final reason that should be mentioned for this has elements of irony. To some degree, engineers were hoist with their own petard. Their very professionalization and increased competence threatened to make them less relevant on the battlefield. Operators of a well-oiled, smoothly running machine, highly educated in the necessary specialized knowledge, they were called upon to do their job and they did it with increasingly predictable results. Commanders could no longer manage a siege themselves, and took less interest in sieges and fortification. Engineers were again perceived to be more technicians than regular soldiers, the experts of "industrial" war but not fit for the battlefield and consequently robbed of the prestige that always adheres to it among military men. The forces acting toward the marginalization of engineers came from within and without the corps. The ever present delicate balance between the savant and the soldier was often disrupted. Some engineers devoted themselves even more to the details and even the arcane aspects of their art, playing a kind of "glass bead game" in exercising their intelligence on the niceties of defilading calculations, attempting to quantify and mathematize the strength of fortifications, and occasionally pondering more scientific methods of analyzing the properties of matter to help them in construction. Even among those who saw themselves primarily as soldiers, there were some who felt that they should have no contact with troops and be spared the daily trouble of commanding rank-and-file soldiers.

Others, for whom the military aspect of their vocation was dominant, strove for greater integration into the army and participation in its operations. They coveted positions on the general staff, considered to be an ideal place for their talents and training, and suggested new and original military roles for engineers. This desire to integrate more closely into the army was not always received sympathetically by regular line officers—not a surprising reaction in view of the long history of the difficulties engineers encountered in their attempts to be accepted as true soldiers. Montalembert at one point even suggested that the military engineering corps was superfluous. Civilian engineers should take over construction of permanent fortifications and siegecraft should be turned over to the artillery. Others, like Saint-Germain, thought highly not only of the intellectual but the military qualities of engineers and abolished the prohibition of their serving on the general staff. Indeed, he even wanted the general staff to be made up of engineers exclusively.[77]

Along with the marginalization that was partly the unfortunate result of their competence and professionalism, there was increasing diffusion into the general body of military officers of knowledge that used to be part of the preserve of engineers or higher general officers. From mid-century on, officers became more learned and more assiduous in studying the sciences necessary to warfare. The exhortations by Errard and the schoolmasters of the sixteenth century to noble officers that mathematics, far from being beneath their dignity, was essential to their careers, were being heeded with enthusiasm. Military education improved and officers became more competent. This military education included parts of what engineers had been learning for a long time. The increasing interest in field fortifications required some competence in the field by regular line officers. Again, this led some, particularly artillery officers, to argue that they did not need the services of engineers, even in siegecraft. Montalembert again was obviously addressing himself to at least some sympathetic ears when he claimed that beyond what most officers could learn in a relatively short time, the esoteric knowledge of engineers was more for obfuscation than use. Military engineers were in the difficult position of being attacked for their qualities as much as for their defects.

5
Desk Jobs

> Let's run to the skirmish and offer them battle
> Let's dispute the field with these modern pests
> Let's attack their clerks and upset their desks.
> Close up, raise the barricades, as we've always done:
> Make way for the brave, dammit! the sword before the pen!
> —Francoeur, in *Vauban à Charleroi,* by Jacques-Antoine, Baron de Révérony Saint-Cyr

Révérony Saint-Cyr was an undistinguished French military engineer, who helped teach some of the first courses at the École Polytechnique, and was also a minor man of letters in the early part of the nineteenth century. He wrote essays, plays and operas (one of which was set to music by Cherubini) that today deservedly gather dust on the shelves of deposit libraries. His play *Vauban à Charleroi* (staged in 1826), from which the epigraph for this chapter is taken, displays a certain self-image of the engineer that is worth examining critically.

In this play, Vauban is besieging a town and simultaneously involved with protecting a damsel in distress and her mother, both fallen from royal favor but not from virtue, from the evil and libidinous machinations of a corrupt royal financial administrator. One of Vauban's siege engineers, brave, intelligent, but poor and without connections, loves the damsel and is loved by her. The play presents an unimaginative and stark contrast between the heroes—engineers who are bluff, brave, and ready chaps—and the villains, who are craven, scheming bureaucrats, moneybags, and court favorites. Predictably, the heroes win: they capture Charleroi and expose the fraudulent behavior of the villains to the king, thus saving the damsel for her true love, who has survived a very dangerous attack on the fortifications thanks to Vauban's technical genius. At one point in the play, when the struggle between good and evil intensifies, one of the heroes, the sapper Francoeur—the name "Frank Heart" is probably not accidental—urges an attack on the financiers and bureaucrats, the intriguers and enemies of Vauban, in the words cited above.

The image of the military engineer as a warrior, in contrast to the clerks whom he will attack and whose desks he will overturn, is rather obvious here: it was a self-image that eighteenth-century French military engineers cultivated assiduously. Even military engineers as academically inclined as the physicist Coulomb, who emphasized that the military engineering corps was a corps à talent, never forgot that it had another mission. "The Corps of Military Engineering must be considered from two points of view: as a corps of talent and as a military corps."[1] Engineers were men of honor and men of action who were not tainted with the shady dealings of men of finance, of intrigue, and paper.

The Birth of a Bureaucracy

Yet in the actual day-to-day practice of an engineering career—even a military engineering career—a great deal of effort was devoted to administration and paperwork. As much as military engineers complained about the paperwork they derided in others, they were both subjects and creators of a world of bureaucratic paper that was an attribute of the modern state, especially in France. Roland Mousnier, in his massive institutional history of Old Regime France, gives to the engineer the honor of being the prototype and first example of the fonctionnaire (civil servant), as opposed to the old class of venal officiers of the administrative and judicial system of the Old Regime.[2] The monarchy had created and relied on the holders of offices in its struggle with the old feudal nobility in the late Middle Ages. But the officiers became more numerous as the financially beleaguered Crown used the sale of hereditary offices as a means to raise money, and eventually they asserted their independence. A major theme of French history in the eighteenth century during the eighteenth century is the struggle of the royal power to control these officiers whose most eminent members made up the noblesse de robe of the sovereign courts [parlements] who had become every bit as troublesome as the old noblesse d'épée. In addition to direct confrontation, the Crown also attempted to bypass the officiers by creating a new category of servants of the state in the form of the commissaires, its directly controlled creatures, who were not only totally dependent but revocable at will. According to Mousnier, the engineers of the Corps des ponts et chaussées, an administrative corps created in 1716, were an entire hierarchy of commissaires at a number of levels who were prototypes for the future civil servant or the fonctionnaire, the Weberian bureaucrat. If one defines the fonctionnaire as someone who is selected on merit, functions within a clear hierarchy with clear criteria for promotion to which corresponds an appropriate salary scale, whose primary and even exclusive loyalty is to his employer, and is subject to dismissal at the pleasure of the king, then the ingénieurs des ponts et chaussées were clearly fonctionnaires. Military engineers had many similarities with their colleagues in the Corps des ponts et chaussées and although Mousnier is more reluctant to see them as full-fledged fonctionnaires, no doubt because of their military obligations, they in fact do possess many of the features of Pierre Gaxotte's "Monsieur Lebureau," future ruler of states and citizens, who appeared on the stage of history around 1750.

The Corps of Military Engineering had by no means reached the mature organization of today's civil service. It was still a corps, with all the connotations of feudal corporatism and aristocratic militarism inherent in the name. But the lineaments of a more modern, more rigidly structured, and more bureaucratic world were clearly visible. And it is this bureaucratic context that must be taken into account to understand the functions and essence of military engineering in eighteenth century France. The fixation on design and engineering science as the defining essence of engineering has led many, with notable exceptions like Lewis Mumford, to downgrade or even miss entirely project management and administration as an important part of the defining essence of engineering.[3] But a more modern administration and management was an essential part of an engineer's work, all the more so in a society where these functions were institutionalized in traditional structures less and less adequate for the needs of the modern state. Technical administration often overflowed into administration pure and simple. The debates over the abolition of the corvée system of forced labor in lieu of taxes for peasants in the latter part of the eighteenth century had extensive political implications and the engineers of the Corps des ponts et chaussées had a very visible position in that debate.[4] It has been pointed out that the designation of state engineers in the civil engineering construction sector in Prussia, a country much influenced by French models in technical education, was Baubeamte [building official] rather than Engineer.[5]

Vauban as Bureaucrat

For examining the bureaucracy of the French military engineering corps it is only natural to begin, as in many other aspects, with Vauban. Even though a bureaucratic structure for French military engineers is discernible with Sully, the minister of Henri IV, in the early seventeenth century, it was only with Vauban and his successors that we see it congealing into its modern form.⁶ An instructive indication of Vauban's thinking on the nature of the engineer's functions is a small book he wrote on the functions of the engineer even before the creation of the Département des fortifications des places de terre et de mer in 1691. *Le Directeur Général des Fortifications*, of which a pirated version was printed in Holland and dedicated to the Prince of Orange, was only 144 pages long in a convenient pocket-sized duodecimo format obviously intended to serve as a working guide for French military engineers.⁷

A closer study of this fundamental work shows a number of attitudes and concerns toward the function of engineers that give an idea of Vauban's conception of that role. He himself says that the book was written to maintain the continuity of practice that is threatened as a result of the heavy losses among engineers during the last war. Here, as at the foundation of the Mézières school, is a primary reason for the replacement of traditional craft modes of instruction with formal, routinized, and written instruction. Historians of technology have long emphasized the importance of tacit knowledge, especially but not exclusively, in the pre-industrial period when low literacy, craft methods, and the secretiveness of traditional structures for artisans led to the transmission of knowledge by apprenticeship over an extended time under the direct supervision of a master in daily work and production. The fabric of human relations that underpinned this mode of instruction was nowhere ripped apart so often as in the military trades.

The manual is supposed to help with measurements, tenders, and estimates and thus to eliminate "crazy expenses and sometimes great knavery."⁸ This sets the tone of the whole book. The primary concerns are clearly the prevention of fraud, the minimizing of expense by rational management and accounting, and the maintenance of central control and supervision over geographically dispersed work sites. From Vauban's book it appears that the conception and design of fortresses is to be kept as much as possible in the hands of the Directeur général des fortifications, the chief official at Versailles with ministerial responsibility for the Department of Fortifications. Planning and designing fortresses, then, is to be centralized. This does not mean, of course, that this official is necessarily sedentary. It is quite apparent that the peripatetic and energetic Vauban, who always emphasized the importance of the knowledge of local conditions in the design of fortresses, sees himself, if not in this role, then at least performing all its technical parts. (In fact, the first person to hold this post was Michel Le Peletier, a former intendant with great administrative talents. But he was counseled on all technical matters by Vauban, who had held the title commissaire général de fortifications since 1678.) Indeed, part of the duties of the Directeur général is a regular inspection tour of the fortifications in his jurisdiction. Nor does it exclude input from the next lower officials in the hierarchy—the Directors of the various regions and the engineers in the fortresses—some of whose most important duties are regular and systematic communications with their superiors.

A printed circular that appeared a short time before Vauban's death regarding the duties of the regional directors of fortification is also concerned almost exclusively with the functions of inspection, verification, and reporting of these officials.[9] They are to "visit" the fortresses in their jurisdiction, "verify" estimates, "examine" the repairs and maintenance to see if contracts have been fulfilled, "attend" bidding sessions for contracts, and "report" to the Directeur général and the intendant any abuses or fraud they detect. They are also to "observe" engineers working for them for a number of reasons: to ensure they are not too friendly with contractors—something which smelled of potential fraud to bureaucrats of the Old Regime; to keep an eye on the moral qualities of their engineers; and to examine their technical competence. They are to find out if anyone lacks the necessary capacity in geometry, drawing, mechanics, and "the other parts they must know" and to point out to them what is necessary for their further instruction while informing the Directeur général of this.

The engineers entrusted with a particular province or region (the directeurs) or a particular fortress (the ingénieurs en chef or chief engineers) are those who are most intimately connected with the actual construction of the fortresses. Vauban says that "their work will be the most difficult of all."[10] The purely technical part of their functions, however, again seems to occupy a secondary place in Vauban's thinking. They will "lay out the plan settled upon by the Director General," certainly an activity that leaves little for engineering creativity in its technical aspect, and their other functions are described at much greater length.[11] They are to attend all bidding sessions for contracts and sign them as representatives of the Directeur général. Good records of all work done are to be kept and signed by all those concerned. Measurements and verifications of completed projects must be carried out publicly in the presence of witnesses. Again, Vauban seems obsessed with fraud, an accusation that he himself was unable to avoid at one point.[12] The engineers are to pay attention to local parlance, technical idiom, and units to avoid misunderstanding and potential friction with local contractors, but they must also convert all units and measurements to standard royal measurements when reporting to their superiors.

A book must be kept in which all work orders and financial transactions are recorded. Not only are engineers required to give an annual report of work done and costs incurred, but they are also required to report every eight days to the Minister on the state of their fortress.[13] Such regular communication is astonishing even by modern standards. (In practice this ideal was never attained, if the repeated exhortations to this effect and the complaints about backsliding are any indication.) It is their responsibility to maintain contact with local authorities such as the governor of the fortress and the local intendant. Keeping in touch with superiors so that they can control the most trivial details of execution, diplomatically promoting the royal fortifications' program within the complex and delicate web of overlapping powers and jurisdictions in the France of the Old Regime, and maintaining scrupulous honesty and maximum economy for a chronically impecunious government are the main elements of the provincial engineer's mission that Vauban considered so difficult.

Sources such as *Le Directeur général des Fortifications*, various instructions sent by Vauban to engineers in charge of fortress construction, and the iconographic evidence give some idea of the day to day work of the fortress engineer at the end of the seventeenth century. Drawings by the engineer Claude Masse show lines of peasants carrying panniers trudging past a well-dressed gentleman to the ditch of a fortress under construction where some of their fellows are working on the masonry revetments of the scarp.[14] Scheduling of work and the surveillance of the work force are of great importance. Thus, conscription of peasants during harvest time is to be avoided and soldiers working on fortifications are to be lodged where they are as close as possible to their work but far away from places where "they could do harm."[15] And, above all, there is the recurring theme of the many reports to fill out for various authorities. Much of it contains models for standard forms that must be completed for various kinds of reports that the engineers were expected to address to the central authorities at Versailles. Thus there is a "Form for the measurement of wood for carpentry to provide information on the different kinds and pieces of wood, their quantity, length, thickness, and volume according to units of Paris."[16]

But surely technical expertise played a major role? The qualifications that Vauban formulated for the first time for recruiting young engineers appear to support this. This is the work where he rejects any kind of favoritism and calls for merit alone in choosing among candidates, specifying that they should have a public examination "not only regarding geometry and surveying, but also on all the other parts of the most necessary mathematics, such as trigonometry, mechanics, arithmetic, geography, civil architecture, and even drawing."[17] Obviously here is a syllabus requiring above normal mathematical knowledge for its time, even though mensuration is considered the basic requirement.

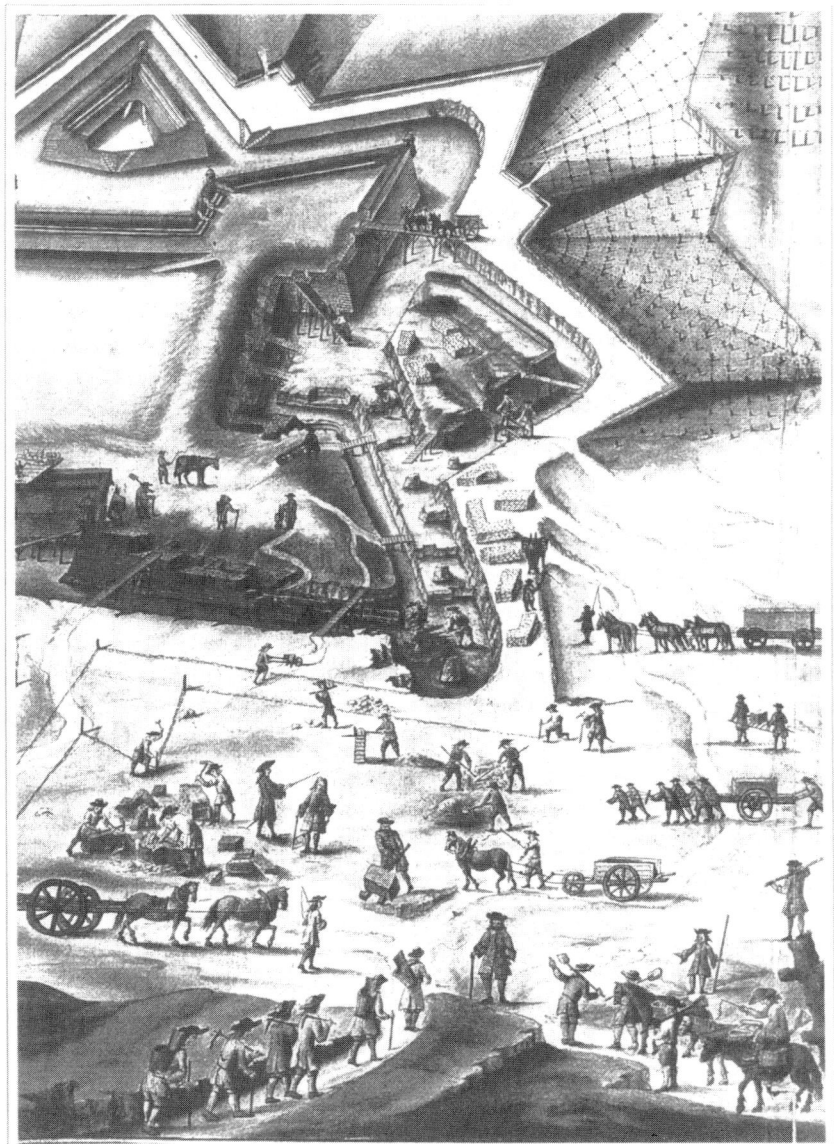

Figure 5.1
A drawing by the engineer Claude Masse (1652–1737) of a work site during the construction of a fortress. A good example of Masse's skill, this has been reproduced often. Source: "Place que l'on bâtit," Bibliothèque de l'Inspection du Génie, Atlas de Masse, ms. in fol. 135.

Science and Administration

To what extent and in what way was eighteenth century engineering scientific? This is a question that will be treated in more detail in a following chapter, but a number of preliminary claims and suggestions can be advanced here. It may be true that one can detect in the work of exceptional engineers like Bernard Forest de Bélidor (1697 or 1698–1761), who in fact was never a member of the engineering corps and was more often than not involved in controversies with it, a clear attempt to present available engineering knowledge in an organized and clear fashion stripped of the mystery of artisanal recipes and tricks, to undertake systematic experimentation to solve engineering problems, and to grope toward the application of some basic scientific principles to engineering construction.[18] In the first half of the eighteenth century, however, in spite of the support he had from the leadership of the military engineering corps, Bélidor's views remained more vision than reality.

Even by the standards of the seventeenth century, the engineer in the field was not abreast of relevant scientific research in institutions such as the Academy of Sciences, nor was he required to have a great amount of it to perform his functions perfectly adequately. He was, of course, not required to be manually adept at the particular technical skills of the various craftsmen employed on the site. Nor does it seem that the average engineer was required to have the responsibility of actually designing the entire fortress, that often being the responsibility of the Directeur général, nor all of its individual parts, that often being the job of the contractor. What was required, beside some familiarity with the practice of construction, was an accountant's and taxonomist's mastery of sums and terminology. In this sense, of course, the engineer was adhering more closely to a Baconian ideal of science as perceived by the majority of his contemporaries and even by some academicians like Duhamel de Monceau rather than a science defined by Whiggish hindsight.[19] The engineer was perceived by Vauban as a man capable of managing the actual construction of fortifications in the seventeenth and early eighteenth centuries. These fortifications required "...infinite care, perpetual activity, great steadiness, much experience and good sense concerning all earthworks, wood, [and] stone, perfect knowledge of the capacity of workers, of all the different kinds of materials and their prices."[20]

The ingénieur des places (later called the ingénieur en chef) described in *Le Directeur général* dealt not only with superiors but with subordinates such as the Ingénieurs en second, Architectes, Appareilleurs, Conducteurs, Inspecteurs des travaux (future engineers in training), Contrôleurs des fortifications, down to the Piqueurs and

Chasse-avant who watch over the workers. The fact that architects are included among the subordinate personnel of engineers provides another clue that the managerial and financial responsibilities of engineers are primary, while those aspects of their work that could be called technical according to the standards of their time are relegated to others. Many of the fortress engineers were, of course, architects themselves originally, but although their technical skill was considered an extremely valuable additional qualification, the fact that architects were considered their subordinates seems to imply that management is their main function.

On the whole, then, in examining the beginnings of the institutional evolution of French military engineering during the eighteenth century, one is struck by the preponderance of accountancy over technicity in the peacetime role of people who were called the ingénieurs des places. It is true that technical knowledge begins to get more and more important as an attribute of the engineer, and it eventually becomes more important in fact as well as in rhetoric. Even this, however, occurred to some degree because of managerial and financial pressures. In *Le Directeur Général des Fortifications*, Vauban clearly indicated his desire for "particular" contracting rather than general contracting in fortress construction.[21] The scale of fortress building was increasingly surpassing the financial and technical resources of general contractors. Therefore, the state, through its agents the engineers, had to assume the role of general contractor and to deal with and organize the work of many subcontractors. This not only put more demands on managerial skills and accountability; by degrees, it also put more demands on technical knowledge. To ensure uniformity for plans, for example, Vauban specified a color scheme that was to remain the standard throughout the century.[22] Obviously intended for ease of reporting and to facilitate administration, the measure led to better maps and plans that elicited other cartographical innovations like contour lines and the mathematical use of these lines to calculate angles of fire and cover toward the end of the century.[23]

All this leads one to believe that the military engineer in France at the end of the seventeenth century is as much the descendant of the sixteenth-century Florentine provveditore di fortezze described by Lamberini as the offspring of the marriage of the artist-architect of the Renaissance and the military officer, in spite of the fact that he often was either an architect or a military officer in the late seventeenth century.[24] The provveditore di fortezze was the intermediary between the civil and military hierarchy of the state who had a special responsibility for fortifications, including the architect who planned the work, on the one hand and the workers, both skilled and unskilled, who executed the plans on the other. Lamberini's Florentine provveditore di fortezze, is somewhat different from the Venetian official of a similar name at about the same time described by Hale.[25] Whereas Hale's officials are nobles elected for a short term to serve as financial and administrative officials to execute the decisions of the political power and who are the social and hierarchical superiors of engineer-architects, Lamberini's provveditore is the indispensable junction between the civil and military hierarchy in the form of an inverted pyramid which tapered down toward him and the regular pyramid of the technical and laboring hierarchy which spread out and below him in the form of a regular pyramid.

> He was the key figure of the building site. In this double pyramid, the provveditore of fortresses occupies the strong but flexible hinge. He was a person who deserves particular attention. To him was entrusted the whole weight of the perfect functioning of the building site, from the technical, economic and human point of view. Generally of humble origins, the provveditore had to know how to read, write and keep accounts. To him they [the architect and the military and civilian administration] trusted the registers of the ingoings and outgoings of the building. Above all, he had to have great experience of men and materials, since it was his role to provide them. He was sort of a factotum, with a thousand brains and a thousand resources; as cunning as a peasant, and as courtly as a minister; conciliatory and servile with his superiors, equivocal with the architect, inflexible with his subordinates; compassionate toward the marrioli he recruited, so long as they followed him; their furious persecutor if they fled him. In short, a true creature of Cosimo [de Medici].[26]

These were duties, along with those of architect and soldier, that were eventually to become those of the military engineer of the seventeenth century and later. Even among the subordinate staff, the function of control and surveillance of relatively unspecialized and unmotivated workers was of the greatest importance for Vauban. Thus, the master mason must be not only a "faithful man knowing masonry" but also someone who "never loses sight of the masons' hands, for the greater part of them are extremely careless in the arrangement of materials either through ignorance or knavery—something that only happens when they are not instructed by someone they fear."[27]

An instruction for chief engineers by Le Peletier circulated for comment two years after Vauban's death essentially follows the guidelines laid down in Vauban's earlier work.[28] It is clear that at least at the beginning of the eighteenth century, most of the final design decisions were left to the Directors, a limited group of about two dozen people that shrank to about a dozen before the Revolution. With the disappearance of Vauban and the domination of the Department of Fortifications by d'Asfeld, whose technical competence was minimal and who did not have a technical adviser of the stature of Vauban, most of the actual design work seems to have shifted toward the regional Directors. Thus, the chief engineers would lay out the ground plan for any work only "in the absence" of the Director. A marginal annotation by Tarade, the Director of

Fortifications of Alsace, indicates that they are to conform strictly to signed scale plans and profiles made by the Directors. In cases of difficulty and doubt, such as the solidity of foundations, they are to communicate with the Director to get his opinion. For the rest, the procedures and duties are essentially those recommended by Vauban. Even the tip that masons "normally are inclined to do bad work" sounds familiar. The chief engineer is essentially a surveyor, a reporter, and a site manager. The Directors advising Le Peletier do express some concern about the number of interim reports requested, which they feel are too frequent, take up too much time, and are sometimes useless. The reports to be given to the intendants are a waste because they "pay no attention to them." The requirement that local military commanders also attend certain surveys and estimations [toisés] is also useless since "they have no idea of the task to which they are attending."

Another anonymous manuscript memoir on the duties of the chief engineer dating from around the same time bears the mark of Vauban himself or someone who had absorbed the old marshal's style:

> The king, having wanted to subject us in estimates to forms that are as of little use for his service as they are inconvenient for the engineers whose limited time should be spared as something very expensive, [and] although one sees very clearly that the regulations concerning the formalities that the king wants observed in estimates do not lead to the result for which they have been made, one must nevertheless carry them out as a thing that comes from our master, whom we must serve in his way and not in ours.
>
> Thus, when estimates have to be made, one of the senior officers of the fortress will be called, according to the regulations, and he will be made to sign on the initialed register given by the intendant to transcribe the estimates. At the end of every month an estimate of all the work that has been done will be made [and] will be sent to the Minister. A copy will be given to the intendant and another to the contractors and this estimate, which will be as close to the truth as possible for the works in progress, will serve to settle approximately the money that can be given to the contractors as the work proceeds.[29]

The subordination of the chief engineer to the Director is evident. It is while the Director is working on "his project" that the chief engineer will gather information on the qualities and prices of local materials. Indeed, this function of becoming familiar with local conditions to enable higher authorities to make decisions is considered important and is emphasized on more than one occasion. It will also come in handy for the engineer in order to make more informed judgments at the local bidding sessions for construction contracts. Before the Director "leaves," the chief engineer, "beyond what he knows by his own capacity regarding the usage of each piece of fortification will beg [the Director] to instruct him more fully." It is "in place" of the Director that the chief engineer will attend bidding sessions with the intendant and he must make no reports directly to the Minister bypassing his Director. This was emphasized even more strongly for the subordinate engineers, the ingénieurs subalternes, in another directive from Le Peletier, the Directeur général. The "principal instruction" that they needed was the recommendation that they have "a perfect deference and complete submission to the orders which will be given to them concerning the service of fortifications by the Engineer Director."[30]

Also noteworthy is the importance given to drawings. This is not a surprising feature. The role of drawing in the development of engineering in the Renaissance and thereafter, as well as its centrality in the education of engineers in Vauban's time, is well known.[31] Drawing was still very much the badge and the ultimate instrument of the engineer at the turn of the eighteenth century. Copies of drawings of work to be done should be given to the contractor, "for no matter how capable or how experienced one is, it is not for a regular engineer ever to work without plans, both because of the bad example one sets to subordinates and because one can make blunders that one only notices when it is too late to correct them."[32]

But there is another reason for using drawings as much as possible: they are also the most concise and eloquent method of communication with ministers and officials. With good drawings one will explain all one's estimates and proposals infinitely better than all the speeches one makes and all the memoranda one writes. Thus drawings—a fundamental element of the by now traditional world of engineering—are also fundamental for the bureaucratic world in which the engineer lives.[33]

The discomfort with some of the administrative trammels of the Old Regime that already shows through above, is no doubt the reason for a greater flexibility of the anonymous author of another memorandum concerning the functions of the chief engineer than is visible in other texts.[34] Although subordination must be maintained and the order of command respected, he is prepared to concede that in some cases, for purposes of efficiency, the chief engineer can make decisions on his own that even involve expense. Thus, changes can be made in the original constructions by adjusting cut and fill of the earthworks if it is found that when built they are exposed to potential enemy fire. (It seems from this that defilading was done on the work site and not calculated from drawings as was supposed to be the case toward the end of the century.) Similarly, it is often better to make decisions about foundations on the spot. There is also a more tolerant attitude to the relations between engineers and contractors. It is conceded that ultimately, rather than scrupulous and continuous control, the only really effective guarantee of probity is the character of the engineers themselves and therefore great care should be taken in choosing them.

The Military Engineer at Work

Yet the efforts at control from Versailles persisted throughout the century, even under the relatively easy-going and paternalistic regime of d'Asfeld. A 1738 circular by his technical assistant, Ramsault, puts forth in great detail office procedures and filing practices of the engineers.[35] Ramsault first gives details of the "manner in which one must report to the Director [general] and the Minister with an indication of the obligatory times and circumstances [for this]." There is a standard procedure for an engineer arriving at a new posting for taking up his functions, beginning with a detailed inventory of the papers of his predecessor. He must keep bound volumes under 12 different headings, such as amount of wood kept in storehouses and firefighting equipment, and send duplicate copies of their contents to the Director of his region on paper of a standard size (paper size 11.5 by 7.5 pouces, slightly longer but narrower than standard French writing paper today) folded once with accompanying drawings of the same size.

There follows a filing system for correspondence and work orders as well as all drawings. Standard scales for the various types of drawings are specified. Thus, the "plan directeur," which is the principal map of the fortress and must never leave the engineer's office has a scale of 4 pouces to 100 toises and plans of individual buildings have a scale of 1 pouce per toise. Atlases for four different categories of maps, plans, and papers are required. For papers not required for transmission and inspection by superiors a system of cartons is specified—five in all—and for current work, four cartons, different in color from the previous ones, are also specified.

What were the kinds of things that filled the current working cartons of a Director? The first carton was to contain letters concerning his service and it is possible to get some idea of what this was thanks to a surviving extract of ministerial correspondence with the engineering district of the Languedoc which spans the period from the administration of Le Peletier in 1698 to the ministry of Saint-Germain in 1776.[36] Languedoc was untypical in some ways because it was a pays d'état (semi-autonomous province) which had little military activity, and in which the military engineers often performed some engineering functions for the local estates, who also contributed to the costs of fortification in the province. The Director was remunerated an extra 1500 livres per annum for his pains. (Their regular annual pay fixed by the Saint-Germain ordinance ranged from 9,000 to 12,000 livres.) But there seems to have been little to do for the average engineer. Thus, in 1763 the retiring engineer supervising the Canal des Deux Mers was not replaced; it was felt that his duties could form part of the duties of an engineer assigned other duties.[37] The question of expenses and the need for economy is a persistent theme and squabbles with the local and municipal authorities take up a fair amount of the correspondence. Occasionally, engineers are asked to sell surplus and unnecessary equipment.[38] Even more frequent is the theme of encroaching civilization on the fortresses of a peaceful, prosperous, and growing province. There are many requests for permission to build in disused fortifications, to widen streets and erect houses near the walls, and to cut doors from dwellings onto the ramparts. The latter request was usually denied, even if it was recommended by great personages. The Minister sharply reminded the Duc d'Uzès that this kind of permission, which the duke had apparently given, was an attribute of the sovereign alone, and could not be transferred to any private person.[39] Inhabitants of the city of Carcassonne were permitted to plow the land near the walls but more usual was the order to root up gardens that were encroaching on the glacis and sometimes even the covered way and outworks. Regulations going back to Louis XIV and often renewed were invoked to forbid gardens and planting near the walls.[40] Enforcing them was complicated by the fact that garrison officers had certain privileges in pasturing animals and having gardens in some parts of the fortifications. This privilege was often abused and considered very dangerous because the officers often did not remove weeds and bushes which began to grow in the fortifications and undermined the masonry. The problem of écorchements (degradations of masonry walls due in large part to the invasive effects of vegetation) was serious and one of the major preoccupations of fortress engineers.

Yet the frequency with which the injunctions against gardens and planting near the fortifications recurs indicates that it was common and that the military engineers were fighting a losing battle against growing urbanization and the long-term effects of peace on French soil, not just in the Languedoc but even in the frontier provinces. The evictions of squatters from disused military buildings is another indication of this.

The reminders to engineers to do their duty, submit their reports, and be regular in their correspondence also indicates that the tidy and orderly world planned in Versailles was resisted by sloth, incompetence, and individualism. On one occasion, it is true, Versailles seemed to be overwhelmed by paper and there is permission to give an abridged report, but much more often one finds the minister remarking that "all the chief engineers do not observe article 42 of the ordinance of 7 February 1744 to give an account every 15 days by annotated memoranda of the execution of works with which they are charged."[41]

While d'Asfeld was Director General, there are a number of reminders that he wanted younger and less experienced engineers to exercise themselves in drafting and design by drawing by submitting projects of fortresses to the Director General.[42]

D'Asfeld, who may have got this idea from Cormontaigne, intended this exercise both as a means of exercising his engineers in their duties and to provide insight into their capacity when decisions for promotion came up. He went so far as to demand regular submission of such projects with the annual reports of Directors. Submitting theoretical projects of fortification, then, was a method that could attract the attention of the Minister and open the road to a higher rank in a notoriously congested corps. Here is evidence that not only bureaucratic pliability but competence in design was valued and encouraged by the senior leadership of the corps. Yet the evidence is conflicting. Some young engineers did attempt to take advantage of this and submitted proposals for new designs for fortification. But there is no evidence that the practice was widespread. There is, on the other hand, evidence that it seems to have been disliked by many of the senior Directors who often ended up judging the proposals. This seems to be confirmed by occasional complaints by the Minister of War: even after the foundation of the school at Mézières, the Minister was concerned that younger engineers were not getting enough practice in the surveying that they had been taught in the school and enjoined the Directors to keep their subordinates busy on such activities whenever they had free time and to submit their plans and maps to Versailles.[43] Two years later the Minister complained: "It has come to my attention that several chief engineers have refused on different occasions to [give] the young engineers under their orders the instruction that they need in the practical side [of their profession]." He warned them that if this recurred, he would take the matter up with the king.[44] On the whole it was the service du détail that in practice occupied most of the time and energy of younger engineers throughout the century, even though there are signs that they were chafing at this more and more after the Seven Years' War.

Although Languedoc may not have been altogether typical, the impression that persists is that the service du détail of the working engineer was fastidious and involved squabbling and litigation as much as new technical work. That the technical work was often left to contractors who sometimes proceeded as they themselves best saw fit, can be inferred from a repetition of the injunction to give contractors the plans and drawings of the projects they had to execute. On a number of occasions it had not been possible to attach responsibility to the contractor for buildings constructed because he had apparently not been given any plans.[45] From now on they were to be given these plans so that they could be "well executed."

Nicole Garcin has done a detailed study of the fortress of Longwy near Luxembourg whose site was determined by Louis XIV himself, planned by Vauban, and built between 1680 and 1684.[46] Garcin has published the *Instruction Generalle sur la*

Fortification de Longvuic by Vauban, which provides an excellent view of Vauban at work and the nature of military engineering and fortress design in the late seventeenth and early eighteenth century.[47] Robust practicality and common sense pervades the document: not only does he specify in detail the tools that need to be assembled but he does not forget to insist that butchers and cooks should be dispatched to the site to be ready to feed the workers and soldiers when they arrive. Although mentioning the possibility of adding an extra story to a barracks he advises against this because "a thousand quarrels will arrive every day" among so many men of different units crammed together. Better make smaller barracks and disperse them.[48] There are detailed descriptions of the various features of the fortifications, from the main gates through the demilunes and bastions, tenailles, powder magazines, barracks, down to the sentry boxes and latrines. All descriptions refer to scale drawings (that are unfortunately lost). Detailed estimates of expenses are given for all aspects of the work: labor, materials, cartage, and the cost of expropriating local residents. He even recommends specific people who could help Choisy, the engineer who commanded the fortress of neighboring Thionville. He asks for two good appareilleurs (master stonecutters) and a fontainier (specialist in water supply) to come from Paris.

In the even more bureaucratized world of his successors, Vauban's reports would have gone into Carton No. 4 of the fortress engineer's office (*Mémoires particuliers concernant la fortification, Manoeuvre des eaux etc.*). This carton was supposed to contain "Memoranda that the engineer could write on the fortress, on the knowledge he had acquired of its environs, the control of water, emplacement of camps, and generally all that which is related to war, to defense and attack, where one will add to these memoranda all the plans and drawings relative [to these subjects]."[49]

Almost a century later, Augustin de Filley, whom we have already met assisting his father at the siege of Mons in 1746, was chief engineer at neighboring Thionville and wrote a series of reports on Longwy that would have gone into Carton 4.[50] It is worth listing the titles to get another view of the military engineer's functions in peacetime:

On the situation and the usefulness of Longwy in relation to the frontier.

Critical Examination of the Fortification of Longwy to make known its advantages and defects.

On the wells and cisterns . . . at Longwy and of the quantity of water that they can furnish.

On the most suitable means that can be found at Longwy to put men and munitions under cover from [mortar] bombs in case of a siege.

The state of the buildings presently at Longwy to lodge troops.

Wood required to fraise [fence] earthworks, form retrenchments, and to palisade the covered way.

Inventory of the ovens and mills available and what they can bake and mill in 24 hours.

Miserable Details

Most of these memoranda are essentially inventories that were part of the familiar routine of the service des places.[51] The work is scrupulously done and seems to rely on the kind of estimates for food and munitions necessary for a fortress that were standard from the time of Vauban on. This is the service de detail, which was still the overwhelming portion of the military engineer's lot in peacetime, even as the average engineer became better educated and more competent in design. In the 1770s, it was a matter of some controversy that deserves the historian's attention. The controversy illustrates the tension between the boringly mundane and the dangerously dramatic that is a normal feature of military life, not just that of military engineers, and the tension between an aristocratic ideal of leadership and one that could be called a bourgeois ideal of administration. Military engineers, increasingly more aristocratic in their social origins as the eighteenth century wore on, yet attuned more than most people to technical problems felt this tension more than most. At the same time there was a tension between the bureaucratic world of their daily lives and the martial vision they held of themselves that is displayed in such a crude form in Révérony's little play.

One view of the world of detail and the military engineer's relation to it appears in a 1776 memorandum to the reforming Minister of War, the Comte de Saint-Germain, by Jean-Claude-Eléonor Le Michaud d'Arçon (1733–1800).[52] D'Arçon portrays military engineering as a laborious, serious, and unglamorous profession, where painstaking attention to detail is required and the natural youthful impulses to flashiness and display must be sacrificed. In his opposition to the amalgamation of the military engineering corps with the artillery corps, he writes that, as a result of such an amalgamation, young engineers would

abandon their workshops, flee from their offices, disdain the art of construction, scorn the indispensable little cares of maintenance of fortresses, lose the temper for work and planning, to give themselves up without reserve to the glittering functions of the artillery, which will always appear to them a very brilliant profession, because in fact it is. Thenceforth, no more engineers. The plans of dispositions and projects that require so much practice and calculation, their execution, the estimates, the appraising, the means of solidity, the laborious research of the best possibles.... All that, as one would expect, would be done superficially and consequently badly, at least by young people, who as a result would have contracted the seductive habit of not examining very closely delicate operations that demand details and that require sustained application. It would then follow from this general tendency of youth for parade and clashing arms, in a new amalgamation, that the artillery would always be served very well but the art that exists in the hands of military engineers would immediately vanish.[53]

Here d'Arçon is bringing in a theme that was becoming increasingly common in the French army in the second half of the eighteenth century, and would have struck a sympathetic chord with Saint-Germain. A growing sense of professionalism among army officers resulted in an emphasis on the importance of the mundane and even tedious details of army duties and training over the more spectacular and dashing aspects of a military career that had been traditionally associated with the army, especially in the eyes of a nobility that loved luxury and display. The good officer was perceived more and more to be not just a charismatic leader, but one who conscientiously stayed with his troops in winter quarters and drilled them, instead of flitting off to court or to the nearest exciting city for most of the year. Military reformers, especially prone to soul searching after the disasters of the Seven Years' War, were concerned that many officers in the army were frivolous, inclined to idleness, and loved luxury too much. These Stoic themes, common in France before the Revolution were almost inevitably accompanied by a new conception of the role of the officer and even the nature of heroism. The army began to believe, as David Bien puts it, that

Officers at all levels had to know what they were doing, and they had to be willing to spend time teaching and training the soldiers in the stylized and accurate formations [that were characteristic of eighteenth-century battle tactics]. Honor was now not so much the great act and episodic heroism, as simple work and the acceptance of subordination.... Living with dull routine was the new heroism.[54]

Implying that engineers possessed the qualities of this new heroism to a superior degree was certainly a common self-image of engineers, but the suggestion that artillerists, because of their "flashy" and "noisy" profession, possessed it to a lesser degree was, of course, an unwarranted jab. D'Arçon hoped that engineers would no longer just plan and inspect but, figuratively at least, would also build their own fortifications. The dignity of planning and conceiving would be united with the activity of execution of the work.

In a discussion of whether civilian constructeurs should build fortifications and military engineers merely inspect them, d'Arçon had answered this question with an unequivocal "No." Rather interesting, in view of d'Arçon's sensitivity to points of honor and rank, is the care he takes in reassuring his colleagues that undertaking these apparently mechanical tasks of civilian architects and engineers will not ultimately detract from their honorable status. Turning over the actual details of design and supervision to constructeurs or ingénieurs des ponts et chaussées may be the path of least resistance; admittedly the cares of construction and maintenance "have something repugnant [about them] for a corps of officers," but "these cares are made indispensable by the usefulness of the matter [and attending to them] . . . would be a necessary sacrifice."[55]

There were a number of serious drawbacks to such an abdication of apparently demeaning duty. Military engineers would still have to supervise the construction of siegeworks in wartime—the thought that here, too, civilians could work under enemy fire is dismissed as unthinkable—and consequently would need the skills and experience of construction work. They would lose these through inactivity if they surrendered them in peacetime to civilians. Moreover, since in engineering details condition the general plan, unfamiliarity with details would gradually make the military engineers dependent on civilians, another situation that needed merely to be stated to indicate its unacceptability:

> . . . since there exists a necessary link between the details of construction and the analysis of their projects with the general arrangements, it is evident that the military engineer who would make such arrangements without considering the means of execution would necessarily find himself dependent on a builder, who would not be a soldier.[56]

Finally, it is more economical to use officers (by definition honorable men and, it is implied, less subject to the temptations of fraud that always have and always will accompany any construction project) because they are already on the payroll and duplication is expensive.

A distinctly contrary view is presented by Louis Le Bègue du Portail (1743–1802), a young military engineer who was later to become the chief engineer of the American army in its War of Independence.[57] For du Portail, construction of fortifications proper was a distinctly subsidiary function of the art of fortification and military engineers should not waste their time paying too much attention to it. It should be the affair of architects and draftsmen—a category of people distinctly different from the military engineer. The construction of fortifications is "gross construction"—the engineer knows enough about it to supervise the work and leave the details to others.[58] Attacking the existing situation where military engineers are required to design various structures on the grounds that this will make them learn their trade, he asks that "someone explain to me a little what relation there is between the science of fortification and these miserable details."[59] Details, that is, drawing up the detailed design projects for construction, are to be avoided.

The attitude of du Portail on the matter of "details" and their importance for the military engineer, although it undoubtedly had some resonance among engineers of his generation who were living through the intense debates on French military doctrines in the 1770s, was nevertheless a minority opinion. There is little hint in the surviving correspondence of senior officers that construction is considered an unimportant part of their duties. Usually their attitudes were closer to those of d'Arçon—calling for more attention to everyday duties and suspicious even of excessive novelty in design. When Fourcroy de Ramecourt, the unofficial head of the corps, later talked of experience, it is clear from the context that much of this is gained in the daily humdrum duties that du Portail despised.

The "spirit of detail" discussed above raises questions that make it crucial for understanding the changing mentalities and attitudes of French eighteenth-century military engineers. There are several ways in which the word 'detail' can be interpreted. Originally it appears to have meant, as in the phrase "service de détail," the details of maintenance and logistics of a fortress. This was closely related to the bureaucratic duties of management and administration that have been discussed in this chapter. But there appears to be something of a semantic shift that led to ambiguity and confusion in the 1770s. For some, like d'Arçon, details meant a greater attention to the technical aspects of construction that had hitherto been abandoned to outside contractors and craftsmen. For others, including du Portail, these technical aspects were a trap and a distraction from the true profession of the soldier. Here was a parting of the ways between two views of military engineering. Both views had emerged, so to speak, from file folders and the desks that some military engineers claimed they were so fond of upsetting. These two directions—the flight from details perceived as petty into the domain of military tactics and strategy, which will be discussed below, and the attempt to understand construction at a more fundamental level and to undertake detailed design—appear and probably only could appear on an established and dense bureaucratic background of reports, memoranda, projects, and plans that were routinely required by the administration of the corps of military engineering.

Returning to Carton 4 discussed earlier, one sees that among the details of cisterns, wood, and ovens, the first two memoranda Augustin de Filley drafted were more than just inventory and accounting. Following the models laid down by Vauban, they required an analysis of geographical, historical, and military factors to evaluate the strength of the fortress as well as its usefulness in the general system of national defense.[60] It appears that military insight—always considered the preserve of the general of genius—was to be reduced to written reports, domesticated, and kept in carton no. 4.

The first memorandum—on "the situation and usefulness" of the fortress—was essentially an analysis of the strategic situation, and it follows the models laid down by Favart and Fourcroy de Ramecourt. Longwy was part of a chain along with Montmedy and Sedan which blocked the French frontier from attacks from the fortress of Luxembourg and at the same time it could serve as a jumping off point for French attacks against Luxembourg, southern Holland, and the Ardennes region. The history of recent military actions was recalled, the most recent being that of a French army that supported Maurice de Saxe in his successful attack on Maastricht in 1748. Augustin de Filley indicated the difficulties besiegers would have in building lines of circumvallation because of the terrain around Longwy and which would make it vulnerable to harassment by a relief army.

The second memoir, on the detailed advantages and disadvantages of the existing fortress, begins with a workmanlike description of its features with references to a detailed plan. Longwy was almost a regular hexagon perched on a height with a hornwork to protect its most vulnerable side. The usual maps made with the cartographic techniques and conventions of the eighteenth century were misleading because there was a strong slope on the plateau on which the fortress was situated which led to an extreme difference of elevation among the bastions of 54 pieds. Augustin de Filley pointed out that Longwy had been built according to Vauban's earliest ideas, with concave instead of straight flanks, and it is of interest to note that he does allow himself some respectful criticism of Vauban who, it was claimed, had allowed himself to be carried away on this point but had come to realize the error of his ways later in life. The fortress was also a little too cramped (162 toises per front of fortification instead of the usual 180) and without the counterguards that had been built later on in most of his fortresses either by Vauban or his successors. He believed that Vauban wanted to make additions to the fortress and was aware of the incompleteness of its works. But little had been done on Longwy since Vauban: Filley enumerated the newer works that had been built and, aside from one of them, considered them inadequate or useless and one almost hears the sigh of envy when he writes "It would be greatly desirable that finances were as lavish and construction as cheap as they were in the times of Mr. de Vauban." He felt that the ability of the fortress to repel attack was poor because such elementary precautions as counterguards in front of the bastions and a redoubt in the gorge of the hornwork had not been built.

Filley had also inspected the ground outside the fortress and found that it was very hard and extremely difficult to dig trenches in it, which was a disadvantage for potential besiegers. Longwy was a strategically vital fortress and had many advantages, but defects in its construction had not been eliminated even though they could have been remedied with little additional work. This left the fortress extremely vulnerable in a condition which was worse than useless, for it could be seized easily by an enemy who could quickly make it much more difficult to recapture by taking the relatively simple measures that the French had neglected.

More often than not reports like Filley's were an elegant but sterile exercise, occasionally attracting a superior's attention because of literary or graphical elegance. Some of the counterguards that Filley so urgently requested for Longwy were only built in 1829. As the century wore on, with increasing numbers of Mézières graduates entering the corps, it is more than likely that peacetime service for military engineers was not only onerous but was perceived to be such to an ever increasing degree. Their design work involved a fair degree of well-established expertise derived from standard rules of military architecture—calculating fields of fire, lines of defense, and fronts of fortification—that nevertheless was not mathematically sophisticated unless it was the techniques of defilading that were developed after mid-century. It produced recommendations for construction but the construction itself, which was usually relatively simple in design, was often simply not done for financial reasons.

Yet dull office routine had other effects on the evolution of military engineering beyond merely inflicting boredom. Reports on fortresses required military engineers to provide information that today would be considered part of military staff work. This stimulated the reflection on military doctrine that du Portail urged as the proper vocation for the military engineer. The problems of design and construction, along with the attendant need to keep a tight rein on costs, promoted not just more bureaucracy but more involvement in construction and more attention to the details of the science of construction—something the rigorously educated engineers in the second half of the century were increasingly qualified to do. In a context of increasing frustration, the sometimes conflicting tendencies pushing toward a more military, a more scientific, or a more managerial role for the military engineer were growing stronger at a time when the country and the corps were entering a period of crisis.

III
Crisis

With all due caution toward indulging in fashionable exaggeration, the situation in which military engineers found themselves in the last quarter of the eighteenth century can be fairly described as a crisis. This was also true for France as a whole. Gillispie (1980, pp. 3–4) has put it as follows: "The juncture that brought Turgot to office [in 1774] was more crucial than a change of ministry or even reign. Either it was the beginning of the revolutionary movement or the last chance to avert it." As soldiers and as trusted servants of the Old Regime, military engineers were affected by the general crisis that was tearing the country apart. As technicians, they were subjected to stresses peculiar to them that simultaneously threatened the practice of their art and their very institutional existence.

There are three aspects of this crisis of military engineering at the end of the Enlightenment. First, there are what may be called the structural aspects of the crisis—that is, those tensions and ambiguities in engineering as a profession that have existed since the beginning of engineering, and perhaps technology itself. They exist today and will no doubt continue to exist forever. Then there are the contextual aspects, those that were present in the milieu in which French military engineers worked at a time when revolutions were occurring in America and brewing in France. Finally, there are the contingent aspects, among which the main one was the irascible personality and ideas of the great bête noire of the military engineering corps at the end of the eighteenth-century: the marquis Marc-René de Montalembert, well-connected cavalry general, officer of the royal bodyguard, diplomat and confidant of ministers, and enormously prolific writer on fortification.

The interest in this story resides in the significant ways in which military engineering in France was involved in or reflected primary changes not only in French but Western European society in general. A society of estates was giving way to a society of classes, the role of the traditional noble elite was being challenged and it was responding to this challenge in diverse and interesting ways. The army, because of its central role in the state and its intimate bond with the nobility was a major theatre of struggles between various sections of the nobility, between nobles and commoners, and among those who held different views on what the role of an officer should be. Military engineering provided an influential template for the type of officer who eventually prevailed in this contest. A nascent professionalism among engineers both military and civil, appeared earlier than we have hitherto believed and influenced not only the nature of engineering but also the professionalization of military officers, which military engineers themselves were. A distinct profession of engineering was emerging in which military engineering was the central sub-profession that had catalyzed a broader movement in other fields of engineering, as well as in the army. It was influential because of its priority as a technical bureaucracy with a strong sense of corporate cohesion, administrative and technical efficiency, and its obvious link with the military.

Warfare was undergoing major changes and its most articulate practitioners were paying less attention to siege warfare and fortresses, which was the privileged locus of the military engineer's activity. Indeed, some were even arguing that fortresses should be ignored and for others this had the implication that military engineers were useless. Added to the residual snobbery against technical soldiers that had always been present in the army, this posed a threat to the military engineers' self-image, prestige, and even their institutional existence. Military engineering and fortifications remained important in the more mobile warfare of the future but the siege warfare that characterized Louis XIV's wars disappeared and led to a reconfiguration of the role and responsibilities of engineers in the army.

No less interesting for the general issues it raises is the transformation that science was undergoing at this time of general crisis. It was the period of the "chemical revolution" associated with Lavoisier, when work in pneumatic chemistry was transforming chemistry and integrating it more closely with the quantitative model of physical science that had been associated with the rational mechanics of the classical Scientific Revolution of the seventeenth century. Related to this was the rise of a mathematical physics. The traditional experimental physics often resembled a natural history of physical phenomena, where description and classification played a greater role than analysis. In the nineteenth century this would change with the increasing mathematization of diverse fields such as electricity and magnetism. It is significant that some of the pioneers of this movement were military engineers. Some engineers were also laying the groundwork for what was to be called engineering science, which is also often attributed to the nineteenth century. Engineering science—the scientific approach to both the natural and artificial world in order to facilitate design by technologists—involved increased quantification and the use of mathematical theory and less reliance on rule-of-thumb methods. Military engineers like Charles-Augustin Coulomb, although prophets in the wilderness even for most of their brother engineers, were influential in this movement. It is both interesting and significant that Coulomb was also considered a pioneer of a true mathematical physics with his work on electricity and magnetism. It was presented in 1789, the same year Lavoisier published his revolutionary *Elementary Treatise of Chemistry* and the French Revolution began. Hindsight suggests the crisis through which France and the military engineers were passing naturally culminated in the Revolution, and this has often led historians to see this crisis through the glass and with the categories of the Revolution. A richer and more accurate view can be obtained by viewing the crisis on its own terms.

I have chosen to focus the analysis of this crisis of military engineering by describing the polemic between the military engineers and the maverick Montalembert, documenting in as detailed a fashion as possible the warp and woof of this polemic to provide relief to the general analysis of the crisis of military engineering. Moreover, the case of Montalembert is itself part of the crisis as well as its illustration. A brilliant aristocratic critic of the corps of military engineering, he achieved some things that were unique and never done before or after, such as meddling with the monopoly of the corps over the design of fortifications. He was a beneficiary of a kind of patronage—royal and ministerial favoritism—that looks quaint today. Although many of his ideas were naïve and confused, some of them were forward looking and original, so much so that they were taken up by European military engineers in the nineteenth century, with the notable exception of the French. (Some doubt is legitimate, however, whether even in France the military engineers were as impermeable to his ideas as some of them proudly proclaimed.)

Unlike Gribeauval in the artillery and Guibert in the line army, who succeeded to a great extent in their reforming efforts and left their mark on Revolutionary and Napoleonic warfare, Montalembert failed. His story needs to be examined carefully and in detail to understand why this was so in spite of his many initial advantages. The episode of Montalembert not only completes the story of the military reforms in France on the eve of the Revolution, but it also illuminates the better-known and more successful attempts at reforms in artillery and tactics. His case is also different because it continued into the Revolution when the political context had changed radically. It will show that the categories of "revolutionary" and "conservative" when applied to engineers are slippery and need to be reexamined. The following seven chapters in this, the central part of this book, will deal with each of these dimensions of the crisis of military engineering in turn.

Chapter 7 looks at engineers as officers in a royal army where the traditional justifications of the nobility for the exclusive right to be officers were being questioned and revised. The example of the engineers was perhaps only one reason for this but it was an important one.

Chapter 8 describes the military thinking toward the end of the Old Regime regarding tactics and strategy as well as the possibility of a military science that would diminish the element of chance in war. Here, too, military engineers served as both inspirers of some of the ideas while at the same time their necessity was being questioned.

Chapter 9 examines the beginnings of engineering science.

The next three chapters deal with Montalembert. His early career and his attempts to become a metallurgical entrepreneur are the subjects of chapter 10. His attempt to build a fort at the island of Aix according to his new principles is described in chapter 11. The fierce polemic over his methods with the head of the military engineering corps, Fourcroy de Ramecourt, is presented in chapter 12. The controversy between Montalembert and the military engineers lasted for more than a quarter of a century, and in the end they had successfully fought off his challenge. By that time they had been transformed as a military and technical community and had come to resemble today's military engineers much more closely than they had previously.

This part will begin, however, with chapter 6, which will discuss the ways in which engineering has stayed the same. Prominent among them are the permanent tensions inherent in the nature of engineering. The chapter will discuss the problems engineers, particularly military engineers, have always encountered and which have shaped their actions, their attitudes and their professional lives. The amphibious nature of engineering whereby it partakes of the concrete and the abstract, theory and practice, and science and art was complicated by the difficulties military engineers encountered as technicians who were officers and who worked in the very different situations of war and peace. Throughout, engineers always sought stability and order—things that were as elusive and fragile then as they are now.

6

Engineering and Its Discontents

In the past half century, the ethos of the engineering profession has dissipated. Engineering has expanded vastly to include first science and then society. Even where engineering has stayed focused on design, the meaning of design has evolved from things to code, from matter to reified consciousness, both the expansion of the profession and the redefinition of design have led to diffusion of the professional core. The center has not held.
—Rosalind Williams[1]

Engineering is at the interfaces of the concrete and abstract, of theory and practice, of science and art, of discovering and settling, and of innovating and controlling. Military engineering lies at the boundaries of war and peace, of matters military and technical, and of direction and execution. Engineers are not mathematicians, although a mastery of mathematics has been one of their attributes in modern times and they have contributed to new fields of mathematical endeavor (dimensional analysis, numerical and graphical methods, descriptive geometry). Nor are engineers the proverbial "horny-handed empirics" with superb manual skills and raw technical intuition, although the better ones share these qualities. To the sciences that the ideal architect/engineer was to master, Vitruvius added an impressive number of the manual building crafts, without which the true engineer was incomplete. The article "Ingénieur" in the Supplement to the *Encyclopédie*, written by the military engineer Flachon de la Jomarrière, calls for a truly encyclopedic range of competence in the crafts of construction for the military engineer.[2] Individual engineers are sometimes tugged more in one direction than in another, and that is why engineers like Coulomb have gained respectable places in the history of mathematical physics as well as in what today is called engineering science. That is why James Brindley, a superlative semi-literate millwright, has entered the Victorian pantheon of engineering created by Samuel Smiles.

Hélène Vérin has devoted some brilliant pages to discussing the "new space" in which the engineer of the Enlightenment worked.[3] There is a fundamental and irreducible tension between the plurality and diversity of the material world and the generalizing and abstracting tendency of the intellect in attempting to understand and manipulate that world. The engineer is in the cross-fire between the demands of attention to the particular (client, artifact, process) and the general (laws of nature). The amphibious and protean nature of engineering is the cause of stresses and centrifugal forces in engineering that are both fruitful and vexing. This uncomfortable state is inevitable and inescapable for the engineer. The future offers no hope for resolving these tensions because of the perpetually increasing richness and diversity of the artificial world that engineers themselves create. Comprehending and manipulating this artificial nature requires perpetual enlargement of the engineer's domain and skills. Vauban was surely right when he wrote this: "Engineering [le Génie] is a craft that exceeds our strength. It contains too many things for it to be mastered by a man to a sovereign degree of perfection." One result was the increasing specialization of engineering (something that was noticed in the eighteenth century and was rejected by some, as the example of the polytechnical ideal initially espoused by the founders of the École Polytechnique shows). Another was the porosity of the professional boundaries of engineering to other professions and crafts. Yet another was the dynamism of these processes. In recent decades, new fields of engineering have often been accompanied by controversy, sometimes to the extent that their right to call themselves engineering fields has been contested—the most notable recent example being software engineering.[4] In the France of the 1770s new engineering specialties, such as that of naval architects (ingénieurs-constructeurs de la marine), were also emerging, and these perennial features of engineering were accentuated by the environment in which engineering, and particularly military engineering, was practiced.

Engineers are, of course, technologists. As technologists, they deal with matter and its transformations for human purposes by the application of energy through tools and technique.[5] However, engineers are only a subset of the community of technologists. The farmer, the artisan, the sailor, and the miner are technologists, but they are not necessarily engineers, even in those cases where their activities fall within a framework constructed by engineers. The distinction between engineering and technology in general could be said to reside in the complexity and scale of engineering projects—a complexity and scale that has led to the interaction of engineers not only with individuals but with societies and governments as well. In this sense engineering deans keen on attributing a lengthy lineage to engineering are right. It is at least as old as civilization, and Stonehenge and the Pyramids can be legitimately called works of engineering.

Engineering as we know it, however, is more recent. Although its rudiments can be traced back to the time of Galileo, its foundations only began to coalesce toward the end of the eighteenth century. By then a number of additional features that made engineering modern had appeared. One was the use of the findings of the natural science that arose from the classical scientific revolution. Another was the creation and use of an engineering science that dealt with matter and energy from the perspective of their usefulness in design.[6] This new engineering science was informed and inspired to some extent, but not entirely, by modern science. Before the acceleration of its development in the nineteenth century, Galileo, Bélidor, and Coulomb were laying the groundwork for Poncelet, Sadi Carnot, and Rankine. Often incorrectly consigned to the jejune and derivative category of "applied science," engineering science sprang from what Ken Alder calls a "middle epistemology" and was created in the tension between theory and practice.[7]

Yet another distinctive feature of engineering is social: the growth of a profession of engineering. By the middle of the eighteenth century, there were already well-established bureaucratic organizations in France for what today would be called military engineers, civil engineers, naval architects, and even mining engineers. True professions were gradually emerging from these organizations—professions that were characterized by esoteric expertise, a distinct sense of corporateness, and responsibility to impersonal entities like society or the nation. A related feature that is both social and intellectual was the increasing specialization of engineering. Relatively underdeveloped in the eighteenth century itself, and driven more by administrative and old-fashioned corporatist considerations typical of the social landscape of the Old Regime, they would flourish in the following century. A final feature is the rise of what could be a called a self-consciously dynamic aspect of engineering that is a characteristic of professionalism as defined by Samuel Huntington. Huntington argues that "the ordinary skill or craft exists only in the present and is mastered by learning an existing technique without reference to what has gone before" whereas professional knowledge is intellectual in nature, capable of preservation through writing, and has a history.[8] If this kind of knowledge has a history, it also has a future and is continually being expanded by the members of the profession whose leading members at least consider this a primordial aspect of professionalism. In a sense, craft knowledge is conservative, while professional knowledge is not. This rejoins the view of Vérin on the new kind of engineering knowledge being created during the Enlightenment. Departing from an etymological analysis of the word 'engineer', Vérin emphasizes the importance of the engineer's capacity for inventing new material things. Engineers, with the help of mathematics, use the innate power of the human mind [ingenium] to obtain new material effects to satisfy human needs.[9] Unlike craftsmen, engineers are in the business of novelty within the domain of the material world. This is exemplified in Bélidor's hope that, with the aid of mathematics, engineers could design new things and escape the rut of routine that sage tradition tested by failure forced them to follow. Whether this means that it is revolutionary will be addressed in a later chapter, where I will argue that there is always a dialectic between novelty and the constraints of the material world in its concrete and financial aspects.[10]

All the aspects of modern engineering that have been described lead to its being an inherently fluid and fragmented profession, no matter what the period. The fact that it is always engaged with a material world that is already diverse and multiform in its natural state but is continually transformed by the technological activity of engineers, who create an artificial world alongside it, inevitably leads to changing functions and modes of operation. The existence of an engineering science, expanding and changing as it encounters new problems, lends dynamism and instability to the mental world of the modern engineer. The scale and complexity of engineering projects that is another feature of engineering inevitably brings the engineer into contact with social concerns and interests. The fact that engineering is a modern profession injects another element of instability that will be examined below. In a recent article, Rosalind Williams, from the privileged vantage point of an administrator and educator at the Massachusetts Institute of Technology, has argued that engineering is currently undergoing a radical change to the point of "dissolution" of the traditional core of engineering.[11] For her, engineering has "expanded vastly to include first science and then society" and the result has been a transformation to the extent that we can now speak of a "little e" engineering and "big E" engineering—the former traditional, the latter radically new. Williams urges us to look again at new words like 'reengineering'—a word many engineers have a tendency to dismiss as nebulous concoctions of spin doctors—and try to understand why such words have emerged. She thinks that the increasing closeness of management and business with engineering and the increasingly dominant position of computer engineering and technologies where machines manipulate symbols and language instead of inarticulate matter have led to unprecedented transformations of engineering. This has made life difficult for traditional engineers and for historians of technology. The former deny that William's new engineering is real engineering; the latter have to construct entirely new analytical frameworks.

Thomas Hughes' book *Rescuing Prometheus* strikes a similar chord.[12] Building on his previous work on technological systems—exemplified best by the systems of power distribution that appeared in the Western world at the turn of the twentieth century—Hughes sees a new kind of system and new kinds of engineering of systems emerging after World War II. Once again, MIT, the cynosure of American and even world engineering, assumes a prominent place. In projects such as the SAGE air defense system and the Atlas missile, new "postmodern" systems were put into place, where the old disciplinary boundaries broke down and the relationships of engineers with government, business, and society took on a new cast. Jay W. Forrester, one of the driving spirits at MIT behind the SAGE system, went on to develop the idea of system dynamics, which he hoped could be used to engineer entire societies and cure poverty, slums, substandard education, and other critical problems of the United States in the 1960s. His disciples literally took on the world in developing computer models commissioned by the Club of Rome to examine the "predicament of mankind."[13] It does seem that the ambitions and scope of activity of engineers had increased and moved into areas that formerly were those of the businessman, the politician, the general, and even the prophet. Frank P. Davidson, chairman of the Steering Committee of the System Dynamics Group at MIT, spoke of a new kind of engineering that was increasingly touted as the indispensable engineering of the future. The name given to it was macro-engineering.

Broadly speaking [macro-engineering] implies the study, preparation, and management of the largest technological undertakings of which human society is capable at any given time. Characteristically, macro-engineering involves "hardware"—the construction of pipelines, railroads, airports, dams, bridges, refineries, rocket bases, new cities. But it would be unwise to define "large-sale technology" too narrowly in terms of physical plant and equipment. Communication systems, for instance, while physically and financially impressive, connote highly developed elements of "software"—the research, plans, methods, regulations, and procedures that govern the design, building, and use of the installed network.[14]

However, even Davidson mitigated his tone of boosterism for macro-engineering with the caution that it would not "be right to regard our own era as unique in its dependence on large-scale technology." Indeed, many examples from antiquity show that large-scale technology was a defining trait of engineering even before it could be considered modern. Similarly, it is fair to ask whether the dissolution of engineering that Williams detects in the postwar period has not been present at least as a possibility and a threat—to a greater or lesser extent depending on the era—from engineering's inception, and particularly in the modern era. With the increasing size and complexity of its fields of activity, engineering has tended to expand, with more or less success, into new domains and has elicited resistance, attempts at assimilation by other professions, or fission into new specialties. There has always been a certain fragility in what Williams calls the ethos of engineering. Depending on the context, this has always led to tensions within engineering about its role and its purpose. Even at the height of engineering's self-assurance, during the heroic Victorian period, there were muted struggles between the conception of the engineer as a gentleman entrepreneur and that of the engineer as a technical working man.[15] They were caused in part by the rapid expansion in numbers of engineers and the increase in engineering work. In eighteenth-century France, the domain of engineering was still relatively restricted and usually fell within the domain of the activities of the state. Although the examples of Riquet's canal uniting the Atlantic and the Mediterranean and the Duke of Bridgewater's canal heralded an increasing involvement of engineering with private capital, engineering's main purpose was still the enhancement of state power—fortification and siegecraft, shipbuilding, metallurgical enterprises, and the production of weaponry. Today engineering impinges on all aspects of our technological society, and it has displaced many other technological activities. However, the increasing diversity of its activities has merely accentuated the recurrent tensions inherent in it; it has not created them.

The tensions within French engineering were all the more acute because they were the infantile diseases of a profession in its infancy. As a profession, engineering fits in well with the model of professions proposed by the sociologist Andrew Abbott, who has presented a systems view of professions and who argues convincingly that there is no single Platonic model of professions. They do not have a unique historical trajectory of development through various stages of immaturity and consolidation to reach ultimate stability that in its general features is identical for all professions. For Abbott, "the central phenomenon of professional life is . . . the link between a profession and its work."[16] A profession cannot establish this link, which Abbott calls its "jurisdiction," unless it finds it vacant or fights for it. This has the implication that one cannot study any profession in isolation; it is part of a system of professions, whose elements continually interact. Any displacement or vacating of the jurisdictions of a profession leads to an expanding wave of consequences, which affects the system as a whole and not just a particular profession. This is particularly true in a profession's nascent state, a condition that best characterizes engineering in the late eighteenth century.

It might be argued that engineering and military engineering were not yet professions but only bureaucracies in the eighteenth century. It may be premature to speak of a military engineering profession—as opposed to military engineers—not only during the Renaissance but even during the time of Vauban. With the indispensable help of Louvois and Louis XIV, Vauban certainly created an effective organization and bureaucracy of military engineering. This bureaucracy was an essential and even preponderant element of the profession as it began to emerge in France. Without a bureaucracy, engineering might not have become a profession. Yet there was more to the profession than bureaucracy. If bureaucracy provided the skeleton on which the profession grew, it did not entirely control its growth, particularly at a time when engineering and its bureaucracy were subject to many conflicting forces.

There are a number of other implications of Abbott's model of professions. First, professions depend on the context in which they exist—both the context of other professions and the social context generally. This view undermines fixation on the Anglo-American archetype of the free profession consisting of self-regulating, independent members. Professions in different countries will be different, and Abbott himself points out to the radical differences between professions in English-speaking countries and France, where the role of the state was always greater in their development. Next, professions are never stable. The tidy, static view of a profession dissolves into one of a dynamic struggle along shifting epistemological, social, and administrative boundaries. This has special pertinence for engineering since a major component of the "work" of engineering is technology and technological change has been accelerating throughout the modern era.

Engineering professionalism has always been a knotty subject for historians and sociologists. In 1969, Perrucci and Gerstl argued that the engineering profession in America was quite different from the traditional professions such as law and medicine for a number of other reasons.[17] For one thing, it dealt with social rather than individual problems. The boundaries of the engineering profession in the United States were much more porous than those of the medical and legal professions, and the professional community was fragmented in a number of interesting ways not discernible in other professions. Engineers were more fragmented among technical sub-specialties—when Perrucci and Gerstl wrote their books, 24 distinct engineering specialties were taught in American engineering schools. There were also fissures among engineers with basic degrees and those with graduate education, between engineering educators and engineers working in industry, and between independent consulting engineers and engineers working in large organizations. According to Abbott's model, such diversity and fluidity is normal in all professions. Instability of the engineering profession is clearly visible in France in the eighteenth century even within the confines of a vigorous bureaucracy. What needs to be explained is not professional fluidity and diversity but the relatively greater instability of French engineering, particularly French military engineering, during the Enlightenment.

There were clearly cleavages within the French engineering community, such as those between the Corps des Ponts et Chaussées, the ingénieurs-géographes, and the military engineers, not to mention the smaller groups of engineers whose existence made the professional landscape so untidy and deplorable for the founders of the École Polytechnique.[18] Within the important subgroup of military engineers, there were further divisions that have been described earlier—divisions between the ingénieurs des tranchées and the ingénieurs des places, between those who held military commissions and those who did not, and these had given way to less obvious ones between those who saw the engineer as primarily as a soldier and those who saw him as a technician and builder. These cleavages differ from those we see today, but their presence was and still is a feature of professions and even more so of the profession of engineering. Thus, during the eighteenth century, military engineering was subjected both to forces of cohesion and consolidation as well as forces of dispersion and even annihilation—forces always present in professions generally and in the engineering profession especially.

The bureaucratic aspect of engineering has attracted attention in the United States, where engineers did not easily fit into the model of the free professional that was normal for traditional professions like law and medicine. Although some heroic figures (e.g., John Smeaton and John Jervis) may have fit this model in the early years of the Industrial Revolution, fewer and fewer engineers did as the nineteenth century drew to a close.[19] Both Edwin Layton and David Noble have examined this aspect in some detail. Layton has described the tension between the Anglo-American notion of the free professional and the demands of bureaucracy for American engineers in the early part of the twentieth century.[20] He sees this tension as having been painful for at least a certain group of engineers and that an incipient "revolt" of engineers against the demands of business and its bureaucracies ultimately failed. David Noble went further.[21] He saw a dialectical relationship between the modern business corporation and engineers. The corporation, more often than not based on modern technologies, was created in large measure by engineers, and was often managed by engineers. On the other hand, the engineers were themselves a recent creation for the benefit of corporate capitalism, which had a primary role in molding the profession for its own benefit. For both Layton and Noble, bureaucracy—either welcome or not—is an integral part of the modern engineer's existence.

Bureaucracy, however, although a cardinal element of modern capitalist enterprise, was around before capitalism. Indeed, managing large numbers of men and large amounts of money was as fundamental a feature of engineering as involvement in scientific and technical activity. Even before the building of the pyramids, bureaucracies of some sort were indispensable for this activity. The engineer is as much a creator as a creature of modern bureaucracy. The pressures and tensions exerted on French engineers in their bureaucratic context were different from those exerted in the context of modern corporate capitalism. They formed French engineers and left their mark to this day. Until recently, the French engineer was more étatiste than capitalist in his thinking, more a product of the state than of business.[22]

In the last quarter of the eighteenth century, however, engineering was becoming more than a bureaucracy and it was showing clear signs of developing into a profession. The profession was surrounded by conflict—a normal feature according to Elias and Abbott.[23] It was not yet fully formed and retained much of the old corporatist structure of the Old Regime, but it was more developed than scholars like Huntington and Vérin are willing to concede. Vérin has argued that the ingénieur du roi, the military engineer of the Old Regime, was just that—a personal attendant of the monarch whose raison d'être was service to the royal person in a technical capacity.[24] The ingénieurs des ponts et chaussées, according to her, had a more abstract allegiance to a broader collectivity, being members of the administration of the finance ministry. What both had in common, however, was a strong ethos of service to the state, whether originally in the form of personal fealty or administrative subordination. Perhaps Vérin states her case too strongly. The military engineers as well as the rest of the army were also evolving toward a more depersonalized and bureaucratized relationship to the state during the Enlightenment, even as the ideals of a military caste that emphasized service and sacrifice remained a powerful undercurrent in the ideology of French engineering professionalism.

The eighteenth century, the century of bureaucrats and of the prolonged birth of engineering science, saw the appearance of many features of engineering that are still visible today. The independent engineering artist of the Renaissance was not replaced by the modern engineer during the so-called "Second Industrial Revolution" of the late nineteenth century (the revolution of electricity and industrial chemistry that Noble sees as the crucible of modern engineering). This was already happening a century earlier in France, and it can be seen in fortification as well as in shipbuilding and in state construction projects.[25]

Not just bureaucracy but instability, tensions within engineering, and friction with other professions, have been around as long as modern engineering itself. They provide the background on which the particular histories of engineering in particular contexts are written. The interaction of context with the deep structural features of engineering gives the particular physiognomy of the profession at a particular time. Certain contexts are of particular interest. The case of France merits attention for a number of reasons. France was the locus of many changes that made engineering truly modern and truly a profession in the eighteenth century. It illustrates the central importance of military engineering not only in the building of the profession of engineering in general but in its impact on other professions, such as those of the army officer and the civil servant. Finally, examining this context illuminates certain aspects of engineering that are sometimes only visible during cardinal points in engineering's existence but are always present, such as the tendencies to "dissolution" that seem so exceptional and unprecedented in contemporary engineering. The following chapters will attempt to describe the interaction of the particular context of eighteenth-century French military engineering and the fragile but enduring core of engineering.

7

Soldiers, Nobles, and Engineers

Have confidence in me rather than in all the engineers of the world, and you, as well as the king, will be all the better for it. There are people born to command and these sorts of gentlemen are created only to execute the orders they are given.
—the Duc de la Feuillade, in a letter dated 1705[1]

The changes that had been transforming European society since the late Middle Ages affected all its institutions and its social and occupational groups. The nobility and the armed forces were particularly affected and so were military engineers, both as soldiers and as nobles. There were complex and dynamic interactions between these changes. One aspect of the crisis in late-eighteenth-century France was that they appear to have accelerated significantly during the eighteenth century.

Birth, Merit, and Talent

The French nobility, like the nobility in the rest of Western Europe, had been originally defined by its function: fighting. In theory, the purpose of the warrior nobility was to provide protection and security for the other orders of society against external and internal threats. Along with this came a certain code of honor, probably as much observed in the breach than in the practice, that attributed a certain moral superiority to nobles. With the end of the Middle Ages, an increasing number of nobles were no longer involved in military pursuits nor could they be considered virtuous by any stretch of the imagination. Ellery Schalk has argued that toward the end of the sixteenth century in France two things undermined the claims to moral superiority and guardianship of the nobility.[2] One was the wars of religion, with their attendant horrors, which had undermined the view of the nobility as virtuous. Some nobles had abandoned Roman Catholicism, and many of them had participated in atrocities.

Moreover, an ascendant class of commoners was demonstrating many of the virtues that were supposed to be the traditional marks of nobility. Another was the increasing number of non-military nobles, as wealthy and talented commoners were ennobled for their services to the monarchy. It was then that the nobility began to do something it had never done before: justify its right to social and political preference. The main justification was that nobles were noble by virtue of their birth and nothing else. Education became increasingly important for some nobles, but it was a mark of, rather than a justification for, nobility. By 1789, only about one-fourth of noble families followed the career of arms.[3] In the last quarter of the eighteenth century—the years that are the focus of the discussion here—there seemed to be a surge in the attempts of nobles, particularly in the sovereign courts (the parlements) and the army, to assert the privileges of birth. This has appeared in hindsight as one of the major triggering factors of the Revolution. The perception of and the resentment toward nobles' emphasizing the racial idea of exclusiveness were later used against them, with devastating effect, by the Third Estate.

Another feature of late-eighteenth-century French nobility was that it was extremely fragmented. It is impossible to speak of a homogeneous nobility or a feudal class, as old-style Marxists would have it. There were great differences in occupational pursuits, wealth, education, and length of pedigree. Guy Chaussinand-Nogaret has brilliantly presented a sophisticated portrait of this group before the Revolution and convincingly argued that the trauma and suffering of the Revolution created an imaginary monolithic character that it never really had before the Revolution.[4] Moreover, the wealthier and better-educated nobles, who shared many of the values of the Enlightenment, were not averse to mixing their blood with that of wealthy and educated commoners. The values in question were increasing secularism and religious tolerance, an environmentalist view of man and society, a rejection of ascetic values, an acceptance of commerce and industry as respectable and legitimate activities, a rejection of arbitrariness and blind adherence to tradition in religion and government, a respect for utility, and a degree of solidarity for common aims that transcended narrow parochial, corporatist, class, and regional loyalties. Titles and money flowed across the boundary between the two groups that were on the point of fusing into a new category: the notables, who were to rule France in the first half of the nineteenth century.[5] Molière's upwardly mobile bourgeois gentilhomme of the seventeenth century had his counterpart in the eighteenth-century gentilhomme bourgeois whose lifestyle was sometimes difficult to distinguish from that of his commoner counterpart. Court nobles of Versailles, with their enormous wealth and privileged access to posts in governmental administration and the army, were a world away from the poor rural country gentlemen [hoberaux], who were sometimes hardly better off than their peasants and were finding it ever more difficult to pursue the only trade they knew or tradition allowed them to pursue—the trade of killing for one's king and country—because of the price of commissions. This "nobility of the sword" was challenged by a "nobility of the robe"—the judicial and administrative elite of the nation who dominated the sovereign courts [parlements] and the royal administration. Along with the cleavages of wealth and culture, there were also the cleavages between town and country and the different regions of France.

There were a number of possibilities for rising from the status of commoner to noble, including intermarriage, purchase, government service, and military service. As France and its bourgeoisie became wealthier at the end of the century, there is evidence of increasing exclusiveness by the nobility. However, recent research shows that the concept of an "aristocratic reaction" that has been a staple of general histories of the French Revolution is somewhat overblown.[6]

Birth was still the official criterion for membership in the nobility, and it was more or less enforced for admission into the parlements and the officer corps of the army. In the army, enforcement became more rigorous in the 1770s and particularly after the famous Ségur ordinance of 22 May 1781, which required four degrees (generations) of nobility or an honorable and extended paternal career in military service in order to be accepted as an officer.

At the same time, many nobles increasingly turned to other justifications for their status that were more consistent with the ideology of the Enlightenment, such as utility or education. In the army, the apparently reactionary Ségur ordinance, which was apparently directed against commoners in such a blatant and wounding manner, is considerably more complex when its precise requirements and context are considered. In fact, the Ségur ordinance, reactionary though it was even by the standards of the eighteenth century, was a reversion to the idea of nobility as a nobility of function and attempt to improve the professionalism of the army. It is an event that is part of a complex phenomenon related to the modernization of armies, the growing technical sophistication of warfare, and the beginnings of military professionalism. Although it is a reassertion of hereditary status, it is much more than that. Surprisingly, it has many modern and reforming aspects. The apparent paradox of reform and reaction has been brilliantly analyzed by David Bien.[7]

In 1750, a royal edict created a noblesse militaire whereby, depending on one's rank and length of service in the army, commoners could acquire nobility. This consciously created a military nobility whose purpose was to attract and especially retain the class of poor hobereaux who formed the backbone of the lower ranks of army officers and ostensibly to reward those commoners with a proven military vocation with noble status. In practice, a relatively small number of commoners benefited from this edict, nor was it primarily intended for them, but it did postulate a concept of a military nobility composed of those who were already noble military officers (with no account of their degrees of nobility) and the small number of commoners who could acquire this nobility after devoting themselves to military service.

The very concept of a military nobility only appears with the edict.[8] The adjective 'military' would have been a pleonasm during the Middle Ages, when nearly all nobles were soldiers. In an influential book that appeared a few years later, the chevalier d'Arcq argued that the true nobility, and the mainstay of the monarchical state, was the military nobility.[9] For him, there were two fundamental aspects of this nobility, along with the fact that it was to be considered the preeminent nobility. First, its sole business was war, and it was to be barred from any commercial activity. (Commercial activity was being defended by some, on the grounds that it would alleviate the poverty of most nobles for whom access to the army was difficult because they could not afford the price of the commissions for most ranks nor the lifestyle expected of an officer.[10]) D'Arcq suggested that venality of commissions be abolished and financial aid be given to military nobles to make their families larger, their sons better educated, and their daughters more marriageable; apart from this, however, military nobles were to eschew wealth and lead a Spartan existence that would make them better soldiers. Indeed, poverty would be worn proudly as a badge of their sacrifice to the monarch and the country. Second, the military nobility was to be open to commoners who served honorably for a prolonged period. The sons of such commoners would have access to the officer corps and could eventually acquire noble status for their service to the state if they maintained the military vocation in their families. (This was a road fraught with difficulty and accessible to few. During most of the eighteenth century, the number of true commoners, as opposed to fake nobles or the recently ennobled, who were often looked down upon by nobles with longer pedigrees as not being true nobles, hovered around 5 percent in the army as a whole.[11] Among artillerists and engineers the figures were higher.) Vauban had presented similar ideas in *Idée d'une Excellente Noblese et les Moyens de la distinguer par Générations*.[12] Although he accepted the idea of ennoblement for services to the state in administration and even finance, he wanted differentiation within the nobility on the basis of generations of nobility and rejected the access to the nobility by purchase. He also emphasized the primacy of the military nobility and argued that only through exceptional military services people should be allowed to jump over the rungs of the generational ladder.

D'Arcq's ideas resonated strongly with the many military nobles influenced by the pre-Romantic Rousseauist ideas of the simple, uncorrupted life and the Enlightenment ideals of the importance of education and usefulness to country. A proper military education was an extended affair from early youth and military families were the best places to implant the virtues of honor, sacrifice, and merit in their sons. The army was a place of serious lifetime service to the state and not a hobby for the rich and uncommitted to be neglected during long periods of leave and abandoned after a few perfunctory years of service.

The Ségur ordinance of 1781, clearly impregnated with the ideas of the edict of 1750 and the Chevalier d'Arcq, was the work of reformers in the army. Proud of their nobility they certainly were, but in many other ways they were far from reactionary in a simple mechanical way. The vast majority of the committee which worked to reform the army in the late 1780s defended the monarchy and emigrated.[13] But of the nobles elected to the États-Généraux on the eve of the Revolution, the most liberal and reformist were members of the military nobility, and they were the motors of the reform movement in the army before the Revolution.[14] The paradox has been explained by Bien, who reminds us that reform was not a milder relative of revolution but often distinct from and opposed to it. The military reformers wanted to reform the army but had no desire to reform the nation. The Ségur ordinance was aimed not so much at commoners, never a massive presence in the officer corps in any case, but against other sections of the nobility, particularly the recently ennobled and wealthy nobility attracted to the secular prestige of the noble calling of arms and whom the reformers perceived as being incapable of being good officers at a time when such officers were badly needed. Birth and pedigree were not justified by invoking tradition or divine purpose, rarely were they justified by the "genetic" argument that innate martial characteristics were passed on to offspring. The primary argument was based on education and Enlightenment ideas of epistemology and psychology. Knowledge came from the senses and the surroundings, character had to be imprinted at a early age on pliable spirits, and this was best done in an atmosphere that was imbued with military attitudes and traditions. Even before formal schooling became increasingly important, the small circle of the family was to exert an indelible influence on the future officer. Birth was merely a gateway to the acquisition of those characteristics that forward-looking and patriotic officers were finding to be essential in a modern army with modern technologies—merit and talent.

'Merit' as it had been originally used in military parlance in the Old Regime had meant seniority.[15] An example of this was the Ordre du Tableau, instituted by Louvois under Louis XIV, whose purpose was to regulate the replacement of an army commander in case of absence or incapacity. In such an event, the officer would be replaced automatically by appointing the most senior serving general officer in the army to take his place. It was meant to avoid bickering over precedence based on birth or personal connections. In this sense, it undermined the principle of birth in the army and was initially rather unpopular. But the principle of length of honorable service as a criterion for promotion, seniority, also became a part of merit and became increasingly more popular, especially among reformers. It soon came to be associated with assiduity to one's duties, time spent with the troops, and punctual execution of duties, including the more tedious and time-consuming ones such as drilling troops and attending to administrative details. Staying at one's post and putting up with the dreary life of provincial barracks and the multiple details of military administration in an increasingly bureaucratized army was considered to be a new form of sacrifice as noble as that of heroism on the battlefield. It was almost the very opposite of the medieval ideal of the independent warrior knight. Military reformers pushed for reducing the length of leaves (and the practice of being able to buy extended leaves from poorer brother officers by the wealthy) and minimum periods of service in a given rank before promotion to the next rank. In this way they succeeded in forcing wealthy young nobles who had purchased commission into learning something, at least, about military life. These colonels à la bavette (colonels with a bib) were perceived as a symbol of everything that was wrong in allowing frivolous, wealthy young men with no aptitude or commitment into the army to indulge their appetites for display and status.

The reformers accepted, albeit this does not seem to have been popular with the people whose interests they usually articulated, the direct admission into senior ranks of the army to high court nobility. In this, their resistance to wealth and sometimes dubious competence gave way to their sense of noble solidarity and deference to the higher nobility. But they fought for and finally succeeded in getting (under the reformer Saint-Germain) the abolition of the venality of commissions. Here again, the reformers ran up against wealthy nobles who often had shorter pedigrees. Both in the importance accorded to merit and the insistence on old military stock that shines through in the Ségur ordinance, there was a fear and rejection of wealth and opulence. These were firmly believed to enervate military fiber, sap morals, and dissolve solidarity within the ranks. They were also supposed to weaken one's physical constitution and undermine one's tolerance of hardship. In this regard, there was a deviation from one

Enlightenment attitude toward luxury. Voltaire mocked Christian asceticism and believed that luxury embellished the world, stimulated the economy, and increased employment. Rousseau, however, was against luxury and wealth, preaching the virtues of the simple life and hardiness that was more congenial to the army officers.

There is a final element in what could be called, following Corvisier, the "triangle" of qualities that sometime competed and sometimes complemented each other in determining what the reformers considered to be the ideal army officer. In addition to "birth" and "merit," there was an increasing emphasis on "talent." Although seventeenth- and eighteenth-century dictionary definitions of talent defined it as an exceptional gift of nature, by the 1770s there was evidence that it was being distinguished from the more romantic concept of genius (intelligence and creativity above and beyond the norm that was an innate and lucky gift of nature to the individual) and coming to mean something that could be acquired and learned. Bien argues that the dictionary meaning was behind the times, particularly in the army.[16] In this, he seems to be supported by Corvisier, who also mentions the importance of talent in the reforming military ideology of the eighteenth century.[17] In fact, talent was often associated with wealth, because it could be attained by a good education that was often costly and beyond the means of the poor country nobleman with a numerous family. The good classical education of a wealthy bourgeois who perceived it as desirable qualification for an officer was regarded with suspicion by more traditional officers as promoting disputatious and sophistical attitudes out of place in a fighting army.

An example of a talent was the coup d'oeil so often mentioned by military theorists and educators. While some felt that this ability to pierce through the smoke and confusion of the battlefield and grasp the lie of the terrain and the positions of forces was a rare and innate gift reserved for great captains, others felt that it was gift more widespread and could be cultivated with serious effort and exercise. Opposed to the Romantic idea of the power of raw, untrammeled and unique genius welling up in the individual, was the idea that talent could be acquired by hard work, discipline, and study.

Such a view was consistent with many of the views of military engineers and is demonstrated in d'Arçon's desire to protect the young engineer from the flashy and brilliant temptations of the spectacular and the facile in military life (attributed particularly to artillerists, who in fact had a similar educational tradition to engineers). It appears also in Du Portail's claim, in spite of his contempt for "miserable details," that the feel for space that is the prime characteristic of the good commander comes through long training and practice. When Montalembert and Guibert denied that the etymology of the word used first by Vauban to describe military engineering—'génie'—had anything to do with genius since engineering was a rather dull and mechanical profession, many engineers might have been inclined to agree, at least in private. (They would likely have also agreed with Edison that genius is 99 percent perspiration and 1 percent inspiration.) For them, much of military engineering was dull and fastidious work involving attention to little details, but the devil (or the enemy) was in the details and it was no mean feat to grasp them. It required long and arduous training, not just innate qualities, to develop the talent—the character and the intelligence—to do this.

The increasing agreement on the importance of talent—meaning education—for a military career is demonstrated in the foundation of the École militaire in 1750. It was particularly welcomed by poor nobles, who often had not been able to afford a good education and who were intended to be its main beneficiaries.[18] The military engineering and artillery corps, with their obvious emphasis on competence and talent, along with the fact that commissions in them were not purchased, were becoming more attractive, as the increasing proportion of nobles taking up a career in military engineering indicates. In the "cascade of contempt" that has been mentioned as a feature of the French royal army, military engineers had been located toward the bottom for some. The Duc de la Feuillade, after ignoring Vauban's advice and before going off in 1705 to his (unsuccessful) siege of Turin, to the Minister of War of Louis XIV, could write "Have confidence in me rather than in all the engineers of the world, and you, as well as the king, will be all the better for it. There are people born to command and these sorts of gentlemen are created only to execute the orders they are given."[19]

De la Feuillade's haughtiness does not seem to have been shared by the king. The attitude toward military engineers had bee improving since the time of Louis XIV and probably before. Proportionally, engineering officers had a better chance of being promoted to general officer rank than regular infantry officers.[20] True, among them were remarkably capable people, including Vauban, and the statistics only appear to include those with military rank, which was not the case for a substantial number (almost half) of them. Yet while disdain for engineers did not entirely dissipate—Montalembert was to take advantage of it in his struggle with the corps—it appears to have diminished substantially and been mixed with a considerable amount of respect for their education ("talents"). Here the engineers benefited from the increasing importance attached to talent by all sectors of the army.

By the middle of the century, there was an increasing conflation of talent and merit, qualities that had earlier been seen as conflicting and even opposed to some degree.

The qualities of insight and superior abilities arose from stoic subservience to servitudes, both intellectual and moral. Engineers saw themselves as a corps à talent in which talent had to be nurtured and continually and patiently burnished through disciplined and hard work and proper education. This was an attitude entrenched more deeply and earlier in the corps of military engineering than in the rest of the army (with the usual exception of the artillery) but it gradually was taken up by the rest of the army, particularly after the "military awakening" [reveil militaire] that swept the French army and French society after the disastrous defeat of Rossbach at the hands of the Prussians during the Seven Years' War.

The fusion of talent and merit was facilitated by the idea of a good education that was espoused by military reformers. This differed from the classical education of the gentlemanly bourgeois and the honnête homme that was dispensed in the classical collèges of Old Regime France.[21] For soldiers, mathematics rather than classics had always laid claim as the prime matter of education from the time of Errard and even before. The importance of science and mathematics during the Enlightenment gave it additional prestige. It was fundamental at Mézières and the artillery schools, and it was beginning to play an even larger role in the education of other officers as well. At the École militaire and its various avatars in the provinces, mathematics occupied a major place in the curriculum. Sixteen of the 31 professors at the École Militaire were teachers of mathematics, a subject that took half of the class time each morning in the 1770s. David Bien refers to a veritable "mania for mathematics" in the French army.[22] Royal scholarships enabled poor nobles to acquire this kind of education more easily. Aside from its obviously practical benefits in arms like engineering and the artillery, it was perceived to have moral virtues that set it apart and in contrast to the cursus of the classical colleges, which inculcated skill in rhetoric and disputation. Mathematics, even if it had relatively limited usefulness for regular infantry officers, would nevertheless provide the astringent mental discipline that would form the character and the mind of a good officer. Straining for long hours on difficult and dry problems would be the equivalent of a hair shirt requiring the development of character to support. Instead of a ready fluency and ability to excel in captious and specious argument, so useful to the legal gentlemen of the robe, a rigorous training in mathematics would enable the future officer to think clearly and concretely. The spare and unadorned logical thinking that would be the product of a good mathematical instruction would protect against entrapment in thickets of sophistry. Officers reading Laplace in their spare moments in Alfred de Vigny's *Grandeurs et Servitudes Militaires* may strike the modern reader as highly incongruous but they were not entirely a product of an author's imagination.[23]

A Corps à talent

In many ways, engineers were at the cross roads of the transformations that were skewing the triangular base of birth, merit, and talent that was the underpinning of the reform movement in the French army in the last decades before the Revolution. Commissions in their corps had never been venal, advanced technical education had always been a requirement, and much of their work, even on the battlefield, was administrative and bureaucratic. Their status as officers had never gone unchallenged. Their usefulness, at a time when siege warfare appeared to be in decline was also questioned; even their very efficiency at conducting sieges made them appear less necessary, or at least made only some of their duties the narrow purview of specialists—specialists who did not necessarily have to be military men. They increasingly offered a model for a new kind of officer for reformers and an aberration for others, generally more attached to the perquisites of birth alone or imbued with a vision of a traditional kind of army that was being destroyed by the reform movement of the military awakening of the Old Regime.

For those who valued its achievements, engineering had a double prestige in the French army of the Enlightenment—it was both military and scientific. This had a special appeal to many of the reforming officers of the army at the twilight of the Old Regime who were developing a more professional attitude to their calling. Samuel Huntington, in his analysis of military professionalism, cites three fundamental characteristics of professionalism that began to be displayed among officers of Western European armies in the nineteenth century—expertise, responsibility, and corporateness.[24] In fact, professionalism in France seems to have appeared somewhat earlier, toward the end of the eighteenth century, and nowhere earlier than in the military engineering corps. Within the corps, expertise had long been important, indeed it was the initial raison d'être of the corps, even before the regular line officers had begun to feel that expertise and technical mastery were as important, if not more so, than individual qualities such as bravery, self-sacrifice, and loyalty to one's commanders. Long before the trauma of the Seven Years' War, Vauban had insisted on a "novitiate" for his engineers and tests of their ability before admission to the corps.

Engineers qua military engineers had always been soldiers. What prestige the military architect of the Italian Renaissance had had as mere builder and fortifier had been ephemeral and had owed much to the association with arms and the noble occupation of war. By the time of Vauban, architects working in the military field were becoming military engineers, along with soldiers who were learning the architects' skills of construction as well as siegecraft. The descendants of the master masons of the Middle Ages with a close connection to things military had been transformed into contractors and were also absorbed into the military body of military engineering. (The descendants of the ingeniator had mostly become gunners and were providing one of the groups of the nascent artillery corps.) Although its necessity was sometimes contested by people within and outside the corps, building expertise for fortification had been assimilated and militarized. So had siegecraft.[25]

In addition to expertise they acquired through practice and instruction, military engineers were getting more exposure to science. Science was gradually becoming increasingly important in the corps since the time of Bélidor and was yet another mark of prestigious expertise during the Enlightenment. A small minority of military engineers were making a substantial name for themselves in scientific work. The process of integrating science into military engineering was raising questions about what kind of science and how much science was necessary, but it was clear that science was an indispensable part of the expertise of the engineer.

Engineers were also part of a corps. This corps was not an artisanal corps of the corporate world of the Middle Ages, but a distinctly more modern entity that looked forward to the great state corps of the French state today. As members of such a corps, they thus fall into Huntington's category of a bureaucratic profession rather than an associational profession.[26] It was, as Huntington claims for the army officer corps in general, both a bureaucratic profession and a bureaucratic organization.[27] The personal relations between the engineer and the monarch were still real and easily visible at the time of Errard in the title ingénieur du roy, but ever since Vauban the corps mediated between the political power and individual engineers.[28] This aspect of corporateness that Huntington includes in the trinity of professional characteristics is almost inevitable in the kind of bureaucratic entity fashioned by Louvois and Vauban, with its hierarchical and territorial divisions. The responsibility of engineers to something more than themselves or to a superior of the moment was also making them a profession. A concept of France as a space to be dominated and defended for the sake of the king and the country, loyalty to the values of service and sacrifice that was part of the official ethic of the nobility, whose members they either were or aspired to be, promoted a concept of stewardship and responsibility.

In light of the work of Bien, Huntington's view that "prior to 1800, there was no such thing as a professional officer corps" needs to be modified.[29] Reformers in the French army immediately before the Revolution were quite clear that they wanted to build an officer corps with the qualities of expertise, responsibility, and corporateness, which Huntington sees as the hallmarks of professionalism. If the military profession in its modern Huntingtonian sense was not yet fully formed in France, it was well on the way to becoming so in a process that was clearly conceived. Furthermore, in the case of the military engineers, one could almost speak of a profession that was even more advanced in its development. They were already a corps, had their own school of advanced instruction giving them specialized esoteric knowledge, and had never had the proprietary and personal bonds with their office and the soldiers that occasionally came under their direct command that the military reformers were trying to abolish in the line regiments. It is indeed plausible—and an interesting subject for further research—that military engineering provided a model for this movement toward professionalism in the French army. However, if military engineering was a model, it was also a target for some other officers of the army, both those who were pushing toward professionalism and those who were resisting that thrust. Those who believed that mere birth was necessary and sufficient for an officer were hostile to professionalism and would have seen engineers as a prime exemplification of the movement toward it. But even those who wanted to professionalize the officer corps, like Guibert, sometimes questioned the *kind* of expertise the military engineers had.[30] From their beginnings, their legitimacy as a *military* corps had been questioned and some still continued to attack it. Also very galling to the military engineers was the view by many officers that they were mere executants who should be subordinated to either line officers or artillery officers. They were barred from the general staff for which they thought they were particularly suited.[31] The same impulses driving reforming officers toward professionalism and reform sometimes conflicted with the older tendencies to the same end among engineers. Yet the very conflicts themselves that multiplied and became more acute in the French army are themselves a sign of the onset of professionalization. As Norbert Elias has put it, "if one attempted to work out a general theory of the genesis of institutions, one would probably have to say that the initial conflict is one of the basic features of a nascent institution."[32]

Saint Germain's Ordinance of 1776

The Comte de Saint-Germain, a "pure soldier without any idea of social reform,"[33] was a member of the often impoverished provincial nobility that traditionally provided most of the officers in the French army below the rank of colonel. His military career had been an illustration of the frustrations endured by professional soldiers without the right court connections. Because of his inability to purchase a colonel's commission he had initially followed a career of a soldier of fortune for various European powers. Then, after a period of brilliant service during the Seven Years' War, he had left the army as a result of court intrigue and his own irascible personality, and had gone to Denmark, where he became the Minister of War and tried to reform that country's army. Aging, ailing, and crotchety in retirement, he had devoted himself to farming and an occupation of old soldiers that is almost as common as the publication of memoirs: dispensing advice on military reforms. One of his publications had attracted the attention of Louis XVI's reforming minister Turgot, who wanted to prevent the post of Minister of War going to a member of the court nobility at Versailles and who also wanted a reformer to remedy abuses in the army, particularly the financial ones. Appointed by Turgot, surrounded by a number of capable aides and administrators, Saint-Germain initiated a flurry of reform that resulted in no less than 98 major army regulations being promulgated during a brief ministry that lasted less than 2 years.

Among Saint-Germain's actions during this ministry were the improvement of living conditions for common soldiers, the streamlining of military bureaucracy and administration of logistics, and the reorganization of the French army into 16 permanent territorial divisions commanded by serving professional soldiers. These officers had command of all military units and installations in their respective divisions in peace as well as war, and hereditary noble commanders who had hitherto occupied ceremonial military posts in their traditional lands were subordinated to them. Although he would dearly have liked to abolish outright the purchase of commissions and to replace this system with promotion by merit, he was not able to do this in spite of support from junior officers. Nevertheless, he did initiate the eventual phasing out of venality of commissions by successive reductions of their price. Furthermore, a military career was made easier for poor nobles by the government's providing hundreds of scholarships in provincial collèges for those contemplating a military career. (Napoleon Bonaparte was one of the beneficiaries of this system.)

At the technical level, Saint-Germain recalled the artillerist Jean-Vacquette de Gribeauval (1715–1789) and appointed him First Inspector General of the Artillery to finish his interrupted reforms in the materiel and organization of the French artillery that was to make it the best in the world at the beginning of the Revolution.[34] The tactical ideas of Guibert were studied, and some of them began to be integrated into standard army regulations and procedures.[35]

Perhaps the best-known of Saint-Germain's actions, and in hindsight the most controversial, was his attempt to make the officer corps a preserve of the nobility and the sons of former officers. Like many of Saint-Germain's acts, this one must be viewed in the context of his primary concern for military efficiency and only secondarily in terms of social considerations. Not only did he think that it was unjust for someone wealthy, whether bourgeois or noble, with no military family traditions, to buy his way into the army over the head of a poor noble, prevented by law from engaging in most commercial and industrial activity, who was expected by popular prejudice and caste tradition to follow a military career; it did not produce good officers. In the view of the old soldier, military competence was acquired in a military ambiance and dedication to military professionalism was nurtured in families of military professionals. Although by no means a social reformer—indeed such ideas would have horrified him—he was very much the military reformer.

Saint-Germain was one of those who explicitly described the military engineering corps as a "corps distinguished by its talents" in his reorganization of the military engineering corps.[36] The "Ordonnace concernant le corps du génie" was therefore not the most radical of Saint-Germain's acts and did not attract much attention outside the engineering corps and the army. Although it does institute the recruitment policy based on noble birth that applied for the rest of the army, many of the other provisions were intended to consolidate and rationalize existing practices. It fixed the number of military engineers in the French army at 329. This was a reduction from the 400 specified by a regulation toward the end of the Seven Years' War, and it reflects one of the major concerns of Saint-Germain's ministry: cutting expenses.[37] The French military engineering establishment throughout the eighteenth century had fluctuated between 268 and 390 engineers.[38] This reflected the needs of the largest standing army in Western Europe, and at the end of the Seven Years' War the decision to raise the number to 400 had been taken as the result of the realization that a pool of trained engineers could not be formed as easily as additional infantry forces. It was claimed that, even in peacetime, a number of engineers sufficiently large for immediate service in war had to be maintained in addition to those fulfilling regular peacetime duties.[39]

The high point of 390 engineers had been reached in 1772; the perennial financial problems that were a constant constraint on the conduct of the French monarchy led to the cuts of 1776. Even with the smaller numbers, the Corps of Military Engineers was overstaffed for its peacetime duties and increasingly even for its wartime functions. The example of Guibert shows that it had become fashionable to denigrate siege warfare that was coming to be perceived by many in the army as outmoded. In spite of the forced retirement of 25 engineers in 1777 as a result of the ordinance, the lack of any mandatory retirement age (and automatic pension) made promotion painfully slow as senior engineers—a physically hardy lot it seems—hung on to their posts for dear life.[40] Furthermore, the number of territorial administrative units—the *directions*—was reduced from 23 to 12. Consequently, the number of the most senior posts in the corps—the Directors of Fortifications—was accordingly reduced and this further aggravated the stagnation of ranks in the corps du génie. During the Revolution, an engineer claimed that only a third of the engineers had achieved the rank of captain after 30 years of service and that it took a further 30 years for promotion to colonel.[41] Thus, it was impossible to become colonel before the age of 70!

A first glance at the new Corps of Military Engineers reveals a bureaucratic structure that looks surprisingly modern. Serving under 12 directors, assigned a certain geographical area with a certain number of fortresses ranging from 8 to 24 (172 fortresses in all), were a number of "chief engineers" in charge of "ordinary engineers."[42] Within each direction there was a further subdivision of engineers into engineering brigades—units that could be called upon to participate in sieges in case of war. At Versailles, a thirteenth Director was attached to the Minister of War to deal with ministerial correspondence concerning the engineering corps and to act as an on-the-spot adviser to the Minister on engineering matters. Although this directeur was in theory a primus inter pares among his fellow Directors, in practice, his proximity to power made him an unofficial head of the corps. Fourcroy de Ramecourt was appointed to this position.[43]

But in addition to titles of technical seniority, engineers held a military rank ranging from sub-lieutenant as cadets at the military engineering school at Mézières to maréchal de camp (the lowest of general officer ranks). One of the aims of the 1776 ordonnance had been to impose a rough correspondence between what one could call technical rank and military rank. The previous lack of symmetry in this regard that had been a major cause of friction, both between the army as a whole and the Corps of Military Engineers as well as within the corps itself, was finally eliminated.

The coordination of military and technical rank was perhaps the major achievement of the ordinance. Closely related to it was the decision to make the corps of military engineering "royal." It now enjoyed the status of the other combat arms of the army and was closely integrated into it. Military engineers had finally become true officers and truly royal—a demand they had voiced forcefully on the eve of the reorganization. Two other demands were also satisfied. Promotion henceforth was to be based exclusively on competence and not on seniority. It was also decided to set up a committee on fortification, much like the assemblée des ponts et chaussées, that was to be composed of representatives of all ranks that was to meet regularly to discuss new projects. Evidently the aim of this provision was to permit an impartial and more effective examination of new ideas that would minimize the weight of favoritism and technical incompetence by administrators in the Ministry of War and promote the better circulation and development of new ideas in the domain of fortification.

Integration and rationalization of ranks were in fact attained. Indeed, the trend of the previous decades was in that direction in any case. But other aims of the ordinance were not put into effect before the Revolution. For example, only two exceptional engineers, Le Michaud d'Arçon (1733–1800) and André-Joseph de La Fitte-Clavé (1740–1794), were appointed over the heads of more senior colleagues before the Revolution.[44] An engineer such as the scientist Coulomb apparently did not rate this favor in spite of his classic papers on soil mechanics and the strength of materials. A more probing examination of the human reality behind the formal bureaucratic structure shows that senior ranks often remained a cozy family affair. Six of the thirteen Directors were themselves the sons of Directors, indeed three of them had begun their careers under fathers' orders. Another three were close relatives of engineers. None could be called young—their average age was almost 59 years and they had been in the corps an average of 38 years. When the Revolution came, 13 years later, all but five were still in their posts, and of these only one had retired voluntarily (at the age of 68). The others had died in service in their late sixties and seventies. It was only in 1791 that the government released the logjam of promotions by retiring all but two of the current directors, whose average age was 71.[45]

It is also unclear to what extent the committee on fortifications ever became a viable institution before the Revolution. All the evidence seems to indicate that it did not. It was never mentioned in the Montalembert controversy, where it could certainly have played an obvious and natural role. Yet it did meet. In a letter to Fourcroy de Ramecourt, by then the unofficial head of the Corps royal du génie a year before Montalembert built his fort at Aix, the Director of Fortification Dajot mentions the meeting of the committee that lasted from 16 to 23 February 1778: "Nothing could be better imagined for the good of the service than these committees. May the Minister value them sufficiently to not succumb in future to the dreams of Court Vaubans."[46]

It is entirely possible that Montalembert was among the "Court Vaubans" to whom Dajot referred. But it was only during the Revolution that the Comité des Fortifications came into its own.

One change in the Corps that was nowhere mentioned in Saint-Germain's ordinance, and which may have had much to do with the faltering progress of a committee of fortifications, was the appearance of a single effective head of the corps in the person of Fourcroy de Ramecourt. The regulations had stipulated a Director at Versailles, but

Saint-Germain is alleged to have said that this was done to "enlighten [the Minister of War] on the administration of fortifications and not to give himself a master."[47] In fact, however, Fourcroy de Ramecourt became pre-eminent in the corps and remained in the Versailles directorship until the Revolution. The question remains open whether this was due to his adeptness at intrigue, as suggested by some historians, or because he was the whole-hearted choice of his fellow directors, as Condorcet claimed in his eulogy.[48] Nevertheless it was Fourcroy de Ramecourt who came to be considered the unofficial chief of the corps and his election to the post of académicien libre of the Academy of Sciences in 1785 is doubtless an indication of this status.

The passage of a new charter for the military engineers that finally made all of them officers and truly royal and brought them under the command of a sole chief who was one of their own and who had direct access to the Minister of War did not bring stability nor contentment to the corps. There were obviously Directors who resented Fourcroy de Ramecourt's preponderant influence and said so in memoranda to the Minister in the 1780s, which called for further revisions of the ordinance.[49] In the same memoranda they protested against the system of promotion by merit as leaving too much room for subjectivity and favoritism and called for a return to promotion by seniority. Others resented the tighter control by general officers in charge of the newly formed military divisions over engineers and asked that engineers be allowed to report directly to the Minister of War.[50] Complaints that the military engineers were insubordinate and uncommunicative with field commanders indicate that even the tension between integration and the autonomy of the corps had not been resolved.[51]

The upheaval in French society in general was reflected in the institutions of the French state and military engineering. Even if there was some externally apparent bureaucratic tidiness, belied by bitter struggles such as that with Montalembert, the working role and epistemic space of military engineers were not rigidly delineated and there were forces both external and internal that were expanding, reducing, or changing the occupational domain of military engineering. It was a situation that Fourcroy de Ramecourt and many of his colleagues found difficult to tolerate because they felt it undermined the effectiveness of their corps and made its very existence vulnerable. While military engineers were continually dealing with change as part of their professional life, they needed to do so in a systematic and orderly manner to complete construction projects, direct sieges, and maintain a far-flung network of fortifications. The conservative art, as fortification was often called, implied the commitment to defense of a status quo whose main features were accepted even though improvements could be considered. As military officers and privileged members of the Old Regime, military engineers were committed to stability as a social virtue. They were also committed to it as the condition for professional legitimacy and technical efficacy.

At the end of the Old Regime, the Corps royal du génie, unlike the Corps des ponts et chaussées and in spite of an apparent confirmation and elevation of status, was still in an ambiguous position. The loss of many of its former functions to the Corps des ponts et chaussées, such as the responsibility for many harbor works in militarily sensitive sites, aggravated the perceived lack of institutional solidity.[52] This was only feebly compensated by the control of the corps des ingénieurs géographes that had been given to the military engineers by the 1776 ordinance. To compound its difficulties, the financial and political crisis that was racking the army and the country as a whole brought further blows to the corps. A special Conseil de Guerre created in October 1787 to reform the army called for the decommissioning of numerous fortresses.[53] Whether willingly or not, Fourcroy de Ramecourt went along with this idea and raised a storm of protest from the junior ranks of the corps. His reputation as an administrative despot was reinforced at a time when he was locked in a ferocious polemic with Montalembert. The dry administrative regulations and ministerial correspondence give ample indication that as an institution the Corps royal du génie, notwithstanding its proud designation as a corps à talents, was on the defensive, divided, and vulnerable. It was not only a beneficiary of the trend to professionalism; it was threatened by it as well. Integration into the general professionalizing movement in the army portended changes in its own movement toward professionalism.

8
Thinking Wars

… the art of war is like that of medicine, murderous and conjectural.
—attributed to Voltaire by John Frederick Charles Fuller[1]

Interest in warfare as a subject of intellectual analysis in the eighteenth century was not new; it dates back to antiquity. There had been a revival of war as a subject of discourse and study in the Renaissance partly because of an interest in antiquity and partly because of the beginnings of a political science and humanistic studies. The belief that human nature was universal, unchanging, and consequently could be studied, suggested that war too had invariable principles. Machiavelli's strong philosophical belief in an unchanging human society and principles of war led him to depreciate technical innovations such as artillery and to emphasize valor precisely at a time when technical innovations were beginning to make an impact on warfare.

In eighteenth-century France, there was another spurt of study both on the subject of war in general and the mechanics of waging war. Under the former rubric, war was studied for its place in a polity, its relations to a given society, and its morality. Under the latter, an attempt was made to attain a rational understanding of military activity and to develop a science of warfare that would facilitate victory. Interest in antiquity and classical history continued to inspire French writers on military subjects throughout the eighteenth century.[2]

But there were additional factors that molded French military thinking during the Enlightenment. Perhaps most importantly, the impact of science on the intellectual climate in France, particularly after the triumph of Newtonianism after about 1730, was felt in the military sphere as well. Even before this, and even before the dominance of Cartesian scientific ideas, science had a major influence on Raimondo Montecuccoli (1609–1680), who enjoyed a reputation in the eighteenth century as great as that enjoyed by Clausewitz in the nineteenth.[3] Montecuccoli's scientific reading included a preponderance of works in the Hermetic and Paracelsian tradition as well as books on fortification by Italian Renaissance fortifiers. Although one may be tempted to dismiss this as proto-science, it is nevertheless clear that the attempt to distill general rules from experience ran parallel to Montecuccoli's own attempts to create a theory of warfare.[4] Another influence was the persistent legacy of seventeenth-century neo-classicism in the arts. Based on Aristotle's *Poetics*, this view posited the existence of rules and principles for different arts, such as drama and painting, and saw it embodied in the exemplary work of great geniuses whose work served as a model for further artistic creativity.[5] A third factor was the sobering and thought-provoking French military defeats, particularly during the Seven Years' War, which stimulated the best military minds to reflect on the nature of their profession. Finally, as the belief in progress in human affairs grew stronger, there was a modification of the view that the lessons of history were universally valid and atemporal: changing military technology could play a role, even a decisive one, in warfare and had to be taken into account in any analysis of war.

Yet at the beginning of the eighteenth century the Chevalier de Folard (1669–1752), a soldier and classical scholar who was a translator and commentator of Polybius, still looked to antiquity for inspiration in his writings on battlefield tactics.[6] Having little respect for new technology and firepower, more enamored of the Greeks than the Romans, he advocated the importance of shock combat with the arrangement of troops in compact and dense masses that would conquer because of their vigor and spirit channeled to best effect in column formations.[7] Folard, who had epistolary relations with some of the great commanders of day, such as Maurice de Saxe and Marshal Belle-Isle, is therefore among the first and the most influential of that school of tactical doctrine associated with the ordre profond, one of the main camps in the military debates at the end of the century.[8]

For the philosophes of the Enlightenment, war was a scandal and an aberration, associated with bloodshed and irrationality, and rarely a fit subject for reflection other than as an object of contempt and condemnation. The misadventures of Candide and the harsh judgments of the Swedish warrior-king Charles XII by Voltaire are a fair indication of his attitude to war. And he was not alone: From Leibniz to Kant via Diderot and Condorcet, eighteenth-century intellectuals were pacifists. The abbé de Saint-Pierre's sweetly utopian book *Projet pour rendre la paix perpétuelle en Europe* reflected a strong and widespread desire for peace.[9]

But it was not only sensitive souls who displayed a disinclination to deal with war as a subject of polite intellectual discourse; many soldiers agreed with them that war was certainly irrational and subject to the vagaries of chance, genius, and courage, so that the idea of reducing warfare to scientific principles was futile. Only long personal acquaintance with war gave the necessary experience to acquire expertise and some limited competence.

Marshal Jacques-François Chastenet de Puységur (1655–1743)—a writer, a former military tutor to the Dauphin, and a former Minister of War—made the following rueful admission in the preface of his book *Art de la Guerre, Par principes et par règles*:

> We have no school where one can be instructed in the military art, no teacher who teaches its fundamental rules. Worse, people are almost persuaded that such resources are useless, that war can only be learned in the tumult of camps and the movement of armies, as if all arts did not have certain rules and a theory founded on solid principles, without which not only cannot one hope to succeed in them, but one should not even expose oneself to practice them.[10]

Puységur quite consciously set about remedying this defect:

> It is to disabuse soldiers of a prejudice so contrary to their advancement and their own interest that the author determined to share with the public the knowledge that he had acquired in the art that he had practiced . . . ; he has proposed to develop in a concrete and palpable manner this theory of field warfare that exists independently of practice and to explain everything that is associated with it in a manner so detailed and easy to understand that by merely studying it, and without leaving one's chamber, every attentive and moderately intelligent person can afterwards put himself in a condition to apply true principles to all the proper movements for troops. . . .[11]

Here one can see an example of the general admiration of the Enlightenment for science and the scientific worldview. But the sudden increase in military publications in Europe that was centered in France can also be attributed partly to the disasters of the Seven Years' War, which inspired much reflection by the more thoughtful French officers. The first *carrière ouverte aux talents* was the literary one, and officers with a flair for writing could attract the attention of the public and superiors in the army. As the brilliant military theorist Jacques-Antoine-Hippolyte de Guibert (1743–1790) put it, "to innovate or to attach oneself to innovators has become a way of [making one's] reputation and fortune."[12] Between 1748 and 1756 there was a fourfold increase in the rate of publication of military books over the rate of publication from the beginning of the seventeenth century until that time. And between 1756 and the Revolution there was another fourfold increase to a level that would be maintained throughout the nineteenth century.[13] These publications dealt with both the technical aspects of waging war and the theory of war in general. The most interesting, while dealing primarily with narrowly technical issues, also had fundamental implications regarding the nature of general strategy and war itself. They were published in a context of intense and often acrimonious controversy that eventually led the Minister of War Ségur to forbid officers to publish without his permission.[14] The three most important foci of controversy were on the nature and use of artillery, the marshaling of troops on the battlefield, and the architecture of permanent fortification. The first two issues have attracted considerable attention from military historians and the last is the subject of this work. A quick survey of these debates will cast light on their interaction.

The Gribeauval System

Two broad categories of reasons have been proposed for the victories of French armies during the Revolutionary Wars—sociopolitical reasons and reasons having to do with military technology and organization. Until recently the former were considered more decisive. The liberation of the human potential of the nation, French nationalism, and revolutionary fervor were supposed to have contributed to the élan of the Revolutionary and eventually Imperial armies which swept from Lisbon to Moscow and kept Europe in a state of turmoil for 20 years. Along with increasing evidence that the two categories are not nearly as distinct as might appear at first glance, the second category has also gained more consideration.[15] The royal army inherited by the Revolution, even when shorn of most of its officers and modified by revolutionary governments, formed the solid basis for the victories of Revolutionary France. The basic drill book of the French army embodied in the ordinance of 1791, which was used by the Revolutionary and Napoleonic armies and stayed essentially intact until 1831, was basically the result of the pre-revolutionary debates on military matters in the French army.[16] The pre-eminent feature of the armament of this royal army was the artillery that had been thoroughly reorganized a few years before the Revolution by Gribeauval, who died the year the Bastille fell. The British military historian Fuller has argued that Gribeauval's system of artillery had effects that were comparable to the introduction of the bayonet in European armies about a century earlier.[17]

In the century before Gribeauval's final reforms there were almost twenty major changes in artillery organization indicating the kind of perpetual ferment that was feature of the military engineering corps as well.[18] The general aim of these reorganizations was also similar to that seen with the military engineers—the standardization and integration of men and materiel into the army. From a loosely organized semi-civilian corps of technical experts as late as the early years of Louis XIV's reign, the artillery was transformed into a regular combat arm in its own right with its own professional officers who were specially trained at schools set up for that purpose.[19] There was also an acceleration of an already existing trend to standardize equipment. Louis XIV's chief of artillery Jean-Florent de Vallière (1667–1759) had already established uniformity in reducing the number of kinds of guns by replacing all the long, medium, and short types of cannon for a given caliber with a single medium-length gun. But Vallière's guns were heavy and difficult to move on the battlefield, where they were still used sparingly and generally in positions from which they were not moved throughout the battle. Vallière did not believe in specialization and believed that the same guns and gun carriages should be used in fortresses, in the field, and in coastal batteries. As a veteran of Louis XIV's wars where siegecraft played such a central role, his design philosophy of artillery tended to accommodate the needs of siege warfare rather than field battles. His guns were sturdy, heavy, and overpowered, with the main consideration being range rather than mobility.

Vallière bequeathed his position to his son, Joseph-Florent de Vallière (1717–1776) who continued his father's policies until the spectacular rise of Gribeauval.[20] Gribeauval, of relatively humble background, whose advancement in the French army had been slow, had gone to fight for Maria Theresa and distinguished himself at the epic siege of Schweidnitz where he had helped hold off the great Frederick himself. During his years in the Austrian service, he had come to appreciate the advantages of mobility for artillery and to respect the lighter guns used by the Austrians.[21] Like Montalembert, Gribeauval had attracted the attention of the Duc de Choiseul, ambassador to Vienna and later all powerful minister to Louis XV, and he had risen directly from captain to lieutenant-general in the army after his return to France. Choiseul put him to work to reorganize French artillery after the disasters of the Seven Years' War and Gribeauval's reforms were radical, leading to bitter opposition from some officers. After Choiseul's fall from grace, Vallière fils resumed his place for a short time and Gribeauval was dismissed. A new ordinance for the artillery inspired by Vallière père returned the old ways and, in words reminiscent of Fourcroy de Ramecourt, declared that "nothing was more contrary to the good of the service as innovations."[22] But this turned out to be only a brief interlude and Gribeauval returned in triumph when the reforming Saint-Germain assumed the Ministry of War in 1775. By the time of the Revolution his reforms were essentially complete.

While keeping Vallière père's basic range of calibers of (24-, 16-, 12-, 8-, and 4-pounders), Gribeauval lightened his guns considerably, which led to an increase in the recoil and a decrease in the range. Thus, the lightest field cannon, the 4-pounder, decreased in weight from 575 to 300 kilograms and in range from around 1,500 meters to 1,000.[23] A series of tests at Strasbourg in 1764 (favorable) and Douai in 1772 (unfavorable) gave conflicting results on the ranges and efficacy of the new cannon.[24] The bitter quarrel between the bleus (partisans of Gribeauval) and the rouges (partisans of the old system of Vallière), like the quarrel over the perpendicular fortification of Montalembert, eventually spilled over in the Academy of Sciences and the military and scientific press.[25]

In addition to new kinds of cannon, Gribeauval standardized gun carriages and ammunitions boxes, developed improved gun sights for his shorter cannon, and reorganized the command structure and promotion policies of the artillery regiments.[26] In 1789, in a move fraught with major tactical implications, the French army created a horse artillery, which was intended to move about the battlefield and intervene at key points.[27] After witnessing the cannonade at the battle of Valmy in 1792, Goethe wrote that it had ushered in "a new era in world history." He was not only right from a poetically political point of view; unwittingly perhaps, he was also right from the narrow technical point of view of artillery tactics. And there was some connection between the two. Gribeauval's achievement is more than any individual innovation by itself. It is the creation of a coherent and integrated system of both animate and inanimate elements designed to fit together and perform at the highest efficiency. Consistent with this aim, Gribeauval showed himself to be an energetic patron of Honoré Blanc, who was attempting the standardization of the manufacture of small arms, an effort that had some influence on the eventual development in the nineteenth century of the so-called American System of Manufacture using interchangeable parts.[28] Rosen sees it also as an example of McNeill's "command technology"—a technology created to serve purposes neither attainable by existing technologies nor even envisaged in all their ramifications by users of these technologies.[29] The battles for which Gribeauval designed his system had not yet been fought and they implied tactics and strategy of unprecedented mobility. The system only achieved perfection in the early battles of the Revolution as young and talented commanders like Napoleon Bonaparte realized its full potential.

For the strategy and tactics of the early eighteenth century, the supporters of Vallière were surely right and it is hard to reproach them for rejecting a system that was designed for a world that did not yet exist. They were believers in artillery as an art, not a science, and defenders of artisanal and aristocratic traditions. Gribeauval's followers saw artillery as a science and were meritocratic in their attitudes to hierarchies. While the rouges believed that their art was based on and justified by unchangeable principles, the bleus felt that science and reason were tools that could be used to innovate and develop a new kind of artillery and artillery system.

Gribeauval triumphed partly because of protection in high places but also because of undeniable competence in his profession and substantial support from the majority of artillery officers. It is said that Saint-Germain made his decision about recalling Gribeauval after having solicited the opinion of the artillery corps and having found overwhelming support for Gribeauval and his system. Thus, although in a sense Gribeauval was an outsider by his having temporarily left the French service, he was nevertheless a bona fide member of the corps with solid support and collaborators like Tronson du Coudray and the du Teil brothers who implemented and helped to defend the system.[30] And no matter what one might be tempted to say about the Old Regime and what might logically appear to be its most reactionary component, the army, the quarrel of the bleus and the rouges and its outcome demonstrates that it was capable of fundamental reform with far-reaching social and political implications.

Ordre mince and Ordre profond

Marshaling troops on the battlefield has been a constant concern of generals and a source of arguments among military writers from the time of the Macedonian phalanx and Roman triplex aciem. The order of marches and deployment on the battlefield itself particularly attracted the attention of great captains like Maurice of Nassau and Gustavus Adolphus as changing technologies, notably the use of pikemen and firearms, posed problems that were unknown for the relatively undifferentiated feudal host. The general trend from the Renaissance on was toward subdivision into smaller units acting together on the field of battle. This in turn led to problems of articulating the various units of an army so that they worked in unison and could change from a mode of marching to attack or retreat with a minimum of confusion. Elaborate geometrical patterns such as those depicted in the *Encyclopédie* illustrate the various combinations of maneuvers that became part of the standard drill manuals and officially accepted battle orders. Even though simple, the various combinations of these geometrical patterns provided one of the reasons for the study of mathematics by military officers.

Eighteenth-century French developments regarding these problems are of particular interest because of a belief, which has recently been successfully questioned, that the tactics of Revolutionary armies arose out of both drawbacks and advantages of using large numbers of relatively untrained recruits and a shortage of professional officers. Inspired by revolutionary ardor and patriotism, willing to suffer heavy casualties, French armies attacked in dense columns that smashed their way through the conventional lines of enemy soldiers without indulging in the complicated ballet of battlefield maneuvers that were a characteristic of professional armies of the eighteenth century. This view neglects the intense ferment of military thought during the

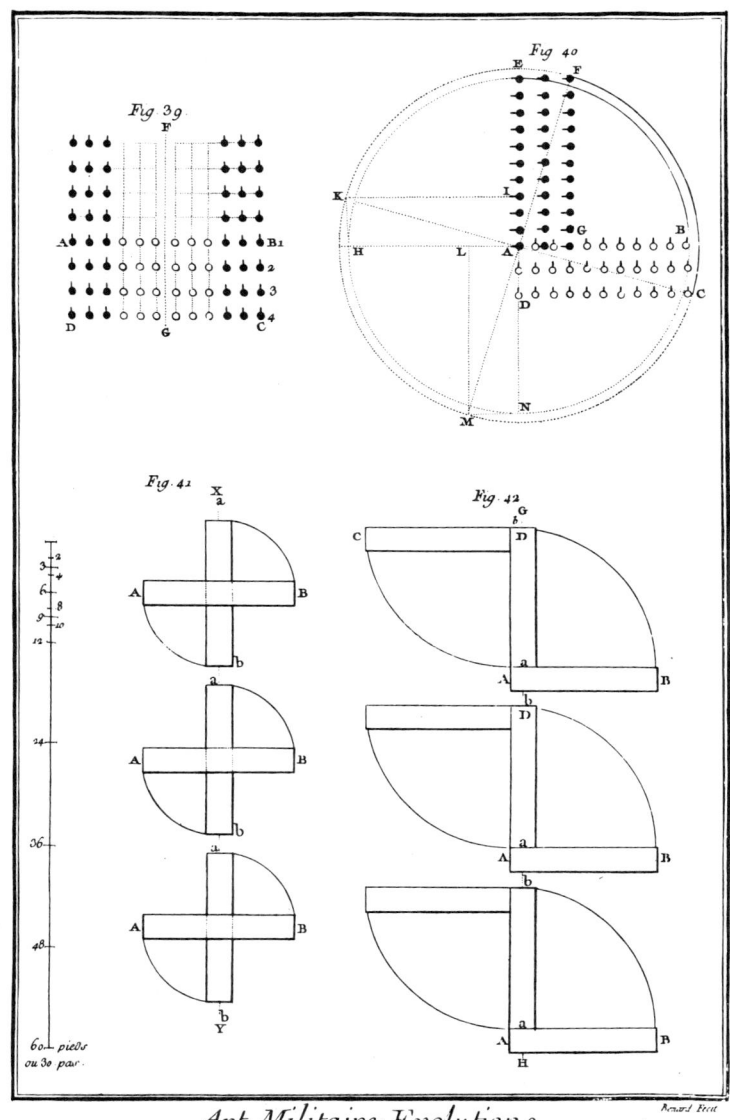

Figure 8.1
Plates from the section on the *Art Militaire* in the *Encyclopédie* of Diderot and d'Alembert illustrating military "evolutions" of the eighteenth century. The plate above shows how a troop of soldiers breaks up into two parallel columns twice as deep as the initial troop (fig. 39), and how a troop can wheel to face an enemy in different directions pivoting about the top left corner of

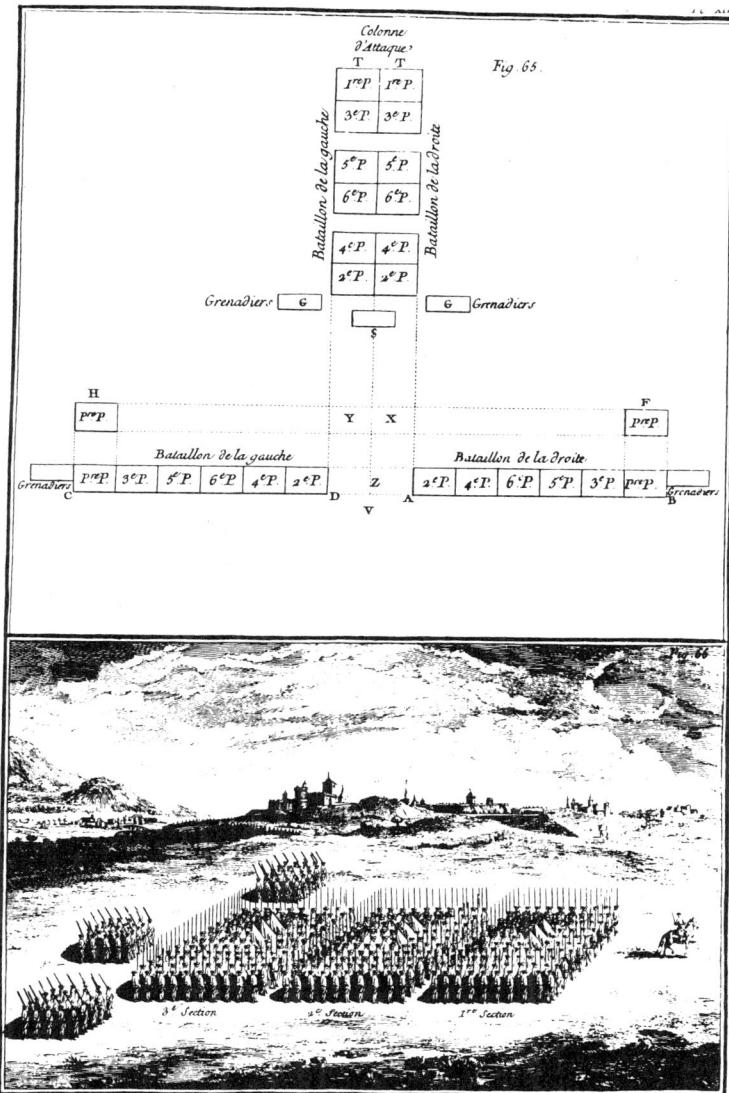

Art Militaire, Evolutions.

the troop (fig. 40) or around its center (fig. 41) and about two successive pivots to end up pointing in the same direction but displaced to the left (fig. 42). The plate above shows how two battalions drawn up in ranks maneuver to form a column.

Enlightenment in France and the fact that the battle orders of French armies had developed in large part before the Revolution. True, the Revolution did add even more dynamism and vigor to these maneuvers for a variety of political and social reasons but its soldiers were not nearly as unskilled nor innocent of tactical doctrines as some historians have implied.[31]

In fact, the idea of the column for attack was a relatively old French idea that goes back at least to the Chevalier de Folard. A number of disciples, of whom the best known was François-Jean de Graindorge d'Orgeville, baron de Mesnil-Durand (1729–1799), vigorously defended the so-called ordre profond against the partisans of the ordre mince, long and thin lines of infantry that could use firepower to maximum effect.[32] The essential tactical issues underlying this controversy was whether the combination of bayonet and flint-lock musket was a shock weapon or a fire weapon.[33] Two major arguments were adduced in the favor of a columnar attack over the advance of a line—one technical and one psychosocial. First, the defenders of the ordre profond had little faith in firepower. In this there was definitely some justification that went far beyond a general admiration for antiquity and contempt for modern military technology. Even the Prussians under Frederick, known among European armies for the most rapid rate of fire and heavy use of musketry were amazingly ineffectual because of the technical limitations of the grossly inaccurate unrifled muskets and inattention to careful aim, which was often justified because of the quality of their weapons. The French officer Rostaing reported the case of 2000 killed by 500,000 shots (250 shots per fatal casualty) in the battle of Chotusitz in 1742.[34] The other reason was a belief that by national temperament the French were vivacious and active, impetuous, and unsteady, unlike the Dutch and the English. This was a common claim; it is encountered in arguments by military engineers regarding the suitability of casemates in French fortresses. Hence, it was thought that French troops were superbly suited for the attack, and that this quality would best be exploited in columnar attacks.

The ordre mince had its defenders, especially among admirers of Frederick the Great, and their hand was strengthened by the defeats of the Seven Years' War. In 1774, the Baron de Pirch, a Prussian in the French service, drew up new regulations for battle order that emphasized the importance of long lines of infantry exploiting the firepower of this configuration.[35] But this did not settle the issue. Partisans of the ordre profond were not silenced and the issue was tested empirically with inconclusive results in large-scale maneuvers in Northern France at Bayeux and more importantly at Vassieux in 1778. The best French general of the Seven Years' War, Marshal de Broglie, defended the ordre profond, while Guibert engaged in a literary duel with Mesnil-Durand.

Guibert, while not wanting to rely on fire exclusively, called for better training of troops in fire principles with practice in loading, aiming, and firing weapons. Yet he was considerably more pragmatic than the systematist Mesnil-Durand, admitting the usefulness of columns in certain situations and calling for training for rapid deployment from column to line and vice-versa. The result was what was sometimes called an *ordre mixte*. Guibert was extremely influential in the group of officers around Saint-Germain and later, just before the Revolution, was the guiding spirit in the Conseil de Guerre set up in 1787 to advise the Minister of War on new procedures and field manuals.[36] It was these procedures that formed the basis of French combat practice when the Revolutionary Wars began.

In general, one can distinguish between a group of systematizers of the military art such as Mesnil-Durand and Pirch whose interests were focused almost exclusively on battle tactics. More pragmatic thinkers like Maurice de Saxe and Guibert, were also the ones who were more attentive to broader implications of tactics and thought about general political and social issues and how they could be related to military policies. Guibert in particular called for a more energetic and aggressive type of warfare that belittled both complexity of maneuvers on the battlefield and the importance of fortresses.

In addition to the more narrowly tactical matters of maneuver, the other primary question was on the ideal size of armies. Frederick the Great still fought controlling an entire army on the battlefield but armies were growing in size during the eighteenth century. In 1760 Marshal de Broglie was already experimenting with the idea of a division—a part of the army consisting of units of cavalry, infantry, and artillery that was capable of operating independently but in conjunction with other divisions and was strong enough to withstand the initial assault of an entire enemy army and keep it at bay until reinforcements arrived. Saint-Germain formalized the practice, and it was to serve Napoleon well in his wars.

It is difficult to determine where military engineers stood in these tactical debates. One is forced to rely on the opinions of a few, albeit prominent, spokesmen. D'Arçon defended the ordre profond when he clashed in print with Guibert in defending Mesnil-Durand.[37] Although he later established cordial personal relations with Guibert, d'Arçon seems to have maintained his views on tactics. One might expect engineers, members of the armes savantes, to emphasize the decisiveness of technology in warfare—a view that was generally associated with partisans of the ordre mince. Instead, d'Arçon endorsed a brand of tactics—the ordre profond—that depreciated the importance of firepower and saw battles in terms of the shock of columnar masses of troops using the bayonet. Besides the relative ineffectiveness of firepower, one cannot rule out the possibility that there was some affinity between those supporting what was allegedly a French mode of fighting with those supporting a French way of defense and fortification.

Military engineers' interest in tactics was welcomed, indeed encouraged and even demanded, by Guibert himself. He felt that with the growing importance of field fortifications, demonstrated during the Seven Years' War, engineers should know something about the subject when constructing fieldworks in conjunction with regular line officers.[38] Fieldworks had been a major concern of military engineers from the middle of the century when the chevalier de Clairac had published a book on the subject with the encouragement and support of Cormontaigne.[39] Lazare Carnot, a military engineer and a defender of Montalembert, was also a supporter of cold steel and shock tactics when he held his influential position of de facto military planner for the Committee of Public Safety during the Revolution.[40] In his book on the principles of fortification, commissioned by Napoleon, Carnot emphasized the importance of raw courage and stubbornness in defending fortresses.[41] He too called for a defense that would rely more on sorties by the garrison. Gone were the days when overly dogged resistance was considered ungentlemanly. If Louis XIV thought that "playing the hero at the wrong time is the way to becoming a laughingstock," Napoleon now wanted his troops to fight beyond what was reasonably possible and to display the kind of tenacity that he had himself witnessed in the Spanish defenders of Zaragoza.[42] The trend had already begun during the first sieges of the Revolution. The politicians in Paris had reproached their commanders, including and especially their engineers, for following the old rules of war and elegantly surrendering when the outcome was no longer in doubt and could only be staved off with a useless expenditure of blood and treasure. Officers who followed this line of reasoning found themselves summoned to Paris and accused of treason.

These views on the role of fortresses went beyond the merely mechanical aspects of tactics, where in the final analysis the views of military engineers could not have been particularly influential or relevant. Although in the eighteenth century such views were still subsumed under the name of tactics, they entered into the domain of what today would be called strategy. Was it possible to develop a science of winning wars that went beyond the mere science of the details of material equipment and its use? The study of battlefield formations and tactics had looked at the art of war from the limited perspective of the eighteenth-century battlefield but it later expanded to a broader vision of war and a science of war.

A Science of War?

It has been suggested that one of the reasons for Louis XIV's love of fortification was the alleged control and predictability it could bring into military operations.[43] But fortification was not merely a tamer of chance in war generally, it also served as a model for what one day could be accomplished on the battlefield. The regularity and (at least limited) predictability of operations was an ideal that field commanders could also strive toward. Puységur's faith that one could train a good general from books was explicitly based on the example of fortification.

> I propose . . . to demonstrate that without war, without troops, without armies, and without being obliged to leave home, by study alone, with a little geometry and geography, one can learn all the theory of field warfare from its most minor to its most important parts. And [I propose to do this] in the same manner as Marshal Vauban, by the theory locked in the books he left us and by the practice he established conformable to it, teaches us the art of fortifying, of attacking and defending fortresses, which is taught daily by people who have never gone to war nor managed the fortification of fortresses.[44]

Puységur's charitable attitude to mathematics' masters would not have been shared either by the majority of soldiers or military engineers. It displays, however, his firm belief in the possibility of a science of war based on principles that can be taught in the form of rules. In spite of an extended case study of army operations between the Seine and the Loire, Puységur falls in the category of the systematizers of warfare whose science Clausewitz would have considered as imperfect and even misleading if uncritically accepted. Yet Clausewitz too admitted that siege warfare had provided the germ of the science of warfare:

> Siege warfare gave the first glimpse of the conduct of operations, of intellectual effort; but this usually revealed itself only in such new techniques as approaches, trenches, counterapproaches, batteries and so forth, and marked each step by some such product. It was only the thread needed to link these material inventions. Since in siege warfare that is almost the only way in which the intellect can manifest itself, the matter usually rested there.[45]

Other writers were more skeptical on the possibility of creating a science of war. Maurice de Saxe (1696–1750), perhaps the best eighteenth-century military writer aside from Guibert, was also a great military commander who spoke from experience gained fighting for Eugene of Savoy in his youth and for Louis XV at the peak of his career. Unlike most eighteenth-century military writers (except for Guibert), he appreciated the importance of morale in warfare and advocated the importance of vigorous pursuit after a battle. His own successful campaigns in Flanders during the War of the Austrian Succession, which saw the capture of many fortresses, led him and many others to question seriously the efficacy of the existing permanent fortifications and the accepted manner of warfare.

Although Maurice had written his modestly titled *Mes Rêveries* in 1732, his book was published posthumously in 1757.[46] (Puységur's book, also posthumous, appeared in 1748.) Even the chevalier de Folard, who de Saxe admitted had attempted to go beyond blindly accepted tradition and custom, had failed:

> War is a science covered with shadows in whose obscurity one cannot move with an assured step. Routine and prejudice, the natural result of ignorance, are its foundation and support. All sciences have principles and rules; war has none.[47]

Maurice disclaimed any "desire to establish a new system of the art of war," but he went on to provide a comprehensive treatise that examined the details of army organization, battle formation, and armament, as well as the more "sublime" aspects such as field battles, mountain warfare, and sieges.[48] His distinction between the "principles" of war—the details of training, armament, marshaling on the battlefield—and the "sublime parts" of war, the operations of whole armies in different situations—is roughly that between those aspects that can be subjected to rules and those which depend on the genius of a general and are more difficult to prescribe.

Saxe is quite original in his thinking and impatient of the trammels of accepted practices. He proposes a host of new ideas—more comfortable and practical uniforms, new combat units (such as the legion), new types of weapons (e.g., a breach-loading carbine and a light cannon that could be wheeled by two men). He also proposes resuscitating some old ideas, including cavalry lances and body armor. Clearly he believes that war is not stuck in perennial patterns that admit of no improvement or modification on theoretical principles. In this sense he marks a major departure in military writing. Puységur's work was more a codification and rationalization of existing practice. But in spite of the freshness of his thought, much of Maurice de Saxe's writing is still limited to what Clausewitz would later describe as the first stage in the development of military theory:

> Formerly, the terms "art of war" or "science of war" were used to designate only the total body of knowledge and skill that was concerned with material factors. The design, production and use of weapons, the construction of fortifications and entrenchments, the internal organization of the army, and the mechanism of its movements constituted the substance of this knowledge and skill. All contributed to the establishment of an effective fighting force. It was a case of handling a material substance, a unilateral activity, and was basically nothing but a gradual rise from a craft to a refined mechanical art. It was about as relevant to combat as the craft of the swordsmith to the art of fencing. It did not yet include the use of force under conditions of danger, subject to constant interaction with an adversary, nor the efforts of spirit and courage to achieve a desired end.[49]

The works of de Saxe and Puységur appear to have attracted a considerable readership and become standard components of the libraries of all soldiers interested in a theoretical discourse on war. Saxe's *Rêveries* were reprinted three times in 1757 alone, and again in 1761 and 1763, along with two German and three English editions. Most eighteenth-century military authors, even when admitting the existence of a "sublime" part of war that escaped systematization by "principles and rules" nevertheless provided comprehensive treatises on war that saw it as a mechanical system, reminiscent in many ways of attempts to achieve the same result in natural philosophy. Like the Cartesian system, it possessed the attractive features of intellectual elegance and coherence. At the same time it left a sense, considerably more obvious in matters military than in cosmology and the natural world, of incompleteness. There was a certain artificiality, irrelevance, not a little implausibility, and an inability to take into account the active power of human passion in combat. Two figures whose military writings have been given special attention by Azar Gat provide excellent illustrations both of considerable originality in late-eighteenth-century military thought as well as the dead end which it had reached by the end of the century.

The Paroxysm of Military Rationality

Henry Humphrey Evans Lloyd (1718?–1783) was an Oxford-educated clergyman who eventually became a Russian general after having served in the Austrian, Prussian, French, and British armies, not to mention the Young Pretender's army in Scotland.[50] This romanesque character, who had gone on two espionage missions off the English coast for the French and wrote a detailed analysis of defensive plans of that same coast against a French landing for the British, is said to have received the compliment of having his lodgings ransacked for papers by British agents after his death in the Hague. On a number of occasions, he is supposed to have served as an engineer: with the Pretender's army and with the French in the siege of Bergen-op-Zoom.[51] His excellent geographical and cartographical talents seem to have been an asset in his career as a spy and to have conditioned his most original ideas on military affairs. A man of great intellectual breadth, he wrote on economics, demography, and politics as well as military matters. He was read attentively, and although Napoleon, one of his readers, thought his work an "absurdity," some of his ideas, such as the concept of a "line of operations," which he popularized if he did not invent, lasted for a long time.

Although Lloyd wrote on the organization of armies and petty tactics (where his ideas are merely variations of better-known French ideas on the subject), as well as military psychology (where he attempted to apply the hedonistic-materialistic sensationalism of Helvétius), his most original ideas were on the conduct of operations rather than the organization of armies. Seeing the army as a great machine and fortification and artillery as nothing but mathematics, Lloyd saw military operations in terms of lines on maps.

The first Cassini survey of France was completed in 1744 and the second, considerably more substantial mapping was launched in 1747. Among many illustrious members who supported the company set up by Cassini to perform this task was Montalembert. Josef Konvitz argues that the project was an important element of a major rationalization and reorganization of French administration of the transportation infrastructure that also saw the creation of the École des Ponts et Chaussées and the arrival of Daniel Trudaine at the head of the department of the Corps des ponts et chaussées.[52] By 1789, the project was completed, as was an extensive secret topographical atlas of Prussia and her neighbors in 1780, which was followed by the foundation of the Ordnance Survey in Britain in 1791.[53] Movements of armies could now be represented and planned more effectively on maps than ever before and Lloyd's "line of operations" is one of the first graphical-military concepts for field operations to appear on the map tables of generals.

As a soldier of the eighteenth century, Lloyd was aware of the cardinal importance of supply for armies that adhered to a civilized form of warfare and relied on depots rather than plunder for sustenance not only in victuals but also in the ever increasing supplies of munitions as firepower became more important. For primitive armies, like the oft-mentioned and much maligned Tartars, space was supposed to be neutral and it was asserted that they could move in any direction Nature permitted because they lived off the land, but for European armies space was not isotropic and lines of operations were conditioned by man-made logistical nodes. The good general was dependent on his supply depots and his strategy was always to keep his lines of operations protected and open while at the same time menacing those of his enemy. War thus became strategic maneuver and battle was often forgotten. The Austrian army in particular was fond of this concept of military operations, and it may have been this that excited Napoleon's contempt for Lloyd's ideas. But for Lloyd and a significant number of eighteenth-century generals who thought like him, war was tamed by mathematics and cartography and if properly conducted was a clash of minds with less effusion of blood.

Like Lloyd, the Prussian officer Adam Heinrich Dietrich von Bülow (1757–1807), brother of the von Bülow who distinguished himself in the Napoleonic wars, was equally peripatetic, fascinated by geometry, and had a strong whiff of the bizarre in his character. Politically radical and francophile to the point of treason, von Bülow developed a system of military thought that was a culmination, indeed an exaggeration, of the mathematical-mechanical streak in eighteenth-century military thinking.[54] He died in prison after being arrested by his government and declared insane. One of Clausewitz's first works was an anonymous and highly critical review of von Bülow's work, which he reiterated in his treatise *On War*.[55] Von Bülow's mathematical hobby horse, which he rode in his map games while sublimely ignoring anything but geographical factors, was the "base" of magazines and reserves. The word 'base' is quite explicitly taken from geometry: it is the base of an imaginary triangle whose vertex is the military objective under attack and whose area consists of the supply routes safely controlled by an advancing army. The enemy must aim to attack the sides of this triangle to penetrate to the rear of the attacking army, cutting it off from its "base." The optimal shape of this triangle from the point of view of the attacker is determined: the narrower the triangle the more vulnerable it is to enemy attack. Therefore the minimal vertex angle is 90° and preferably the angle should be greater, an obtuse angle.[56] Drawings accompanying the analysis reveal a military Euclid who announces the triumph of reason over war: "War will be no longer called an art but a science . . . everyone will be then capable of understanding and application" and "the sphere of military genius will at last be so narrowed, that a man of talents will no longer be willing to devote himself to this ungrateful trade."[57] The result would be the elimination of small states with limited resources leading to larger, more powerful and invulnerable states, which would make war unviable and lead to perpetual peace.

Von Bülow's predictions coincided with the rise of Napoleon, surely a "man of talents" in the military domain, and war was becoming fiercer than ever before. Lloyd's and von Bülow's tidy map-room warfare, which had never really existed in anything approaching the pure form they postulated, was disappearing rapidly. Its end had been predicted by Guibert, a decade before Lloyd's writings and almost a generation before those of von Bülow.

Jacques-Antoine-Hippolyte de Guibert (1743–1790)

Although Guibert came from a military family, had been trained by his father for a military career since his childhood, and had had some military experience, he died just before the Revolutionary Wars, which exemplified many of his ideas and justified

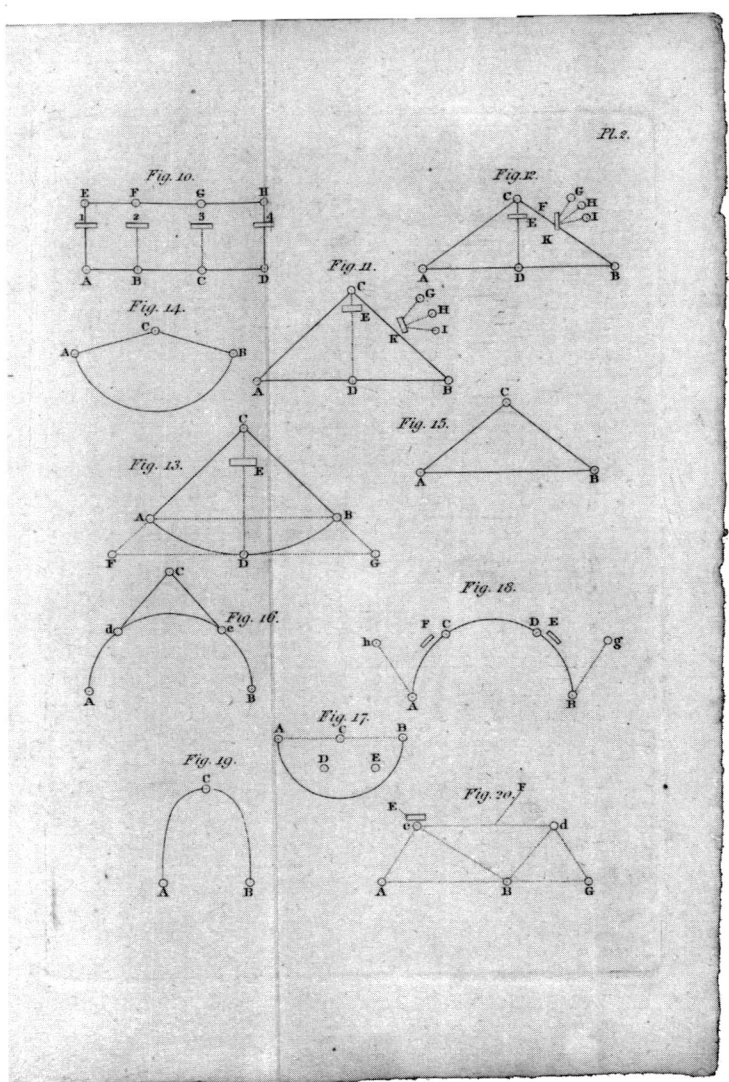

Figure 8.2
This illustration in von Bülow's work on geometrical warfare shows the concept of the base of operations. For a successful attack, the angle ACB must be greater than 90°. C is the object of the attack; AB is the base of operations; E is the attacking army; G, H, I, and K are the defender's field armies and fortresses. Source: Bülow 1806.

many of his fears. He is therefore remembered not for his military career, as was Maurice de Saxe, nor even for his work as an administrator at the Ministry of War reforming the French army in the last day of the Old Regime, but for what he wrote. Handsome, dashing, and confident, lion of salons, lover of Mademoiselle de Lespinasse, and precocious man of letters, Guibert even attracted the admiration and sympathy of the old pacifist Voltaire. A member of the Académie française after 1785, Guibert had literary talent, which manifested itself not only in his plays but also in his controversial works on military subjects. He was truly, in his own words, a "soldier and philosopher." The work that made his reputation, the *Essai Général de Tactique*, appeared when Guibert was only 27 years old. Although Guibert was a constitutional monarchist, a revolutionary tone suffuses the book beyond individual passages such as the one where he hoped for a "prince who will bring about this great revolution."[58] The *Essai* is not dedicated to a noble protector or an influential personage, but to "my fatherland (patrie)," and Guibert talked of "citizens" who will form the armies of the future and revolutionize warfare under a leader of genius and audacity who will break the old bonds of tradition and "leave behind him all the so-called barriers and carry war into the interior of states, to the capitals themselves."[59] No doubt he saw himself in the role that his early death prevented and Napoleon would one day assume. Like Maurice de Saxe and others before him, Guibert repeated the now standard Enlightenment complaint of writers on military science—that there was no military science. Although the subject of war was changeable, unlike mathematics, he felt nevertheless that it should be studied. Interestingly, from a man who was a prophet of the nation-state in arms, Guibert felt that differences in warfare due to national character, although real in earlier times, were being attenuated, and that as Europe became more homogeneous Europeans would become more similar in their ways of fighting. A universally valid science was possible. Military science for him obviously fell in the category of those "arts and exact sciences" whose "progress arises from discussion and even errors, while in the case of ethics, metaphysics, and all the sciences of opinion, [their] writings only augment doubt and ignorance."[60]

Didactic works such as those of Puységur and Folard, however, were judged useless by Guibert. Although he left the outline of a major work where he would examine the military and political systems of Europe and continue with "complete course on tactics," he limited himself to a "general essay" on tactics, which was never followed up by the complete work he promised.[61]

Clear from the outset and all throughout his book is Guibert's desire to innovate and the belief that innovation is possible and desirable in military affairs. According to him, the previous works had spent too much attention to describing and codifying practice in the implicit belief that war had naturally evolved to perfection and that the science of war involved rational description whence rules and precepts would automatically flow. The similarity of "military constitutions" in Europe were not proof of such an evolution and a sign that these constitutions were based on unshakable and common principles. Rather they were the result of indiscriminate imitation of practices that were based on routine and unthinking tradition.

A problem of central importance for Guibert was to reconcile the humane tradition of the Enlightenment and its abhorrence of war with the value he himself attached to the subject. War was bad because it interrupted the "communication that should exist among peoples," yet if it did come "let not the reprisals based on the laws of nature [that are] violations of the so-called laws of war be called barbarism."[62] War, fought as it should be, would be terrible, and Guibert's new science would only make it more terrible. Nature, it seems, did not always have the smiling face some philosophes imagined it to have, and it had rights that deserved respect equal to those accorded Humanity. Like the Chevalier d'Arcq, Guibert was against luxury, which some Voltaireans saw as a positive and civilizing influence, and saw it as having a detrimental effect on the martial capacity of a nation. Unlike Condorcet, who hoped that technology in the form of gunpowder weapons would give civilized Europe an advantage against the more primitive and physically vigorous peoples that perennially threaten sedentary civilizations that have lost their will and capacity to resist barbarian threats, Guibert put his trust in moral factors to stop the hitherto inevitable cycle of degeneration of states. Civilization did not necessarily bring military decadence. Proper selectivity of the civilized values and the maintenance of primitive virtues could lead to powerful and invincible armies. Alas, Europeans had put their faith in such things as fortresses, a symptom of weakness and political decadence. Thus, he regretted the destruction of a military aristocracy. He felt that Russia, because of its barbarism, had had an opportunity to become great, but had lost it because of Peter the Great's adoption of the wrong kinds of civilized values, which led to luxury and decadence.[63]

But knowledge was not necessarily bad and enervating for a soldier, even though there existed in practice an inverse correlation between the arts of civilization and military progress:

It was not the arts and sciences that made the military art of the peoples of antiquity decay; it is not the arts and sciences that prevent it from progressing today. General enlightenment should on the contrary improve this art along with the others. It should make tactics simpler and more scientific, troops more knowledgeable, generals better. It should put method in the place of routine, combinations in the place of chance. If, while all the other sciences are being perfected, that of war stays in its infancy, it is the fault of governments who do not attach enough importance to it, who do not make it an object of public education, who do not direct men of genius toward this profession, who let them glimpse more glory and advantage in the frivolous or less useful sciences, which make the career of arms a thankless career in which talents are outstripped by intrigue and prizes distributed by luck.[64]

What would be necessary to maintain the humanizing benefits of civilization and the defensive virtues of barbarism is a revolutionary transformation in the entire political system. It was necessary to create a new science of morals and politics, which would be encyclopedic, not a science of details but of relations and basic principles. Guibert saw an organic relation not only, as did Clausewitz, between politics and war, but also between politics and morals. The government not only had a duty to diffuse enlightenment of the proper kind but also to maintain morals. In words that would have found ready resonance with the poor rural and military nobility resentful at seeing venality of commissions being used by wealthy commoners and nobles of the robe to annex their traditional career of arms, Guibert called for a prolific warrior caste—"a military nation in the bosom of the nation."[65] Maintained by the state, it would live and train like Spartans. Its members' physical hardiness, courage, patriotism, and relative independence of supply trains would make them better and more mobile soldiers. They would move more quickly than other armies and ravage the land they crossed. The wars they fought would be different:

> To look at that [eighteenth-century wars and particularly the campaigns in Flanders during the War of the Austrian Succession] from the point of view of philosophy and humanity, it might be lucky that, either [from] the effect of fortresses or established routine, wars are thus waged in little operations, alternate fortresses taken and retaken, instead of conquering and ravaging, as they did formerly. But looking upon the military aim, the art of war has undoubtedly lost from this, since its effects are less grand, since ultimately they do not fulfill the first and unhappy purpose that they should have—that of doing the greatest possible harm to the enemy and promptly deciding the quarrels of nations.[66]

Here was a principle that ran directly counter to Montesquieu's injunction, invoked by the engineer d'Arçon, that in peace nations should attempt to do as much good to each other as possible and in war they should attempt to do the least possible damage to each other.[67] It also ran counter to the argument of fortifiers and engineers who argued that fortress warfare—ruinously expensive—led to shorter wars because states could not afford protracted wars of that kind; indeed, they would avoid them in the first place for the same reason.

Guibert and Fortifications

Guibert's emphasis on moral values and morale makes his skepticism on the decisiveness of artillery and fortifications understandable. He felt that there was too much artillery and too many fortresses. These slowed down an army on campaign and led to attitudes of timidity that undermined the offensive and ultimately made nations more vulnerable militarily. Guibert was annoyed at the presence of both Fourcroy de Ramecourt and Gribeauval on the Conseil de Guerre, feeling that they were unnecessary in discussions of most general military matters and that they were not particularly competent in their own specialties.[68] He succeeded in badgering Fourcroy de Ramecourt to agree to the decommissioning of a number of secondary fortresses, much to the indignation of most of Fourcroy de Ramecourt's subordinates in the Corps of Military Engineers. If Guibert became more sensitive with time to the importance of artillery, he was consistent in his stand on the uselessness of most fortifications.

He was prepared to concede that Vauban and Coehoorn were superior to the fortifiers of antiquity and the vast majority of their contemporaries: their "knowledge was founded on the considered concourse of all the branches of mathematics. The others, deprived of geometry, were miserable routines."[69] Moreover, Vauban and Coehoorn had perfected the method of attack on fortresses—a view that Montalembert was to repeat indefatigably—but they had not improved the defense of fortifications. They had even fallen behind. Why? Possibly it was the decline in courage that fortification fostered or ignorance of the principle that the best defense is offense. For Guibert, the only correct way to fight a war was offensively. The mentality of the defensive was always bad and besiegers had always won over the besieged. Even worse, Vauban and Coehoorn had imparted great and undeserved prestige to siege warfare by their personal exploits. The "spirit of imitation and mania" and the genius of the great fortifiers had "swept along all opinions."[70]

In any case, fortification was merely an auxiliary science and had played too large a role in warfare. Fortifications were symbolic of the trust in material barriers rather than moral forces, but the ultimate basis of military excellence was a good army composed of good soldiers. He rejected the idea of fortified lines and the argument that Vauban's fortified lines in Flanders had saved France during the critical period of the War of the Spanish Succession: if the French army had been good, there would have been no need of desperate measures to save France. The argument that fortresses prevented devastation of the countryside by invading armies was also specious: they only did that for today's armies; the armies of the future as envisaged by Guibert would not be deterred by them. Because of fortifications, "wars have become more ruinous and less scientific (savants)": army size had increased because of the necessity of defending fortresses in wartime and garrisoning them during peacetime. "In countries covered with fortresses, like Flanders, war took on a character of routine and softness, which is certainly not that of genius."[71]

To add insult to injury, Guibert contested the etymology of the word 'génie', arguing that it came not from 'genius' but from 'engines'.[72] Vauban's label for his art was alleged to assimilate engineers to the dull mechanics they really were and not the creative artists they thought they were. Genius is what was needed by the good officer and it is something that was both innate and something that could then be sharpened by study and experience. Above all, it was mental and it was difficult for genius to thrive in the army:

Everything concerning genius and generals is so delicate to discuss! Everything concerning genius necessarily has a systematic imprint, and announcing systems is already putting yourself against an infinity of people who condemn them without reading them. It is even more dangerous to reason about [things] concerning generals.[73]

Although Newton could attack Descartes without fearing accusations of blasphemy, it was more difficult to have a true intellectual debate in the army because the military profession was fearful of genius and respected seniority.

Was there no place for fortifications and military engineers in Guibert's scheme of things? There was: the art of field fortifications. It was an art purely mechanical, but which engineers should not be allowed to practice without a knowledge of tactics, and it could be acquired by a regular line officer in 6 months at a good school. Even so, field fortifications were be minimized for they went against the spirit of offensive warfare. Both engineers and tacticians should learn about each others' respective specialties. But, he complained, engineers had not accepted the view that they should know about tactics.

In fact, Guibert did have a policy on fortifications that gave them a limited but integral place in his scheme of war, which is similar to that of Napoleon.[74] Fortresses were to be kept, but they were to be both far fewer and far stronger. They should be "places d'armes, points for stores and support, bastions whose curtain is a good and maneuverable army."[75] And temporary fortresses [places du moment] would be built during a campaign. Ultimately, however, it was the fate of armies that decided that of fortresses, and fortresses were clearly seen in an auxiliary role to highly mobile field armies.

Guibert's voluntarism and emphasis on morale and physical hardiness led him to downplay the importance of topography as well as fortresses. In fact, topography and terrain would be less important than they were now if armies were more aggressive and mobile. While praising Bourcet and his topographical staff work, he argued that this was only the initial basis of a theory of reconnaissance. Obviously skills in reconnaissance were vital to any commander, but topography and cartography by themselves would not provide it. The best way to acquire this art was constant practice from youth and participation in sports and travel. The real knack of reconnaissance was not merely being a good topographer; it was a "present from nature, and the instinct of genius."[76] The correct way to do reconnaissance was not only to take into account the terrain but to be able to envisage the movement of troops over this terrain. Like many others, Guibert believed that the "science du coup d'oeil" belonged to rare geniuses.

The author of the *Essai Général de Tactique* was clearly the most original military thinker of the eighteenth century, and he was remarkably prescient regarding the kind of war that would be fought in the future. If he lacked the sophistication (and the frequent obscurity) of a Clausewitz, he is one of the few who shared a common awareness of the importance of morale and politics in waging war. The blending of Enlightenment values and raw primitivism was skillful. Here was a writer who believed in vigorous free trade and communication between peoples, thought France had reached its natural boundaries and should avoid territorial expansion, believed in the positive value of lumières (enlightenment) for a society, and also admired a military aristocracy, acknowledged the virtues of the barbarian horde, and believed that war should be waged without mercy or respect for the law of nations.

Guibert's science of war concentrated on tactics and downgraded technological features like artillery and fortification. He was never really clear, in spite of emitting bloodthirsty noises, about how to integrate the all-important moral aspects into his science of war, nor did he ever, unlike Clausewitz, seriously try. Although mobility is a cardinal feature of his idea of war, he paid only limited attention to the terrain over which his mobile armies would move. For Guibert, the science of war was not the science of dominating space, as it was for du Portail, the military engineer who was closest to his way of thinking: it was the science of harnessing human energy to wreak maximum destruction on an enemy.

It is difficult to ascertain to what extent Guibert's ideas were shared by serving officers in the French army, in spite of their apparently widespread application in the Revolutionary and Napoleonic Wars, but it is clear that he was widely read and a focus of active controversy. Obviously, his ideas on fortification were of particular interest to military engineers.

Louis Le Bègue du Portail (1743–1802)

It may be an oversimplification to characterize the debate between Guibert and the engineers as a clash of opposites, between proponents of a more modern offensive warfare and a more traditional defensive warfare. Yet it is true that Guibert's writings reflect major differences between a significant group of regular line officers and military engineers. They reflect a belief that there is a radical difference between the offensive war of movement and the defensive and static warfare supposedly espoused by the engineers. Moreover, they are a symptom of the persistent disdain by regular line officers for engineers that is confirmed by the defensive tone engineers often adopted when asserting their claim to be true officers and the long delay it took for the military engineers to get their own Corps of Military Engineering designated as a royal corps in spite of innumerable proofs of their valor. The desire to be genuine officers often went so far as to downplay the importance of their own technical achievements and capacity. This comes out on a number of occasions.

The most striking is in a memorandum written for the Minister of War by du Portail, who has already been discussed in connection with his views on the importance of "details."[77] In this memorandum he promoted a new science of engineering that was at the same time a science of war. Du Portail's science of engineering was a far cry from Bélidor's. Du Portail denied the fundamental nature of the rift between engineers and regular line officers as well as the distinction between the radically opposite natures of offensive and defensive war. While du Portail's opinions were his own, the presence of a number of copies of his memorandum in the archives, and the resonance of his ideas in other contexts, suggest that they represented a point of view that was popular among engineers, especially among the younger ones whose recruitment was more carefully restricted on the basis of noble origin. It also indicates another aspect of the malaise from which the military engineering corps was suffering around the time when Montalembert began his attacks on it.

In 1773, when he wrote his memorandum, du Portail, who had graduated from Mézières in 1765, was a captain in his corps. He spent some time in the service of "eastern powers" before leaving for America to become George Washington's aide-de-camp and the head of American military engineers in their War of Independence, which had ended with the siege of Yorktown under his direction. After returning to France with the rank of major-general in the American Army after the peace of Versailles, then spending a few months in Naples, he transferred to the infantry with the rank of maréchal de camp. During the Revolution, du Portail would serve Louis XVI as

Minister of War from October 1790 to December 1791. Although he had authorized soldiers to participate in politics and join political clubs, he became suspect to Revolutionary politicians and was forced to flee France and exile himself in America after the downfall of the monarchy. Struck from the list of émigrés after Napoleon came to power, he died en route to France from America.

In short, du Portail was not a typical military engineer. After transferring to the infantry in 1788, he was not even an engineer. One suspects that he was discontented with his corps, in which promotion was slower than in the other branches of the army in the years before the Revolution. Yet neither the ideas nor the tone of his memorandum were abnormal or far-fetched: discontent was a widespread sentiment in his corps.

What was the cause of this restlessness? One thing seems clear: du Portail did not like being called an engineer. In a section of his memorandum called "On the title of Engineer," he writes

This originally honorable title has been prostituted to so many people that it is almost debased. The builder of roads and bridges, the geographer, the shipwright, the machinist, the instrument maker, etc. have taken hold of it.[78]

For du Portail, this was unacceptable. All these people with different occupations, who are not only the engineer's social inferiors and have usurped his title to bask in his prestige, gave a false idea of what the true engineer did. He wanted to forbid the use of the title but, realizing that this would be unrealistic because the public had become accustomed to the present usage, he suggested that military engineers be called "officiers du génie" (officers of military engineering). Thus there would be no confusion about the true functions of military engineers.

What were these functions? The answer to this question took up most of du Portail's memorandum which was divided into two parts—"the object of the science of engineers" and the "constitution" of the corps and its relations with other parts of the army. The latter subject required that one have a clear idea of the former—the true essence of the military engineer must be understood before one could talk about organization and relations with others. This had not been understood, and du Portail set out to remedy the situation. That his "science of engineers" was not that of today's engineers is perhaps not surprising, but it was also not the science of engineers as envisaged by Bélidor, Coulomb, Prony, and Navier.

With the approach of an encyclopedist, du Portail attempted to find the links between his science and other sciences. One of the capital points of his argument was that ignorance of the true relations between the sciences disfigured the military sciences as a whole. He believed that the science of engineers was truly a military science, composed of tactics and fortification. The aim of the military art was "not to leave to chance or moral causes alone the outcome of a battle."[79]

Contrary to many soldiers, including military engineers, du Portail saw continuity rather than a radical difference between what he felt were these two intimately linked branches of the military art[80]:

> One sees . . . that everywhere defense is linked to attack, that if one wants to attack with a decided superiority on one side, one must necessarily reduce oneself to the defensive on the other and that this defensive must be all the stronger by the advantages of the terrain since it has lent (so to speak) more men for the attack.[81]

Even in pitched battles, terrain often made it necessary to resort to field fortifications. One always tried to use as effectively as possible the obstacles that nature itself gave us. Improving these natural obstacles and creating artificial obstacles to strengthen defense was the aim of fortification. Without a good defense, one could not mount a good attack and as much in strategy (du Portail does not use the word) as in tactics there was always a natural combination of mobile armies and fortresses, of attack and defense. The ideal warrior should thus be both tactician and engineer. Invoking the example of antiquity, du Portail asserted that the Roman legionary had used the spade as much as the sword and that Caesar had been equally good as an engineer and as a leader in battle.

Since the invention of gunpowder, however, war had become more terrible and more complicated. Rare were those who could be both good tacticians and good engineers; specialization was necessary, and no longer would superficial knowledge of a few general principles suffice:

> *To know roughly;* there's one of those expressions invented by ignorant presumption. The world is thus full of people who know everything roughly, and who, if one considers it well, know nothing at all. Perfect ignoramuses are a hundred times more preferable, at least they have no pretensions, they take advice, but our *rough savants* have no doubts on anything. They are the flail of the arts and sciences.[82]

One could no longer do without specialist engineers in planning military operations. The science required was not only complex and mathematical; it was also in large measure tacit and not accessible to the uninitiated by purely formal instruction. Assuming the tone of a secretive craftsman instead of an enlightened enemy of secrecy and tradition, he addressed himself to "those people who are not of the craft [métier]." They would never be able to acquire the knowledge they need exclusively from books. There was a reason for this:

> . . . there circulates in the corps [of military engineering] a traditional science that is given to one another, but which does not pass to outsiders. This science of traditions is the result of the cogitations and of the experience of the whole corps since it has existed. It is the assemblage of

the most profound reflections, the most ingenious ideas, the most subtle observations. It has never been written down and it is good that it has not. One does not find it in books, so that he who only learns fortification in books, only learns the gross elements. . . .[83]

There was therefore a collective memory of the corps, rich in the experience of its individual members both living and dead, which one assimilated by osmosis in doing one's daily work. These arguments would recur again in the polemic of military engineers and Montalembert. No matter what experience any individual had acquired in fortification he could never hope to match that of the corps of military engineering. Thus there was a technical knowledge, sometime bought with blood, sometimes more somatic than cerebral, which the engineer, like the artisan, acquired with his hands and his work, which modified and tempered scientific knowledge and made his knowledge superior to simple bookish knowledge obtained from "books on fortification," which were "very good for giving a false idea of this art."

In war, one must have a specialist—an engineer, and not a rough savant—participating in the command of armies to take care of the indispensable complement of tactics. In modern warfare the science of these specialists was a "complicated science combining geometry, the use of weapons, tactics and many other things. . . ."[84] Citing Puységur, du Portail mentioned that geometry was also the foundation of tactics.[85]

But the importance of mathematics in general and geometry in particular had led to fundamental misconceptions about the work of engineers. Knowing that geometry was a fundamental element of his science, familiar with his role as a builder of fortresses, and seeing the elegant ground plans in the books on fortification that du Portail despised, the layman had come to the erroneous conclusion, so dangerous for the prestige of the engineer, that he was nothing more than an architect. Maybe this was so in the past but now the engineer was primarily a soldier. The engineer was not just a fortifier, and fortification was not just construction. It was actually part of a science of military space—something fundamental to military science where war took place in space and one needed to understand and exploit this space for victory:

> The most difficult part of the art of fortifying, that which truly requires talents and experience is applying fortification to the terrain. To grasp rapidly, at the sight of any place whatever, the fortification of which it is susceptible, to present its effect and to be able to foretell its strength—it's certainly only the very skillful engineer who can do that. And the grasp of him who is only a little versed in the theory as in the practice of this Art is far short of this.[86]

Here du Portail expresses a view clearly put forward by Vauban himself: the strength of fortifications does not come from adherence to some system but from adjusting them to the constraints of the natural site. The engineer as topographer, capable of reconnoitering and studying terrain, should have the upper hand over the builder whose almost mechanical functions could be left to the architect.

This is far from Bélidor and Coulomb. It is much closer to another concept that occupied a prominent position in military writings of the time—the concept of the *coup d'oeil militaire*. All writers of the time considered the coup d'oeil—the penetrating and synthesizing glance—to be the supreme quality of the good general. For many, the coup d'oeil—the capacity to judge terrain serenely and rapidly in the chaos and disorder of the battlefield—was an innate gift of nature. It was de Saxe's "sublime" part of the art of war and not reducible to rules. But du Portail argued that one could learn and develop this capacity, and that it was precisely the engineer who, by his training and his natural functions (as defined by du Portail) was best placed to acquire it.[87] Therefore, the engineer, who was trained to master the terrain intellectually, was as useful in attack as in defense.

Engineers should therefore not be excluded from the general staff as they had been by a scandalous formal order, but should be admitted to it as indispensable members.[88] Du Portail denied any attempt to acquire for engineers the exclusive right to be on the general staff, but he did go so far as to say that "the corps of military engineering is the only one whose members must be considered generally capable of becoming skillful soldiers and can be employed with confidence."[89] In spite of his denials, he undoubtedly saw them as the essential officers of the army.

While du Portail demanded the right of engineers to serve on the general staff, he wanted to deny the right of artillery officers to position batteries during sieges, something commanding generals were allowing more and more. Sensitive to the possible reproach that he was being inconsistent in asking that engineers be admitted to the high command of armies on the grounds of their specialized knowledge and denying a right to another corps that was also a corps of specialists to do what might at first glance seem a natural function, du Portail argued that artillerists were mere technicians in the narrow sense of the word. While the engineers were generalists and theoreticians, applying their science to specific situations, artillerists were people who mechanically exercised a trade whose effects were always independent of the situation where it was employed. If one accepted the demands of the artillerists, why not accept the demands of common fusiliers, who are also mere manipulators of an instrument, the same rights to participating in the high command of military operations?

This vision of the engineer as the martial theoretician par excellence thanks to his geometrical knowledge and his topographical experience led du Portail to renounce a hotly pursued demand of most other engineers: the right to command troops during peacetime. In rejecting the amalgamation of the artillery and engineers (something that occurred for a short period during the Seven Years' War and was then countermanded to the relief of most engineers who had vigorously rejected the idea), du Portail said that one of the great drawbacks of this amalgamation would have been that a relatively small number of engineers, all officers, in the new corps, would be obliged to occupy themselves with details of administration and command of ordinary soldiers of the artillery corps.[90]

That would simply not work in view of the "repugnance which people who work with their head have to devoting themselves to purely practical matters," and it would only be a burdensome distraction for those who "meditate on their art and try to discover its principles" in their study "away from camps."[91] With the exceptions of sappers and miners, unsuccessfully demanded as troops by the engineers from the time of Vauban, du Portail did not want to occupy himself with the common soldier. Even the sappers and miners, he recommended, should be mostly left under the control of noncommissioned officers in most peacetime situations. Having the right to command troops was a right the engineer should have in order to affirm his military rights as an officer, but du Portail did not want to have much to do with the actual details of commanding them.

Following this train of argument, du Portail also rejected another demand that had been voiced by Coulomb and other eminent military engineers: expanded responsibilities in the construction of bridges, canals, and roads. The need to keep military engineers active in peacetime was all the more acute at a time when the Corps de Ponts et Chaussées was successfully taking over more and more of these functions even in areas where military engineers had traditionally operated, such as naval bases, military roads and canals.[92] No, said du Portail. One must abandon all this because the military engineer is primarily a soldier and must always meditate and calculate war and terrain. These construction activities would be distractions.

People had acquired a false idea of the true functions of engineers because of what they had done in the past. Like Vauban in his famous eulogy of the art of the engineer, du Portail deplored the scission between the "trench" and "fortress" engineers. There must be amalgamation; however, unlike Vauban, who felt that the most difficult part of the military engineering was construction and this was the aspect that should be emphasized in the education and recruitment of engineers, du Portail felt that the engineer's role in combat was central. This should be developed and it was for this that the engineer should be educated. The right kind of military construction, he felt, directly derived in a rather straightforward and mechanical way from the engineer's qualities as the intelligent warrior of modern times.

Because of the distinction between the sapper-engineer and the constructor-engineer,

It's been believed that the engineer's craft is a double one, that service in peacetime or in fortresses was different from field service, that one could be fit for one and not for the other.[93]

In fact,

The war service of engineers is to attack and defend fortifications; that which is called the service of fortifications is to make new ones. I [can] only see there absolutely the same art. He who knows how to make the best fortification design for any place whatever is apparently he who knows best the effect of each part in particular and who knows best how to combine the parts among themselves and with the terrain. It will be he again therefore who, if you give him a fortress to attack or to defend, will have grasped the relationship of the works, the influence of them on each other, the strength they lend each other mutually, etc.[94]

The construction of works was therefore a subsidiary function of fortification, and the engineer should not be too involved in it. That should mainly be the affair of architects and draftsmen, with whom he should be careful not to be confused. The construction of fortifications was "basic fortification" [fortification grossière]; the engineers knew enough about it to manage the work and leave the details to others.[95] Du Portail deplored the present situation, in which engineers were obliged to draw up plans for various buildings on the grounds that this would help them to learn their profession. He asked: "But tell me a little then what relation there is between the science of fortification and these miserable details."[96] A few pages later, he wrote:

... if the military engineering corps of France is not careful, let me be permitted to say, if it perseveres in burying itself in those miserable details that it calls the service of fortresses [service des places], it will have soon lost that recognized superiority that it has had until the present over all those of Europe—a superiority which it owes only to the fact that it was created before them and which it can only maintain in always keeping ahead of them in the establishment of the proper laws for training skillful soldiers.[97]

One must never forget, he wrote, that *"It is the art of fortifying* that constitutes the engineers and not the *art of construction."*[98]

Once again it is clear that the science of engineers of du Portail was not the science of engineers of Bélidor. But it was the science many engineers practiced in the early Revolutionary and Imperial wars. The science du Portail proposed left many questions regarding its details, but its very vagueness clearly shows that almost a century after Vauban had coined the word 'génie' the concept was in a state of flux, even among military engineers. It was in this situation that they had to face the onslaught of Montalembert, a well-connected and outspoken amateur, who questioned their science and their competence to practice it. The skirmishing over the frontiers of their science was brought on as much by a defensive reaction as by any hegemonic visions. The imperialistic tone of du Portail betrays an underlying insecurity. In spite of proud traditions of their corps, their superb training, and their proven valor on the battlefield, engineers were threatened overtly and more subtly in their status, in their claims to competence, and even in their identity by fellow soldiers and outsiders. This was all the more unsettling in an environment where society was becoming politically turbulent and the army was questioning its own mission and the traditional means of fulfilling it.

9
Science and Military Engineering

. . . at the time *La Science des ingénieurs* appeared, builders, ordinarily guided by blind practice and mostly unfamiliar with mathematical knowledge, were reluctant to rely on the principles that these sciences furnished to the art of building. It was therefore necessary, to give some prestige to the theories that learned men wished to establish, to present them as free of uncertainty and to thrust aside as much as possible [the uncertainty] that arises from the impossibility of representing physical circumstances exactly by means of calculations. But the study of mathematics is now so widespread, and its application to physics has been so perfected, that it has become on the contrary necessary to caution builders against the drawbacks of a too absolute confidence in the results to which it leads. The question of soil mechanics, in particular, no matter how exact its analysis presently is, is not much clearer, as to the practice of construction, as it was in the time of Bélidor.
—Navier, in his notes to Bélidor[1]

Of the two new sciences discussed by Galileo in his epochal book of 1638, historians of science have concentrated on the second science, the science of kinematics of objects moving in space.[2] Yet the first science, which was essentially a science of the strength of materials, was equally revolutionary, and from the viewpoint of the engineer it was central. Galileo was keenly aware of the problem of scale-up, which was and has always been fundamental for engineering. Galileo knew that maintaining isomorphism when expanding scale could lead to structural failure.[3] This was well known to engineers and was one of the primary reasons for engineering's conservatism. A traditional technology relied on the proven and the tested. Failure, always costly and dangerous, was and to some extent still remains an excellent teacher to discover the limits of a certain design. But understandably builders were reluctant to resort to it deliberately in a systematic fashion; they usually encountered it unwillingly on occasions when they summoned sufficient courage or arrogance to violate the traditional injunctions of their craft.

An archetypal distinction between the modern engineer and the traditional craftsman is the former's attempt to resort to intellectual manipulation of analytical principles to predict the behavior of objects in situations where no previous experience existed. It can be said that Galileo's clear conception of the problem and the possible solution, relying on the mathematical analysis of the properties of materials, inaugurates this modern tradition. An example of this is Galileo's classic analysis of stresses in a cantilevered beam, which provided an exact quantitative rule for what had been only intuitively suspected earlier: the resistance of a beam to breaking was in direct proportion to its thickness and not its overall volume. Although Galileo made a slight slip in the actual ratio, which was corrected by Mariotte some 30 years later, the essential idea was correct and was the model for similar attempts to analyze the strengths of materials. Similar attempts to analyze the laws of nature as they manifested themselves in the artificial materials and the constraints imposed by the technological activities of humans occurred in other domains such as hydraulics and soil mechanics. Several members of the French Academy of Sciences, including Mariotte, Parent, and Pitot, were active in this area. They made experiments to ascertain basic constants for different kinds of materials and to verify some of the theoretical derivations. Later mathematicians (including Euler and the Bernoullis), fascinated by the complexity of the problems, continued this work.

The non-homogeneous nature and complexity of materials that served the engineer made the problems of analysis difficult and solutions were few and often erroneous. The whole domain remained recalcitrant to the efforts of the scientific engineers and the mathematicians with inclinations to practical problems until there was an improvement both in mathematical and materials testing techniques as well as the generation of adequate quantities of experimental data. Theories of elasticity put forth by Navier, Cauchy, Poisson, and Barré de Saint Venant in the nineteenth century finally put the science that Galileo had foreseen on secure foundations.[4] It is not accidental that historians are better aware of Galileo's work in kinematics, where simplifying assumptions could be made more easily to permit the theoretical analysis that made spectacular headway with Newton and Euler. These spectacular results were a constant source of emulation and justification for the scientifically inclined engineer who struggled with more intractable and uncontrollable matters. Most engineers either labored on in the tested tradition, often in ignorance of the work of the Galileos and Mariottes, or abandoned the attempt. Of those who did not, the most outstanding was Bernard Forest de Bélidor.

Bélidor's Science of the Engineers

Bélidor's *Science des ingénieurs* and *Architecture hydraulique* are great classic engineering texts. They are quite distinct from other texts of the time addressed to engineers and architects, which concentrate on global design. Bélidor's texts deal with the details of design and construction, and the primary approach is from a theoretical and analytical point of view. For the time, this made them unique and even revolutionary outside the forum of the Academy of Sciences. Yet there is a certain ambiguity about them: at the same time that Bélidor preaches with great eloquence a gospel of a new kind of engineering, engineering tradition is respected and codified. Bélidor is too aware of the functions of his profession and the limitations of his colleagues to call for an outright severance with traditional practice. This pioneer of engineering science knew that there was more to engineering than science. As a practical engineer he would argue that the proof of the pudding for engineers ultimately boiled down to a completed project and the details of reporting, estimating, and tendering were what ultimately counted for successful completion.[5] Even while calling for a new kind of engineer with a new kind of intellectual tool kit with all the fervency of a prophet, Bélidor concludes by enjoining engineers to assimilate Vauban's practice in drawing up tidy specifications and records.[6]

In 1720, Bernard Forest de Bélidor (1697 or 1698–1761) was appointed teacher of mathematics at the newly founded regimental artillery school at La Fère, where his teaching and his work attracted international attention.[7] He became a member of the Royal Society of London, the Royal Prussian Academy, and a Corresponding Member of the Royal Academy of Sciences in Paris. His correspondents were practical scientists such as Pitot and Bouguer until he was finally elected an associate member of the Academy in 1756 (5 years before his death and 9 years after Montalembert's election).[8]

Bélidor was best known for his texts in mathematics and engineering.[9] His first book, *Nouveau cours de mathématique*, published in Paris in 1725, appears to have been a résumé of his lessons at La Fère. There were a number of French editions of this work as well as German and Russian translations.[10] This was followed by his classic *La Science des ingénieurs* (1729), a practical manual for artillerists (*Le Bombardier françois*) in 1731, his great *Architecture hydraulique* published between 1737 and 1753, and his *Dictionnaire portatif de l'ingénieur* in 1755. *La Science des ingénieurs* and the *Architecture hydraulique* enjoyed many editions and translations. They were reprinted in France as late as 1830 and 1819 respectively, with notes by Navier, and a German translation of *La Science des ingénieurs* appeared as late as 1857–58. These two works are considered fundamental by modern historians of engineering.[11]

On a number of occasions in *Nouveau cours de mathématiques*, Bélidor reiterates that he wants to be "useful, clear, and interesting."[12] What could well have served as a motto for the Enlightenment and its pedagogy reflects an attitude that conditions the content and style of the book. Acknowledging the well-known justification that mathematics is an excellent way to sharpen the intellect for abstract studies such as physics and metaphysics, he nevertheless addresses himself to his main constituency: those who want "to serve with distinction" in the corps of artillery and military engineering. He therefore exerts himself in applying mathematics wherever he can to subjects where its utility will be apparent. His approach is pragmatic, elementary, and operational. The elementary nature of his text appears in his practice of routinely including, along with the traditional theorems, propositions, proofs, corollaries, and remarks, a number of solved practical problems or illustrations for each subject covered. Typically they are the kind that are germane to the experience of soldiers. Grenades and gambling are used in examples of solving simple simultaneous algebraic equations.

Bélidor suggests more advanced works for those of his students who have the interest and inclination. He goes so far as to suggest Malebranche's *Recherche de la vérité* and to reassure officers that it is not unseemly to be seen with a book of metaphysics in their hands. Malebranche's book, he argues, can be more useful than Caesar's *Commentaries* because it teaches methods of exact reasoning. Along with the works of Descartes, those of some minor authors, and the *Mémoires* of the Academy of Sciences, he recommends Mariotte, Rohault, Varignon, and the marquis de l'Hôpital. (He suggests that students might find l'Hôpital "a little abstract" and recommends de Crousat's commentary.)

Not surprisingly, there is no trace of concern for rigor or even of respect for the traditional format of Euclid. Bélidor proudly asserts that he has given more information in 12 propositions than Euclid had done in the 72 of his first two books. And he summarizes in 16 propositions "the most beautiful discoveries of Archimedes on the sphere, the cone and the cylinder."[13] Bélidor was at all times acutely aware of the lack of mathematical sophistication of the audience he was addressing and attempting to convert to his views, and this can be seen in Bélidor's more famous book, *La Science des Ingénieurs*.

La Science des ingénieurs is significant for its theoretical approach to engineering problems, although even here Bélidor takes pains to reassure mathematically backward readers that they can always resort to simpler rules and tables for determining specifications. Subjects are treated with a generality that is absent in earlier texts. The subject areas Bélidor defines are readily recognizable by today's engineer: strength of materials, soil mechanics, and theory of structures, albeit presented at a level that today might be considered primitive, are distinctly delineated.

La Science des ingénieurs, which was intended to be one book in a series of four, is itself composed of six parts paginated and referred to as separate books.[14] The first book, "On the Theory of Masonry," concentrates on the dimensions and profiles of retaining walls, an obviously useful topic in the construction of fortresses with massive earthworks and masonry envelopes. There is nothing more complicated than simple statics and algebra in this book, although Bélidor attempts to apply statics and algebra to the problems of soil mechanics as well as to the stability of elementary structures. The assumptions he makes in analyzing masonry retaining walls (homogeneous materials, perfectly solid foundations etc.) and soils (they will slide along a plane of 45°) are often fallacious, although not worse than those of later pioneers in the field. They do, however, give him answers that he can compare with current practice, including that of Vauban.

In the second book, Bélidor deals with the stability of arches, drawing heavily on the work of predecessors, especially the mathematician La Hire. Again the assumptions are faulty, but they tend in the direction of overdesign, so that the practical results are perfectly acceptable in ordinary building construction. The military engineer Grenier would use Bélidor's methods in his refutation of Montalembert published in Fourcroy de Ramecourt's *Mémoires sur la fortification perpendiculaire*. The third book is devoted to a description of materials of all kinds used in construction. This is a catalogue-style description along with some theoretical principles based on the chemistry of the time. Construction of military and civilian buildings is the subject of the fourth book. Along with descriptions of tests that he had performed on the strength of wooden beams to verify relations between fracture stresses and the material's dimensions, Bélidor's main emphasis here is on the arrangement and design of various buildings. What today would belong to the domain of the architect is even more clearly pronounced in the fifth book, which deals with decoration of buildings. This book is nothing more than a description of the various kinds of classical orders and ornamentation, a staple of architectural treatises from the time of Vitruvius on. The last book gives examples of project descriptions and offers of tender for civil and military building projects. Concrete examples are taken from Vauban and from other eminent engineers.

Bélidor remarks that the sixth and final book of *La Science des ingénieurs* is really the most important for any engineer, and all that has gone before is merely preparation. This remark points to an important tension in what could be called Bélidor's program for engineering. On one hand, Bélidor wants to create an "engineering science," and for him this means the mathematization and more particularly algebraization of engineering. Mathematics must provide not only the rules of engineering but also its language.[15] This will allow mathematics to be used as an instrument of research for discovering new truths about engineering in domains such as construction. Bélidor would have avoided using algebra if he could have, he says, but the rules of the science can only be understood by people speaking and understanding this language. Eighty years earlier (about the time of the death of Descartes), algebra was hardly known. "But I hope," Bélidor writes, "that the day will come when mathematicians [géomètres], physicists, engineers, and architects will think in more or less in the same way."[16] On the other hand, Bélidor wants to codify the traditional practice of French engineering to save engineers time and effort by presenting in a rational and efficient fashion the rich and varied knowledge that working engineers usually acquired laboriously and unsystematically through the hazards of experience and apprenticeship:

> . . . since experience in the art of construction is the rule that can be followed with most assurance, especially when only works that have been successfully executed are being imitated, I have thought that the surest course was to report exactly the plans, profiles and elevations of the most approved buildings that have been made in the new fortresses. Because, since those who designed them can reasonably pass as masters of the art, it can be presumed that they did the best that could be done and that one cannot be mistaken in following their example, leaving to the prudence of those who will have [buildings] constructed to make the changes that they will judge to be appropriate.[17]

Yet this is the same person who writes

> . . . structures . . . must draw their solidity from the rules of the art rather than from the abundance of materials, since if one knew the mechanics appropriate to this subject, edifices would be build that would be a lot more daring than most of those that do so much honor to past centuries: one would work surely and one would not notice a certain timidity that is rather common in modern structures.[18]

The same dichotomy in Bélidor's thinking is evident in his approach to the work of Vauban. He professes great admiration for him, and he reprints a table Vauban is said to have circulated in the 1680s giving the thickness of retaining walls with a standard slope of 5:1 at different heights.[19] However, he remarks:

> When one is accustomed to act according to the principles of mathematics, one gets finicky. Unless everything that is presented to us as right is absolutely evident, the mind is not satisfied and that which appears indubitable to everyone often is a subject of great concern to mathematicians. I have been for a long time in this state concerning the general profile of M. de Vauban.[20]

The Marshal's standard profiles must be good, for they work. But what if they are overdesigned? And Bélidor applies his mathematical derivations (the subject of the first part of his book) to Vauban's suggested profiles and finds that, indeed, for revetments

10 feet high, the thickness is almost double that sufficient for safety. For 20-foot-high walls there is a 25 percent safety factor. At 30 feet there is a margin of $1/8$ for safety, which drops to $1/19$ at 40 feet, $1/21$ at 50 feet, and only $1/58$ at 60 feet. At heights above this, walls with Vauban's suggested dimensions are dangerously unstable according to Bélidor's calculations. Yet he is aware that Vauban's forts do not collapse, and he knows the reason why. Most of the walls are rather low—in the range where Vauban's prescriptions lead to overdesign—and even for the higher walls, the assumptions used about soil cohesion in deriving the theoretical thickness of walls do not apply. The soil in practice is packed hard, layered, and reinforced with brush and trees, so that it does not slide smoothly to give a natural slope of 45°, which Bélidor uses in his calculations.

Why then bother to use theory and to publish these results? The answer points to Bélidor's other great aims in addition to the codification of practice: the mathematization of the engineer's language and the inculcation of habits of thought and judgment that Bélidor believes are those of the mathematician. He wants to "put people of the craft [gens du métier] in a condition to examine things with precision and by ways that lead to truth and whose principles can serve for a quantity of other related subjects that would be related."[21]

Bélidor writes primarily for technical practitioners, such as engineers and the artillerists he taught at La Fère and those who "came to the Arsenal of that fortress, to convince themselves of what they had heard me say about the strength of timber."[22] His role is that of a translator and popularizer of research and methods that were generally restricted to the mathematicians and scientists of the Royal Academy of Sciences. Thus, he is a Moses coming down from a mathematical Sinai not merely with laws but with a language in which to read, understand, and use those laws, and to make new ones. Speaking of his research on the strength of materials (in this case, wood), he mentions the work of the academician Parent. Bélidor thinks Parent is the only person to have published on this subject (in the *Mémoires* of the Royal Academy of Sciences), "but in too elegant a manner to be understood by everyone."[23]

Bélidor is aware that his attention to detail, the elementary nature of his calculations, and his painstaking and sometimes repetitive working out of examples could "bore clever people who would find only insipid [things] there," but he "[begs] them to consider that a book like this one is not made for them."[24]

In another almost apologetic aside to the more mathematically inclined reader, Bélidor says "It will be said to me perhaps that [accounting for the impact of cannonballs on walls as well as calculation of simple static loads] is trying to examine things too physically, but in a subject like this one, everything has to be taken into account."[25]

The first two parts of *La Science des ingénieurs* deal almost exclusively with mechanics and strength of materials from a very theoretical viewpoint, but there are some theoretical sections in the fourth part. Practitioners had done good work before Bélidor, and sometimes their instincts were sound, "but they have only had vague feelings concerning [the reasons for sloping retaining walls and] that cannot be otherwise when things are not considered in their principles."[26] Bélidor aims to change this by introducing practitioners, especially in the first and second parts of his book, to principles expressed mathematically, which will give certainty and illuminate experience. The remaining parts of the book deal with job procedures and collections of the best methods of practice. Here Bélidor relies not just on academicians but also on his own experience of engineering. Speaking of the merits of fresh or salt water in mixing mortar, he mentions opposing points of view and finally says that he cannot pronounce a fixed opinion on the subject "not having done the experiment."[27] Talking about foundations and the various precautions that must be taken to obtain a good understanding of them before building, he continues:

Aside from precautions that must be taken to have a perfect knowledge of the foundation on which one wants to work, it is good to question the workers of the locality: some will always be found to whom good sense and the continual habit they have of working in the same place will have led to remarks and reflections about which one should be apprised. Often these people give more knowledge in half-an-hour than one could acquire by long and laborious researches.[28]

But even at the level of individual parts of the book, the tensions within Bélidor's approach are apparent and even explicitly brought out by Bélidor himself. Thus, part two, on the theory of arches, begins with a theoretical analysis of arch stability couched in algebraic language. But even before he does so, Bélidor is careful to reassure the impatient reader that after this first part he will include a section that "includes the rules for finding the thickness of the piers of vaults of all kinds by the mere calculation of numbers, for the understanding of those who do not know algebra."[29] In this section, he says, "I have eased up a bit on the great precision that is required in geometrical matters, but always in favor of practice, that is to say in erring on the side of solidity of the edifice that one would want to construct...."[30]

Using the methods of La Hire, Parent, and Pitot, with some modifications, Bélidor derives a rather straightforward algebraic expression to give the thickness of the piers supporting an arch.

In the second section, designed for those "who do not know algebra," Bélidor goes into almost excruciating detail, while carefully avoiding any hint of an algebraic symbol, telling which dimensions to multiply together and set apart for subsequent arithmetical manipulations. He aims to get the practitioner to proceed in four steps. The result is mostly simple arithmetic, the most complicated operation, as Bélidor had promised, being the extraction of a square root. The arithmetical operations are in fact based on a simplified versions of his earlier algebraic expression. Two terms are omitted, which actually gives a slightly larger, and consequently safer value for the thickness of the pier. This is an illustration of his earlier remark that he had eased up on precision, but always in favor of greater safety. This is perfectly good and acceptable engineering mathematics, although Bélidor does not say so.

What strikes the modern reader most is not the pragmatic engineering analysis with assumptions being made in such a way as to increase safety factors, but what implicitly appears to be the almost childish aversion to algebraic symbols on the part of his intended readers. The engineers of the early eighteenth century seem to have had a mental block against algebra. A stereotype still common today among some engineers and the public is that of the engineer as someone always consulting tables and handbooks. Another stereotype, distinct from the former but often assimilated to it, is that of the engineer living in a world of mathematical formulas. Bélidor certainly seems to have desired to see engineers become men of formulas, but in the meantime he wanted them to have recourse to tables. He failed in his primary aim: engineers started to think in terms of formulas only toward the end of the eighteenth century. Vauban's tables, such as those regarding the thickness of retaining walls, were considered almost sacrosanct.

Bélidor's contesting the validity of Vauban's tables has already been mentioned. So has his reprinting of them as a code of building practice that was not perfect, perhaps, but would serve as a necessary standard to ensure uniformity in a far-flung, geographically dispersed, and intellectually heterogeneous group of men who were members of a vital governmental service. But even his own engineering research is couched in terms of tables. Thus, after developing a method for calculating the resistance of retaining wall without additional parapets and without supporting buttresses, he prints a table of results.[31]

> Since there are people who could find themselves puzzled in using the rules I have taught on the subject of revetments of banks and ramparts for lack of a good understanding of the reasons on which they are established, I have thought that it would be appropriate to give a table that would dispense them from making long and laborious calculations in which it is always dangerous to be mistaken, unless one is very attentive. . . .[32]

He tells how he had assembled a staff of three "very intelligent" calculators who had done the work under his supervision. Similarly, in the fourth part of his work, as an afterthought and at the persistent urging of friends, he says, he adds a number of tables for the benefit of those who would fear "the length of calculations and of abstract operations that must be done."[33]

Obviously, part of the aim of providing tables is to facilitate the work of the engineer by eliminating tedious calculations, but part of it is to help people unused to anything but the very simplest calculations at all, and eliminate error for those for whom algebraic manipulation was a mysterious and frightening experience.

La Science des ingénieurs was endorsed not only by the Academy of Sciences but also by senior officers in the military engineering corps, and he enjoyed the support and encouragement of the marquis d'Asfeld, the head of the corps, in publishing the book.[34] Bélidor requested and received help from military engineers while writing it. He also included very flattering testimonials from eminent military engineers. It was not only, however, for his colleagues in the Royal Academy of Sciences or for the leadership of the military engineering corps that Bélidor was writing. It was also, and mainly, for the working engineer. What was the reaction of this increasingly numerous and important group to Bélidor? To what extent was his program accepted?

A direct answer to this question is difficult, and even the indirect evidence is scanty and relatively unexamined at present. But a preliminary hypothesis can be attempted. Interesting, for example, is the reaction of the military engineer Nicolas Buchotte to Bélidor's work.[35] In a preface to a work on military engineering he had written for his two sons who were intending to go into engineering, Buchotte expressed surprise that, although works on siegecraft and fortification, both manuscript and published, existed, there was no complete "instruction pour le service des places."[36] Architects, artillery officers, and even infantry officers possessed manuals of their crafts, but according to Buchotte military engineers did not. There was Bélidor's *Science des ingénieurs*, but Buchotte felt that it was of little use to military engineers. According to him, the first two parts of Bélidor's books were simply summaries of treatises on mechanics, of which many scientists, notably Varignon, had already given superior and more complete examples. The same was true for the section on architecture and decoration. Buchotte appeared to think that the primary subject matter for the engineer was to be found in Bélidor's sections on drawing up project proposals, job specifications, and tenders, but these he found verbose and disorganized. By implication he found fault with Bélidor's sections on engineering materials and with the design and layout of such specialized structures as barracks and powder magazines.

Buchotte wrote that the method of organizing his own treatise was to collect his practical observations and the memoranda he had written on engineering subjects and arrange them in the "most natural order." He also would include the "principles and remarks" of appropriate members of the Royal Academy of Sciences "on [explosive] mines," plus those of Vauban and Vallière (inspector of the artillery). It appears from Buchotte's description that his main interest lay in codification and not in the mathematization of engineering. He wanted to provide "all the instructions an engineer might need to serve in a fortress"—in short, a corpus of knowledge that is a distillation and to some extent a rationalization of the best practice.[37] Collection, rather than manipulation and expansion of knowledge in what might be called the portfolio tradition, is what he had in mind. Furthermore, this knowledge dealt mainly with the managerial-administrative aspects of engineering rather than with design.

Buchotte was not alone among military engineers in his low opinion of Bélidor's work. Cormontaigne fulminated against the system of fortification proposed by Bélidor. It was, he said, a

shapeless cloth extracted for the most part from that which one sees as most extraordinary in school authors who have wanted to meddle in writing on this science, which practice alone can cleanse of all that the ideas of the study can engender [that is] monstrous. One will see an example of this and at the same time of the usual reef of those who abandon themselves too much to the search for new discoveries in this field without having previously studied with method and practice this great art, in passing gradually from a simple well-conceived [bien médité] system to another well arranged and consequently better conceived, always based on the received principles of building fortresses and approved by the true masters of the art.[38]

It seems, then, that Bélidor encountered serious opposition not only among artillerists but also among engineers in spite of support from exalted protectors such as d'Asfeld and Marshal Belle-Isle.[39] It seems also that this opposition focused mainly on Bélidor's departure from traditional methods of doing things and his desire to resort to mathematical methods in solving engineering problems. Bélidor was guilty of novelty not only in his proposed solutions to technical problems but also in his methods of solving them. In spite of a brilliant military record that included the construction and demolition of fortifications, he was considered a member of the tribe of professors and study-bound theoreticians, rather than a practical man acquainted with the way things were done. Like most prophets, Bélidor was without honor in his own country. There can be no doubt that his work was known, but it is probable that only the practical tips that came out of this book were absorbed. The project of an engineering science as defended by Bélidor stagnated because of the inherent difficulties in the enterprise, the mathematical limitations of most engineers, and the adequacy of existing engineering methods for most engineering needs.

Bélidor's true intellectual descendant, the engineer and scientist Coulomb, who grappled with many of the same problems as Bélidor in his monumental memoir to the

Academy of Sciences in 1773, was also an isolated figure in the Royal Military Engineering Corps.[40] Coulomb too attacked—only much more successfully and in a more sophisticated fashion mathematically—the same problems that Bélidor treated in the first, second, and fourth books of *La Science des ingénieurs*: soil mechanics, the statics of structures, the theory of arches, and the strengths of materials. Like Bélidor, and unlike Euler and the Bernoullis, Coulomb saw engineering problems from an engineer's point of view. Yet many of his ideas were ignored in his own engineering corps for 30 years or more.[41]

Frézier and a Geometry of Engineering

Besides Coulomb, there were other military engineers who shared Bélidor's view of engineering in a general way and made major contributions to science and engineering. Not as well known as Coulomb today, they nevertheless merit recognition. One such person, who was actually Bélidor's senior and much respected by him, was the explorer and military engineer Amédée-François Frézier (1682–1773), who like Bélidor is supposed to have studied with Lahire and Varignon. Besides being an inventive practical engineer, Frézier published a *Traité de stéréotomie* in Strasbourg in 1737.[42] This was reprinted in a number of editions and translations and is considered a major text in the early history of descriptive geometry.[43] According to Frézier himself, he had three classes of readers in mind. The first was composed of "the few men of letters, amateurs of the [practical] arts, who, being initiated in geometry, only look for the theory of the productions that merit their attention: such is the science of *stone cutting* in architecture."[44] Frézier's bookseller had informed him that this was the part of his work that attracted most of his readers who had preferentially bought the first volume of his work, which dealt with this material. The second group of readers consisted of

> artisans, carpenters, cabinet makers, master stonecutters [appareilleurs], and some marble masons [marbriers] who on the contrary only need instructions for a servile practice, which they consider the secrets of their profession. These not being in a condition to understand the reasoning, based on geometry, they skip them as useless [things], which only confuse them. They only want very detailed descriptions of tricks [traits] all digested and circumstantial for the execution of the work that they have to do, like those one finds in the books of Deran and de la Rue, particularly the latter, whose engravings are more expressive, bigger and better engraved, without being bothered with demonstrations nor if there are more general and correct ways of operating, like the ones found in the last two volumes of my Stereotomy.[45]

This group, one might say, is the opposite end from the first on a simple theory-practice axis: workmen and skilled craftsmen merely interested in operational techniques and recipes without any desire to know the underlying principles. The third group included engineers whose social status and technical expertise Frézier desired to consolidate. This group consisted of

> those who are interested in the science and the art of *stone cutting*, as one of the principal objects of the occupation [état] they have followed, such are young civilian architects, some to put themselves into a condition to oversee private buildings, others for the public ones, like roads and bridges, and finally military officers destined to preside over fortifications, formerly distinguished from architects by the name of *Engineers,* which a number of civilian occupations have latched on to by an abuse that muddles the professions.[46]

No doubt Frézier was at least somewhat disappointed that he had not penetrated to his second category of skilled artisans. He was not unwilling to present the results of his work stripped of the underlying theoretical considerations in terms that they could understand. There is a stoic resignation in his words:

> If nevertheless some artisans without theory and private persons who build in places where only stone cutters without knowledge of their trade (those called *hammers without heads*) are found, seem still to want a second volume for the setting out of the most common traces predigested, I will gladly employ the little leisure that the occupations of my profession for this work of the study, which the weakness of an already advanced age renders laborious and bothersome, to furnish them the aid of the instructions they need for the execution of that which they wish to construct, persuaded of that noble maxim of the Romans that we are not born only for our own pleasure but also to make ourselves useful to society. *Non nobis, sed reipublicae nati sumus. Cic.*[47]

But, as Frézier says explicitly, he was writing for architects and engineers; the architect Blondel had urged him to do so.[48] It was they, in theory, who should have been most able to appreciate the following:

> Practices alone are enigmas that do not enlighten the mind and do not put the observer in a condition to imitate that which he has seen done if ever there is a change in the job, because it may be of a different nature from that which he has seen executed, although similar in appearance. All that is necessary for that is only a little circumstance more or less.[49]

Frézier's engineer, like Bélidor's, was supposed to be capable of adapting to new and unseen circumstances by virtue of his possession of a science of the engineer. For Frézier, one of the most important parts of this science was coupe de pierres—literally stone-cutting, but already recognizable as a nascent descriptive geometry, which the great mathematician Gaspard Monge (1746–1818) would later consider the language of the engineer. Knowledge of principles was the key to innovation, breaking the crust of tradition, escaping the rut of routine—in short, the precondition of doing creative rather than imitative engineering. Yet Frézier, like Bélidor, is a pioneer. While he points the way to the future and will eventually be followed by others—if not in the precise details of his science of coupe de pierres, then in his belief in the necessity of principles for doing creative engineering—he is almost alone and it will be some time before the essence of his views will be assimilated by the engineering community. For the average engineer, the assimilation of tried and true techniques and recipes was a much greater priority than developing methods that made one independent of the knowledge of such techniques and permitted safe and effective departures from tradition.

Frézier, like Bélidor, had some influence in the Corps of Military Engineers. His books, like Bélidor's, were available at the library of the school of military engineering that began to function at Mézières around the middle of the eighteenth century. But, as with Bélidor, it is highly doubtful that the more advanced techniques of his science of stonecutting were ever fully assimilated by the majority of French military engineers. Just as for Bélidor, a major reason must have been that the more difficult elements of his books were not necessary. In its details, not much was changing in stonecutting. The stonemasons who worked on fortification projects, and no doubt the vast majority of engineers who supervised them, were probably among those about whom Frézier complained that they were merely interested in learning the how and not the why of their craft.

Two points are worth repeating about Frézier's enterprise because they cast light on the nature of engineering then and now. First, Frézier's science (unlike Bélidor's, which is algebraic) is geometrical. It provides no solutions for structural problems. Its usefulness lies in fabrication by permitting the most economical and efficient stonecutting, mostly in those cases where complicated shapes are necessary, and it is useful also for improved graphical techniques such as shading and drafting. Drawing had always been important for the engineer, and graphical methods of solving engineering problems fitted in well with this tradition of graphical techniques. More intuitive, if less quantitative, it had always been part of the engineer's tool kit. Indeed, graphical methods have been used as substitutes for analytical methods, and many engineers have preferred them because the loss in precision in comparison to analytical techniques is compensated by the clearer vision of the physical situation and an improved grasp of the viability and safety of the solution with no significant loss in accuracy.[50] Frézier, Chastillon, and other engineers manifest a growing concern for a more standardized depiction in their graphical documentation. Ordinary perspective and the projections used by architects were considered to be inadequate for the needs of engineers, whose role as constructors, bureaucrats, and soldiers required different approaches. In this sense, Frézier could be seen as continuing the traditions of graphical experimentation that had appeared with Brunelleschi and as giving these traditions a new impetus in a different direction: improved capacity for the design of detail drawings and construction procedures. Second, this is a science of details: it does not deal with holistic design. The element of Frézier's concern is the stone block or the spiral staircase, not the fortress as a whole. Something much more complicated than angles of fire is involved. Associated with this is the implication that engineers must be more intimately involved in the details of construction that were formerly left to the artisan. The dialectic between conception and execution that during the Renaissance led architects to become more involved with design and less involved with construction was reversing direction. For some military engineers at least, involvement with the details of construction as well as with the overall plan was necessary for successful building. Frézier's views would be taken up and developed by Chastillon, and his work in stonecutting would set the stage for Gaspard Monge's work in creating modern descriptive geometry.

With the arrival of Gaspard Monge at Mézières in 1764, the techniques of coupe de pierres would be generalized and developed, and they would come to occupy a significant part of the curriculum.

Chastillon's Engineering Drawing and Monge's Mathematics of Engineering

Monge, a traveling merchant's son from Beaune in Burgundy, came to Mézières at the invitation of Antoine-Nicolas-Bernard du Vignau (1716–1795), the second-in-command of that school, who had been struck by the quality of a map Monge had made of his native town and had engaged him to teach in the Gâche (the annex of the Mézières school where students, along with some selected youths of humble extraction who were training to become artisans, were taught the practice of stonecutting and carpentry).[51] In spite of a brilliant academic record at an Oratorian college, Monge seems to have been hired to serve as a draftsman and an instructor in the practical aspects of the engineer's art. Yet his obvious brilliance soon led to his becoming an assistant to the abbé Bossut, mathematics instructor, protégé of d'Alembert and future academician. On Bossut's departure, Monge became his replacement and eventually (March 1772) his official correspondent at the Academy of Sciences. While teaching at Mézières in the 1770s, Monge wrote a series of brilliant mathematical papers that led to his election as a full-fledged member of the Academy as an associé géomètre on 14 January 1780, once again moving into Bossut's place after the latter had been promoted to pensionnaire. The requirement which this entailed of 5 months' residence in Paris each year, as well as additional duties teaching a course in hydrodynamics in the capital and appointment as examiner to naval cadets in 1783, weakened his links with Mézières and he resigned his position there at the end of 1784.

238 Chapter 9

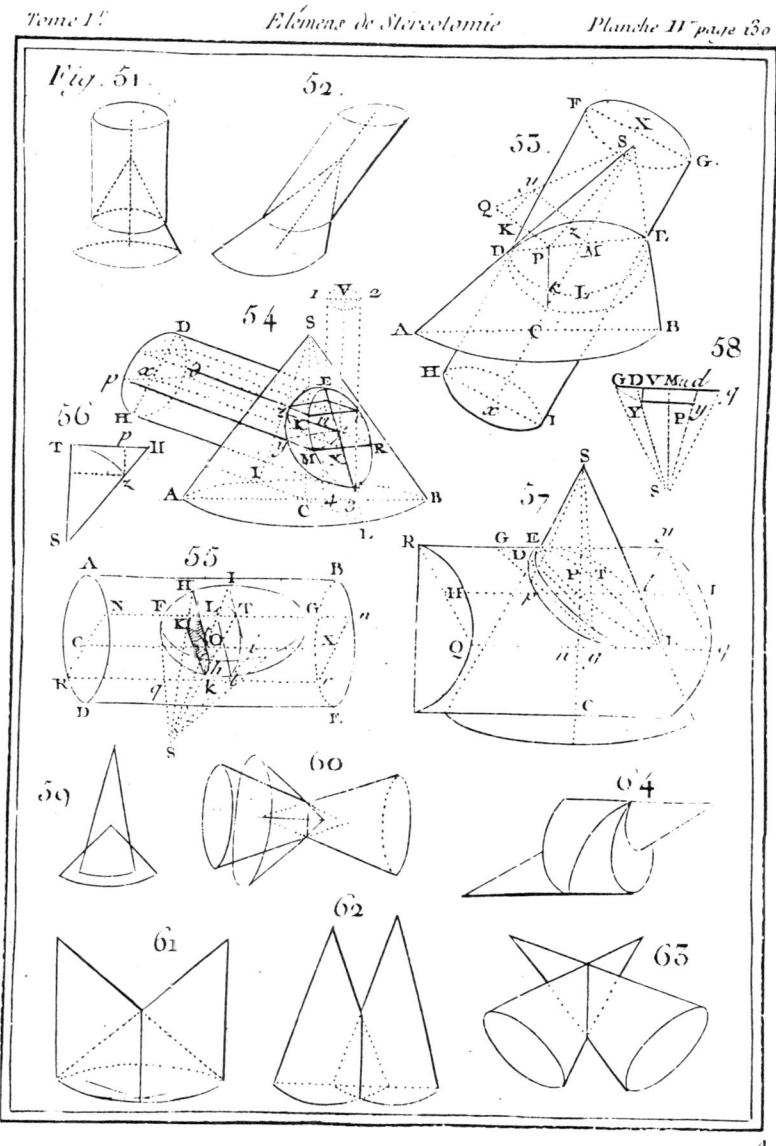

Figure 9.1
This illustration in Frézier's *Éléments de Stéréotomie* gives an indication of the complexity of the drawings in stereotomy texts that preceded Monge's development of a modern descriptive geometry. Source: Frézier 1760.

During his 20 years at Mézières, Monge had moved from a lowly assistant in the student workshop to the chair of mathematics as well as physics (succeeding the abbé Nollet on the latter's death in 1770) and eventually to a position at the summit of scientific life in Paris. A number of Monge's students at Mézières, including Charles-Marie de Tinseau, Montalembert's defender Lazare Carnot, and the great engineer and future theoretician of ballooning as well as aide to Lavoisier, Jean-Baptiste Meusnier, went on to achieve great reputations in mathematics and engineering.[52] Along with Coulomb and Pierre-Louis-Georges du Buat, they have helped to give Mézières the great reputation for scientific excellence that has fascinated historians. Later, in his courses on hydrodynamics in Paris, he also taught the future Jacobin Jean-Henri Hassenfratz (another aide of Lavoisier) and Gaspard Riche de Prony (a great civil engineer, a covert ally of Montalembert, and a future director of the École des ponts et chaussées).

The presence of Gaspard Monge at the military engineering school at Mézières has been one of the claims to fame of that institution. It was at Mézières that Monge began his work on descriptive geometry, differential geometry, and analytical geometry. It was also there that he, independent of Cavendish, burned oxygen and hydrogen to produce water. And it was no doubt his experience at Mézières that made him a logical choice when his former students Carnot and Prieur called on him to help with the organization of the École Polytechnique during the Revolution.

By all accounts Monge was an exceptional and inspiring teacher, and one must take very seriously his possible influence on graduates of Mézières in the last years of its existence. Nevertheless, Monge's active scientific work suggests that his influence was greater among an elite of exceptionally bright students with pronounced scientific interests than among the average graduate of the school. The average graduate would certainly have been exposed in a masterful way to advanced mathematics and physical sciences, but the integration of such sciences into design practice is not a necessary consequence of this.

Recent work, without detracting from Monge's mathematical genius, which deservedly made him one of the great mathematicians of France at a time when it was rich in mathematical and scientific talent, has redirected some attention to the founder of the school, chevalier Nicolas-François-Antoine de Chastillon (1699–1765). In 1748, Chastillon was the person who had impressed the Minister of War, the Comte d'Argenson, with his proposals for educating military engineers. Those proposals had led to the founding of the school at Mézières, where Chastillon was based as the engineering officer. He was a scion one of the great dynasties of French military engineers so lovingly described by Anne Blanchard.[53]

Thanks to Chastillon, a more general and mathematics-based approach to graphical techniques as design tools was taught at Mézières and began even before Monge's arrival. Bruno Belhoste's undeservedly neglected study has illuminated Chastillon's role in the transition from relatively unsystematic and particular graphical techniques to something that can be recognized as the first stage of a true descriptive geometry which was subsequently developed and formalized by Monge.[54] Belhoste has clearly shown that Chastillon made a fundamental but little noticed transformation in the education of engineers by advancing the graphical analysis of stonecutting, which was so important for Frézier, as a fundamental part of military engineers' education. Chastillon was undoubtedly influenced by the work of Frézier as well as by the importance that engineers had always attributed to drawing. In a manuscript treatise on shadows in geometrical drawing intended as a manual for students, he wrote:

> Since the beginning of the School of Military Engineering at Mézières, particular attention has been paid to giving young officers the principles of drawing and to do so with method. Facility of expression with a pencil and the understanding of drawing lead to good construction of works in fortresses, of the trace and relief of fortification and, in general, everything related to them. [Drawing] of the works relative to siege and field warfare train the power of observation [coup d'oeil] for the reconnaissance of fortresses. In a word, it is drawing, understood and applied to all objects, which in part characterizes engineers.[55]

Not only does this passage indicate drawing as the elemental art of all aspects of the engineer's work (both overall design and detailed design of construction); it also reiterates drawing's function in the education of one's technical intuition. Drawing helps the engineer acquire a practical feel for design and helps the military officer acquire the coup d'oeil considered such an important part of the eighteenth-century commander's talents. The rest of the treatise, however, concentrates on the adaptation and modernization of drawing techniques to the specific problems of stonecutting—in other words, the use of drawing in detailed design.

The redaction of Chastillon's treatise coincides with a significant overhaul of the teaching at Mézières. The syllabus of 1754 gives the order of instruction as fortification, decoration, and construction. Cutting stone and wood, which were building techniques of specialized artisans, were relegated last in this succession. Even then, Chastillon considered them to be highly important for the reasons mentioned—inculcation of a practical sense of design and the handling of spatial elements. At the time of his death, however, the order of instruction had been inverted: the student began with "the elementary practices of drawing and making drafts of stone cutting and carpentry, perspective, drawing shadows, architecture and fortification."[56] On one plane this marks an increasing concern with the fine detail of design and the finer points of construction rather than what could be called macro-design. This suggests that for some engineers attention to detail signified not merely attention to administrative details of daily management of a fortress but also the details of construction hitherto left to contractors. D'Arçon was a defender of this view in his memoranda to Saint-Germain in 1776. On another plane it implies the persistence of a strong belief in the importance of drawing in design and the lack of penetration of Bélidor's ideas into the domain of construction. On yet another plane, it is a reminder that the mathematical and algebraic instruction at the school, which took place in the salles de théorie and was quite substantial by this time, was neither the only nor most important influence on the intellectual development of young engineers. The salles de pratique were also important.[57] Because of the need for more engineers felt at the end of the Seven Years' War, the ordinance of 4 December 1762 had raised the regular complement of the military engineering corps from 300 to 400 members, and this had led to an increase in enrollments from 30 to 50 students.[58] The instruction offered in these salles de pratique became more formal and Chastillon's treatise was one of several appearing after 1762 to provide veritable texts for regular classroom instruction.

If Chastillon's insistence on drawing can be seen as rather traditional, his approach to the act of drawing is original. The engineer is to be a draftsman and his drafting is to be of a different kind from that of architects. While perspective drawing is useful and is undeniably important, the drawing par excellence of the engineer is geometrical—the use of projections on various planes.[59] Shading is also important not merely for aesthetic reasons but for making the drawing clearer. Here he comments on the difference between architectural and engineering drawing. Engineers draw projections of objects as they are illuminated in nature, whereas architects shade their drawings to make them more pleasing to the eye or simple to understand for their clients. They also position the various projections in no regular pattern, whereas engineers are to have a standard procedure of positioning views. Their procedure allows someone inspecting the drawing to see the object in any projection as it would be seen in nature illuminated by the sun at a standard elevation of 45°. Chastillon clearly distinguishes the two modes of representation: by architects for laymen with the intention of pleasing and accenting desirable features and by engineers reporting to superiors initiated in the techniques of reading their drawings and desiring to know the exact appearance of the object in nature.[60]

The idea of eliminating all subjectivity and presenting objects as they really are with the help of orthogonal projections and shading leads to the problem of coordinating

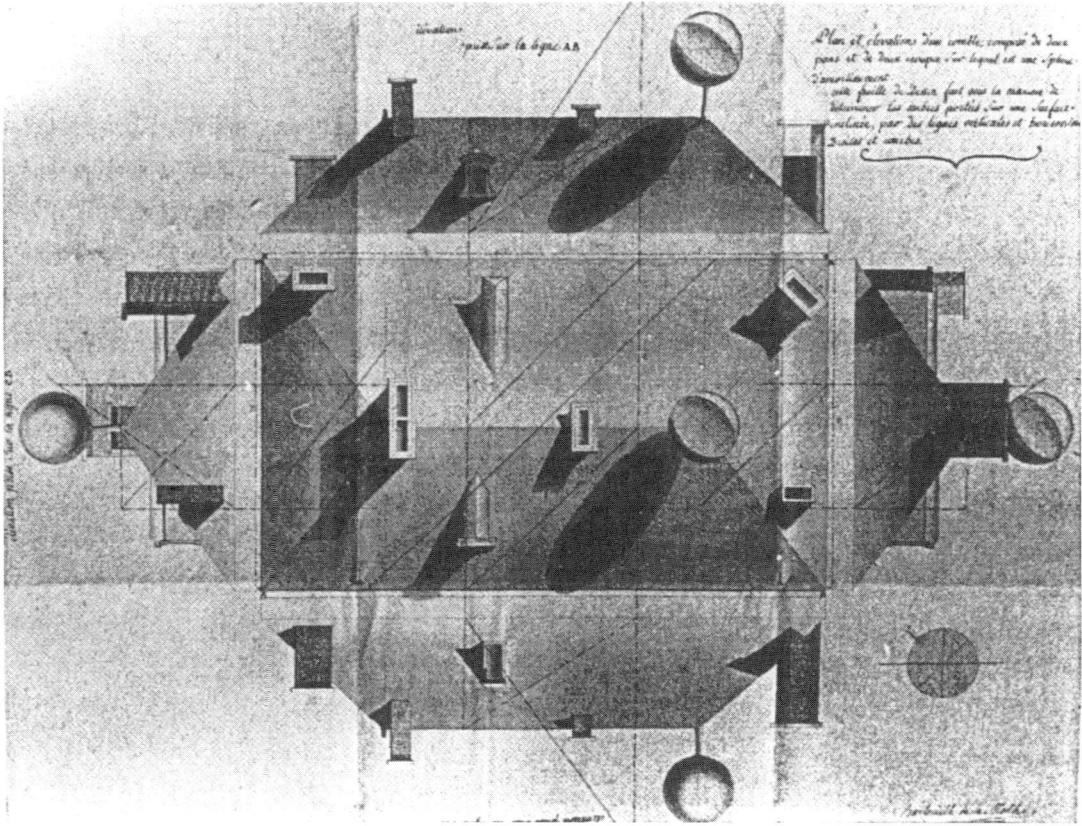

Figure 9.2
Drafting of a building with appropriate shadows cast by the sun according to the procedures specified by Chastillon. Source: Belhoste 1990.

the various projections with each other. It also leads, as in drawings for stonecutting, to problems of rays cutting objects at different angles in different projections, which is the stuff of descriptive geometry. By standardizing and rationalizing graphical presentations, Chastillon moves a great step forward from the working drawings of masons and artisans. Yet it is ultimately Monge who takes up these problems and develops the general consequences. While Chastillon had developed a general graphical method, he nevertheless continued to deal with concrete objects like lampposts. When Monge wrote his own *Treatise on Shadows* in 1768, he did not go from the particular to the general as did Chastillon, discussing lampposts and houses; he dealt with abstract

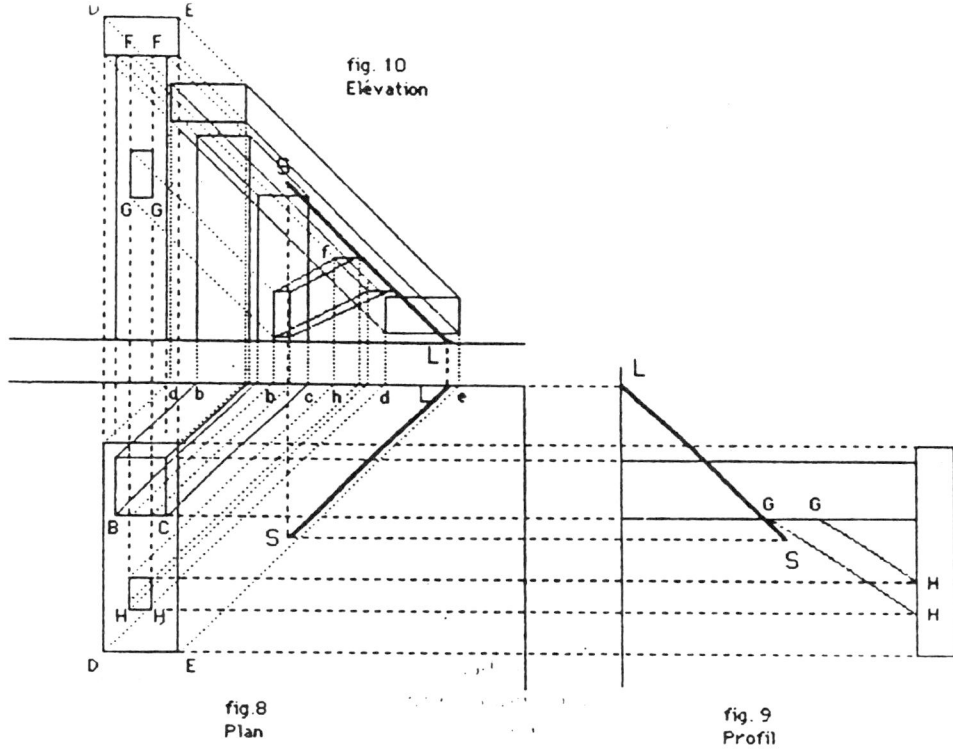

Figure 9.3
Determination of the shadow of a jib on a wall by the ray of a lantern SL. Reconstructed from Chastillon's procedures in Belhoste 1989a.

shapes, such as the shadow of an ellipsoid on a cone. And while Chastillon's drawings still resembled architectural drawings with his separation of profiles and elevations from his plan view, those of Monge, with projections attached to each other and folding down along this line of attachment (called the ligne d'élévation by Chastillon and the ligne de terre by Monge, are clearly drafts of descriptive geometry.[61] The evolution of the practical art of stonecutting into the science of descriptive geometry has been magisterially described and analyzed by Joël Sakarovitch.[62] There were, however, other problems that contributed to the creation of a modern descriptive geometry and they fell even more in the domain of military engineering than stonecutting.

Drafting for stonecutting and design of architectural elements in construction were not the only subjects in which instruction became more formalized after 1762.

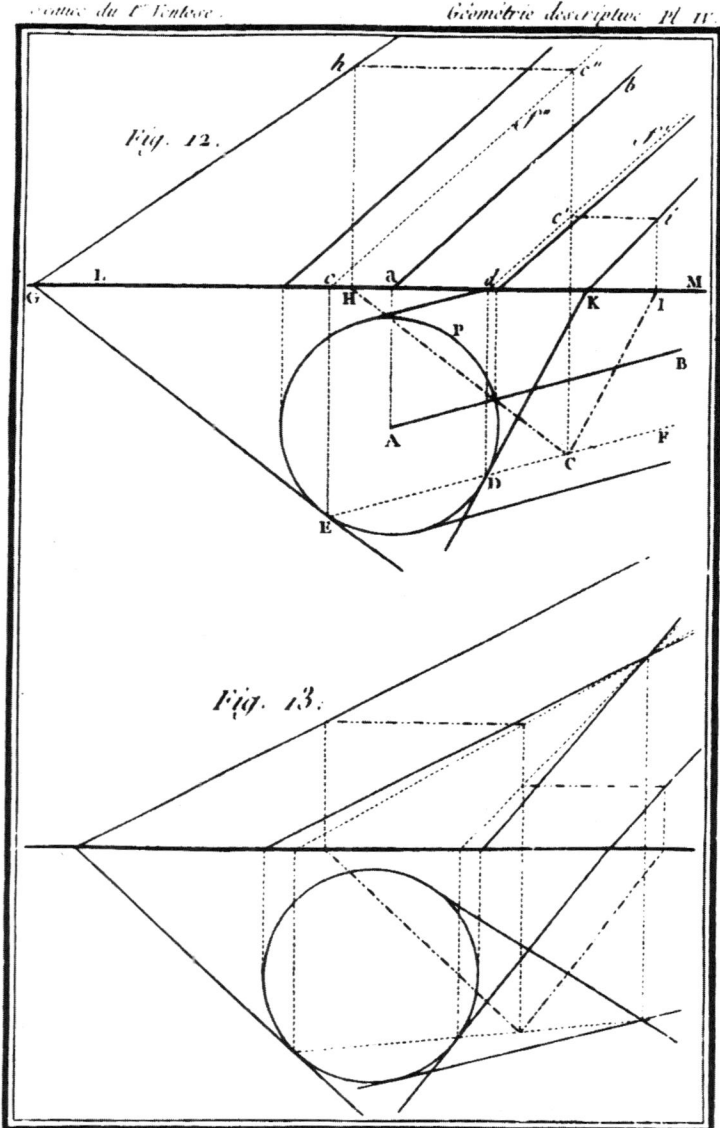

Figure 9.4
A figure from Monge's *Géométrie Descriptive* (1794). Note the severe sparseness of Monge's technique as opposed to Frézier's (figure 9.1 above). The two problems considered here by Monge are drawing a tangent plane to a point on a cylindrical surface whose projections in the horizontal and vertical plane are given (fig. 12) and drawing a tangent plane to a conical surface

Chastillon also was apparently the first to attempt a graphical analysis of the problem of defilading [défilement]. Here again Monge outshone him and may have dazzled subsequent historians; indeed, Monge's meteoric rise from a humble assistant in the workshop and drafting room has been attributed to his brilliant solution of this problem.[63]

Most people examining books on fortification from the Renaissance up through the eighteenth century and beyond, then and today, are attracted to the elegant symmetries, the elaborate geometrical patterns, and the beautiful shapes of the traces or ground plans of classical artillery fortifications. Star-shaped fortresses, triangular demilunes, and pentagonal bastions, often in baroque profusion, project an image of frozen rationality imposed on nature—a tidiness that almost inevitably, albeit subtly, conveys a sense of order and security as well as aesthetic satisfaction. According to A. Rupert Hall, constructing the patterns of traces required more mathematics of the military engineer than that required of the master mason of the Middle Ages, but not incomparably more.[64] For the military engineers, however, the patterns of the trace or the horizontal projection, which so fascinated the layman, were not the important ones in a fortress. Instead, military engineers concentrated on the profile of the fortifications. For Simon-François Gayvernon, military engineer and professor of fortification at the École Polytechnique during the Directory and Consulate, the profile or vertical projection was the primordial geometrical form.[65]

The profile had to satisfy three requirements. The different works of fortification had to "command" (overlook) the external works in the direction of the countryside. The amount of cut and fill in the profile ideally had to be roughly equal so that construction would be most economical. Finally, the fortifications had to be properly defiladed to protect the defenders from enemy fire. This latter condition involved fairly sophisticated mathematics by the end of the eighteenth century, considerably more than that of Professor Hall's military engineers between 1400 and 1700.[66]

The *Oxford English Dictionary*, quoting a definition taken from a mid-nineteenth-century military dictionary, says that "the object of defilade is so to regulate the relief of the parapets in covering masses, that the defenders may be perfectly covered by them from the view of the enemy." The problem of defilading is trivial if the fortification is on perfectly flat terrain. One simply has to raise parapets above the surrounding

whose horizontal projections are given (fig. 13). Source: Gaspard Monge, *Géométrie Descriptive, Leçons données aux Écoles normales,l'an 3 de la République; par Gaspard Monge de l'Institut National* (Paris: Baudouin, an VII [1798 or 1799]).

countryside to a height that will cover the defending soldiers and ensure that the fortifications in the next line of defense are even higher, so that inner fortifications always "command" outer ones. But the situation becomes more complex when the terrain is irregular and there are commanding heights around the fortress where the enemy can establish batteries. With the exterior ground level at some points higher than the ground level of the fortress, the enemy can see and fire into it. The problem is then to minimize the area visible and subject to direct fire.

Traditionally, military engineers laying out a new fortress had determined the heights of walls by sighting from the area they wanted safe from enemy fire toward the highest points within artillery range outside the fortress. A plane passing through the highest points on the surrounding terrain that were within artillery range as well as the interior of the proposed or existing fortified area was called a plan du site (site plane or site plan) and all defensive works were constructed with respect to this basic plane. The plane of defilade was generally parallel to the site plan but some distance above it. Chastillon was possibly the first to attempt this operation on paper rather than on the actual site. Even though Chastillon thought that the best school of the military engineer was attentive study of existing fortifications, and that "the study of fortresses is the best book," he had written his treatise because he hoped that his techniques made the execution of tasks "so simple, that if ever the persons entrusted with them had some conception of the art of building, it would be easy for them to complete [them] in the absence of engineers, too few to follow all the work, and reduced to this sole means of achieving their aim."[67]

The extensive building program of Louis XIV had aggravated this problem of supervision and the related problem of communication between those who conceived and those who executed. Like Gayvernon more than 40 years later, Chastillon thought that "the most scientific part of fortification is a good understanding of the relief" and therefore special pains had to be taken to get the profiles of fortification right. Written shortly before his death, the techniques for defilading were not, however, of much use, since they dealt merely with the case of perfectly flat terrain around the fortress. It is likely, however, that students were also taught defilading in irregular terrain because a few years after Chastillon's death, his second-in-command du Vignau drafted an exercise on the practical aspects of military engineering that included a solution to the problem of defilading in irregular terrain.[68]

Du Vignau's text, then, is the first to deal with the general case of defilading in irregular terrain. He does so by making use of a map marked with the elevations of significant points of the area under consideration. Picking a point within the fortress and joining it to two of the highest points within artillery range outside the fortress, Du Vignau constructs a plane. If all points of the terrain are below this tangent plane to the countryside, then the plane can serve as a site plan which can be taken as the effective ground level of the fortification and on which one can design fortifications as if they were on perfectly flat terrain. However, it may turn out that this initial tangent plane cuts intermediate points of the terrain, in which case the process is repeated with a new tangent plane being drawn. After a number of trials a tangent plane that does not cut the terrain at any intermediate point is obtained. This was a time consuming, fastidious, and graphically messy process. Monge was the first to eliminate these difficulties with an elegant and general solution that in principle could remove much of the trial and error from the process.

Monge constructed a tangent cone (whose apex was the point within the proposed fortress) to the terrain and then found the trace of this cone on a vertical plane cutting the exterior terrain. The tangent plane to this trace that included the reference point provided a site plan on which one could construct properly defiladed fortifications. More elegant and general than the previous ones, the method Monge uses is what we now recognize as the techniques of descriptive geometry to draw lines and construct planes. Essentially his method reduces to the classical problem in descriptive geometry of constructing a plane tangent to a cone. There is no record, however, that this method, which Monge never published and may have only started to teach after 1775, was widely used by students at Mézières, or that it was ever considered a fundamental design technique by the military teaching staff at the school or by Monge. The method first appeared in print only in 1796 in a publication by Horace Say (1771–1799), a professor of fortification at the École Polytechnique.[69]

But another method, even better from the engineer's point of view, was developed around this time (ca. 1777) by the military engineer Meusnier, perhaps Monge's most brilliant student at Mézières. Meusnier's method of defilading used contour lines on topographical maps.[70] Contour lines for hydrographic work had been suggested by the geographer Philippe Buache (1700–1773) and later by Marc Ducarla (1738–1816).[71] The latter presented a paper on the subject to Académie royale des sciences in 1771, and it is likely that Meusnier, a future academician who undoubtedly followed the activities of the Academy closely, would have been aware of this development.[72] From the point of view of descriptive geometry, contour lines can be seen as the intersections of horizontal planes at given constant levels with the terrain. Muenster's solution to the problem of defilading is original and is easier than Monge's. On a map with contours, taking a line to be defiladed, Meusnier graduates it according to the different

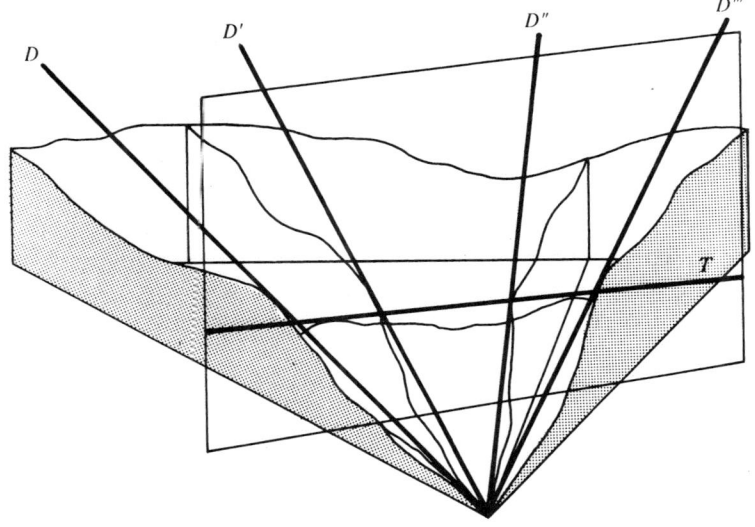
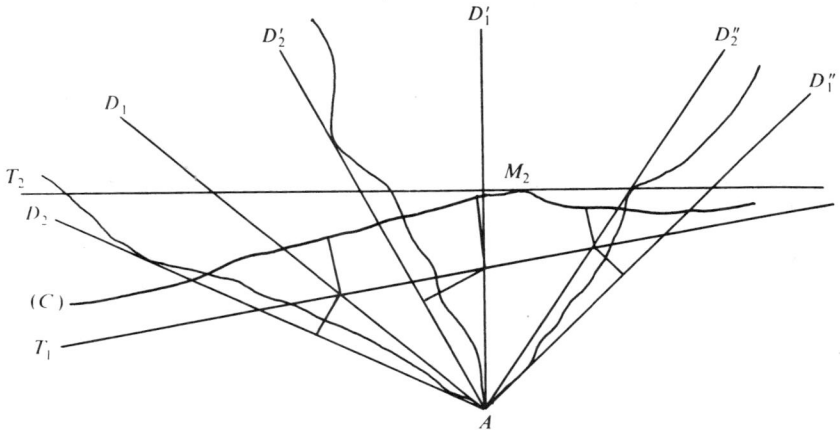

Figure 9.5
Block diagram (above) and projections in Monge's method of defilading. Source: Belhoste 1992.

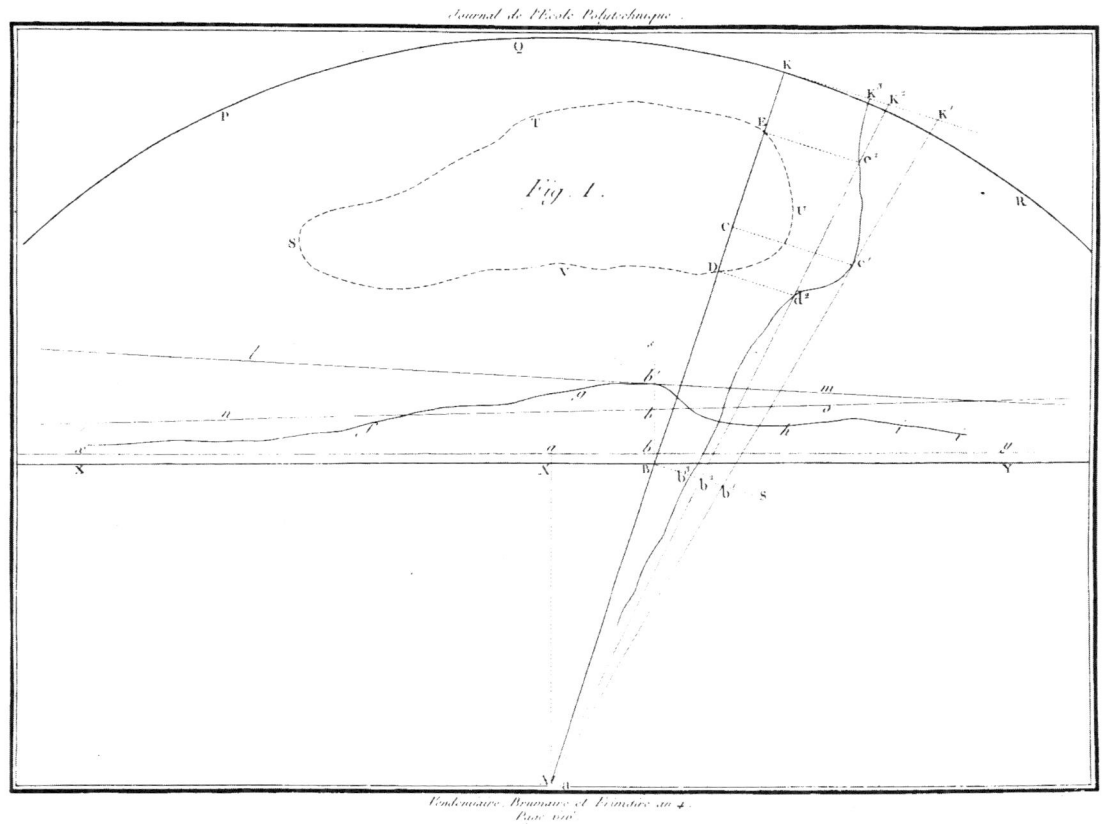

Figure 9.6
Horace Say's solution of a problem in defilading fortifications. Source: Say 1796.

elevations along this line. He then draws tangent lines from each point of given elevation on the line to each of the contour lines of the same elevation. Of all the lines from the graduations on the defiladed line to the corresponding contour, the line that make the smallest angle with the defiladed line itself gives him the tangent plane required for a site plan. Yet the applicability of this method was limited because of the quality of cartography available to engineers.[73] Contour lines were more a feature of school exercises than maps even in the late eighteenth century. It seems that, even though Meusnier's method was quickly recognized as the best and developed in the nineteenth century by F. Noizet and others, it could only achieve its full potential with the routine use of modern topographical maps.[74]

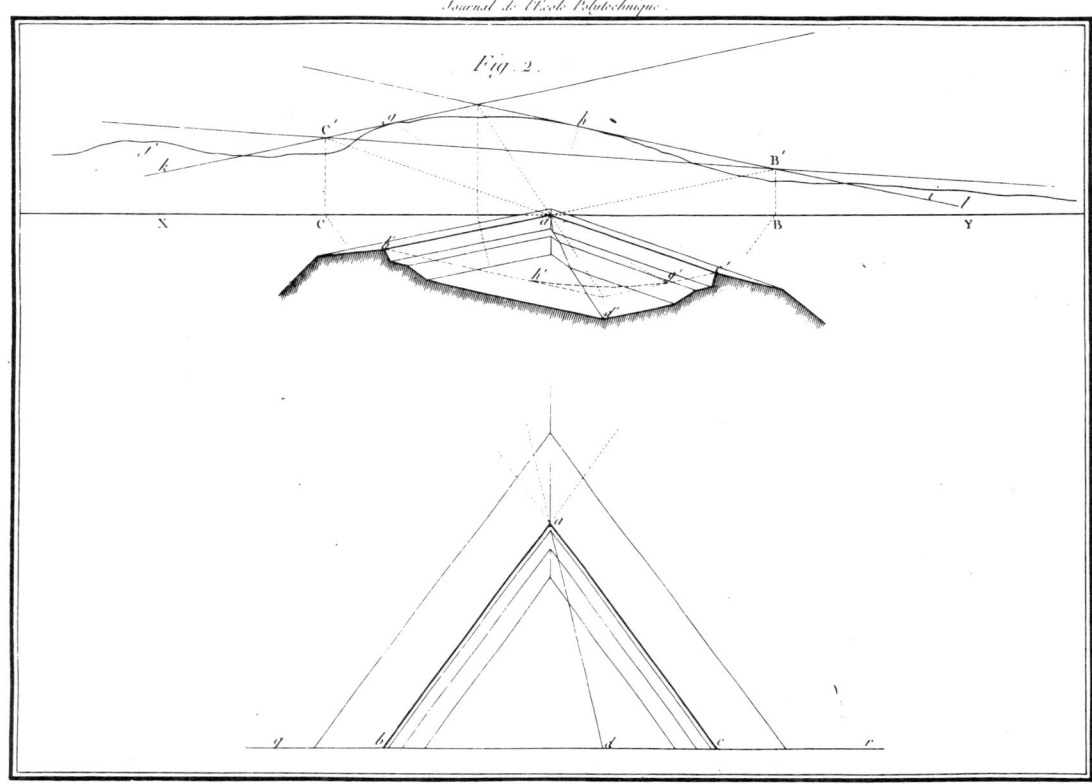

Figure 9.7
Say's solution of a problem in defilading fortifications, continued from figure 9.6.

Monge's later behavior seems to confirm the suggestion that his interest in the technique of defilading seems to have been restricted to its being one example among many of more general problems in descriptive geometry involving the behavior of tangent and secant planes with different kinds of surfaces. Presenting his lessons on descriptive geometry at the École Normale during the Revolution, Monge mentioned it in passing without any development in detail in his lecture on tangent planes to various curves.[75] Indeed, he managed to confuse the plan de site and the plan de défilement. Monge's solution to the problem of defilading was a solution of a problem in descriptive geometry while Meusnier, in spite of using descriptive geometry as a tool with great mathematical virtuosity, was solving a problem in practical engineering. Notwithstanding

Monge's keen interest in practical problems and the inspiration they provided for his more abstract mathematical work, he considered this mathematical aspect of his work central. The practical military engineering problem of defilading served as a point of departure for more abstract and general considerations in descriptive geometry and ultimately in differential geometry and analysis. It was in these last two areas that he submitted his papers to the Academy and established his well-deserved reputation as a mathematician. Monge's geometry—and the geometrical school that derived from it—was as abstract as anything in Euclid.

Monge's attitude is illustrated by another problem he tackled during his time at Mézières, which also arose from the practical problems of engineers. This was the problem of calculating cut and fill. Du Vignau's method merely involved partitioning the various excavations and earthworks into regular volumes that were computed. An attempt was made in the design of the fortress to ensure that the amount of cut roughly balanced the amount of fill that could be obtained from the excavation so that transport costs would be minimized. In practice, this meant a series of guesses at the depth of the ditch (one of the major sources of fill), calculating the cut and the necessary fill, checking if they were close, and repeating the calculation with another estimate for ditch depth if they were not. The process was repeated until rough equality was obtained. This was a problem more tedious than calculating defilade and whose ease of solution depended in large measure on the practical experience and "feel" the military engineer had for the situation. Monge deals with an even more complex problem: to minimize the costs of transport when moving earth from an area of cut to an area of fill. He treats the problem as a theoretician: given an area of cut and another of fill he needs to minimize the sum of the products of volume of earth times the distance each volume is moved.[76] Various complicating situations are added, such as the existence of a river with a bridge at a given point that the workers must cross with their loads. The problem is mathematically complex and was only finally solved in 1886. After dealing with the problem in two dimensions (assuming flat areas rather than volumes), he turns to the case of three dimensions, which leads to highly novel considerations on curves and surfaces in space. René Taton considers this part of the work, quite removed from the practical problems of engineers in spite of being inspired by them, to be one of the more original and significant developments in eighteenth-century mathematics.[77] He quotes Condorcet, who in his report on Monge's paper argued that it provided a contrary example to the better-known case of highly abstract theories leading eventually to highly practical applications, such as the famous example of Apollonius and Kepler. Here, practical problems had led to highly abstract and novel theories. It is almost

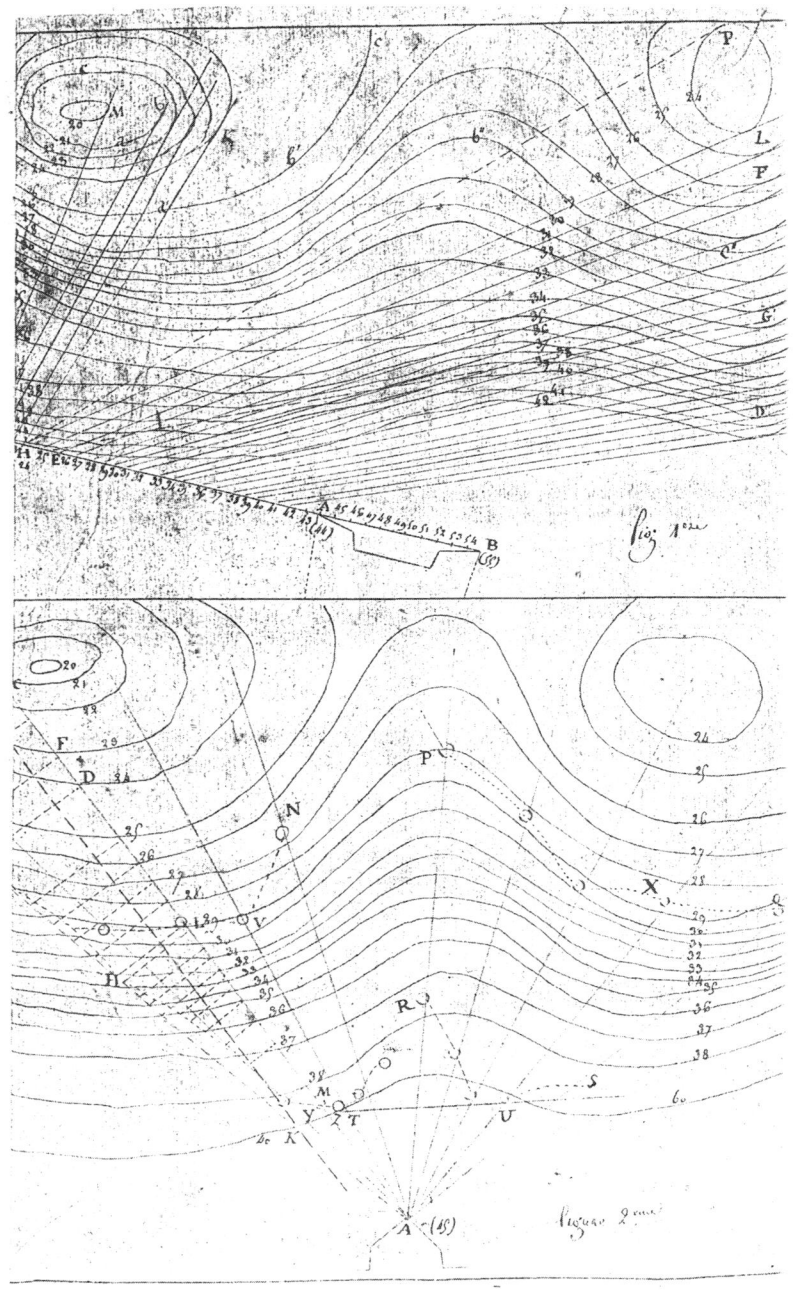

unnecessary to add that even those engineers who understood Monge's paper never used it in any practical manner.

There are two clearly discernible streams in defilading analysis in the eighteenth century. One is that of Chastillon, who is interested in solving particular problems of design. It continues with Meusnier, who, although he can be considered a great mathematician, is primarily an engineer in his mentality. In the nineteenth century it is represented by people like the military engineer Noizet, who writes that "fortification drawing, however, is not immediately deduced from the principles of descriptive geometry . . . even though in both cases the basic principles are the same and cannot vary."[78] In spite of Noizet's respect for Monge, a veritable cult figure for the polytechnicians who entered the military engineering school in the nineteenth century, he considers Monge's methods—rightly so—less useful and more cumbersome than Meusnier's. Monge's methods, the second stream, continued to be taught at the École Polytechnique by people like Say, and historically they were undeniably inspired by the problems of military engineering. But their format was that of an abstract mathematical problem rather than a practical exercise. The former stream can be characterized by the search for graphical efficiency, the latter by the search for intellectual elegance and generality.

Conversely, the sophisticated approach to the design problem of Meusnier is also inspired by the descriptive geometry of Monge, but there is a radical difference in approach. In most cases, Meusnier succeeds in creating a method to perform his defilading analysis in a single horizontal projection by using contour lines, which not only give a clearer vision of the terrain but also permit a superior method of defilading. A metaphor that is appropriate for an engineer who was also one of the pioneering mathematical analysts of balloon flight, as well as the first to present an analytical, rather than merely descriptive, use of contour lines may be useful to characterize Muenster's approach. Meusnier ascends from the terrain into the heavens of abstraction to get a better view of his terrestrial engineering problem from which he ascended and to which he will return. On the other hand, while Monge's method has its roots in a practical design problem, it quickly shoots up into the heavens of abstraction, there to stay and

Figure 9.8
Meusnier's method of defilading is much simpler than Say's and shows what could be a practical exercise and not a drafting room problem. Defilading exercises in military schools in the nineteenth century followed Meusnier's model. Source: Meusnier, Jean-Baptiste, ms. "Mémoire sur la Détermination du Plan de Site," [1777], AIG Art. 21, Sect. 1, §1, carton 6, p. 12.

to bear its true fruit. One can use the abstract science to solve engineering problems—finding a site plan can be seen as finding a tangent plane to a cone. And the fact that very rapidly descriptive geometry did become an abstract science (in the period between about 1760 and 1790) can easily lead to the belief that Monge's work is an application of an abstract science to a practical engineering problem. Historically, however, this is not the case: Monge was following a path first laid out by Chastillon, from which he eventually was to deviate in original ways. On the other hand, Muenster's practical engineering problem was solved in an eminently practical engineering way that strictly speaking had no need of descriptive geometry.

To what extent was this mathematics inspired by engineering assimilated into practice? The answer can be obtained by examining a fortification exercise for students during the early years of the Revolution, when instruction had been stripped down to essentials. This text contains a complete step-by-step analysis of the design of a fortress with a quick résumé of the history of fortification in France and the predictable endorsement of Vauban's methods as the best.[79] There is little discussion of the traces or even a description of methods of constructing their ground plans. After listing the principal systems (those of Errard, de Bar-le-Duc, Marolois, Pagan, De Ville, Coehoorn, and Vauban) that have been described "in several books," the reader is referred to those books and told that the aim of this treatise is "to give young officers principles applicable to any system." The bulk of the rest of the treatise is devoted to calculations of relief, defilading, and cutting and filling.

There is no mention of Monge, Meusnier, Du Buat, or even Du Vignau, but Chastillon comes in for high praise and is credited with developing "principles and simple methods" for analyzing problems of defilade. Indeed, Chastillon is credited with having come up with the idea of contour lines although they are never used in the treatise. What is striking, however, is that defilading the site is done with the implicit assumption that the surrounding territory is flat. The more complex task of defilading for irregular terrain is not attempted. There is little evidence in the surviving archives of the military engineering corps that the exercise was done often, either in the classroom or on the job.[80] Although the maxim that "absence of evidence does not imply evidence of absence" should make one careful about categorical assertions that defilading was not done very often, it does seem safe to suggest that it may have only been an occasional exercise of technical virtuosity rather than a routine procedure.[81]

Further support for this suggestion comes from the virtual absence of contour lines and even plans with indications of elevations in the eighteenth century. Including indications of height as a standard feature of plans and maps was proposed as early as 1762 but was only formally recommended by the Comité des Fortifications in 1796. Even then, according to the engineering historian Augoyat, it took a long time to become accepted.[82] Contour lines became a regular feature of maps in the military engineering school at Metz only in 1802. At the École Polytechnique, the descriptive geometry method of two projections was used to indicate altitude.[83]

There are a number of possible reasons for the scant amount of evidence for widespread use of sophisticated defilading techniques. Working drawings have a slimmer chance of survival than other kinds of documents. Detailed information on terrain around a fortress and the levels of various features of the fortifications would have been very valuable for potential enemies and was therefore guarded very carefully and possibly even destroyed in emergencies. In practice, there was not much fortress construction after the graphical techniques of defilading were developed and most of what there was the construction of maritime fortifications where the trivial case of flat surroundings applies. Proper attention to siting fortresses, also much emphasized in design exercises, would have avoided placing them within range of guns on commanding heights. While the exercises rightly admit that this is not always possible, it probably was possible in most circumstances.

A course of fortification at the military engineering school of Metz during the First Empire and after the foundation of the École Polytechnique implies that defilading was of the simplest kind.[84] The author of the course does not mention Monge or Montalembert, and Carnot is mentioned only to reject his system of fortification. The great authorities are Cormontaigne, Noizet Saint-Paul, and the émigré engineer Bousmard. Noizet Saint-Paul and Bousmard had been students at Mézières under Monge and, with Gayvernon, were the only ones to publish works on engineering as taught at Méziéres.[85] Yet the author criticizes the methods of defilading taught at Mézières as bookish and not particularly useful exercises. Meusnier's is considered the best method and probably was becoming more realistic as an option with the heroic efforts of Napoleon's cartographers.[86] Say, expositor and developer of Monge's techniques of defilading at the École Polytechnique, is treated with a certain degree of condescension as being an impractical schoolman. It seems clear that the magnificent tools fashioned by Meusnier had to await the popularization of the new techniques of topography—contour lines—which they required.

But the value of the whole exercise was becoming more and more questionable. In practice, the final touches of defilading earthwork fortifications would most likely be done on the site with actual leveling and sighting. Cover against howitzer and mortar fire could not be guaranteed by defilading in any case. Monge's interest in defilading as the application of theoretical principles may have been more realistic than that of those who claimed it was a useful technique of design. One feels that the public flaunting of defilading analysis (without much or any discussion of its actual details) by military engineers was more useful as demarcation from the profane and tribal reminder of youthful exploits in the drafting room.

The author of the exercise of 1792 also rejects Bélidor's results for the calculation of retaining walls, explicitly mentioning that his results lead to overdesign for walls 80 feet high. Vauban's recommendations are considered quite adequate and have proved to be valid by the longevity and resistance of walls designed according to his specifications. Furthermore, he argues for the importance of "young officers" getting as much experience as they can so as to

get used to training the informed glance [coup d'oeil], to combine according to a different location the different ways of disposing the trace and the relief of a work and to judge the effect it would produce in different cases so as to achieve the aim one has in mind most simply and exactly, without being obliged to have recourse to ruler, compass, and level. One must know how to do without these things in certain circumstances that occur often in war. The celerity with which one is most often forced to work requires that an officer make his decision on the spot, and he will succeed in [doing] this only to the extent that he will have trained his judgment in advance by reflection and experience.[87]

Once again, as suggested by Chastillon, graphical techniques and graphical analysis of engineering problems seem to be important not so much for themselves as for a propaedeutic purpose: to prepare the military engineer for the art of war. This became explicit and central in Louis Le Bègue du Portail's conception of the military engineer as an expert in a science of space applied to war. The military engineer was also a soldier, and his science was not only an engineering science but also a science of war.

The Sciences of Military Engineering in Eighteenth-Century France

Although the nineteenth century is usually considered the time of the beginnings of a truly scientific engineering, it is clear that the groundwork for and even many features of this kind of engineering were being created from the time of Galileo on.[88] Bélidor's *Science des ingénieurs* and the *Architecture hydraulique* are milestones in the literature of engineering science, and his books were treated as such by engineers of the eighteenth century.[89] Yet it is likely that Bélidor's books were of more ornamental, inspirational, and symbolic than practical use for the working military engineer. This no doubt started to change with the establishment of the school at Mézières; the school increasingly provided the education necessary to absorb the work of Bélidor and those who were following in his footsteps. But only with the creation of its successor institution—the École Polytechnique—was that part of the legacy of Mézières concerned with analytical mechanics and mathematical design taken up in earnest as well as developed and enriched.

There are a number of good reasons why the fledgling engineering science was not pursued with great vigor by the average military engineer, or even by many senior military engineers, with the notable exceptions of Coulomb and a few others. As Navier, with the benefit of a solid mathematical education at the École Polytechnique and the absorption of Coulomb's improvements on Bélidor, pointed out in his own edition of Bélidor's *Science des ingénieurs*, Bélidor's work was premature.[90] While praising Bélidor's faith in the power of analytical mathematics to solve the problems of engineering construction in the field, and agreeing that such a pioneering faith was necessary even if unjustified at the time, Navier pointed out the inadequacy of empirical research on strength of materials and the unsophisticated and often inaccurate nature of the theoretical models that Bélidor had used in his work. This was obvious not only to Navier but also to military engineers of the eighteenth century whose daily work taught them to be dubious of elegant but inadequate mathematical analyses of engineering problems.

Nor were such analyses particularly necessary. The type of construction of fortification, usually involving revetments and earthworks, was well known, and the empirical tables passed down by Vauban were quite adequate in practice. Generally, what was involved in construction practice was nothing like, say, the dome of Soufflot's church of Sainte Geneviève (today known as the Pantheon). The unprecedented size and design of the Pantheon forced engineers to go beyond the regular dimensions and forms, but more typically they had to deal with standard profiles and designs that had been tested on numerous occasions since the Renaissance. For the fortresses surrounded by water and rivers where hydraulics was important, as was often the case in Flanders, the construction was often complex, but here too there was a rich tradition that could be relied on to give adequate design in most cases.[91]

Antoine Picon argues that the military engineers were both ahead of and behind their time.[92] On the one hand, some of them attempted to penetrate an opaque and recalcitrant matter with the tool of analytical mathematics to create an artificial nature subject to laws they discovered, understood, and used. But the fruits of their efforts only blossomed in the following century and were of little use in their own time. On the other hand, they had considerably less opportunity than their competitors, the ingénieurs des ponts et chaussées, to actually construct innovative structures. In 1789, the budget for public works in France has been estimated at 32 million livres.[93] At the same time, 12,000 leagues of roads were being maintained and an equal number were projected or under construction; 336 bridges of various sizes had been constructed by 1765 in the rather backward generality of Auvergne alone; many great bridges had been built, and dikes and canals were being built and maintained. Most of this activity was going on during the reigns of Louis XV and Louis XVI, at a time when major military engineering construction projects (with the major exceptions of Metz and Brest in the 1740s and Cherbourg in the 1780s) were few and far between. From a low of 620,447 livres (excluding salaries) in 1718, funds allocated to fortifications rose to an average of 2,521,905 livres in the period 1764–1769 and to 3,700,000 livres in the decade after 1777.[94] Since the expensive works at Cherbourg were being built in this period, this indicates skimpy funding at the best of times. Their vigorous program of construction not only gave the ingénieurs des ponts et chaussées considerably more opportunity to practice their skills but also ample possibilities to experiment and advance their craft empirically.[95] If they lagged behind the military engineers in theoretical education in mathematics and mechanics, and if even fewer of them appear to have contributed to theoretical advances in engineering science, their actual achievements were remarkable, attracting the admiring notice of travelers like Arthur Young.[96]

Thomas Kuhn wrote some insightful pages on the transformation of the physique expérimentale of the eighteenth century to modern physics, with its strong mathematical and analytical bent.[97] Physique expérimentale could be called a natural history of the inanimate world. It was a Baconian, often local science, concerned with detailed description, classification, and systematization of phenomena quite different from those treated by the physics that we know today. Acoustics, much of optics, electricity, and magnetism began to emerge from the cabinets of curiosities and the lecture halls of itinerant lecturers and were recast in mathematical form with general laws that made them indistinguishable from the rational mechanics and astronomy of the Newtonian Revolution.

Kuhn argues that the delay of the methods of the classical scientific revolution penetrating the congeries of isolated collections of curious details was due partly to the necessity of accumulating an initial critical mass of experimental observations and the development of qualitative theories that could then be mathematized. Partly, however, it was also institutional and individual factors that resulted in early French leadership in this movement. He argues that, after the introduction of the teaching of physique expérimentale at Mézières by teachers of great talent such as Nollet and Monge in the 1760s,

> Baconian subjects [chemistry, electricity and magnetism, acoustics, etc.] increasingly penetrated the education of French military engineers. That movement culminated in the establishment during the 1790s of the École polytechnique, a new sort of educational institution at which students were exposed not only to the classical subjects relevant to the arts mécaniques but also to chemistry, the study of heat, and other related subjects.[98]

Kuhn's suggestion that the emergence of scientifically trained engineers from the Mézières school was a stimulus to this movement is plausible. The school produced graduates who had the mathematical training and the opportunity to grapple with the complexity of matter in their daily work, which provided the right combination of conditions to create a physics that was more than description and classification of phenomena.[99] The slow emergence of modern physics from the context of Mézières and what in many ways was its successor—the École Polytechnique—has a counterpart in the emergence of engineering science from the artisanal, the architectural, and the earlier, more qualitative, engineering traditions. Thus, the development of the new or transformed sciences of electricity, magnetism, and chemistry occurred simultaneously with the growth of engineering science as we know it today and it was closely associated with the French military engineering corps and its school. What is noteworthy in both these movements, for our purposes, is that neither culminated in the eighteenth century. In the period considered in this work, the average military engineer remained a Baconian in his methodology and a man of detail in spite of possessing all the intellectual tools to move into the realm of engineering science and modern physics, where some his most eminent colleagues had already established solid outposts.

A serious obstacle to the whole-hearted adoption and development of a modern engineering science among French military engineers in the period immediately preceding the Revolution was the lure of incorporating all or parts of military engineering science within a more general military science, as sketched out by du Portail. Regular line officers were also wrestling with the problem of the scientification of their profession at a time when science enjoyed great prestige and legitimizing value. Those military engineers who were more attracted to their military roots than their purely technical ones looked on this movement with some favor, but the result was to muddy the concept of an engineering science and to make it almost as obscure as the tentative military science proposed by a number of military theorists, including Puységur and Lloyd.

Another problem was that some military engineers attempted to conflate the engineering science that was growing in the bosom of military engineering with an autonomous science of fortification. The idea of a distinct science of fortification was already embraced by some military architects in the Renaissance with an enthusiasm that was not matched with corresponding success at articulating the precise nature of this science. Vauban never had any illusions about such a science and considered it part of the false posturing of armchair fortifiers and mathematics professors in search of legitimacy and status. His views were generally accepted by most French military engineers. The declarations by Fourcroy de Ramecourt about military engineering being a science (one founded by Vauban, no less) are therefore surprising at first sight.[100] Yet they can be understood in the context of Fourcroy's self-serving polemic with Montalembert. Invoking science to legitimize military engineering and to endow the practice of military engineers with authority was a natural tactic in the Enlightenment. Moreover, the general prestige of mathematics in French intellectual life and its increasing penetration into the regular officer corps—a development that may have been inspired by its importance in military engineering—also no doubt contributed to the prestige of activities attempting to model themselves on Newtonian science.

The conflicting claims about the scientific nature of military engineering (identified as a science of fortification), and about what precisely this science was, gave an appearance of confusion to the whole debate, and it was aggravated by Montalembert's attacks. Yet this confusion, if greater than previously or subsequently, was by no means abnormal. At this particular time and place in its history, military engineering shared the fluidity and tensions of other sciences, most notably chemistry. Bernadette Bensaude-Vincent and Isabelle Stengers have argued in their chapters on Enlightenment chemistry that it was a "conquest of territory."[101] They reject not only the idea of an "essence" unveiled as the science of chemistry progressed to its inevitable triumph in the form of a university department, but even Kuhn's sequence of paradigms. The history of chemistry is more than cumulative progress out of pre-scientific shadows, and it is more than the clash of paradigms. Throughout history, chemistry constructs itself on a territory where it is confronted with heterogeneous spaces it must either integrate or exclude within its own autonomous science. Military engineering, like chemistry at the same time, was also attempting to defend and to expand a territory under attack from other professions and from Montalembert, who incarnated many of these institutional threats and who provided some new personal ones.

Perhaps the belief in engineering's "concreteness" misleads us into thinking that its boundaries are more rigid than those of the sciences, where we might be more prone to entertain the idea that there is a gradual displacement of loci of activity and foci of interest. But these things can also change precisely because of the material challenges and limitations of technical or scientific activity. For Bensaude-Vincent and Stengers, instrumental techniques, crafts, and institutions were the loci around which the form of chemistry was constantly being renegotiated. The same was true for engineering, where one can witness a displacement of interests and practice because of more complex construction, more involvement in that construction, improvements in cartography, and even the temptations of expanding into the domain of tactics and military planning.

10

The Early Career of Marc-René, Marquis de Montalembert

> This gentleman used to be poor, but ever since this [cannon casting] scheme, he has bought a commission in the gendarmes, lands, [and] built a château and a theatre.
> —Marquis d'Argenson[1]

Marc-René, marquis de Montalembert, was born in Angoulême on 16 July 1714 into a family of well-established nobility.[2] Although his family could be described as eminent at the local level and the possessor of a proud name, it does not appear to have been particularly wealthy or influential at Court. He needed, like many others of his kind, to make his way in the world in spite of his noble birth. Fortunately, an eminent patron appeared early in his life. Louis-François de Bourbon, Prince de Conti (1717–1776), was a cousin of King Louis XV, a prince of the blood who could deal on equal terms with the Orléans and Condé families, and Montalembert became an ensign in his cavalry regiment at the age of 18.[3]

No information has come to light so far that tells anything about the details of his early life and education. It is probable that, as for most aristocrats intended for a career of arms at that time, it was rudimentary. The claim in one biographical dictionary that he was educated as an engineer in his youth is probably false.[4] Almost immediately after joining the Prince de Conti's regiment, he fought in the War of the Polish Succession (1733–1738) and was present at the sieges of Kehl and Philippsburg in Germany. These were the first of nine sieges he would often invoke to claim expertise in siege warfare and fortification. In the War of the Austrian Succession (1740–1748), Montalembert again saw action with the Prince de Conti, this time not only in Germany and Austria but also in Italy at the bloody battles of Cuneo and Madone del Ulmo, as well as the sieges of Villefranche, Château Dauphin, and Demonte. He appears to have been liked by Conti and served as the captain of the Prince's personal bodyguard, a select company of 50 noble officers, brilliantly attired and equipped at great expense

264 Chapter 10

Figure 10.1
The frontispiece of volume 1 of Montalembert 1776.

to the Royal purse. The company caused diplomatic headaches because of its showiness during Conti's joint command with the heir to the Spanish throne.[5] Montalembert's connections with a Prince of the Blood put him within the ranks of the court nobility, if not at the highest ranks, then close enough to benefit from the networks of patronage that nobility enjoyed. He seems to have shared Conti's interest in literature and science and it is more than likely that Conti's influence and his interest in the sciences propelled Montalembert into the seat of an académicien libre at the Royal Academy of Sciences in spite of few apparent credentials for that post.

Entry to the Academy of Sciences

The post of académicien honoraire was given to high-born aristocrats and ministers who took an interest in the sciences and the Academy. But the académiciens libres, while not always of the same level of competence as the regular academicians, were normally more involved with science than the honoraires. There was always a large number of army officers and members of the regular clergy with distinct scientific and technological interests who occupied these positions and they were an intermediate grade between the wholly professional academicians and the socially illustrious but not professionally active honoraires. The artillerist and author of engineering textbooks Bélidor, the director of the École des ponts et chaussées Perronet, the artillerist Vallière, the navigator and explorer Bougainville, and Montalembert's nemesis and the unofficial head of the French military engineering corps, Charles-René de Fourcroy de Ramecourt (1715–1791), were all académiciens libres. Montalembert's election was a sign of interest rather than competence and in his particular case it is fair to say that the post of académicien libre represented a lower grade of académicien honoraire.[6]

Montalembert the Metallurgist

Election to the Academy of Sciences, both because of the prestige it represented and the personal contacts it provided in the world of science and the court, no doubt proved useful to Montalembert in the next stage of his career. During the decade after his election, the gentleman from the provinces, who had developed a respectable military career in the entourage of the Prince de Conti and had no previous experience in the management of metallurgical enterprises, became one of the premier ironmasters of France. The story of this episode in his career, although it ended with no apparent continuation, is important for understanding some aspects of his next career as a fortifier.

In spite of its financial and technical failure, it appears to have convinced Montalembert of the possibility of making cheap cannon—an indispensable requirement for his fortifications. And it provides insight into traits of his personality that appear in the future when he presented and fought for his new system of fortification.

When Montalembert returned from his campaigns after the War of the Austrian Succession, Maurepas, who was not only Minister of the Royal Household that included the management of the Academy of Sciences, but was also the longtime Minister of the Navy and the Colonies, set about the task of rebuilding the Navy. In 1750, the Navy had only about two-thirds of the guns required to arm the authorized fleet of 60 ships of the line and 40 frigates.[7] Unlike the army, which obtained its ordnance from state arsenals, the Navy traditionally relied on a number of well-established private suppliers. The area of the Périgord and the Angoumois, Montalembert's native country, was one of the centers of cast-iron cannon production for two reasons.[8] First, it was the hinterland of Rochefort, the major naval base and arsenal which was the prime consumer of its product and relatively easily accessible at a time when the difficulties of transport of heavy artillery pieces (such as the 36-pounder cannon weighing 3.6 tonnes) was a problem that could never be taken lightly.[9] Second, the numerous streams, which provided these transport facilities, could be harnessed for hydraulic power and traversed a well-wooded area that was rich in good quality iron ore. A surge in production of cast-iron ordnance in this area was thus inevitable and Reix des Fosses, an established iron founder in the area, received a contract from Maurepas' ministry for 1,200 cannon in 1748.[10]

But the structure of the French economy during the Old Regime meant that there was no national market for the iron industry.[11] Consisting of relatively small-scale and numerous establishments which generally produced products of poor quality for local markets, the industry had difficulties in coping with massive and sudden orders from institutions such as the Navy. The urgency of the needs of rearmament and the scale of production required was only aggravated when Antoine Louis Rouillé, comte de Jony (1689–1761), Maurepas' successor as Minster of the Navy and Colonies in 1749, increased procurement demands even more and created a seller's market. This may very well be part of the reason for Montalembert, a man with no experience in metallurgy, receiving a contract for the manufacture of cannon. Another reason must have been Montalembert's patrons at Versailles: the prince de Conti, Maurepas himself, and the Duc de Chaulnes (who, unlike Conti, was well liked by Madame de Pompadour). Along with other members of the Court, the Duc de Chaulnes appears to have been involved financially in some way with Montalembert's fledgling metallurgical enterprise. Montalembert's own newly acquired prestige as an academician could only have enhanced his putative capability of successfully running a major metallurgical enterprise. He had already purchased a paper mill with its attendant water resources at the town of Ruelle on the Touvre in the Angoumois in 1750, no doubt with the intention of establishing himself as a maître des forges.[12]

On 19 September 1750 he obtained a contract to provide 800 cannon for the Navy by 1754—a contract that was the second largest in the area after the one for about 1,200 cannon from the well-established cannon founder Reix des Fosses.[13] This was not enough for Montalembert, for he also acquired the smaller contracts of M. de Roffignac and M. Brassac (for 400 and 200 cannon respectively) to give him a contract for a total of 1,400 guns, which made him the largest supplier in the West of France for an amount totaling 1,193,400 livres, a significant portion of the annual French naval budget.[14] He became the recipient of 20,000 livres a month as cash advances from the government as well as important concessions to crown forests in the area. He quickly purchased and made operational a number of blast furnaces and foundries and built a new one at Ruelle, an action that called forth vigorous protests from the town of Angoulême.[15] All in all, he ended up with seven establishments scattered through the Angoumois and began production by the end of 1751.[16]

Montalembert, in his role as a maître des forges, combined both traditional and novel aspects of a dynamic court nobility with a keen sense for exploiting new commercial and industrial opportunities.[17] Undoubtedly he had some indirect familiarity with an industry so closely associated with land and the aristocracy. Yet unlike many traditional aristocrats who were content to lease their properties and mineral resources to others to develop, he took a personal interest in the matter and did not hesitate to introduce new technologies in an industry that at the time was backward by comparison with best metallurgical practice, that of England. Early on, he decided to abandon the traditional method of casting cannon with a core and replacing it with casting without a core and then boring out the center of the cannon (the so-called method of "solid casting" or "massive casting"). This method had been employed in France by the Maritz family for bronze cannon cast for the army, but Montalembert flouted tradition by developing a new machine for the purpose and attempted to apply the method to cast-iron cannon.[18] In addition, Montalembert used two blast furnaces instead of three for casting the larger 36-pounder cannon.[19] At least two furnaces were necessary for the larger cannon. But standard practice, more expensive, was to use the contents of three furnaces, to ensure better quality iron that had undergone a more thorough process of smelting.)

For a time, things seemed to go well. Montalembert had established a new ironworks at Ruelle in addition to the existing works which he had purchased, and at St. Gervais, one of his forges, guns were cast ahead of schedule.[20] In 1752, Montalembert purchased the prestigious office of Lieutenant-General of the King of his native province of Saintonge and Angoumois.[21] He also purchased a commission in the Light Cavalry of the Royal Bodyguards and built a private theatre in his Parisian hôtel.[22]

But soon there was trouble. There were complaints about Montalembert's unorthodox methods, and the first five cannon he delivered to the Rochefort arsenal failed the proofing tests on 13 May 1752.[23] Artillery officers, who acted as inspectors, condemned the iron as brittle and prone to shatter when fired. In addition, they complained that Montalembert was adding extraneous decorative features not prescribed by the official 1689 ordinance for the manufacture of cannon, "in order to give a more pleasing appearance to the guns of his supply."[24] Did this disapproval indicate rigidity on the external appearance of the cannon or was it connected with another complaint that Montalembert's cannon were overweight (a charge that implied fraud, since payment for the ordnance was calculated according to weight)?[25] There were also intimations of fraud regarding glazing of the internal surfaces of the cannon in some cases—a practice that was used for covering impermissible pitting and grooves that were a symptom of structural weakness and propensity to explode.[26]

In view of Montalembert's further career, one can legitimately speculate that both of these possibilities were true. Fraud was an option that does not seem to have been below Montalembert. Not only would ministers of the Crown, as well as his enemies among the military engineers, imply this for his enterprises of cannon manufacture, but he would be accused of this both by his widowed sister-in-law, who took him to court on the grounds that he had misappropriated her dowry, and by business associates in a paper mill enterprise.[27] Similarly, a rejection of standard procedures and a love for the novel, particularly when it was strikingly visible, are another feature of his character that would show itself in the future and in other contexts.

Montalembert had rejected the criticisms of the ordnance inspectors by arguing that they were arbitrary and excessively demanding in a technological situation where perfection was not possible.[28] The ordinance of 15 April 1689 restricted itself almost entirely to visual and tactile inspection of the pieces and required strict conformity with the prescribed dimensions and maximum permissible size of any flaws in the cannon. Montalembert claimed that the artillery officers who served as the ordnance inspectors were demanding even stricter standards for rejection of physical flaws than prescribed by the ordinance on the grounds that the experience of more than half-a-century dictated this. On the same grounds, they were making demands about the quality of the ore and the procedures of casting, something not mentioned in the ordinance, but which the inspectors felt was crucial for determining the strength of the cannon. It was, in fact, standard practice to make rather loose specifications for ordnance on an individual basis for each contract between the state and its suppliers; rarely was reference made to the official ordinance of 1689.[29] Montalembert's insistence (when it suited him) on adhering to an outdated ordinance that was ignored in the light of more advanced practice by the technical inspectors of the artillery is characteristic of his unpredictable style.

Strangely enough, in view of his later defense of novelty, Montalembert assumed the role of sensible conservative and defender of accepted traditions. "I will not plead the cause of ancients," he began, but continued by arguing that since techniques had not changed and that there had been no substantial technical progress between 1689, the date of the official specifications for tests, that "the ancients could say to the Moderns that the present time resembling their own . . . they should follow their example to admit by necessity the defects that do not affect the goodness of the cannons."[30] Thus he argued that the flaws the inspectors found unacceptable were not really significant and did not affect the quality of the cannon, that technology had not advanced, that the older criteria for accepting cannon should be followed, and that the inspectors were needlessly and unfairly demanding perfection from an inherently defective technology. Although his original contract had called for 125 and 175 pieces in 1751 and 1752 respectively, he had cast more than 500 and claimed that more than three-fourths of them should have been accepted.[31] He rejected the concerns of the inspectors for the safety of the gun crews by arguing that "sentiments of humanity" should not lead to "superfluous precautions." The inspectors were accused of wanting "mathematical precision," something that would have required hiring "angels for workers" since one could not expect that men, "especially the men one can get for this kind of work could achieve that kind of precision."

During the first quarter of 1752, only 13 of the 36 guns delivered to Rochefort by Montalembert were accepted.[32] Montalembert's reaction to his difficulties again seems characteristic of his later behavior when dealing with obstacles to his projects for fortification—boldly embracing a novel and relatively untested technical solution to his problems. On 3 February 1753, he claimed that a "new method" would eliminate the defects pointed out by the ordnance inspectors.[33] This "new method" was apparently the solid casting of iron. On 18 June 1753, he met Jean Maritz, who had developed a method of solid casting and then boring for the cannon that were produced in the army arsenal at Strasbourg.[34] Whether this was at his own initiative or at the request of the Navy, which was beginning to take a much closer interest in Montalembert's affairs, is not known. What is known, however, is that Montalembert pursued the idea with enthusiasm. He was not, however, willing to use Maritz's boring machine, which he felt would cost him too much, but proceeded to develop his own boring machine.[35] This was a risky affair, for to replicate a boring machine as good as Maritz's would have been difficult enough, but it was even more difficult for an entirely new material—cast iron—that was considerably more difficult to bore than bronze. Unfortunately Montalembert's machine was not adequate to the task and the cannon he delivered to Rochefort continued to be as defective as before. In the six-month period between the end of 1753 and the middle of 1754, more than 350 cannon were cast, but of these, a seventh were rejected outright on simple physical inspection, and almost the same number needed changes before they could be allowed to be subjected to the proofing tests at the arsenal in Rochefort. There is no indication that the results of these final tests were any better than the first deliveries, which had seen only one-third of his guns accepted.

In January 1753, Montalembert had demanded an increase in his advances from 20,000 livres a month to 40,000 livres a month, but by the end of the year, when he had received 1,200,000 livres in advances (the amount of his original contract), his cannon continued to be rejected and he was desperate.[36] The only other producer in the Angoumois area, Reix des Fosses, was also encountering difficulties, some of which, he claimed, were caused by Montalembert's enticing away his workers and his suppliers. He seems to have copied Montalembert's practice of using only two furnace loads instead of three for casting the large 36-pounder cannon, and from 1748 to 1754 he admitted that he had delivered only 485 of the 1,200 guns specified in his contract.[37] Seventy-one of Reix des Fosses' 103 guns were rejected upon reaching Rochefort, and the intendant there wrote: "Sieur Reix des Fosses is absolutely incapable of carrying out the work that must be done to effect the repairs [to his furnaces and ironworks]; it is scarcely possible to find a more inept and less suitable man to occupy these forges."[38] It was suggested that the lease of Reix des Fosses be canceled and this was eventually done. He was only pressed back into service after the outbreak of the Seven Years' War had made the demand for ordnance even more acute.

Perhaps it was the gravity of the situation, where a high priority was apparently attached to rearming the Navy that enabled Montalembert to stay in business although he had signally failed to satisfy the terms of his original contract. Perhaps Montalembert's performance, bad as it was, was relatively better than that of Reix des

Fosses, the only other supplier for Rochefort. Both could claim their difficulties were caused by an unusually tight production schedule and the large amount of ordnance ordered. Montalembert's connections at Court must have again been put to good use. In any case, the Navy had little choice, for there were no alternative suppliers and the situation was desperate. On 14 December 1753, Montalembert renegotiated his contract for an even larger amount of ordnance than previously (80,000 quintals of iron) in spite of his insistence that he needed a 100,000-quintal contract to avoid bankruptcy.[39] He ended his letter asking for the increase in his contract by again complaining about the rigor of inspection with this comment: "There are very learned people in matters of artillery who propose to establish many new things. I think that this is a case for saying that the best things can be abused."

Jean Maritz had already been sent to Montalembert's establishments in 1754 to install his superior boring machine that was to be used on the cannon cast by Montalembert, whose own machines were to be discarded. By February 1755, two of the new boring machines were erected at Ruelle and work was begun on the immense backlog of cannon that Montalembert had already cast. Casting in sand, rather than clay, as was the case previously, was introduced to improve the quality of the castings.[40] But Maritz complained that the iron from Montalembert's forges was still hard and brittle and the rate of rejection of the guns delivered to Rochefort in May 1755 was still high—51 out of 81 guns were rejected.[41] At that point, when Montalembert was 1,900,000 livres behind in his deliveries, the Navy lost patience and halted all payments.

A new Minister of the Navy, Jean Baptiste Machault d'Arnouville (1701–1794), who was considerably less sympathetic to Montalembert, had taken over from Rouillé on 31 July 1754.[42] On 26 September 1754, Montalembert had written a personal and confidential letter to the Keeper of the Seals, the highest judicial officer of the realm to stave off the inevitable. He claimed that he was a victim. "The unfortunate desire of doing better than had hitherto been done, to perfect an art totally ignored, abandoned at all times to the attentions of the coarsest people of the country—in short, that which should have earned me eulogies served as the means to establish the most unfavorable prejudice [against me]."[43] At the same time that he pictured himself as a misunderstood technological pioneer who got opposition instead of support for his laudable efforts, he also struck the pose of grand seigneur whose delicacy of sentiments and attachment to the state should not be questioned by narrow-minded and socially inferior technicians. "But if the lower orders [le bas peuple] are allowed to give their opinion on an equal footing, then only disorder can result."[44] The appeal did him no good and the government moved to take over the management of his enterprises after it ceased payments. The works at Ruelle were placed under the control of the Crown as a régie (government administered enterprise) in November 1755 and Maritz, who had recently been made inspector-general of naval ordnance was put in charge with the power to order contractors to do what he thought necessary.

At this point, Montalembert's direct active involvement with his metallurgical enterprises seems to have ended. A contract he signed with his brother, Paul de Montalembert, in October 1755, naming him inspector general of his metallurgical operations at the princely sum of 120,000 livres for a ten year period (with free board and lodging that included the services of three horses and two servants) appears as much a sign of abandoning his enterprise as an attempt to protect his revenues.[45]

From this point on, the story is one of either overt or covert litigation. Montalembert claimed that he had not been permitted to execute his contract. The king did, in fact, give him an indemnity of 780,000 livres in October 1771, and in 1774 Montalembert sold his interests to the king's brother, the Comte d'Artois (who promptly sold them to the Crown) for a further 300,000 livres.[46] By 1783, however, he went to court claiming that he had been cheated in the transaction because his occupations in the service of the state (he was apparently referring to his work on planning fortifications in the Colonies) had prevented him from examining the contract of sale attentively. The case dragged on and was continued by Montalembert's daughter and was only finally thrown out of court in 1817, when the Comte d'Artois was an elderly man and heir to the French throne.[47]

There had been mutterings at Versailles about Montalembert's cannon casting venture and, by the end of the year 1755, they seem to have formed one of the primary topics of scandal. The marquis d'Argenson wrote the following in his diary on 11 December 1755:

A certain M. de Montalembert, former page of the prince de Conti, claimed to have a secret for making better cannon for the Navy. He received eighteen hundred thousand livres and has still furnished only seven cannons, but he has purchased lands, has built on them, has made a theatre, and the cannon are not coming. The whole Court is involved in his enterprise, especially the duc de Chaulnes who sees himself as a great designer of machines [machiniste]. . . .[48]

Even allowing for exaggeration and the dislike of the Argenson clan for the prince de Conti, it seems that Montalembert was in trouble at court, but he managed to emerge from the affair unscathed. In another entry in d'Argenson's diary less than a fortnight later one can read the following:

He is under the protection of the duchesse de Chaulnes and through her of Madame de Pompadour. The duc de Chaulnes is himself involved in this affair. This gentleman used to be poor, but ever since this [cannon casting] scheme, he has bought a commission in the gendarmes, lands, [and] built a château and a theatre. However, the Keeper of the Seals Machault has summoned him: Montalembert said that his charcoal was burned by accident, the Minister told him that he deserved to be hanged and it's believed that he could be put in the Bastille within days.[49]

Montalembert did not end up in the Bastille in spite of more rumors of shady dealings and cheating of the other foundry owners whose contracts he had purchased when he undertook his cannon-casting enterprise.[50] It appears that protection in very high places led to his being left alone by the authorities and the ministry of the Navy. Although he was forced to give up his establishments, he was paid for them, and Montalembert was by no means silenced. In 1768, he published a pamphlet that provided a synthesis of three legal opinions he had solicited claiming that he had been wronged, that his furnaces should be restored to him, and that he should be paid damages and interest for the government's action.[51] He provided statistics of the guns he had cast, claiming (inaccurately) that he had fulfilled his contracts.[52] Using his own system of accounting, Montalembert claimed that he was not a debtor of the king who should be expropriated but was in fact a creditor. His contract had merely stipulated a number of cannon, irrespective of quality, and he had not been allowed the contracted four years to replace defective cannon.

In addition to legal arguments, Montalembert claimed indulgence on grounds that had already appeared in his correspondence. He had not, he wrote, received the indulgence that an inventor of a new technology needed and deserved. Furthermore, his performance had not been measured with respect to past practice but with respect to the innovations of Maritz. The four year period to replace defective cannon was "more indispensable in the circumstances of an establishment where all the methods were new, and in which numerous difficulties multiplied at the same time as the uncertainty of the Art, and the risks of the enterprise."[53] New personnel had had to be specially trained. Solid casting of iron cannon had never been attempted before him. The secret of the boring technique had been locked up in the arsenal of Strasbourg and in any case was restricted to bronze cannon. Thus he had both to rediscover and adapt a method already novel.

He claimed there were two reasons for opposition: First, there was the natural distrust of anything new and in particular his new method of using two-furnace castings rather than the accepted three-furnace casting. Second, Maritz was accused of setting impossible standards and at the same time of wanting to steal Montalembert's ideas. The pamphlet confirms the somewhat contradictory impression that appears often in Montalembert's previous dealings with the naval administrators and ordnance inspectors at Rochefort. At the same time he appears to be a partisan of outmoded and crude methods of inspection, challenging any deviation from them that proposed to take into account the progress of the art, he claims to be an innovator attempting to bring about a radical transformation in methods of making cannon.

Montalembert had attacked Maritz indirectly earlier in the august forum of the Academy of Sciences. Indeed one historian has seen Montalembert's pamphlets and paper to the Académie des Sciences as a kind of "Platonic" litigation to take the place of the court cases that Montalembert apparently planned but was forbidden from launching by the government.[54] In a paper published in the *Mémoires* of the Academy on methods of testing cannon, he argued that new methods had to be developed and—in what appeared to be an about face—that simple visual inspections of external characteristics were not enough. More fundamental criteria determined the resistance of a cannon to the stresses of firing. There was, according to him, a single feature—density—that determined this quality and at the same time was easily measured and eliminated the need for the expensive and potentially damaging traditional method of proofing cannon, which involved firing with excessive charges after visual inspection. Instead, the cannon would be simply weighed in air and water and the density determined, thus avoiding any potential degradation of the material through subjecting it to extraordinary explosive stresses. The higher the density, the stronger and better the cannon. Here, we see a penchant for a single principle and a simple answer that would recur in other contexts.

This went against accepted practice, which considered higher density, harder, and more brittle cast iron, as inferior to less dense, softer, and more resilient cast iron. Montalembert claimed that Maritz supported accepted practice because cannon with the latter characteristics were easier to bore with his machine. Thus, he argued, the ease of machining, of prime interest for Maritz, had been allowed to obscure other qualities such as resistance to stress. In fact, Montalembert was wrong. The cannon founders of the eighteenth century were correct in considering a softer iron as being on the whole superior to a harder cast iron in its resistance to explosive stresses. No matter that Maritz was fixated, as Montalembert claimed, on characteristics that permitted easier boring, there was in fact a positive correlation between machineability and resistance to stress.

The response of Maritz, who was not a member of the Academy of Sciences, to Montalembert's ideas is not known. His performance in managing what was now a royal foundry at Ruelle and boring cannon, both those left from Montalembert's furnaces and new ones that were solid-cast, is known. The Ruelle works became the first state-owned and managed works for naval ordnance that marked a significant breach in reliance on private suppliers. Ruelle, founded by Montalembert, thrived under Maritz and his successors and is still in operation as a defense plant today. The French naval ordnance industry was never the same. Although the Seven Years' War found the French fleet badly wanting in cannon, production appears to have improved and increased. The horrendous French losses of ordnance during this war—more than 4,000 cannon in all—are a backhanded tribute to the successful productive efforts of government and private producers.[55]

Montalembert was conspicuously absent from that group of producers. Before turning to the next phase of Montalembert's career it is worth casting a last look at his foray into metallurgy. The naval ordnance industry in France, along with being small, dispersed, and decentralized was technologically backward before Montalembert's entry. Reports by inspectors of his enterprises note his efforts to be novel and candidly admit their own prejudices against his innovations.[56] Other inspectors were aware of other technologies that were being introduced in England, such as remelting and purifying the cast iron in a reverberatory furnace, but hesitated to attempt them. In a report on Montalembert's various foundries, the artillery officer Villars des Brosses was by no means hostile to improvements, but did not forget the human element of technology. There was a cautious tone in his remarks. Speaking of the furnace workers, he wrote

. . . all these people have no other science than old prejudices that have been transmitted to them by their fathers, in which they have been confirmed by practices that they cannot reasonably justify and from which one cannot make them retreat.

Little rays of light are beginning to be diffused today on the whole process [mechanisme], errors are being recognized in many ways, but they will only be able to be corrected gradually and with a lot of tact [ménagement].[57]

There is no such caution in Montalembert, however. He was obviously willing to try new technologies, whether because of natural propensity or as a means of bypassing technical or administrative obstacles such as poor quality iron or overly stringent inspectors. And he paid little attention to the potential new difficulties that could arise in surmounting existing problems. Protected by powerful patrons and sure of his own technical flair, he was willing to launch himself into new ventures even if it meant ignoring bureaucrats and sometimes Ministers of the Crown. And he was even less tender to his employees, who complained about his treatment and deserted him for other employers.[58] The result was failure of a spectacular sort in spite of his construction of a new model establishment at Ruelle and his ambitious attempt to dominate and transform the metallurgical industry of the Angoumois. At a time when there was an urgent demand for ordnance, this led to chaos.

But the chaos led to the intervention of technical experts with innovative ideas who *were* competent in their field. Maritz succeeded where Montalembert failed. By the 1760s, solid casting of iron cannon was an accomplished fact and Montalembert's foundries, under Maritz's management, had turned out 2,300 cannon.[59] A technical innovation had succeeded. While most of the credit must go to Maritz, and the extraordinary condition that existed because of government procurement policies at a time of rapid rearmament, Montalembert deserves indirect credit. More cautious, more modest, and less influential technicians and officials would not have rushed so energetically into a field they knew only too well and whose difficulties were clearly apparent to them. Montalembert seems not to have been inhibited by excessive deference to officials in Rochefort and Versailles nor to established practitioners in the business of metallurgy. If the result was initially chaos and financial losses for the state, the end result was limited but significant technological innovation and the reorganization of the supply of ordnance to the Navy on more modern lines through the establishment of Ruelle. Henceforth the government took a direct role in the production of cannon for the Navy. Thus, Montalembert's short-lived and unsuccessful venture resulted in significant changes that may not have occurred as quickly as they did without his negative catalytic role.

Montalembert in the Seven Years' War

By 1758, Montalembert was in Pomerania with the Swedish army and Pajol, the intendant for the Limousin and the Angoumois, was asking for instructions from Versailles on how to cope with the trail of claims and creditors that Montalembert had left in his wake.[60] Although the whole episode had left a strong whiff of scandal and fraud about Montalembert and had done nothing to improve the technological and scientific reputation that he was defending in the Academy of Sciences, he was not forgotten in the choice of important posts in the Seven Years' War.

Although Conti was deliberately passed over for a military command, Montalembert appears to have maintained his credit at Court—an indication of the credit of his other patrons. In 1757, Montalembert was named French liaison officer with the Swedish army operating against Frederick the Great in Pomerania.[61] From here he corresponded directly in cipher with the Duc de Choiseul, Louis XV's Minister of War.[62] In this position, if we are to believe him, he appears to have exerted as much influence as d'Havrincour, the French ambassador to Stockholm himself, and gained the trust of the Swedish government to such an extent that it wanted Montalembert as Sweden's ambassador to Russia, where he went in 1759 to perform the same task he had performed with the Swedes—French liaison officer with their army.

Montalembert later claimed that it was as a result of his advice and urging that the Russian army attacked and captured Berlin in October 1760, almost bringing Frederick to his knees and driving him to thoughts of suicide. At any rate, Montalembert displayed the talents of a good eighteenth-century diplomat, getting on the good side of the mistrustful Russian commander, prince Soltikov, by such gestures as offering a special carriage constructed in Paris as a gift from Louis XV.[63] It is noteworthy, too, that he was elected a member of the Saint Petersburg Academy of Sciences during his stay in the Russian capital, preceding such worthies as Buffon and d'Alembert in this honor.[64]

By 1761, Montalembert had been recalled to France and appointed commander of the island of Oléron in Montalembert's native region. This island, the largest in the estuary of the Charente, was near the French naval base of Rochefort that had been founded by Colbert at the end of the preceding century. The British Royal Navy had been active on the French Atlantic coast during the Seven Years' War. In 1757, it had attempted an attack on the old Protestant city of La Rochelle, where the English may have even entertained illusions of fomenting a rebellion among French Protestants.[65] The enterprise was a failure, but during the operation, a lone English warship had captured the small island of Aix, about 15 kilometers from Rochefort and the mouth of the Charente. A small fort at Aix, built by Vauban, had surrendered after a single broadside from the English ship.

In spite of its small size (3 kilometers long, at most 600 meters in width), the boomerang-shaped île d'Aix was militarily an important position, for it lay in the roadstead of Rochefort. Ships of the line proceeding up the Charente to dock at the Arsenal of Rochefort were forced to unload their guns in order to navigate the shallow river. When leaving, their guns were also loaded in the roadstead under the shelter of Aix. Thus, the great port built for Louis XIV as an alternative to Brest and which was one of the main naval bases for the French fleet operating in the Atlantic and the colonies in the Caribbean and Canada could be effectively closed by an enemy holding l'île d'Aix.

Fortunately for the French, the English had not stayed, and all the ships that had participated in the disastrous attack on La Rochelle returned to England. Yet in 1761, they had returned and captured the large and important island of Belle Isle in the Bay of Biscay.[66] This island, just fifteen kilometers from Quiberon, roughly in the center of an arc from Brest to Rochefort, could provide an ideal base to harass military and commercial shipping from Brest to Bordeaux. A permanent English presence on any of the

Chapter 10

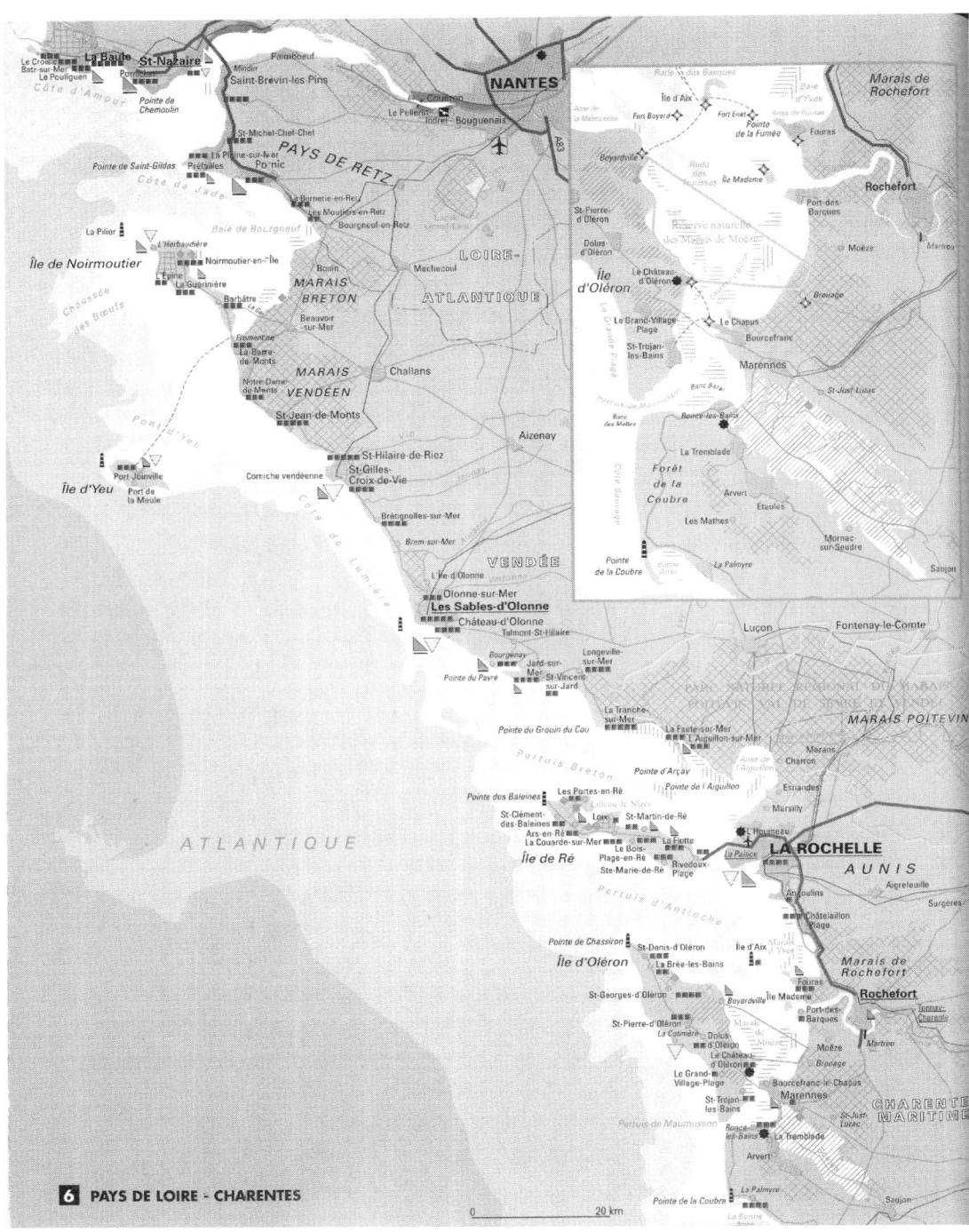

6 PAYS DE LOIRE - CHARENTES

offshore Atlantic islands such as Belle Isle, l'île de Ré, l'île d'Aix, or Oléron would have been a strategic nightmare for French military planners. French commerce moving down the Garonne, the Charente, and the Loire for destinations in Europe and the Americas would have been effectively throttled. Brest and Rochefort would have been in easy reach for a blockading fleet comfortably based on Belle Isle, which could have become a kind of French Gibraltar. Even worse would have been English control of an island even nearer to Rochefort where the Royal Navy's harassment could have become outright strangulation.

Considerations such as these caused the French government to move swiftly to defend the other islands in the Charente estuary after the capture of Belle Isle. One of the moves was to secure Oléron and to do this the main town had to be fortified in haste. It was here that for the first time Montalembert practiced the art of fortifier and supervised the construction of an extensive system of earthworks around the main town on the island and its citadel.[67] The English, however, never attacked Oléron, the war ended, and Belle Isle was returned to France to the immense relief of the French government and the chagrin of influential figures in the British government and Royal Navy. Strictly speaking, Montalembert never held a military command after the Seven Years' War. By then he was almost 50 years of age and had achieved the rank of a maréchal de camp.[68] He had finished his military career as well as that of a maître des forges, and it was to fortification that he would devote the rest of his long life.

Figure 10.2
A map of the Rochefort area including the Charente River and the Ile d'Aix in the roadstead. Source: Massoud and Piboubès 1994.

11
The Challenge of Montalembert

You wrongly criticize my system; it is excellent. As proof I [Montalembert] am presenting another one which differs from it completely.
—Cosseron de Villenoisy, commenting on Montalembert[1]

In 1776, the year of the American Revolution (which would eventually suck France into another conflict with England), Montalembert published the first of what would turn out to be a series of eleven volumes. In that year, there appeared in Paris a richly illustrated and sumptuous volume (produced by the printer of the King's Great Council and the Collège de France) titled *Perpendicular fortification*.[2] The volume had been a long time in coming. Fifteen years earlier, Montalembert had announced a subscription to publish his ideas and had approached the Ministry of War for support for his enterprise. The Minister, after consulting two leading military engineers (Pierre de Filley and Fourcroy de Ramecourt), had asked Montalembert not to publish, appealing to his patriotism to keep his ideas secret for the good of the country. Although it is impossible to say whether the Minister's request reflected genuine respect for his new ideas on fortification or a polite excuse to prevent publication is impossible to say. It is more likely, however, that it was the latter.

But times had changed. A new minister dominated the king's councils—Montalembert's old patron Maurepas, who had returned from disgrace with the accession of the new king Louis XVI two years earlier—and reform and novelty was in the air. The movement for reform in the army that had faltered with the departure of the duc de Choiseul resumed not only in the ministry but among serving officers as well. Four years earlier, Guibert had published his *Essai general de tactique* and had unleashed controversy that was still continuing. Vallière fils, who had dismantled Gribeauval's reforms in the artillery, died in 1776 and the same year saw an intensification of the polemic over Gribeauval's ideas in pamphlets anonymously published by officers attacking and defending his ideas. Gribeauval returned from disgrace to continue his reforms of the artillery as Premier Inspecteur Général du Corps Royal d'Artillerie. The atmosphere seemed propitious for people with new ideas.

What were the new ideas of Montalembert in fortification and what was it that he meant by "perpendicular fortification"? This is not an easy question to answer because during the long quarrel with the military engineers Montalembert's ideas changed and the engineers were right to accuse him of being a moving target. There is a cascade of various schemes of fortification from his fertile imagination that gives the impression of extreme versatility and a lack of coherence in his ideas. His ideas had already changed from the prospectus of his intended book he had presented to Choiseul in 1761. There, Montalembert had claimed that his system of fortification would not only eliminate bastions, but also the covered way and the glacis—a claim he abandoned when he published his book.[3] They were to change further in the years to come, but it is best to attempt an understanding of his essential precepts by starting from the beginning.

Montalembert, like his enemies in the military engineering corps, would have opposed Guibert's attacks on the usefulness of fortification in war and society. Like the engineers, with whom he struggled so long and so bitterly about the best means of fortifying, he felt that fortifications had a positive and "conservative" function in the maintenance of peace, political stability, and civilization itself. The first part of his first book contains the kind of obeisance to fortification as the primary element of security of states and societies that was routine in the writings of military engineers. In the long prefatory chapter of his second volume he reaffirms this and argues that many military disasters of the War of the Austrian Succession and the Seven Years' War were the result of failures of fortifications.[4]

Where he does differ from military engineers, however, is in his belief in the possibility and desirability of impregnability of fortifications. Impregnability is a fortification's ultimate purpose and it is to be judged on this feature. For military engineers, on the other hand, impregnability was of relatively little import. Although incompletely articulated, theirs is nevertheless a distinct concept of optimization of military operations that takes into account an overall economy of material and human resources. As much as by spectacular and bloody clashes in the open field, overall victory is gained by wearing down an enemy, frittering away his resources, and conserving one's own. A siege should force the enemy to spend valuable material and human resources, waste precious time, and permit more facility to one's own maneuvers. Fortresses should only be considered as part of a general system and overall campaign strategy. Concentrating on the impregnability of a particular fortress is counterproductive. If it was achieved only with a disproportionate allocation of resources that could always be directed more effectively elsewhere, it could actually be harmful. There is little of this in Montalembert. It is the narrow domain of military architecture rather than broader considerations that is the focus of Montalembert's thinking.

For Montalembert, the ability of fortifications to withstand assault had drastically declined in recent years. He claimed that part of the reason was a certain moral decadence, in that defenders gave up too easily, but this was partly linked to the correct belief that existing fortifications were not capable of withstanding a vigorous attack. But along with the moral decadence that had led to a weakening of fortification was a technical reason. The power of artillery had increased and had exposed the flaws of existing military architecture to a continually increasing degree. The bastioned trace was an ineffectual and essentially unimaginative modification of feudal fortifications that did not take into account the development of artillery. The kind of balance that had been temporarily achieved between the powers of offense and defense by Renaissance fortification had been broken and now offense was unmistakably superior to defense, unlike the situation in the Middle Ages and antiquity.[5]

That the inferiority of modern fortifications in the face of attack had been accepted without comment even by the practitioners of the art of fortification—the military engineers—and that it was considered a normal state of affairs that a fortress attacked was a fortress taken, was scandalous to Montalembert. For him, the teaching of fortification, which was stuck in a rut of conservatism and unthinking acceptance of established procedures had the rather paradoxical effect of identifying very clearly the defects of the accepted system of bastioned fortification while at the same time inducing a sense of hopelessness about the possibilities of ever remedying them.[6] It was, for Montalembert, a profoundly demoralizing teaching. Ordinarily, a consciousness of defects would have led to remedies and Montalembert argued that Vauban, in his final system at Neuf Brisach, had started along the path to the correction of these defects, but that Cormontaigne and his followers in the Corps of Military Engineering had not only stopped along the path to improvement but even regressed.[7] This was all the more unpardonable because the defects of bastioned fortification were so obvious.

One defect, according to him, was that the basic principle of flanking fire—that the various woks should provide mutual support—had been violated. The faces of the bastions were too oblique to provide effective support to each other.[8] At the same time, the flanks of the bastions, which in Pagan's and Vauban's system were perpendicular to the faces of the neighboring bastions, were unnecessarily far away, thus having their lethal range of fire restricted. Montalembert made the assumption, one accepted by virtually all military men in his time on the basis of experience, and which was one of the cornerstones of his own ideas, that soldiers had an invincible tendency to shoot straight ahead during an attack, that is, in a direction perpendicular to the rampart that protected them. Thus, if two faces of adjacent bastions were not perpendicular to each other, the ditch in front of a bastion would not be covered as effectively as it should be by the fire of the adjacent bastion. Moreover, it would have been more effective to prolong the faces of the bastions until they joined at a right angle—the essence of perpendicular fortification and the feature which gave it its name—because the flanks of the bastion were too far removed and recessed to give effective support to the face of the adjacent bastion. As well, the space between the flanks and the curtain wall was not enclosed within the fortress, thus leaving it outside the fortress, subject to easier occupation by the enemy, and effectively lost to the defense. Here was another argument of Montalembert's: more space was enclosed by his ramparts. A drawback of the bastioned trace, more tortuous and indented than the simpler, cleaner lines of perpendicular fortification, was that it at once left more space outside its walls than it should have and that it was longer and consequently more costly than Montalembert's fortifications in enclosing an area of equivalent size. By retracting and tucking in, so to speak, the flanks of the bastions, the defenders were needlessly limiting the potentially lethal range of their firearms. Walls that were further advanced would enable defenders on them to control that much more of the surrounding area in front of the bastions with their fire.[9]

As for the bastions themselves, their pentagonal form made them more dangerous for defenders subjected to cannon fire. Cannonballs just missing the parapet of one face of a bastion, say, would catch the defenders of the opposite flank unprotected in the rear. A cannonball that did not achieve its aim of battering the face or flank of a bastion could always kill soldiers or dismount artillery pieces on the corresponding opposite flank or face. And the enclosed pentagonal space provided a wonderful "echo chamber" for a ricocheting cannonball bouncing around from one wall to another. The cramped bastion was, in Montalembert's words, a "funnel" for incoming projectiles. It was the mortal vice of all modern fortification.[10]

The first feature of Montalembert's new system that strikes the readers of volume 1 of *Fortification perpendiculaire* is the new form of the ground plan. One is tempted to conclude that the traces of existing systems were the crucial defect and a change of this form was by itself almost sufficient as a remedy. Bastions were to be eliminated and all lines were to be straight with all their angles rentrantes (re-entrant angles) forming right angles. Since the salient angles of the trace were to be a minimum of 60° to prevent crowding at the tip (an empirical rule that was also observed for bastions), geometrical considerations showed that the basic polygon on which such a trace could be constructed had to be a dodecagon at least. Thus, it might appear at first glance, that Montalembert's traces were useful only for rather large fortresses. But as the subtitle title of his first book promised, Montalembert claimed that perpendicular fortifications were also applicable to polygons with fewer sides by "embroidering" them to bring them up to the minimum number of sides.[11]

Along with form, a second feature of the system was a perimeter composed of a multiplicity of continuous defensive lines. The traditional bastioned fortress had only a single defensive line (at the interior limit of its defensive zone). Once the enemy had succeeded in breaking through it, he had penetrated into the fortress and the siege was over. The defensive outworks were not continuous and suffered from their fragmentation. By replacing them with *continuous* and *multiple* lines of defense, the attackers' difficulties doubled, tripled, and multiplied in proportion to the number of continuous defensive lines.[12] Penetrating one no longer meant the end of the siege; another siege started all over again at the next line.

In the actual construction of the multiple barriers of his fortifications, Montalembert called for the separation of the masonry reveting walls from the packed earth ramparts. The masonry walls, which had acted as retaining walls, would now be free standing, much stronger, and better able to withstand battering without the pressure of the earthworks that they normally sustained. At the same time, separating the ramparts from the retaining walls and converting the latter into an independent barrier provided the extra continuous enceinte (perimeter) that Montalembert was advocating. The masonry wall could also house casemates for cannon, a feature of his fortification that would be emphasized even more in his later books.

Casemates were not a new idea in fortification and go back to the earliest examples of gunpowder artillery fortifications in the Renaissance.[13] Vauban, too, had built them and they were a feature of his most developed work at Neuf Brisach.[14] Yet there was little enthusiasm for their use in spite of what appear to be the vital advantages of protection from hostile fire.[15] The accumulation of smoke was considered to be a serious problem and efforts to overcome it occurred throughout the eighteenth century.[16] Although strong, masonry casemates were not invulnerable; they could generally stand up to mortar bombs but they could be reduced to rubble by heavy-caliber flat-trajectory artillery fire and this was a significant consideration in regarding the morale of troops. With all the horrors of the ricochet shot on the open gun platform of the bastion, French military engineers continued to talk of the reluctance of troops to fight from

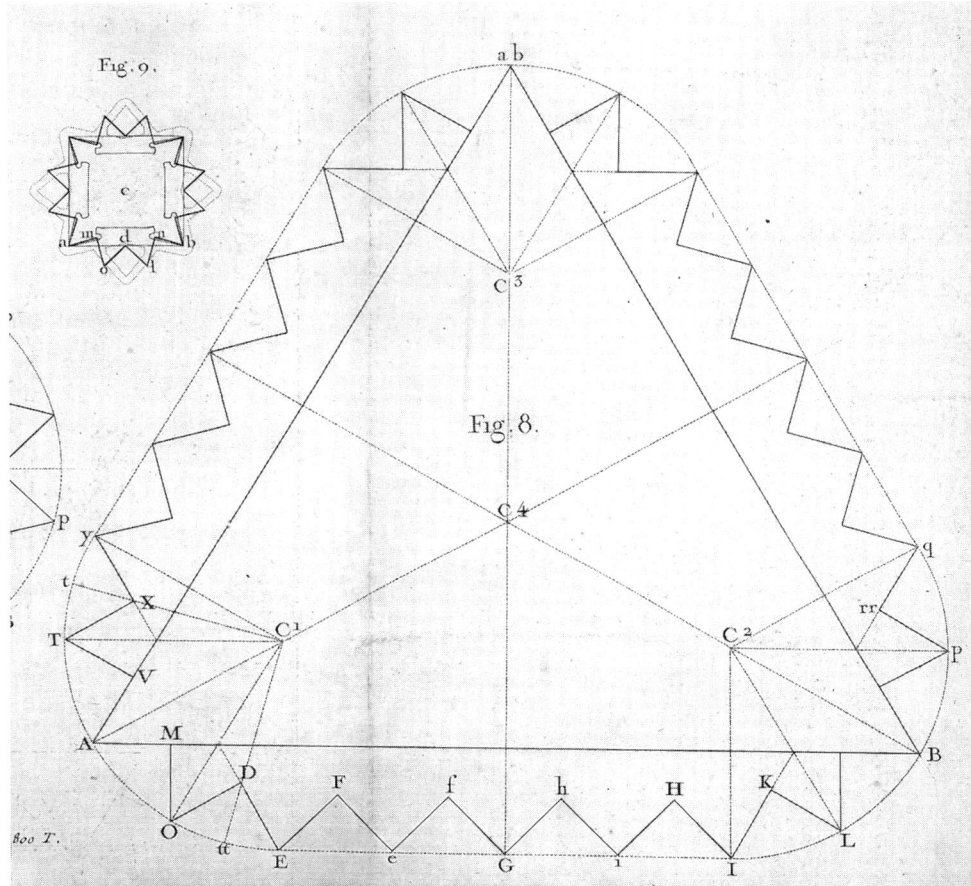

Figure 11.1
"Embroidering" a ground plan. In this illustration, Montalembert shows how a triangular perimeter and a square bastioned fortresses can be converted to plans satisfying the requirements of his "perpendicular fortification": that the salient angles be at least 60° (a requirement that all fortifiers agreed was the minimum to prevent crowding in the salient) and that the reentrant angles always be 90°. Elementary geometry shows that this condition could be satisfied for regular polygons with 12 or more sides (dodecagons). A regular perimeter with fewer than 12 sides had to be reduced in some way to a dodecagon. Irregular perimeters and triangular ones had salient angles of greater than 60° "embroidered" on the perimeter. At corners the connecting salient angles were reduced, but never below 60°. Source: Montalembert 1776–1797. Courtesy of Massey Library, Royal Military College of Canada.

casemates—something considered alien to the French military character that preferred to die out in the open than to be buried in the smoking ruins of a casemate. The embrasures of casemates were also not popular with artillerymen. They were large enough (generally nine feet wide for effective aiming) to be relatively easily hit by an attacker's cannon and showers of lethal flying stone chips were much feared. Montalembert ignored these counter-arguments that appeared again and again; he always insisted that the defending artillery be covered as much as possible.

In his initial ideal system, a double defensive line did not suffice, and Montalembert proposed no less than four successive defensive perimeters. The rampart that enveloped the main body of the fortress was separated by a dry ditch from a simple masonry wall with loopholes that had in front of it a wet ditch, then another earthworks [couvre face], another dry ditch, then a casemated masonry wall, another dry ditch, then the counterscarp with its covered way before the glacis. All in all, an attacker would have to overcome five barriers, all manned by musketeers, and three of them containing cannon, with the unprecedentedly large numbers of cannon in the casemated wall being protected from enemy fire.[17] A bastion could normally hold a maximum of 20 cannon, and half that number was more usual.[18] Montalembert proposed that even small fortresses be equipped with multiple lines of defense. In his full blown trace, the distance from his guntower to the crest of the glacis at a salient angle ran to about 200 toises, almost double the depth of the entire defensive perimeter at Vauban's Neuf-Brisach.[19]

On the other hand, Montalembert's casemated walls were supposed to be able to accommodate cannon spaced a mere 8 feet apart (as opposed to the usual 18 feet), just half-a-foot further apart than the cannon on the crowded gun deck of a warship.[20] They were to run the entire length of the wall and there was thus an increase in the number of cannon in his fortresses by an order of magnitude over those of the traditional fortress.

His new geometrical form of the trace and the radical increase in firepower were the two defining characteristics of his system. Firepower, clearly present but not emphasized in his first volume as much as it would be in the later volumes, would come to displace the shape of the trace as the main feature of his system. While cannon played an important delaying role for the defenders against the attacker in the traditional fortress, for Montalembert, they were supposed to play a decisive role in repelling enemy attacks. The fortress was not merely a passive obstacle by virtue of the bulk and configuration of its material parts, but it was to be an active focus of devastating fire that was to destroy the considerably less numerous attackers' artillery.

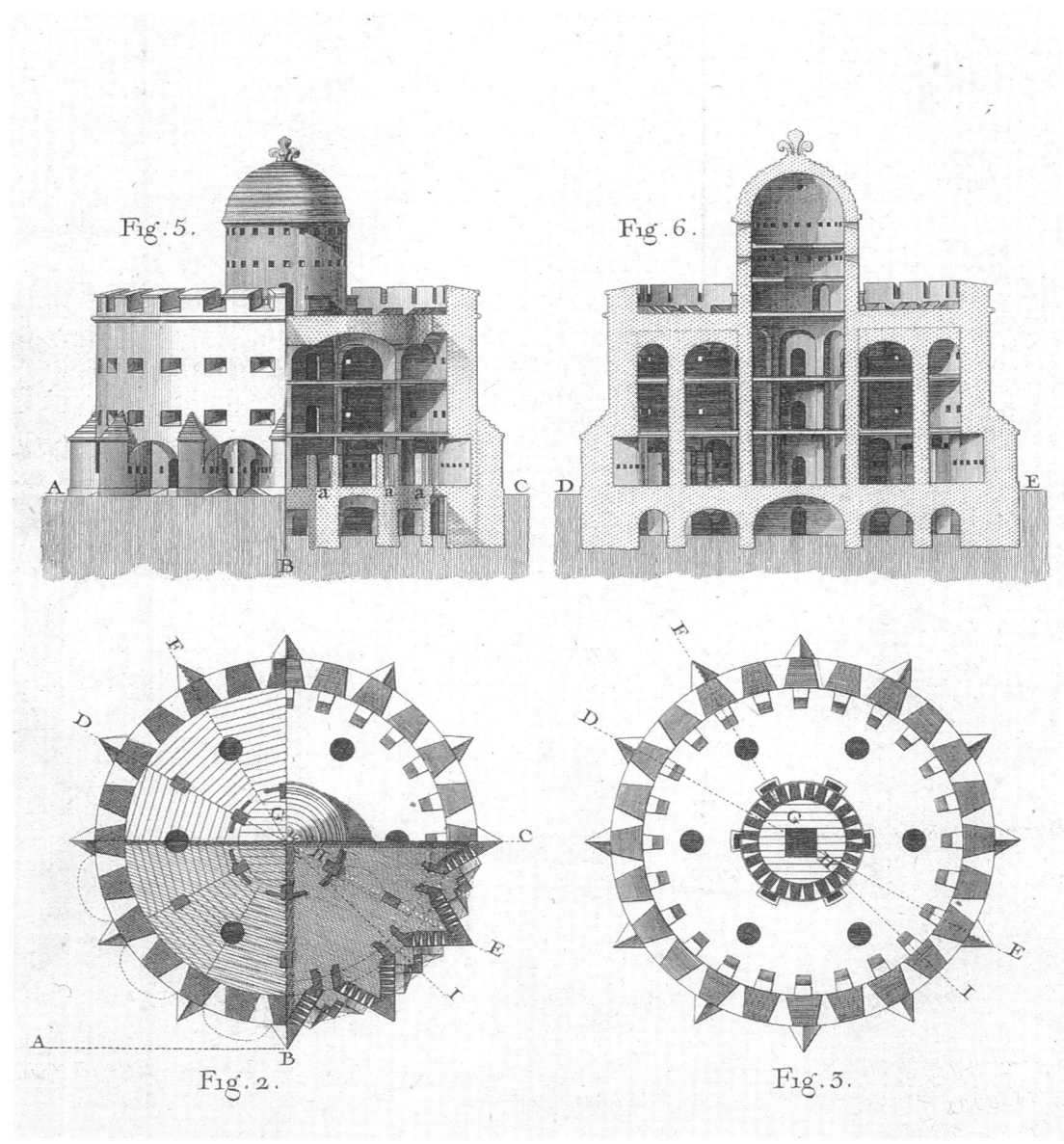

Figure 11.2
An early version of a Montalembert gun tower. Source: Source: Montalembert 1776–1797. Courtesy of Massey Library, Royal Military College of Canada.

An application of this principle is another feature of Montalembert's fortification: the circular multi-story masonry gun tower. Where high cost restricted work to modification of traditional fortifications, such gun towers were to be built at the gorges of bastions to serve as redoubts. In his perpendicular fortification, they were to be located at the center of the base of his triangular salients. Gun towers were to become more and more important in his system—serving not only for proximate defense within the works but also, as the towers became bigger, taller, and stronger, for defense at a distance when its fire was to be directed at the siege works on the glacis.

A feature of Montalembert's system that illustrates both the importance of firepower and military architecture in his scheme of fortification was the caponier [caponnière]. Initially, the caponier had been a communicating path that cut across the ditch at right angles from the main perimeter that was protected on both sides by ramparts and usually open to the sky. The walls were loopholed so that musketeers could also enfilade the ditch to harass an enemy crossing it. Montalembert made them into virtually independent forts, three stories high, accommodating cannon as well as musketeers. Such a caponier, by the intensity of its fire could not only make the ditch virtually impassable but would destroy the besieger's final siege batteries on the lip of the covered way, thus preventing their battering action against the walls. It would be almost impossible to make a breach, and if by miracle it had been effected, an attacking column of troops would be caught in a storm of fire sweeping along the ditch from the caponier. These masonry cathedrals of death had a central nave flanked by two-story arcades housing cannon and musketeers, who not only used the caponier in combat but were also housed there permanently along with their munitions and supplies.

While Montalembert did not like outworks and claimed they dangerously fragmented the defensive perimeter, he did approve of detached forts that were powerful enough to withstand a small siege on their own. Forts surrounding a main fortress, linking together their webs of fire are another cardinal feature of his system. They first appear as an idea in the third volume of his work when he proposed a plan for the fortification of Cherbourg. Vauban had also talked about forts and fortified camps around his fortresses, but Montalembert felt that his own forts would be immeasurably stronger than the bastioned forts of the existing systems. Moreover, Montalembert was against the multiplication of outworks that had been one of the main features of development of fortification in France just before and during the time of Vauban. He felt that they were more harmful than beneficial because they were relatively small, usually poorly coordinated with the main perimeter, and could be easily

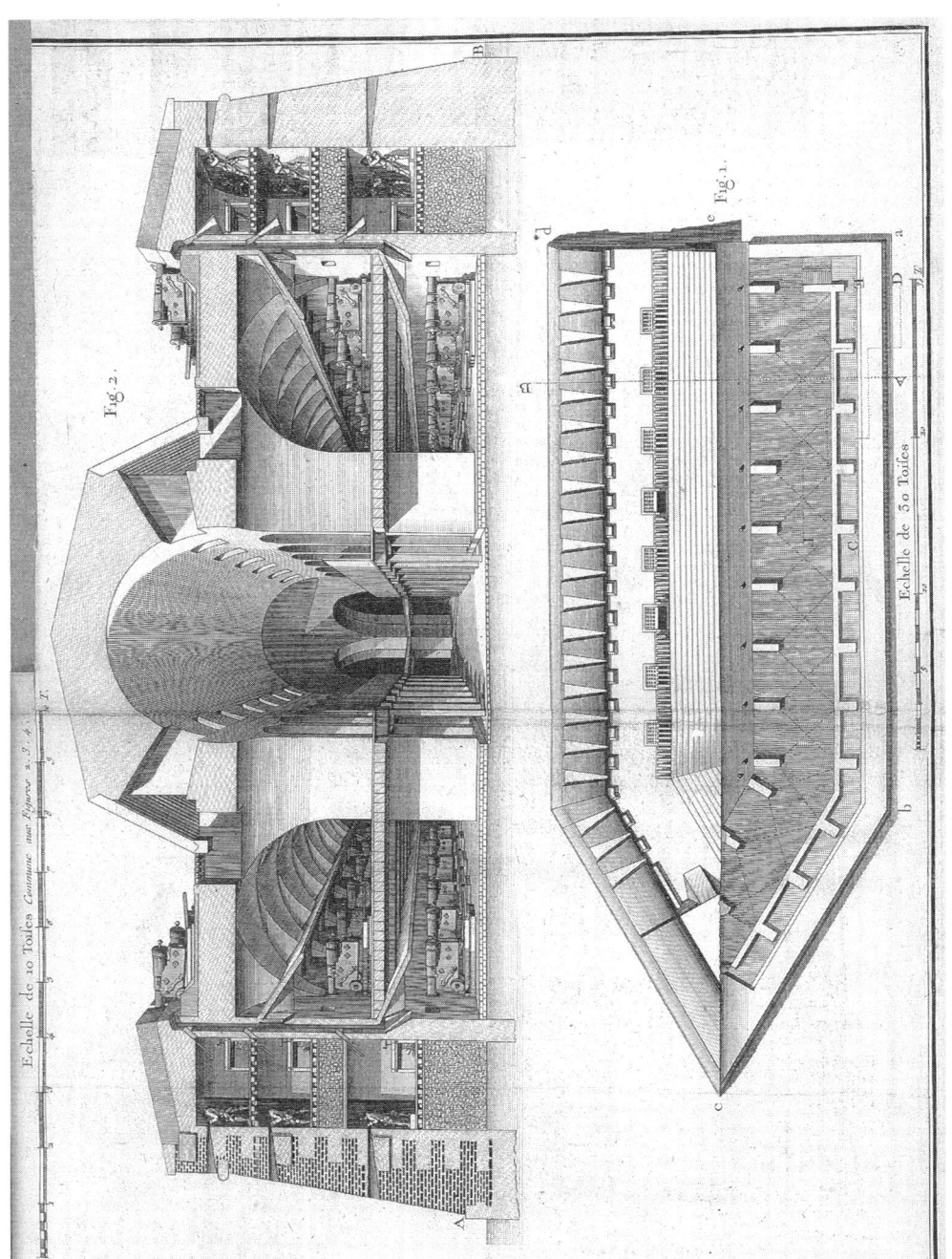

Figure 11.3
Montalembert's Great Caponier, often reproduced in writings on his work. Source: Montalembert 1776–1797, volume 1, plate 13. Courtesy of Massey Library, Royal Military College of Canada.

captured by the enemy, providing convenient attack platforms against the main body of the fortress. A fort, on the other hand, was larger, could put up a self-sufficient defense, and by contributing to a pattern of interlocking fire with other forts could hinder and even stop the advance of enemy siegeworks toward the fortress. They would, he claimed, be as strong as a regular fortress, and he devoted much of his second volume to a detailed discussion of this subject, giving a number of specific theoretical case studies of forts.

It is fair to say that all the elements of Montalembert's system as it developed over a period of more than 40 years are clearly present in his first volume published in 1776. Yet there is also a certain hierarchy of importance among these elements and the order of the hierarchy changed to some degree over time. Within the basic framework of form and firepower, the following enumeration of these elements gives the order of importance they had in his early work:

1. The elimination of the bastion and the institution of the perpendicular trace. Form is initially the primordial feature for Montalembert.

2. Multiple, continuous, and coordinated defensive barriers that follow the basic ground plan. Along with the proper form, the perimeter of fortified areas should have depth and each succeeding layer should be a once self-sufficient and linked with the following one.

3. Separation and strengthening of the masonry revetments from the earthen ramparts. What was perhaps originally the easiest practical way of creating an extra continuous barrier perhaps began to take on extra value because of the possibility of strengthening and converting the simple masonry wall into a line of casemates containing cannon.

4. Multiplication of cannon sheltered from enemy fire in casemates, sometimes along the entire length of a wall. Along with a new form of ground plan, this was his second fundamental idea and was to become even more important and indeed dominant in his later work. It reflected the importance artillery held in Montalembert's thinking and was at the heart of the next two features.

5. The round multi-story masonry gun towers rising well above the fortress that were to provided not only a final line of defense but even to engage in a successful artillery duel with the guns of the besieger located in the siegeworks on the glacis.

6. The caponier, located athwart the ditch perpendicular to the walls, in theory powerful enough not only to prevent the crossing of the ditch but to destroy siege batteries at the edge of the covered way.

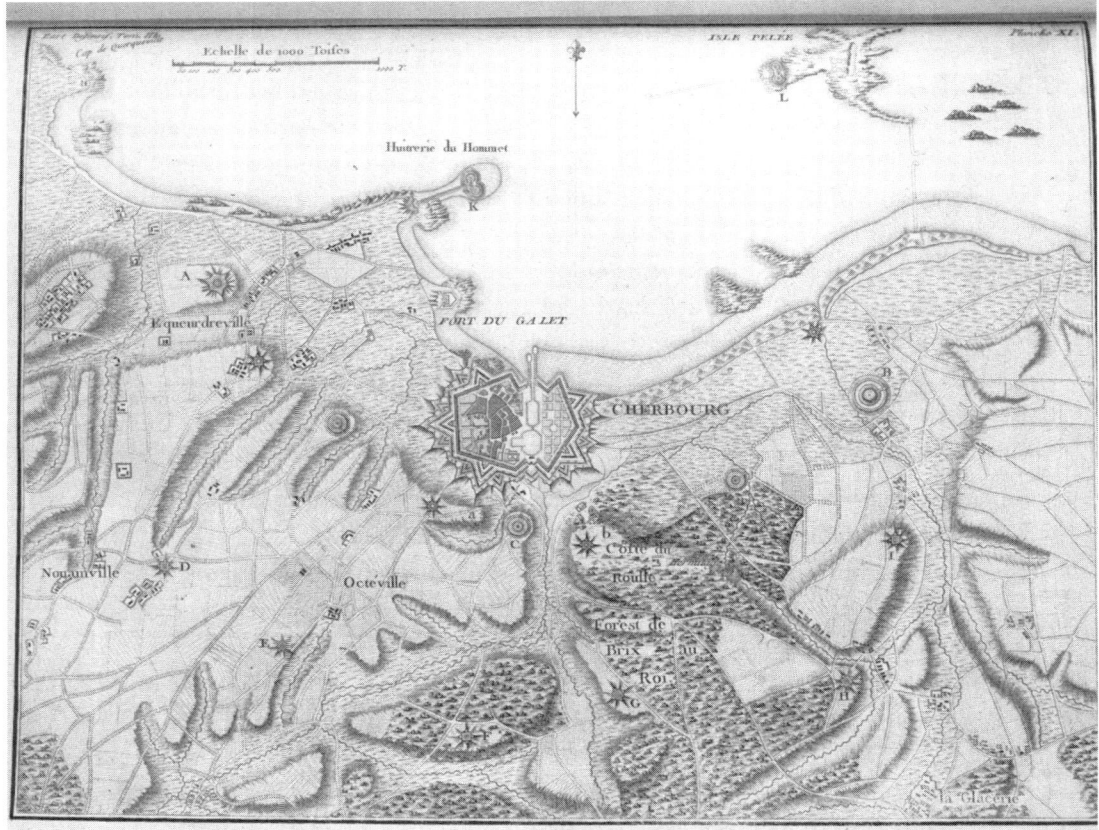

Figure 11.4
Forts proposed for Cherbourg. Note the central fortress, the forts in the harbor, and the forts dotting the countryside. Source: Montalembert 1776–1797. Courtesy of Massey Library, Royal Military College of Canada.

7. A necklace of smaller forts around the main fortress, each capable of sustaining a siege by itself but locked into a mutually supporting web of fire from the other forts to keep the enemy farther away from the main fortress.

None of these features is truly original, in spite of Montalembert's claims. For example, his enemies reproached his circular gun towers for being a regression to medieval forms. Round masonry gun towers were not unknown both in the Mediterranean area and in Northern Europe, where Montalembert very likely had the opportunity to see the towers of Erik Dahlberg (1625–1703), the "Swedish Vauban" who had served three Swedish kings, including Charles XII.[21] The masonry walls, separated from the earthworks behind it, were also considered to be a mere modification of the so-called fausse braye, a feature of early bastioned fortifications, particularly popular with the Dutch, who relied heavily on earthworks in their fortifications, and which had never been popular in France. They had gone out of style by Vauban's time. Everyone agreed that perpendicular traces were desirable and the more one exploited opportunities for perpendicular fire the better.

After the theoretical and systematic considerations of volume one, devoted mostly to a critique of bastions with suggestions for remedial work, geometrical considerations underlying his new system, and the description of an ideal front of fortification, Montalembert had tackled the problems of artillery embrasures in his new fortresses and presented a number of case studies. Instead of large fortresses where his initial theoretical principles could be easily applied, most of these case studies are examples of small forts or redoubts that could be built very quickly with simple materials. There seem to be two inspirations for these kinds of fortifications. One is undoubtedly the challenge of fortifying French colonial possessions that Montalembert had been considering since 1772. (There had been a bitter struggle in various ministerial committees between Montalembert, military engineers, artillerists on colonial fortifications before he had finally decided to publish his ideas.) His suggestions for the "Fort Condé," for example, envisage a square fort with each side (measured along the middle of the external ditch) being 67 toises in length (53 toises along the crest of the main parapet) that could be built in 26 days by 400 laborers and 50–60 carpenters. It would consist of earthworks and some construction in wood—already cut to proper dimensions in advance—for palisades and caponiers. The wood necessary would not amount to more than 26,400 cubic feet—the equivalent of 792 maritime tons—which could be easily transported in a single ship. One cannot help being struck by the coincidence between Montalembert's second and final project for the fort he was asked to build at l'île d'Aix and his proposals for the fort Condé, even though they were radically different in form.[22]

The other inspiration, explicitly mentioned, is the idea of temporary redoubts put forward by the Maurice de Saxe in his *Rêveries* published in almost 20 years before *La Fortification perpendiculaire*. Maurice believed strongly in the importance of substantial field fortifications as well as semi-permanent redoubts occupying strong points during a field campaign capable of sheltering a battalion of troops (about a thousand men). Montalembert used the basic model of Saxe's redoubts to propose a series of his own forts, which he named after branches of the royal family and which increased in size

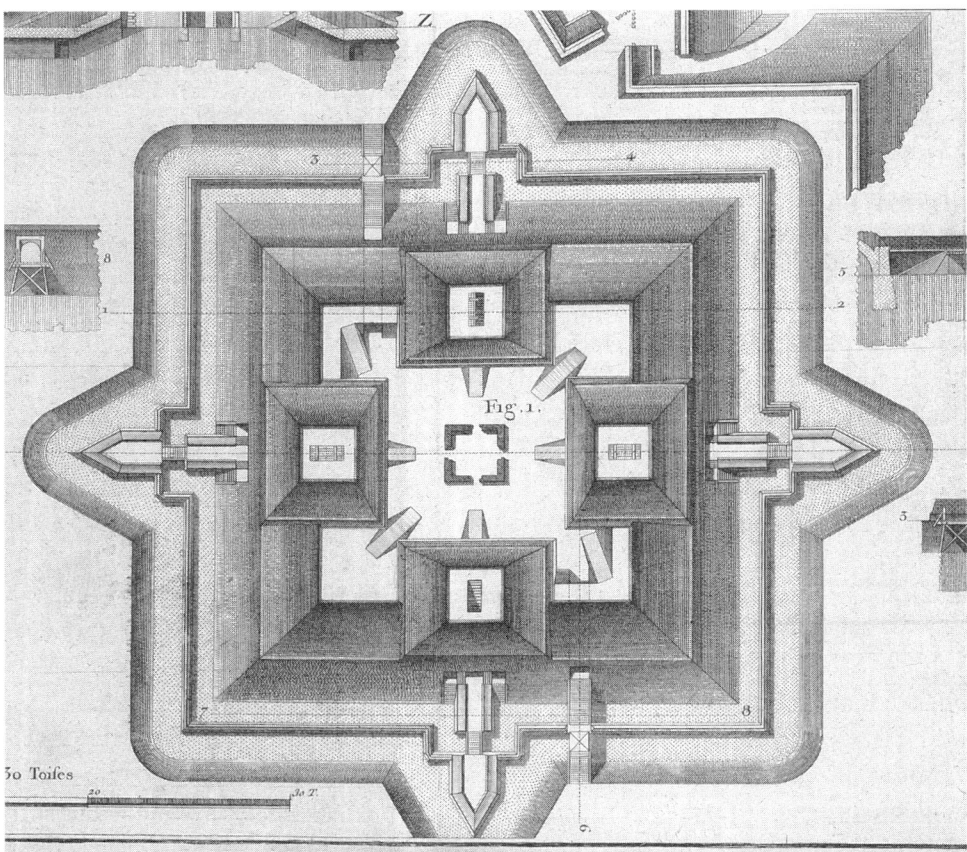

Figure 11.5
The simple Fort de Conti is just a square redoubt with caponiers attached to the middle of each wall and extending perpendicularly into the ditch. Unlike Montalembert's first two elaborations of Maurice de Saxe's redoubts, which were made completely of wood and earth, Montalembert's had a detached masonry wall with loopholes in the ditch before the rampart, and the ditch was widened in order to enlarge the caponier so that it can accommodate some light artillery pieces in addition to fusiliers. Cavaliers (parapeted platforms rising above the fortifications) are built into the wall behind the caponiers. Source: Montalembert 1776–1797. Courtesy of Massey Library, Royal Military College of Canada.

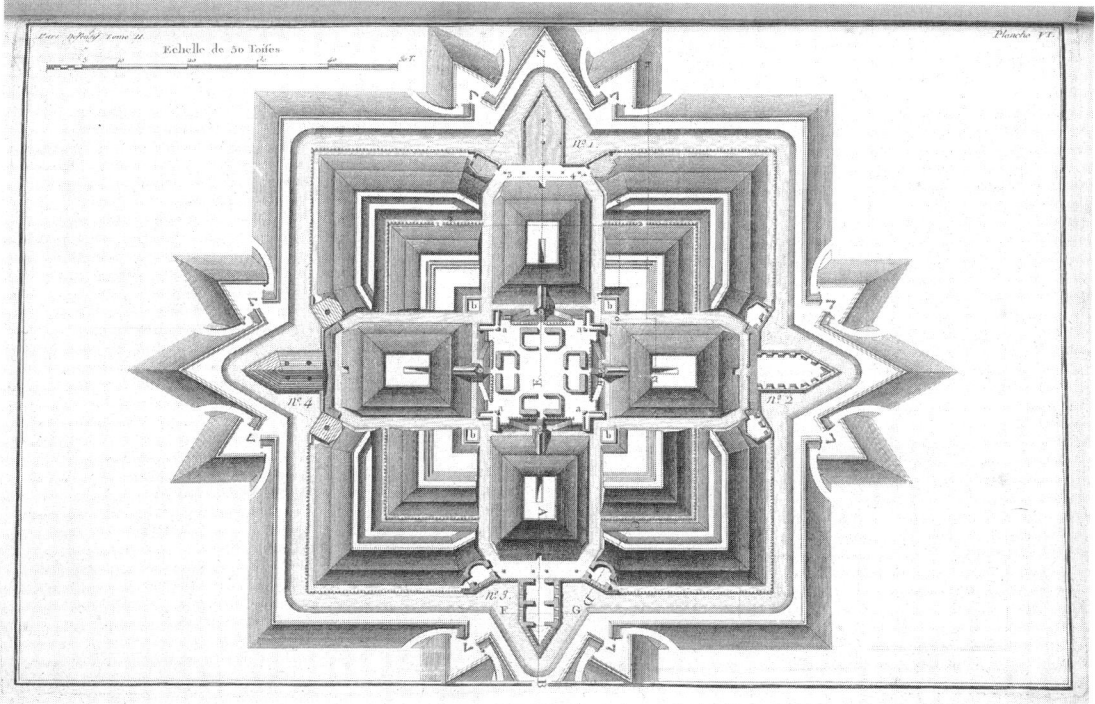

Figure 11.6
The Fort de Condé, based on the same square model as the Fort de Conti, is substantially bigger (though built only with earth and wood) and is intended to be more than just a temporary redoubt; it is supposed to be capable of withstanding a regular siege after being erected in less than a month. The cavaliers, which have galleries for fusiliers at their base, are behind the main walls and detached from them. A continuous wooden palisade is linked to the caponier, which is separated from the main rampart. There is a second discontinuous rampart behind the main rampart, giving four barriers in all. The core of the fort is a raised and palisaded platform with casemated barracks on each corner. A new theme has appeared here. The fort is divided into what are essentially four autonomous and distinct forts that not only provide supporting fire to each other but must be overrun individually by the attacker for the entire fort to be taken. Source: Montalembert 1776–1797. Courtesy of Massey Library, Royal Military College of Canada.

and complexity from the simplest—the Fort de Conti, which like Saxe's redoubts measured 15.5 toises along the crest of the main parapet. They differed from his full-scale fortress that he named Louisville, a dodecagonal fortress with each side between the salient angles measuring 180 toises. Louisville was essentially an application of the ideas put forward in volume 1 of *La Fortification perpendiculaire*. The number of its various defensive lines, moving outward from the innermost enceinte to the glacis is breathtaking:[23] There were 16 defensive lines, and it is little wonder that Montalembert claimed that such a fortress would be impregnable.[24] If one can speak of the primal Montalembert fortress, then Louisville comes closest. The accumulation of additional defenses begins to smack of obsession. The story is different, however, in the case of the smaller forts and redoubts that take up most of the second book.

While the general trace of Louisville can be fairly described as the tenaille trace—essentially a saw-toothed defensive perimeter that was popularized in Germany in the works of the German military architect Speckle during the seventeenth century—the basic ground plans of Montalembert are examples of the polygonal trace that became popular, mostly outside France, in the nineteenth century. Montalembert's forts are mostly squares with caponiers protruding from the middle of each side to provide flanking fire in the ditch. In some cases, the covering ramparts that enveloped the fort and its caponiers gave sharp acute angles and one can speak of a hybrid system with an internal polygonal trace and an external tenaille trace that combines features of both. Thus, fortifiers in the nineteenth century who cultivated the tenaille or the polygonal trace could look back with equal legitimacy to Montalembert as their precursor. This is in fact one of the problems of his legacy.

As Cosseron de Villenoisy, a nineteenth-century military engineer and historian of fortification put it, Montalembert often seems to say "You wrongly criticize my system; it is excellent. As proof I am presenting another one which differs from it completely."[25] There is a cascade of various schemes of fortification from his fertile imagination that gives the impression of extreme versatility and a lack of coherence in his ideas. There are additions and some changes as Montalembert quickly expanded his original single volume to four.

In addition to the initial design of the caponier, several others are added. Cavaliers (elevated firing platforms within fortified works), absent from his first designs, become important in his forts and reflect his concept of what could be called autonomous defensive zones within the fort, which in some ways is in contrast with his initial insistence on the importance of continuous and multiple enceintes. In fact, the continuous enceintes remain for the forts although it seems that Montalembert is now prepared to

Figure 11.7
Detail of the great dodecagonal fortress of Louisville. Source: Montalembert 1776–1797. Courtesy of Massey Library, Royal Military College of Canada.

admit, implicitly at least and only for the sake of argument, the idea that the enemy can penetrate into the interior of the fortress. Here indeed, there is a fragmentation of defensive lines but no decrease in their number.

One of his examples of forts and redoubts, a square fort with cavaliers and casemated walls gives the impression of geometrical delirium.[26] Each side of the square fort is 332 toises long and there is an explosion of covering walls, ramparts and caponiers. As many as ten obstacles (depending on which variant of the design is chosen) confront potential attackers before they come to a central square each of whose sides is the base of an equilateral triangle pointing outward and whose sides in turn each have a caponier.

There is a return to sobriety of sorts with Fort Royal, another large square fort with each side 180 toises long (the standard front of fortification in the classical tradition of fortress design according to Pagan and Vauban). But this is only in comparison with such proposals as that of the Fort d'Artois, described as having a "a mixed line triangular core with an angular base and general counterguard."[27] In Fort Royal, there is a continuous rampart with detached caponiers, and a detached cavalier with its own caponiers behind the rampart. Each corner has a guntower with a number of separate earthen ramparts covering them. The large caponier in the ditch is itself covered by a counterguard, followed by a continuous covering enceinte with casemated re-entrant angles. In the re-entrant angles of the covered way, but separated from it by the ditch, there is a place d'armes with an interior redoubt and attached casemated ailerons with cannon. An interior platform in the fortress has casemated barracks at all its four corners that can cover the spaces between cavaliers and ramparts with lethal fire. All in all, there are 230 covered and 100 uncovered cannon per side (a total of 1,320 cannon, the equivalent firepower of a fleet of about 18 ships of the line), which would have required a minimum of 3,960 soldiers jammed into an area of about 27 acres, most of which was composed of fortifications. This was to bring prompt reaction from critics in both the engineering corps and the artillery.[28]

Another idea of Montalembert's that was widely associated with his name is the string of mutually-supporting forts around a central fortress, which he presents in his proposals for Cherbourg in volume three. Cosseron de Villenoisy is prepared to concede that these proposals are a happy application of Montalembert's ideas that attracted respectful attention in the following century. The main fortifications, although following the principles of perpendicularity, multiple enceintes, and covered guns are stripped of some of the more extravagant aspects of his proposals for Louisville and the unnamed fort in figure 11.8, where the search for ultimate security degenerates into rococo fortification. Perhaps reassured by the surrounding forts, Montalembert made

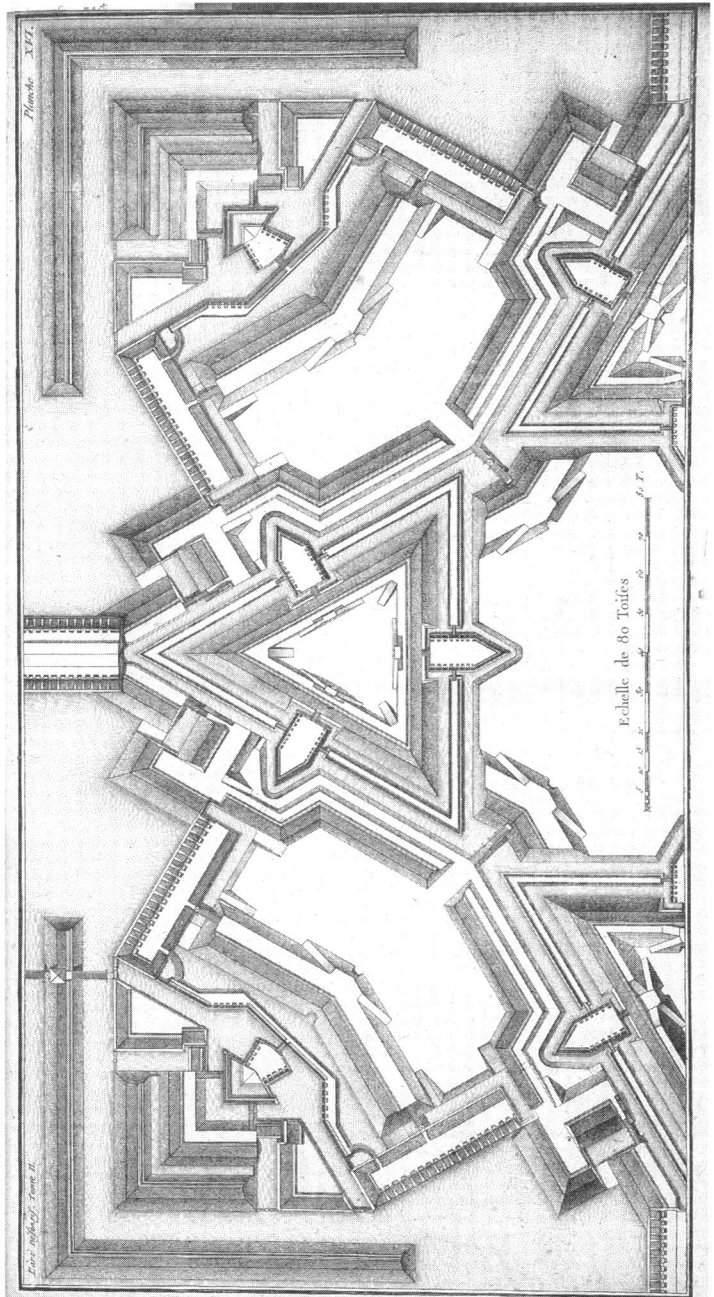

Figure 11.8
This example of a square fort with cavaliers and casemated walls is unnamed. Each side is 323 toises long. The many covering walls, ramparts, and caponiers create an impression of geometric delirium. As many as ten obstacles (depending on which variant of the design is chosen) confront potential attackers before they come to a central square, each of whose sides is the base of an equilateral triangle pointing outward. Each of these sides has a caponier. Source: Montalembert 1776–1797. Courtesy of Massey Library, Royal Military College of Canada.

the proposed main fortifications of Cherbourg simpler and more sober. The forts had the important function not only of keeping the enemy farther away from his main target and thus forcing him to extend his lines of countervallation and the number of troops needed in the siege, but also preventing long-distance bombardment. This was a crucial consideration for naval ports and arsenals with their storehouses of highly flammable supplies.[29]

Attentively reading the first four books of Montalembert's *Fortification perpendiculaire*, one is struck by an almost bewildering fertility of imagination and a torrent of proposals that are undeniably new. This is true in spite of the fact that the diverse elements of these were hardly ever new in their general form. If little known in France—as was the case for the vaulted caponier, for example—they were known elsewhere. But the general exuberance of his imagination could easily raise questions regarding his thought: Was the torrent of ideas evolution or merely incoherence?

Cosseron de Villenoisy detected an evolution in Montalembert's thinking and argued that towers became more daring and more open to the besiegers' artillery with a noticeable diminution in the cover given to these towers. He pointed out also, in connection with this, that the amount of excavation was only half that of the fill in Montalembert's earlier fortress designs while in his later designs these proportions are reversed.[30] (A disparity of these measures was always a drawback in fortress design because in both cases it demanded expensive cartage of earth.) It can be questioned, however, whether Montalembert paid much attention to these aspects or whether he was even conscious of this evolution. Rather than evolution, it is more accurate to speak of an expansion of his ideas. In his books three and four, not originally projected in his initial program of publication, Montalembert continued to add to his already substantial repertory of smaller forts with a growing preference for increasing use of his gun towers and even their use as circular forts. In these volumes he discussed his rings of forts protecting arsenals and ports, proposed long defensive lines at the frontiers of the kingdom, and gave a description of the extensive field fortifications that were constructed on his instructions at the island of Oléron in 1761. In addition to a general theory of fortification he had given enough examples to be considered a universal fortifier.

Yet Cosseron de Villenoisy may have been too harsh in his jibe about Montalembert's inventive fertility. There is in fact some easily discernible continuity in the first four volumes of Montalembert's works, which form the essentially expository part of his oeuvre. All the elements of his future systems are present in these volumes and are discussed at some length. Through all this bewildering wealth of ideas one can detect two themes that grow continually stronger and were present from the very beginning in his thought: the absolute rejection of the bastioned trace and the importance, indeed dominance, of artillery. Whether Montalembert's traces were tenaille, polygonal, or round, there was never a suspicion—even in those cases where his more elaborate caponiers began to sprout covering flanks—that there was the slightest acceptance of the bastioned trace, against which he struggled till his death. And artillery as the primary element of defense was a theme that only increased with time. From his first analysis of the revolutionary effects of Vauban's ricochet shot, which he claimed could only be countered by a stronger and more numerous counter-artillery, to his fort at Aix and his proposals for Cherbourg, the importance of artillery kept on growing. Like the warlike instrument he defended, Montalembert's arguments were relentless, powerful, and concentrated. There is undeniably a certain monotony in this theme, like a siege battery in action at a breach, which recurs over a quarter of a century.

The change in the main title of his works from volume 9 on could serve as the résumé for his system: *L'Art défensif supérieur à l'offensif par une nouvelle manière d'employer l'artillerie et par la suppression totale des bastions*. But even before this one can detect a radicalization of his recipes for the defense. Artillery had become so important that it was the main, if not the sole, characteristic of the fortress. Fortresses were to be constructed for artillery and the disposition of their works was to be conditioned exclusively on their ability to enable artillery to be used to best effect. There was an important distinction here between this idea and conceptions of artillery fortifications since the Renaissance: when flanking fire had always been considered an important criterion in tracing the geometry of fortifications. The geometry of the ground plan had conditioned the direction of fire, but walls had nevertheless retained their function as physical barriers, as passive obstacles to hinder the passage of the enemy. Now, with Montalembert, walls were seen as an active locus of firepower from which artillery destroyed attackers before they could even reach the obstacles. Fortifications were constructed to accommodate, protect, and focus the maximum number of artillery pieces. Safety lay not in stone but in fire. The two were, of course, linked, but Montalembert increasingly gave primacy to firepower. Now the geometry of the ground plan—aside from the criterion of perpendicularity on which he gradually became less insistent—became less important. The evolution of his fortifications toward the circular tower indicates the intention to dominate the entire zone surrounding his fortifications with artillery. He claimed that the besiegers' artillery would never be able to approach the defenders' walls and the inevitable winner of an artillery duel will be the defender, better protected and better positioned with care and science long beforehand. "The strongest wall is that which cannot be fired at."[31]

Though by definition artillery fortifications had long taken account of artillery, mostly this had been in the form of a reaction against the attacker. Furthermore, it was only with Pagan that the primary defensive firepower for fortifications was seen as the cannon rather than the musket. It was, however, Montalembert who emphasized—energetically, obstinately, and even tiresomely—the power of defensive artillery in itself and regarded it as the primary and no longer the secondary element in fortification. It is here that Montalembert's greatest claim to originality resides.

Until 1778, however, Montalembert had not had the opportunity to see any of the plethora of ideas in his books put into practice, in spite of a number of unsuccessful attempts. In that year, however, he finally got his chance. The year after he published the first volume of *La Fortification perpendiculaire*, Maurepas asked Montalembert to undertake an inspection of French coasts and their fortifications in view of the war with England that had been looming since the revolt of its American colonies.[32] This request by the most influential minister at the court of the young king who had come to power three years earlier was a turning point in his fortunes. It marked a period of favor during which Montalembert was able to get some of his ideas put into practice. One cannot help noting the increasing number of coastal fortifications proposed in the second, third, and fourth volumes of *La Fortification perpendiculaire*, published in 1777 and 1778. After visiting Cherbourg, where the Corps of Military Engineers was constructing a new fortified port that was the largest military construction project in France since the end of the reign of Louis XIV, he went on to Saint Malo, Brest, Lorient, La Rochelle, Rochefort, and l'île d'Aix. Apparently his report on the last site attracted the particular interest of Maurepas and shortly after the Minister of the Navy Sartine wrote to Montalembert asking for a proposal to fortify Aix.

Formally requested on 9 January 1778 to propose a design for a fort at l'île d'Aix, he had wasted no time in replying.[33] By 23 January 1778, Montalembert had presented a first proposal. His memorandum, beginning with general strategic considerations on the importance of the island to French naval operations and security in the Atlantic below Brest, went on to propose a masonry fort consisting of three gun towers within a triangular enclosure, which he argued would be operational within a year.[34] The actual design for the fort is identical to a hypothetical design for a coastal fortification that was published in volume 3 of *La Fortification perpendiculaire*.[35] Seven of the twelve and a half pages of the memorandum are simple transcriptions, not always complete, from the book. The memorandum included a small but elegantly drafted and hand-tinted map of l'île d'Aix with a plan and an elevation drawing of the proposed fort, which showed little concern for precision. He also presented a cost estimate of

600,000–700,000 livres that in fact, according to his own calculations later, was significantly less than the probable cost.³⁶

Just as he had done for his plans for colonial fortifications at Pondichery and Mauritius, local topography was secondary in his thinking. What was important was the basic design, which determined whether or not a fortification was capable of withstanding attack. For him, Reason was independent of Nature and dominated it: the accidents of terrain and specificities of location could be subordinated to the theoretical plan.

Figure 11.9
Plan view of original proposal for the Aix fort: two massive masonry gun towers at the base of a triangular fort. Source: Montalembert 1776–1797. Courtesy of Massey Library, Royal Military College of Canada.

304   Chapter 11

In his covering letter, he asked his readers to make allowance for the speed with which he had drafted his proposals and responded to the wishes of the Minister.[37] Somewhat airily he pointed out that his proposal was very short if one skipped over his principles of fortification taken from his book, to which "one can return whenever required."[38] The reader is left wondering whether the elegant graphical presentations and almost cavalier lack of attention to detail are fundamental elements of his lobbying tactics with busy ministers unsure of technical aspects of fortification or the result of haste in maneuvering for the ear of those ministers.

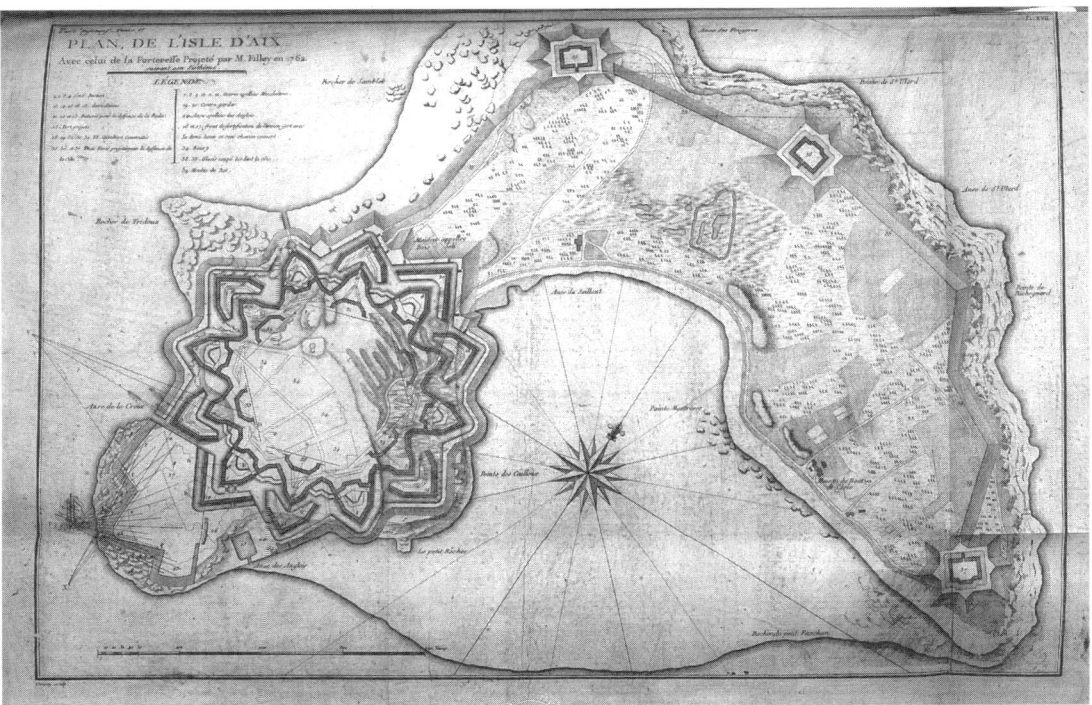

Figure 11.10
A map of the Ile d'Aix by Montalembert. The fortress on the map (proposed by General Pierre Filley and never built) features mezalectres—variants of bastions that also adhered to the principle of perpendicular fortifications. Filley had also called for a number of smaller forts scattered around the island in a very expensive proposal that was unacceptable to both Fourcroy and Montalembert. Source: Montalembert 1776–1797. Courtesy of Massey Library, Royal Military College of Canada.

Montalembert claimed that he could build a fort that would both be capable of defending the island itself from attack as well as controlling the roadstead of Rochefort with its artillery fire. Moreover, he claimed that with his novel methods he could do this very quickly (in time to forestall any British attempt on the island) and very cheaply—an argument that was particularly weighty with a government that was perennially short of funds and facing another expensive war.

Sartine passed on the memorandum to his colleague the Prince de Montbarey, the Minister of War, but by June of that year he had not received a definite reply.[39] On 6 February 1778, France had already renounced its pretense of neutrality in the American War of Independence by signing a treaty of friendship and commerce with the new American republic; diplomatic relations between Great Britain and France were broken off in March, and the first hostilities occurred on 17 June 1778. The anticipated war with England gave new urgency to problems of coastal fortifications, and Montalembert's plans were apparently circulated for more detailed evaluation.

A comment on Montalembert's ideas by Gribeauval, the First Inspector General of the Artillery was hostile: he thought that a simple coastal battery, sufficiently elevated to be safe from the ricochets of a potential attacker's naval guns and protected in the rear, would serve as well as Montalembert's elaborate fort at far less cost.[40] The 200 guns and 1,200 gunners necessary for the proposed fort struck him as excessive. But toward the end of 1778 Montalembert obtained support from a crucial quarter—the powerful Jean-Frédéric Phélypeaux, comte de Maurepas (1701–1781), the most influential minister in the early reign of Louis XVI.[41]

In a letter dated 1 August 1778 that is undoubtedly addressed to Maurepas, Montalembert launched a general attack on the accepted methods of fortification in France, claiming that he "had demonstrated in [his] publications and not been contradicted" that they were not only ineffective but costly.[42] He painted a dramatic picture of the dangerous inadequacy of existing French fortifications: "... one can predict with certitude that, in less than ten years, there will not be a single fortress in a condition to offer the least resistance to the enemy."[43] After this note of panic, Montalembert presented himself as "a zealous servant of the king" who could not remain silent in face of this intolerable situation and mentioned that Sartine had asked him to make plans for fortifying l'île d'Aix and that Montbarey highly approved these plans.

But nothing was being done. Claiming that he had the support of individual military engineers such as Jacques de Carpilhet (1730–1802) and Bourcet, he argued that the administration of the military engineering corps in France made it impossible for good ideas to be accepted. There was "neither coherence in the projects nor discrimination in the work" because individual engineers reigned self-satisfied and indolent in their personal bailiwicks. Meanwhile the unofficial but effective head of the corps at Versailles, Fourcroy de Ramecourt, was isolated from local conditions and only too willing not to interfere with the various Directors of Fortification in the provinces for fear of exciting their animosity and losing his privileged post at Versailles to one of them.[44] It is curious that Montalembert depicts the engineering corps in this letter as being in a state of administrative anarchy. Nor is there a word of reproach to the memory of Vauban. Soon he would be taking the opposite tack by asserting that the engineers were adherents of an outdated technical dogma that was forced on them by the tyrannical Fourcroy de Ramecourt, who kept his subordinates on a leash by controlling favors and promotions.

Montalembert finished his letter with a bald appeal for intervention by Maurepas:

All these things are of the greatest importance, but if they do not have your support, Count, they will not at all be realized. You alone will decide it, if you think it appropriate that you indicate to both ministers what you feel is advisable to do in this regard. Approaches from my side would hinder rather than help. A maxim of state—doubtless the most essential—is Economy Everywhere where Possible and it is proved in a book that is in everyone's hands that my methods are much less expensive and much more solid. It is therefore in the mandate of a good administration to adopt them for use and it is on these grounds that I take the liberty of recommending them to you.[45]

Maurepas, apparently enjoyed attending the private theatre productions at Montalembert's Parisian town house where the charming Marquise de Montalembert acted in operettas and light comedies by her husband, and personal acquaintance counted for much in a society where access to support and power was lubricated by favoritism.[46] But these words just quoted give an idea of Montalembert's dealings with the Court, and there is more that is worth noting here than mere flattery of the minister in indicating his dominant role in the government—a fact well known to everyone. Maurepas, the intelligent courtier with a touch of frivolity who had once been disgraced for his witty and popular verses against Madame de Pompadour was no doubt as sensitive to public opinion as anyone in an age when public opinion was becoming a force in politics. By mentioning that his work was being read by everyone, Montalembert was, of course, also flattering himself, and by declaring that his methods had been "proven" he was implicitly enlisting the influence of science in his maneuvers with the ministry. No rational man of the Enlightenment could easily flout public opinion or science nor could a minister during the Enlightenment be insensitive to the importance of a "good administration." Neither could a minister of the final decades of the Bourbon monarchy ignore the perennial problem of finances. Sometime in the latter half of 1778 the idea of fortifying the Ile d'Aix according to his proposals appears to have been accepted in spite of opposition from artillerists and engineers.

Unfortunately, the war appeared to pose a formidable obstacle to Montalembert at the moment of his triumph. Haste was necessary and the Minister of War announced that Montalembert's claim to get his masonry fort operational in a year was not good enough and in December 1778 requested him to propose some other means of fortifying the island.[47] Typically, he was not at a loss and not slow to respond.

In a memorandum to the Minister dated 14 January 1779, almost exactly a year after his original proposal for a masonry fort, Montalembert proposed the relatively novel idea of a *wooden* fort, an idea he himself had denigrated in the past.[48] As he said later, the idea was "entirely new" and was not even mentioned in the volumes of his *Fortification perpendiculaire* published to date.[49] Montalembert claimed that, if the wood were properly prepared and in place beforehand (a process that he estimated would take 2 or 3 months), a wooden fort could be built in a month and would be strong enough to resist an enemy attack. Even in 15 days, a fortified camp for two battalions of troops to provide security and also work on the fort alongside 500 other workers would be ready and able to withstand any surprise landing. The wood required would be of the most ordinary kind used in the construction of barns, and the larger pieces would not exceed one square foot in cross-section. According to Montalembert, by using a new kind of gun embrasure, four cannon could be installed in the space normally allotted to a single cannon. Once again Montalembert took care not to tire the Minister with details:

It would be very useless to develop this memorandum further. The safety of the operations is evident there, the advantages that would result from them equally so. That suffices to put the Minister in position to determine what he thinks is suitable to do.[50]

A month later, in an even shorter memorandum, Montalembert repeated his call for the wooden fort and pointed out that it was a provisional fort, which could eventually serve as a source of materials for the floors and interior scaffolding of the masonry fort.[51] Indeed, he claimed to have designed the wooden fort with this thought explicitly in mind. His description of the fort as a "vaisseau terrestre"—an expression of which he seems to have been fond and that was to recur in his writings—serves very well to characterize in a nutshell the essence of the Montalembert system throughout its variations and hesitations. Basically his forts were very much like eighteenth-century warships, crammed with artillery and rising straight over their surroundings—the very opposite of the massive classical fortresses of Vauban, huddled in the earth and equipped with relatively few guns.[52]

In plan, the fort proposed by Montalembert looked like an equilateral triangle with a truncated apex and rounded corners. The base of the triangle, which was the front of the fort, faced the sea. At some distance from the front of the fort, there was a palisade with embrasures for cannon to make an additional battery. The fort itself was a three-story wooden structure packed with 150 guns.[53] The second story was covered with several feet of earth to absorb the shock of mortar bombs, and on top of this were additional guns exposed to the sky but behind an earthen rampart.[54] Indeed, in any artillery duel with a ship of the line, this "ship on land" would have the advantage of having more guns than its opponent as well as the advantage of being on solid ground and thus able to fire more accurately. There was an additional support battery on a point to the rear and the left of the fort viewed from the sea, which gave extra firepower. In all, there were more than 180 guns in the fort and its ancillary edifices. It would have been a foolhardy captain indeed who would have dared to attack the fort from the sea.

On the landward side, the fort was considerably less threatening to any potential attacker. Montalembert's plans called for merely repairing and modifying, according to the principles of his "perpendicular fortification," the old Vauban fortifications that had been ruined by the British in 1757.[55] But the modifications were by no means substantial, and there is none of the evidence of ingenuity and care that is apparent in the design of the seaward fortifications, even in the theoretical proposals for land fortresses presented in the earlier volumes of his theoretical works. Montalembert was fixated on seaward fortifications and accorded little attention to those at the back of the fort; almost all of the guns pointed out to sea and the landward perimeter, if anything, was weaker than the old Vauban trace that had consisted of a curtain wall flanked by two half-bastions and protected by a demi-lune in a ditch. On the other hand, all superior officers who were ever connected with the Ile d'Aix project were greatly worried about an enemy landing out of range of the fort's guns and a subsequent land attack on the rear. None of them considered the fort capable of standing up to even a minor assault. To aggravate the matter, the landward fortifications were built only after the seaward fortifications were completed. According to Montalembert's own admission, the landward fortifications were never properly completed.

The construction of the small fort at Aix was to prove the only occasion Montalembert ever had to put his ideas on fortification into practice. His plans for this fort are not the best example of what he called perpendicular fortification—they

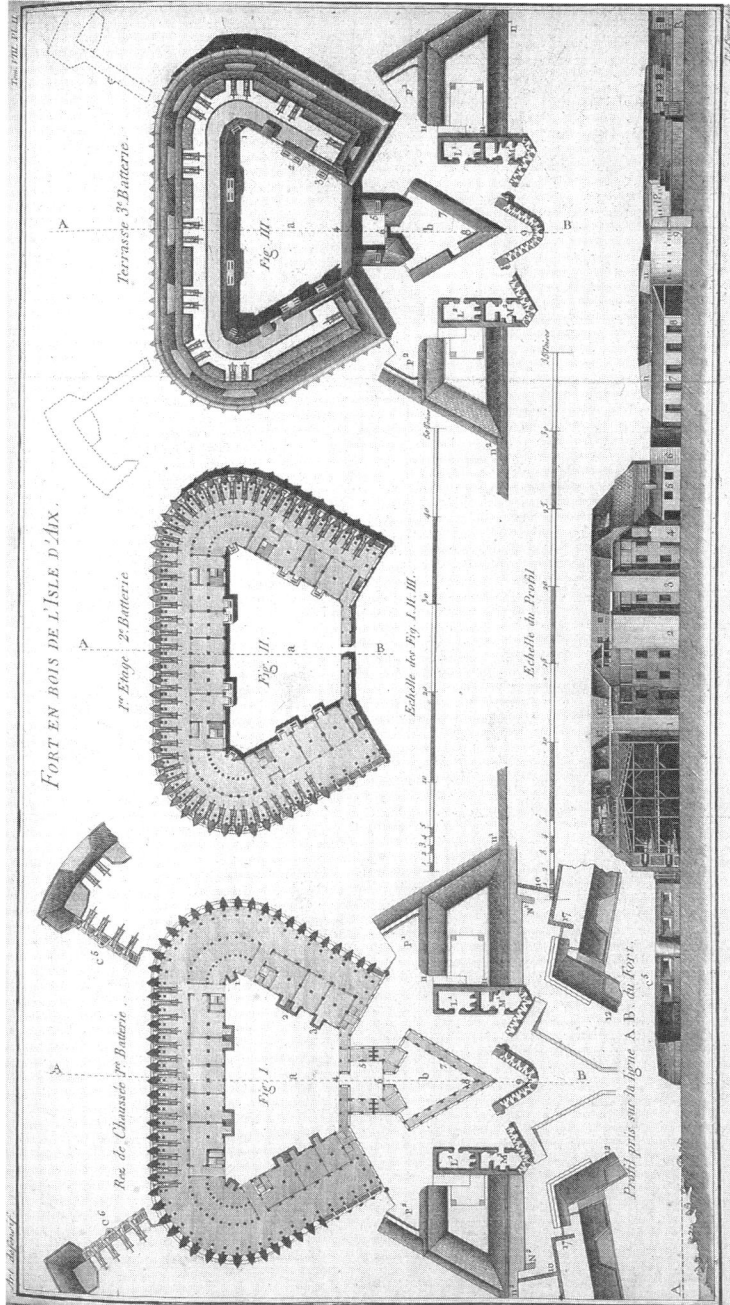

Figure 11.11
Plan and profile views of Montalembert's wooden fort at Aix showing plan views of the first and second stories, and a vertical view and section of the fort along its line of symmetry. Source: Montalembert 1776–1797. Courtesy of Massey Library, Royal Military College of Canada.

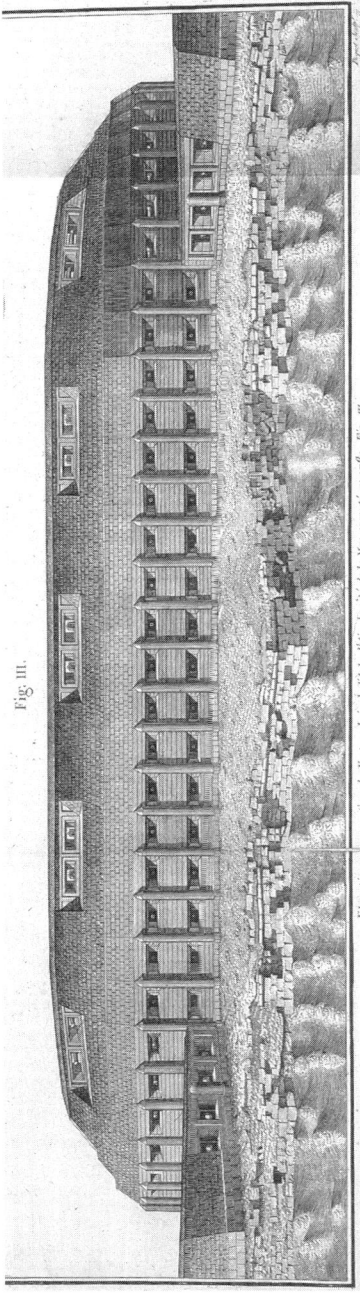

Figure 11.12
A frontal perspective of the Aix fort from the sea. Source: Montalembert 1776–1797. Courtesy of Massey Library, Royal Military College of Canada.

differed from the Louisville model as well as the models for his smaller forts. Moreover, even his ideas for a coastal fortification on an island, which served as the basis for his initial proposal, were rejected, and he had come up with something completely new. In itself, the story of the construction of the fort is a minor episode that often gets no more than a footnote in histories of fortification. But because of its importance for Montalembert's career and the controversy it aroused, it is pivotal in any attempt to comprehend his career. Constructed between 1779 and 1783, it was different from any other fort ever constructed in France for a number of reasons. For one thing, it was made of wood, a material only used on very rare and special occasions in the past. It was also a radically different design from previous forts, lacking the characteristic bastions of Vauban's fortresses. Finally, the corps of military engineering was excluded from any role in the building of the fort, aside from posting observers, who appear to have functioned merely as spies and ineffectual controllers of expenses. This was something that was unheard of and even more surprising in view of the corps' official rise in status three years previously.

By the end of February 1779 the proposal had been accepted, and in a few days Montalembert was on his way to Rochefort. Toward the end of May, he was writing to the Minister of War that work was going well and that 400 workers "some good [and] some bad," were on the job.[56] He thought that the work to prepare the wood would take 3 weeks to a month at most. Carpilhet and Charles-Augustin Coulomb, the two military engineers assigned to the project at Montalembert's request for the duration of the project were, according to him, very impressed by the speed with which the wood was being prepared and were also reassured that the fort would be erected even more quickly. There is a disturbing note, however. Montalembert asks that the troops ordered by the Minister to serve as laborers and guards against a possible English raid should be sent promptly. Work would falter without their presence when the wood was ready and transported to the island. There was also the first sign of more conflict to come between Montalembert and his labor force. Soldiers working in the wood yard had been demanding 25 sous a day for their work, and it had taken an order from the local military commander to impose Montalembert's rates of 12 to 15 sous per day.

Less than a week later, however, Montalembert wrote a memorandum bubbling with enthusiasm on the advantages of wooden forts in colonial wars and naval raids.[57] Formerly hostile to the idea of wooden forts, he now developed a characteristically ardent enthusiasm for them and even suggested that the idea be taken further. He argued that the French navy should consider transporting prefabricated wooden forts during raids on England and its colonies. A prefabricated fort of his design could be transported across the ocean or the English Channel and completed in a month before an enemy could take proper retaliatory measures. Such a fort would be stronger than "three-fourths of the fortresses of Europe" and would be practically impossible to capture. He had visions of one being established after a raid on the Isle of Wight, which would be a "second Gibraltar" that could throttle Portsmouth. The following month Spain was to enter the American War on the side of France with the specific promise of French aid in an attack on Gibraltar, and Montalembert's mention of this fortress may be more than mere coincidence. Once again he appears to have been well informed about the highest levels of diplomatic negotiations by the government, aware of their importance in the consciousness of decision makers at Versailles, and adept at exploiting them for his own ends. It also appears from this memorandum that Montalembert was looking forward to the imminent completion of the work at Aix. According to the memorandum, the preparation of the wood had been completed in only 5 or 6 weeks (he refers to this aspect of his work at Aix in the past tense) and he believed that only a month more was necessary to have the fort operational.

Writing at the end of August, however, the military engineering representative on the site, Coulomb, felt that "this work . . . could still take a very long time," and events were to prove him correct.[58] An accounting of the expenses incurred by the middle of October 1779 estimated that the cost of the fort would be "at least" 450,000 livres.[59] All of Montalembert's correspondence seems to display both irrepressible wishful thinking and a blithe disregard for the logistical problems that were normal in any major construction in eighteenth-century France and were aggravated even more at a large naval base during a major war. But even his failure to keep to his schedule seems to have confirmed Montalembert in his opinion that a fort could be built in a month or 6 weeks.

There appears to be no record that the military engineering corps was formally consulted on the fort Montalembert was asked to build at the Ile d'Aix before construction began.[60] Gribeauval, who apparently was consulted, was ignored.[61] Fourcroy de Ramecourt, who had become the unofficial head of the corps of military engineers by virtue of his appointment as a Directeur attached to the Minister of War at Versailles, however, did not wait long to launch a bureaucratic battle against Montalembert. On 15 September 1779, while the fort was being built, he wrote to Sancquer, the executive assistant to the Minister of War for military engineering affairs, from Saint Malo, where he was in charge of the engineers attached to an invasion force directed against

England.⁶² Disclaiming any sense of rivalry or narrow prejudice of his corps, he wrote that normally

> every military engineering officer could rather rejoice than be tormented with jealousy in seeing executed an example of a system that can only become the laughingstock of soldiers and sailors, both friend and enemies, at the same time being the terror of the people who will think that they have to defend themselves in it. It will be public opinion [la voix public] which will soon assert the rights of true fortification and our knowledge in this matter against the author. But we must regret the funds that are spent on these works.⁶³

Montalembert's ideas were described as "a false theory divested of all knowledge in the practice of defense." Having discussed it with "all the old and good officers I have the honor to command," he felt that the fort at Aix was "repugnant to all notions of defensive and mechanical [méchanique] war."

Some time later, Fourcroy de Ramecourt made a trip to the island to inspect the fort, and on 30 December 1779 he addressed a long and hostile memorandum to the minister.⁶⁴ Fourcroy de Ramecourt appears to have obtained most of his information on the fort from military engineers who had been assigned to Aix at the request of Montalembert himself and who had picked them because he considered them to be sympathetic to his system.

Montalembert had requested and been granted two military engineering officers of his own choosing when he first went to build the fort at Aix. He had asked for Carpilhet, whom he had met on his inspection tour of the French coast at Saint Malo in October 1777, as the first military engineer. According to Montalembert, Carphilhet had expressed support for Montalembert's ideas. He was later to be bitterly disappointed and to claim that Carphilhet had merely feigned support, all the time providing information to Fourcroy de Ramecourt to attack Montalembert.⁶⁵

The other engineer Montalembert specifically requested was Coulomb.⁶⁶ Montalembert had heard from a person who had lent Coulomb his book that Coulomb had praised it highly upon returning it.⁶⁷ It is obvious that Montalembert's choice of Coulomb was motivated by the belief that Coulomb, who was a corresponding member of the Academy of Sciences and a colleague of Montalembert in that body since 1773, could provide the cachet of scientific legitimacy with experts and the public alike. If converted to his new theories, Coulomb's status as an academician would carry great weight in the arguments already swirling around the project. But Coulomb was not converted, and his main role (and that of his successor, Grégoire Seuillet) was that of spy for his superior Fourcroy de Ramecourt, who continued to attack Montalembert in the offices of the Ministry of War.

Coulomb himself appeared extremely annoyed at what he described as an uncomfortable posting in an embarrassing position and requested a transfer: "It would be desirable for me [and] Mr. de Montalembert himself, who had asked for me without knowing me, that we could separate as soon as possible." When sent to Rochefort, he had thought he was going on a military campaign, and "the name of Mr. de Montalembert was not specified."[68] In a later letter he repeated that "Mr de M. asked for me without knowing me and without my consent."[69]

According to Coulomb himself, he was not consulted on the fort, and his only activity, which Coulomb considered highly irregular as well as demeaning to an engineering officer, was to countersign Montalembert's orders for payment. His accounting practice were "absolutely against all the rules and that which is prescribed to us by regulations."[70] Before his colleague Carpilhet left, Coulomb noted,

my commission was easy to fulfill. It was only a question of following, keeping quiet, and being bored. I was never taken aside to have my opinion. But I have since [Carpilhet's departure] been so pressured that I've had to displease or mislead [tromper]. I chose the first [course]. Now I'm not questioned any more but I'm being dragged three times a week to the Ile d'Aix, I am almost always [sea?]sick on the way, and I am absolutely useless for this work in which I do not involve myself in any way and on which I am presently observing the most rigorous silence. . . .[71]

Montalembert confirms Coulomb's attitude. According to him, Coulomb "never wanted to get involved with anything so long as he was employed under my orders at the Ile d'Aix. He only did me the honor of accompanying me on the works, in keeping the most profound silence on what was going on there."[72]

Pressed at last by Montalembert to declare what he thought about a particular feature of the fort, Coulomb had "replied brusquely, while turning his back, that HE DID NOT THINK."[73] There were no personal reproaches by Montalembert for Coulomb, as there were for Carpilhet.

Nevertheless, M. Coulon [sic], full of merit and very knowledgeable, having since become my confrere at the Academy of Sciences, has allowed himself to think there for he has given very good papers there. Only his excessive faithfulness to what he thought was his duty to his uniform had deprived him of a faculty that his talents make so precious and, which, luckily he recovered as an academician.[74]

Yet Carpilhet was not the only one sending information on Montalembert's fort to Fourcroy de Ramecourt. Coulomb's reports on the fort and the activity of Montalembert, whom he refers to as the faiseur, can also be found in the archives.[75] Indeed, he appears to have commented on a preliminary version of Fourcroy de Ramecourt's critical memorandum to the Minister.[76] His comments were harsh:

> ... all the masonry ... which is in loose irregular stone [pierre sèche] will crumble by itself in a short time and in ten years or so nothing will remain of all this work than the regret of the enormous expense and the debris of a project as crazy in invention as in execution.[77]

Armed with information from Carpilhet and Coulomb, Fourcroy de Ramecourt could launch a detailed attack. He made a point of having presented his ideas to his colleagues in the corps and that the memorandum contained "the joint and unanimous observations of 35 engineering officers."[78] In the very first part of his memorandum Fourcroy de Ramecourt reiterated his primary theme, which would reappear: " ... M. de Montalembert uses forms and ideas [that] are absolutely new for me and likely unknown to all the soldiers of Europe."[79]

The accusation of novelty was one that Montalembert would have been glad to accept. For Fourcroy de Ramecourt, however, this had the disadvantage that the best and most reliable method of all—experience—could not be used to judge the quality of the fortification. Since no experience was available, it would be necessary to subject the plans to a logical analysis:

Now since in favor of new proposals in warfare no tests of their success (which are the touchstone of inventions of this kind) can be cited, neither can any equitable judgment be made in advance unless one breaks them down and examines them in detail [to see] if the means used offer at least a sufficient probability to achieve the proposed object by bringing together and combining these means.[80]

Fourcroy de Ramecourt continued to attack Montalembert's "naval citadel" on a number of grounds, arguing that even fortifying the island was unnecessary, that the fort was flimsy and would fall apart under the vibration of its own cannon, that it could be easily destroyed in naval bombardment, that it was expensive (original cost estimates had been exceeded), that the battery in front of the fort would be vulnerable to snipers in the rigging of attacking ships, and that its form would never inspire confidence in the troops manning it.

Montalembert returned to Paris in February 1780 and mounted a vigorous countercampaign against what he called a "procès par écrit"—a trial by writing.[81] He claimed that he deigned to answer the criticisms of Fourcroy de Ramecourt only because he was ordered to do so; otherwise the appropriate reply would be to ask the engineers to suggest a better and cheaper alternative to the fort they were criticizing. Displaying his own flair for bureaucratic infighting, he warned that the Minister's solidarity with Montalembert was as necessary for the Minister as it was for Montalembert:

... I dare to say, my Prince, that it is a matter of your own interest. In attacking me, it is not I alone being attacked. You were right in ordering this work. If you persist in the good opinion you had of it, you will fix that of everyone.[82]

In his arguments Montalembert successively assumed a number of roles. One role was that of the unappreciated expert and individualist confronting the intellectual tyranny of Fourcroy de Ramecourt, who has, he says, adopted "the tone of the master" and treats him as a "schoolboy." Novelty provided another theme and another role: he is the discoverer and the selfless fighter against dull and dangerous inertia. He displays mock satisfaction that his criticisms have finally shaken up the military engineering corps to consider the problem of covering guns from enemy fire that Fourcroy de Ramecourt had made so much of in his criticism. He hoped that the details he would give in his response to the criticisms would "dissipate the vain fears that have been engendered by construction that is so distant from [normal] usages."

Montalembert claimed that Fourcroy de Ramecourt's belief on the vulnerability of soldiers serving the cannon was based on the most unfavorable suppositions. He claimed that Fourcroy de Ramecourt had assumed the ships could come in much closer to shore than in fact was possible and assumed an excessive height of rigging above the deck. But even granting all of Fourcroy de Ramecourt's assumptions, he was wrong because he had made a gross elementary error in calculation. A straight line drawn from the ship's crow's nest on the mast (from where the highest-placed snipers could shoot) passing through the top of the rampart to touch the ground behind the rampart created two similar triangles. The base of the largest was the horizontal distance from the middle of the ship to the point at which a bullet fired from the mast grazing the rampart would hit the ground. The small triangle's base was the horizontal distance from the rampart to the same point. The heights of the rampart being given, the height of the crow's nest above the rampart could be easily calculated from easily available data, and the trajectories were assumed linear. Thus, this problem reduces to a rather elementary geometrical one of similar triangles. Strange as it may appear, the head of the military engineering corps in France had made an error—intentional or otherwise—in this simple calculation. Fourcroy de Ramecourt's calculation had made the area of safety considerably smaller than it actually was. It was, as Montalembert said, "a most astonishing error if not deliberate."[83]

Finally, Montalembert portrayed himself as open minded and willing to put his ideas to the test. If the engineers thought his ideas were bad, let them come up with a new proposal for a fort with covered artillery. For 20 years, he claimed, he had been alone in defending such a system, and now he was being reproached for not effectively putting into practice his principles by the very engineers who had hitherto ignored and even attacked them. Let them suggest something better rather than restricting themselves to sterile criticism was the challenge he threw down and was to repeat on several future occasions. He reproached Fourcroy de Ramecourt for having come to inspect Aix at a time when he knew that Montalembert would not be there and staying only an hour. The implication was that the trip merely served as a seal of respectability for ideas that had been determined long in advance without benefit of hard facts or unbiased analysis. He eagerly took up Fourcroy de Ramecourt's challenge to test the fort's structural solidity by firing off cannon to see if vibrations would damage it.

Montalembert's memorandum displays more verve and sparkle than the plodding pieces by Fourcroy de Ramecourt. After all, Montalembert was a minor author of sorts who wrote and staged plays for his own private theater.[84] But his ally Choderlos de Laclos brought a literary talent of an altogether different order to the memorandum that he submitted on the same subject. Like the military engineering corps, the artillery had also been requested to send a representative to observe the construction at Aix. Although Gribeauval disapproved of the Aix fort as much as Fourcroy de Ramecourt, the artillery officer who was sent as an observer, unlike his engineering colleagues, became an ardent supporter of the marquis. This man was the artillery officer and man of letters Choderlos de Laclos, who may have occupied his leisure moments during his stay at Aix with writing his celebrated novel *Les Liaisons dangereuses*.[85] He was to prove a steady ally of Montalembert.

In his technical reply to Fourcroy de Ramecourt, Choderlos displayed a certain amount of his literary flair and amused himself by expressing mock relief in having Reason (in the form of schoolboy mathematics) on his side in arguing with such a senior officer of such a "justly estimable corps." Laclos does not really bring forward anything new that Montalembert had not already covered; in fact, his shorter piece (10 pages), which consists of a transcription of key points of Fourcroy de Ramecourt's memorandum in one column opposite his own remarks in the adjacent column, benefits from being leaner and less diffuse than that of Montalembert. Just as in *Les liaisons dangereuses*, economy, elegance, and cruelty are combined in the final paragraph:

The summary of M. de Fourcroy de Ramecourt being supported only on the objections that I have contested and I believe entirely destroyed, I think it useless to respond to it and I now finish this memorandum [that] is already too long.[86]

Beginning with the remark that he must reply to Fourcroy de Ramecourt's criticisms because if they are correct he must be guilty either of ignorance or of bad faith, he continues with a cut that must have been painful to the man who though that experience was the touchstone of the military engineer's art:

A consideration which reassures me is that M. de Fourcroy does not rely in any place in his memorandum on his personal experience. Thus, his rank in a justly esteemed corps, the long and useful services that have acquired him the reputation he enjoys, everything, disappears at this point and, thereby raised to the same level as he, there remains nothing between us except reason.[87]

It was therefore reason [raisonnement], rather than experience (which Fourcroy de Ramecourt himself had claimed was the ultimate judge in matters military), that made Laclos the equal of Fourcroy de Ramecourt in the debate. On those grounds Laclos had come to diametrically opposite conclusions. And his point of departure was contrary to that of Fourcroy de Ramecourt as well: "M. de Fourcroy's first motive for mistrust . . . was for me the first motive for hope. As a matter of fact, *absolutely new ideas* seem to me to be necessary to achieve the aim that has been missed till then. . . ."[88] Like Montalembert, he too pointed out Fourcroy de Ramecourt's error in dealing with a simple case of similar triangles, and it was now he who adopted the tone of schoolmaster and took Fourcroy de Ramecourt by the hand to conduct him step by step through a geometry problem: "Let M. de Fourcroy drop two perpendiculars to the level of the sea, one representing the height of the rigging, the other the parapet of the battery, ..." He made even more of the alleged approval of Coulomb: ". . . he of *MM. les ingénieurs* who has most followed the construction at the Ile d'Aix, and who, having merited and obtained by his knowledge in mechanics nomination to the Academy of Sciences, must naturally inspire a confidence that can only give way in the face of contradictory experimental results, for which one can doubtless be permitted to keep on waiting."[89] As for the trials Fourcroy de Ramecourt asked for on the effects of vibration of cannon, and which Choderlos "desired as much as he," Fourcroy de Ramecourt could be "reassured" that preliminary trials of guns firing from the first floor of the fort displayed no problems with vibration. This was attributed to a new kind of gun carriage Montalembert had invented.

Although the exchange between Fourcroy de Ramecourt and Montalembert could by no means be considered to have gone in Fourcroy de Ramecourt's favor, complaints against Montalembert, either by engineers or others, were having some effect. There was a tone of reproach and wounded pride when Montalembert recalled that, having returned to Paris in the month of February 1780 to report to the Minister, the Minister's "great business had not permitted him for several months to deal with these problems," just one of several signs of the Minister's flagging attention for the innovative fortifier.[90]

On 7 October 1781, on the orders of a new Minister of War, Ségur, one of Fourcroy de Ramecourt's challenges was taken up and the wooden fort was tested for stability by firing the fort's cannon in various ways. The results were a vindication of sorts for Montalembert. Not only had the fort not collapsed, but it had shown no visible signs of damage. Moreover, although the trials were intended to study the effects of vibration and not the problem of smoke, witnesses such as the naval officers present said that there was much less smoke and it was less bothersome than smoke on the gun decks of a ship.[91]

The trials were conducted very much like a formal court case, in the presence of senior military and naval commanders including Dajot, Director of Fortifications in charge of the Rochefort area. All had expressed satisfaction and formal minutes (a procès-verbal) had been drawn up by one of the witnesses (Colonel Dyvoley, Director of Artillery) for signature and transmission to Versailles. Montalembert would later read a paper on the subject to the Academy of Sciences, which would include a transcript of the favorable letter to Ségur by the Marquis de Voyer, the military governor of the area, on the problem of smoke.[92] He would publish the procès-verbal and the testimonial letter, along with his own comments in his books on a number of later occasions.[93]

Beyond the fact that the fort did not immediately collapse, the trials in fact did not prove much. And the letter of the marquis de Voyer was not nearly so enthusiastic as Montalembert implied: it was, in fact, rather balanced.[94] He agreed that the firepower was spectacular and that any ship daring to attack from the sea would meet an imposing defense, but he singled out for special praise the battery at the Anse de la Croix that Montalembert had been forced to build by the Minister of War. De Voyer conceded that the fort seemed to provide a greater concentration of firepower with fewer men needed to service the cannon than conventional fortifications. Nevertheless he felt that the fort was also vulnerable to mortar bombs and cannonballs breaking internal pillars, as Fourcroy de Ramecourt had suggested. Most dangerous of all, however, was the weakness of the fort on the landward side. Whatever one thought of Montalembert's ideas, the northern approaches to the fort had to be finished, for the real threat to the fort came from the land.

De Voyer was far from a partisan of Montalembert, no matter what the latter claimed in his correspondence and public pronouncements. This is demonstrated by an incident that is worth mentioning not only because it sheds light on Montalembert's relations with de Voyer but also because it provides insight into another dimension of Montalembert's personality and his approach to the practical problems of construction. A few months after the artillery trials, de Voyer was writing to the Minister of War to say that he was forced to abandon his policy of absolute non-interference on the matter of the fortifications at the Ile d'Aix that had been explicitly imposed on him by Ségur and his predecessor.[95] The shocking state of the health of the garrison of coast guards, who were apparently now quartered in the fort, had led him to do this. He appealed to the Minister's "principles of justice and humanity" to help the "precious class of plowmen, for the most part fathers of families" who were suffering and dying from fevers that had delayed work in the past and forced Montalembert himself to go to Paris for recuperation and medical care.[96] Complaints by workers and soldiers about poor pay and poor treatment by Montalembert recur throughout the entire period of his stay at Aix. The year before, the intendant Meulan d'Ablois had complained about Montalembert's retaining his workers' salaries until the end of the project in order to keep them from running away.[97] To provide for their needs while on the site, Montalembert had been feeding them with poor-quality food whose cost was deducted from their pay. Montalembert's high-handed tone when talking about this "mutiny" indicates another major difference between him and the engineers.[98] His aide Choderlos de Laclos suggested that to overcome a shortage of space soldiers be temporarily required, as they had been for "a century," to sleep three to a bed in violation of the recent regulations introduced by the reforming minister Saint-Germain.[99] The tone of the letter displays a contempt for the comfort of ordinary mortals that Valmont, the hero of his novel *Les liaisons dangéreuses*, would have found quite normal. This contrasts with the injunctions of Vauban, who felt that, in dealing with the work force, humanity was required for reasons of efficiency.[100]

Montalembert's demand that the enceinte of the village be fortified, in addition to the construction of the fort, coincided very well, of course, with the demands of military commanders that the landward fortifications be improved, but it clearly contradicted his earlier claims that the fort would be completely self-sufficient in its defense. Soon he was going so far as to say, on the grounds that the fort had merely been intended as a temporary structure, "I will not be able to answer for any occurrence so long as there is completed only the part of [the fort] in wood solely for the temporary protection of the roadstead—a construction as insufficient as [it is] ridiculous if it must become permanent and devoid of any other protection."[101] This was a crucial qualification of his initial claims that wooden forts built in a few weeks could serve as mini-Gibraltars.

If this did not deter Montalembert from asking for more funds and expanding the project, it seems that his support at the Ministry of War, where the ever present Fourcroy de Ramecourt was in continual contact with Ségur, was fading. Furthermore, Maurepas had died at the end of 1781. There were more insistent demands that Montalembert bring order into his idiosyncratic methods of accounting, which had already aroused the protests of Coulomb in 1779. And work continued to drag. Complaints were voiced about the poor workmanship and materials of the landward fortifications.[102] Latour Dupin, who took over the military command of the Rochefort area from de Voyer, claimed that the landward perimeter of the fort, constructed out of loose stones with no binding of mortar [pierre seche] and on inadequate foundations, was already crumbling and would not last more than 2 years.[103] Thus, the landward enceinte around the fort appears to have been no more solid than the rotting wooden fort itself.

At the beginning of 1783, the hapless Seuillet, who was begging to be transferred away from the island where he had been the engineering representative since the departure of his predecessor Coulomb, was saying that work would be finished on the fort and its protective barrier "this year."[104] By now peace with England was imminent and the original need for the fort was fading. Montalembert was still fertile with plans for improving the fort as well as for the enceinte of the village, but there are clear signs that the Minister was in no mood for any more expense and had lost patience with him. On 16 February 1783, work was ordered suspended on the fort until "a plan has been presented on what appears most appropriate to be done to put this island in a state of defense capable of imposing on the enemy."[105] Although the decision was nominally taken as part of a general decision to halt work on all coastal fortifications in view of the impending peace, the wording was ominous and not a little insulting, for it clearly implied that the previous plans for defense of the island—those of Montalembert—were found wanting. Already Latour Dupin had been soliciting the opinions of Seuillet and Choderlos de Laclos for ideas on completing the fortifications. Montalembert, who had complained to the Minister about Latour Dupin, seems not to have been consulted.

Montalembert was now forced to admit that the fort had some problems. The roof was leaking, and the situation was so bad that troops could not be billeted in the fort. In a delayed and partial concession to Fourcroy de Ramecourt, he argued that the leaks had started after the artillery trials in October 1781.[106] However, in this and another later memorandum he was, as usual, confident that the problem could be easily solved.[107] Furthermore, the irrepressible Montalembert had returned to his original position that the wooden fort, with a few improvements, could last a long time: all that was needed was a roof over the third story of his fort to prevent water seepage.

On 24 March 1783, the fort was ordered demolished. A remark in a letter of Montalembert in September 1783 indicates that the Minister had cut off all correspondence with him on the grounds that his mission was completed.[108]

Montalembert did receive a pension for his services, which he (and his wife) complained in writing was too small. He received no further rewards or commissions from the royal government. His polemic with Fourcroy de Ramecourt and the Royal Corps of Engineers now became public and continued in print and at the Royal Academy of Sciences. But the episode at the Ile d'Aix was finished. The fort was, however, not demolished. A project for fortifying the island according to the traditional methods of Vauban was duly drawn up and submitted but it was never funded by the impecunious government.[109] The fort was still standing and still subject to ridicule by the resident engineer during the Revolution. Montalembert did not forget it either. When another war was brewing with England in 1787, he again wrote to the Minister of War, asking that the fort be reactivated and that the third story be completed. The military engineers came in for abuse and he suggested that the work of reactivating the fort be confided to an artillery officer, his former colleague Choderlos de Laclos.[110] Sometime in 1790, Montalembert again wrote to the Minister of War and stated that his relative, the Baron de Montalembert, had visited the fort and seen that it was in good shape. Again he called for the reactivation of the fort and urged that the third story be built properly. The fort would last "centuries" with proper maintenance and replacement of aging wood when necessary. Even though the Revolution had broken out and Montalembert had once again succeeded in gaining favor with influential people, notably Mirabeau, the Minister's attitude seems clear from a marginal note: "Answer without going into any detail—simply acknowledge reception and that a report would be made to the king."[111] Even during the radical phase of the Revolution, when Montalembert's former draftsman Mandar enthusiastically proposed fortifying the island using his former employer's methods, no action was taken.[112]

The military engineers had the last word and Montalembert's fort slowly fell into ruin and was finally demolished on an order from Napoleon.[113] According to a soldier like de Voyer, the ultimate proof of any military technology in general and Montalembert's fortifications in particular lay in their effectiveness when used. Never attacked by an enemy, this proof was always lacking. Yet one is forced to conclude that if an English force had attacked the island it would have easily taken the fort. This would not have been done by attacking from the sea as in the 1757 attack, for Montalembert had indeed packed an impressive number of cannon into the seaward side of the fort—enough to discourage any ship from attacking, even if his fort was vulnerable to incendiary bombs falling on the roof. But the landward fortifications were never built properly and were considered weak by all observers, including Montalembert.[114] In the nineteenth century, fortifiers felt the need to build a number of additional redoubts and forts on the northern side of the island to prevent an attack on the landward side of the main fort at its southern tip. They also decided, as had been suggested by others before Montalembert, who had rejected the idea as unnecessary, to build another fort to control the roadstead on the sandbank of Boyard 3 kilometers from the Aix fort. Obviously, this was a rejection of Montalembert's view that Aix alone could dominate the estuary of the Charente. On the other hand, the masonry fort built at Boyard during the nineteenth century looks very much like the kind of multistory guntower that Montalembert had recommended. For the original fort on the southern tip of the island, however, the evidence points clearly to the failure of his claim to provide a cheap fort in a very short time that would be much stronger than anything Vauban had built.

Even though the particular project at l'île d'Aix was an expensive fiasco and even though both Montalembert and the military engineers hoped it would serve as a crucial test of the correctness or failure of Montalembert's ideas, the general worth of these ideas remains an open question. There is no doubt that the wooden fort was an improvised response to ministerial requirements after his original idea for a masonry fort was rejected. As such, it was not really representative of the bulk of Montalembert's theories on fortification. After the lengthy and often tedious attempt to follow the memoranda and the counter-memoranda coming to the desks of ministers concerning the wooden fort at Aix, a number of questions both great and small remain. They are the difficult questions concerning motivations and honesty on the one hand and the effectiveness of the new system of perpendicular fortification on the other.

Was Montalembert plainly wrong in advancing his claim that a certain fort could be built in a certain time for a certain cost? If so, did he know that he was wrong and dishonestly hope to extort additional funds and resources once the project was started and beyond the point of no return? Fourcroy de Ramecourt wrote in another context that such tactics were used by many engineers in pushing pet projects. It has been shown that there were many problems beyond Montalembert's control. There was a shortage of resources and manpower, a situation that was aggravated in wartime; there were hostile factions at Court to contend with; and a creaky and complex administration made up of a patchwork of jealous and competing interests where things could go wrong at the best of times and with the best of intentions, even without a whiff of ill will or sabotage. Montalembert was surely a victim of all these factors, yet surely one can argue that no one more than Montalembert himself was aware of these problems and should have taken account of them when coming up with his proposals. As a liaison officer with foreign armies in wartime, a participant (as he himself often pointed out) in nine sieges, the supervisor of the building of fortifications on the island of Oléron during the Seven Years' War, and no stranger to court intrigue, Montalembert should have had a better-than-average awareness of the difficulties any major construction project would generate. Even with the benefit of doubt regarding the inherent soundness of his scheme for fortifying Aix, Montalembert, of all people, should have known that he could deliver on his promises only if things worked ideally.

Perhaps it is here—with the word 'ideally'—that one of the most significant aspects of Montalembert's approach emerges. His tendency to emphasize theoretical aspects in design and construction of fortification leads him to stay at the plane of geometry in design and at the level of optimistic calculation in planning and construction. If geometry ignored topography in design, optimism and illusion seem to have triumphed over sober assessment of reality in planning and managing construction. For a man who had already come to the verge of bankruptcy with his cannon-casting enterprise, this indicates a lack of foresight that is remarkable. It is to be expected that military engineers, who encountered the frustrations and difficulties of construction and administration in their daily practice, would tend to be hostile to what they often called "armchair fortifiers" even if Montalembert had not posed an administrative threat to their exclusive role as fortifiers of France.

When Montalembert presented the results of the artillery trials at Aix to the Academy of Sciences, work on his fort had been stopped, his ideas had been rejected, and he was beginning to turn to the Academy and the public for support in a battle he wanted to win as much as ever. There had been no public reaction by military engineers to his work that was being published since 1776; they had battled with Montalembert in the secrecy of military and ministerial councils. Henceforth, however, the polemic would continue in the open. It began with the publication of Montalembert's fifth volume.

12

The Response to Montalembert

One must . . . agree that these systems [of Montalembert] sometimes enjoy the advantage of being put forward by a sex that, if it can avoid the danger of liaisons, must know all about the art of defense.
—Jean-Claude Éléonor Le Michaud d'Arçon[1]

From 1779 to 1783 Montalembert was too busy dealing with ministers and his fortifications at Aix to publish anything, but the year after the construction at Aix ended he came out with volume 5 of *La Fortification perpendiculaire*. That volume marks a turning point. Originally, Montalembert had not intended to write more than two volumes, but he expanded these to four to discuss detached forts and coastal fortifications when he was lobbying for the opportunity to build a fort of his own design. These first four volumes, which comprise the most systematic exposition of his ideas, were the only ones translated into German. The subsequent volumes are essentially tracts or collections of tracts and lack the coherence of the earlier ones. Yet they are interesting because they reveal some of Montalembert's principles and assumptions in developing his fortification schemes. They also provide valuable glimpses of the mechanics of his struggle with his opponents.

Volume 5 also marks a turning point because it is here that the polemic with the military engineers finally bursts into the open. The indirect but transparent attacks on engineers in this volume intensified in his sixth, which Montalembert billed as a "supplement" to volume 5 and in which he refuted criticisms by military engineers of his construction principles. Volume 7 was yet another round in the polemic, this time with the head of the military engineering corps himself. Four more volumes were published between 1790 and 1796.

Volume 5 had more discussion of artillery, since this was now shaping up to be one of the major criticisms of his work. It was alleged both by artillerists and engineers that crowding too many guns into forts would hamper the ability of the gunners to fire them efficiently. The first part of the volume discussed new embrasures for artillery of Montalembert's own design, that increased the sweep of fire without a corresponding widening of the embrasures and which had retractable wooden "cheeks" to protect the cannoneers from enemy fire.

The attack on the fortification theories of the military engineers came in the latter part of volume 5 in the form of an analysis of the systems of fortification of Bélidor, Cormontaigne, and Pierre de Filley. Montalembert first contrasted the systems of Cormontaigne and Bélidor by printing, with extensive annotations, Cormontaigne's violent attack on Bélidor's proposal to fortify Metz in the 1740s. He rebutted and ridiculed Cormontaigne's theoretical sieges of Bélidor's suggested fortifications and argued that Bélidor's proposals were vastly superior. They were not as good as his own, of course, and they were probably borrowed from someone else, but they were much better than the official technique proposed by the man who, after Vauban, was the master engineer and model of the corps of military engineering. Montalembert must have known that he was hitting a sensitive nerve among military engineers. He was attacking not only a major paradigmatic figure but also a paradigmatic technique: the fictitious siege journal. The argument over the usefulness of fictitious siege journals would be central to his debate with the head of the military engineering corps.

Cormontaigne had applied this technique to Bélidor's fortifications to demonstrate their inferiority, and Montalembert was very sarcastic about the arbitrary assumptions that Cormontaigne had made. The most blatantly unfair assumption was that the besiegers could open up their first parallel and install siege batteries at a distance of 25 toises on the first night of the siege operations. Montalembert's criticisms, for all their venom, are fair.

The fictitious siege journal was a well-established technique in evaluating defenses and was taught at Mézières. The assumption was made that both the besiegers and the besieged were brave, intelligent, and made no mistakes—an assumption that, though admittedly false in virtually all real sieges, permitted rational analysis and comparison of different systems. For Cormontaigne it had permitted quantification in fortification—one of the prime criteria of its "scientific" nature. A brave and energetic commander faced with an ignorant and craven attacker could certainly compensate for defective defenses and upset the predictions of the fictitious siege journals, but such a possibility entered the domain of the human, the unforeseen, and often the irrational—a space that was opaque to the engineer's analysis.

The engineer had to be content with assuming omniscient, flawless, and almost inhuman actors to permit him to judge material structures and arrangements. If psychological considerations did enter into these computations, they were unsophisticated. For example, troops would not defend a redoubt to the last man but would retreat if they saw that they would be cut off from the main fortress. The fictitious siege journal was perhaps a primitive and immature mathematization of a complex and irreproducible reality, but it was all that was possible in the circumstances. If a brother officer in the closed and secret company of the initiated argued that *any* circumstances of military combat, characterized by trauma, fright, and irrationality, precluded mathematization, the engineer might have replied that such a method could at least be used for rough comparisons of fortifications. Moreover, the journal was necessary to establish the engineer's credentials as a scientist. Such a naive view of the nature of science, which smacked more of Pythagorean numerology and the Kabbala than of Euler and Robins, was already discredited among those with a solid mathematical education, including many military engineers graduating from Mézières. But for an older generation, Montalembert was attacking an almost sacrosanct, even if not particularly old, tradition.

Montalembert's other target in volume 5 was the system of fortifications suggested by the recently deceased general Pierre de Filley, who had cautiously welcomed Montalembert's views back in 1761. Filley, too, had abandoned the bastion in favor of the mezalectre, adhered to the principle of perpendicular fire, and even suggested casemated flanks. (See figure 11.10.) This and the hostility of the military engineering corps to Filley's system might have made Montalembert sympathetic to Filley. But Filley had used a project for the fortification of the Ile d'Aix to unveil his system, and perhaps that caused Montalembert to compare his project for Aix with Filley's and to argue that his own was better and cheaper. He represented Filley as a "man of genius" trapped "in the circle of the most common ideas, by the mere effect of habit."[2] Even though Filley's fortifications were based on the principle of perpendicular fire so dear to Montalembert, his system was rejected as defective. Montalembert condemned as useless Filley's square forts protecting the main fortress on the island, somewhat like his own redoubts in volume 2, and he thought them likely to be captured by simple assault. He thought Filley's main battery whose cannon were to be protected with temporary wooden shielding was exposed to fire from the rigging of attacking warships. The mezalectres that had replaced the curtain walls, though much better than bastions, did not provide enough flanking fire. Montalembert's main complaint was the enormous expense, which was around 20,000,000 livres.

Montalembert's reflections on Filley's scheme resort to imagery from fortification to address the psychology of error:

The progress of the arts depends on the truth of their principles. When errors and prejudices consecrated by time take their place, they remain in their original state and it is from their very imperfection that they obtain their merit in public opinion. People persuade themselves that there is nothing better and no one bothers any more about them. Therefore it is only in dissipating error, in attacking principles, that one can cross the bounds in which they are retained. But these bounds always find powerful defenders: the initiates in consecrated uses, all those who have practiced bad methods hold on to them—first by conviction, then by vanity, and they cannot be overrun in this last entrenchment. One must bring them to the tribunal of reason, confront them with their own writings, analyze them, point out their errors, and appeal to an impartial public.[3]

In volume 1, Montalembert's statement "Merit is, so to speak, innate in the corps of engineers" is a tribute that seems strange when one considers his later dislike of the corps.[4] In volume 10 he went to the trouble of indicating all the places he had spoken well of the corps in order to repudiate the charge that he bore a grudge toward the military engineers.[5] However, their initially cool relations descended into vicious exchanges that transgressed the bounds of courtesy and came close to the level of the gutter press of the eighteenth century. Epithets such as "graphomaniac," "charlatan," and "spinner of theories" [faiseur de systèmes] were flung back and forth in a quarrel that ended only with the death of its major participants.

Montalembert attributed a number of sins to military engineers. They idolized Vauban and used him to legitimize their own productions. They tended to be slaves to habit and to reject novelty reflexively. Montalembert put forward a number of explanations for this. He had argued earlier that Directors of the various engineering districts were local potentates, who wanted to hang on to their posts and their procedures at all cost. They prevented change by making the Director at Versailles, supposed to have the ear of the Minister, subservient to them by the threat of undermining his position.[6] Thus this Director—Fourcroy de Ramecourt after 1776—dared not innovate. Later, during Fourcroy de Ramecourt's relentless opposition to his schemes, Montalembert began to accuse him of being a dictatorial chief who imposed his ideas on the entire corps against its will, manipulated the Minister, and acted as a "snuffer" [eteignoir] of new ideas. He claimed that certain engineers had been persecuted for their support of his methods and that others had had their names used without their knowledge in writings against him.[7]

Certainly no engineer came out publicly in Montalembert's defense before the Revolution except for Lazare Carnot. Montalembert later argued that the heavy hand of the corps had reduced Coulomb to silence when queried on his reaction to the system being used at Aix. Montalembert used the examples of Coulomb and Carnot to argue that his quarrel had never been with individuals, most of whom were talented and honorable men, but with the corps. From volume 5 on he pictured himself as an assailant of the pernicious scientific corporatism of the corps, which he contrasted with the enlightened scientific corporatism of the Royal Academy of Sciences. In a favorite analogy, he compared the Corps of Military Engineers to an official corps of explorers who lived on land where they could more easily enjoy the substantial perquisites of their positions.[8] Confronted with the discoveries of a Captain Cook in waters hitherto believed empty, their reaction would be to deny Cook's findings and to discourage any other discoveries that would force them to justify their official mission and take to the sea themselves. Another analogy was with the conservative Faculty of Medicine. If empirics and charlatans consistently saved lives without following orthodox procedures, surely the Faculty would accept them into its midst.[9] Any innovation would raise embarrassing questions about the stagnation of the science the Corps was supposed to cultivate and advance. Better for it to claim that everything had been discovered and there was nothing new to find. It could then piously maintain an established and perfect tradition.

Attacking a corporate guild was nothing new, but it was very clever. Turgot's abolition of guilds had been hailed as a positive step in the diffusion and advancement of knowledge as well as a boon for commerce and the economy. There were many army officers who had felt that the skill of the engineer was not a special one that merited the existence of a separate corps; civil engineers and architects could take over the details of construction that soldiers were not able to do. The implications of Montalembert's attacks were no doubt keenly felt by the military engineers. Their status as a military corps was still fresh and still contested by some in the army. Montalembert also was accusing them of being a corps in the sense of a guild, with its connotations of trade, secretiveness, and selfishness.

How had engineers convinced so many that they were both a distinct and a necessary corps? Montalembert, who obviously thought of himself as a maligned and persecuted Captain Cook of fortification, saw several ways in which they did this. Like all guilds they used secrecy both to emphasize their own importance and skill and to hide their incompetence. Since military secrecy invoked the safety of the country, it could cover any selfish actions by people hoping to extract personal profit. Lifting secrecy would force military engineers to justify their systems in public debate. Indeed, the public would learn that fortification—so crucial to a nation's security—was not a mystery accessible only to privileged initiates but could and should be taught to the public, particularly to officers in the other branches of the army. After all, they had to apply the principles of fortification in their professional careers, and they could also help develop these principles by their experience (as Montalembert's own example demonstrated).

Secrecy also fostered the mystique of expertise. The military engineer presented himself as the sole expert of fortification and its only qualified judge. Montalembert was not against expertise in itself; however, having tried but failed to obtain the blessing of the Corps for his ideas, he challenged its fitness as an arbiter of fortification. A worthier tribunal, he argued in volume 5, was the Academy of Sciences, of which he had been a member since 1747.[10]

Unlike the military engineering corps, which instinctively attacked any threat to its corporate wisdom, the scientific academies were always active, welcoming new ideas, and working constantly to replace the old with the new. Montalembert contrasted static and frozen dogma with vigorous and creative activity.

In 1776, when the first volume of *La Fortification perpendiculaire* was published, Montalembert had been a member of the Royal Academy of Sciences for almost 30 years. His connections with the Academy had doubtless been useful when he launched his metallurgical career, and he had used the Academy as a platform to attack Maritz, the man who was put in charge of Montalembert's Ruelle foundry after it had been taken over by the government. On one other occasion he appears to have used the Academy to promote another business venture (a paper mill).[11] Otherwise he was not particularly active in the Academy, publishing only a few minor papers, such as a description of a stove he had seen in Russia during the Seven Years' War and a method of preventing a cannonball from spinning.[12]

It was an official duty of the Academy to examine its members' books. The members of the commission that examined Montalembert's first book were Yves-Marie Desmarets, comte de Maillebois (1715–1791), Louis-Élisabeth de la Vergne, comte de Tressan (1705–1783), the chevalier Jean-Charles de Borda (1733–1800), and Jean-Baptiste Le Roy (1720–1800). The comte de Maillebois was an eminent military man, commander of the French armies in Bohemia and Italy during the War of the Austrian Succession. The comte de Tressan was an associé libre of the Academy and, like Montalembert, a man of modest scientific achievements. A senior military officer with a taste for letters, he was a member of the Académie française. Borda, educated at the military engineering school at Mézières, was pursuing a career as a naval officer. He was one of the very few persons admitted to the Mézières school without an examination. He had been a member of the committee examining Montalembert's colonial fortification projects. Le Roy, a physicist, seems to have done most of the commission's work and written the final report. His report began with an outline of the contents of the book. For an evaluation of some of its controversial aspects he relied on the opinions of military men. Thus, he did not undertake to

> analyze the many objections of M. de Montalembert against bastions. That would involve us into too many long details. We will be content to observe that the consequences he draws from them against modern fortification are justified at least up to a certain point. For the greatest generals of this century unanimously agree that the art of fortifying fortresses is well below that of attacking them. And the most experienced military men in this domain of war all want to find better means to balance the defense and the attack of fortresses.[13]

But Le Roy did bring up some criticisms and objections: smoke in the casemates, explosions in the caponiers, and the potential expense. More points could have been raised, but he stopped. "Our aim," he wrote, "is not . . . to pronounce definitively on this system that the author still proposes to improve and which has not appeared to us to be based on proofs [that] are tied together well enough to demonstrate its advantages." He finished by saying that he approved of the efforts of Montalembert over the last 20 years (thus confirming that Montalembert's serious interest in fortification began after the failure of his metallurgical enterprises) and that his work merited the approval of the Academy. Only the mention of approval of the Academy made its way into the printed copies of Montalembert's book, and this level-headed evaluation of his work remained within the walls of the Academy. It shows that the attitude toward Montalembert was fair and balanced. It was appreciated that he was trying to solve an important and widely recognized problem in military architecture. His solutions appeared to have some merit, but they would have to be developed and verified before any pronouncements could be made. The imperfect nature of the work was noted, and the attitude that the scientific professional should defer to the military professional is clear. Montalembert's ideas were considered worthy of consideration without definitive and normative approval. Montalembert was no Newton, whose work and program had become orthodoxy, nor was he an exhibitionistic crank like Marat or Mesmer.[14]

Volume 2 of *La Fortification perpendiculaire*, published in 1777 and again examined by the same commission, did not receive a detailed report. The commission contented itself with the remark that this second part of Montalembert's work "is nothing else but an application of the principles and views presented in the first [part]. We think that it similarly merits the approbation of the Academy and publication under its auspices [sous son privilège]."[15]

The next lengthy report on a work by Montalembert (volume 5 of *La Fortification perpendiculaire*) was again written by Le Roy on behalf of a commission that included the Duc de la Rochefoucauld and Buffon. The session of 2 April 1784 at which Le Roy's report was presented was devoted in large part to artillery.[16] The rough notes [plumitif] for the session indicate that there was also discussion of floating batteries, very likely of the kind that had been used unsuccessfully by Montalembert's enemy d'Arçon in the siege of Gibraltar. Le Roy mentioned that roughly one-third of the book he was reviewing dealt with fortification ("particularly of the methods used to eliminate arbitrariness in the manner of judging the degree of strength of a fortress") and the rest with artillery.[17] This time Le Roy wanted to discuss Montalembert's ideas on artillery at length. He questioned Montalembert's claim to have invented an all-purpose carriage that could be used for naval, field, or fortress artillery. He felt that the complexity of such a carriage might adversely affect its operation in combat, but he admitted that the approval of naval officers at Aix, which Montalembert claimed he had received, "should reassure up to a certain point on this subject."

Montalembert's ideas on gun carriages had landed him in the midst of another bitter military controversy: the debate between the "rouges" and the "bleus" over the Gribeauval system of standardization of guns and artillery materiel.[18] Both bleus and rouges were opposed to Montalembert's gun carriages. The conservative defenders of the Vallière system (the rouges), although they defended the use of identical gun carriages for fortress and field artillery, did not want to change the existing types of gun carriages. Gribeauval's defenders (the bleus) called for change, but they wanted specially adapted gun carriages for different combat needs. Montalembert had attacked their new gun carriages for coastal guns.[19] Gribeauval had been one of the more outspoken opponents of Montalembert's ideas for the Aix fort. But the marquis de Saint-Auban, one of the leaders of the rouge faction, had also vigorously attacked Montalembert in print.[20] In short, rouges and bleus were united in rejecting Montalembert's ideas on guns and fortifications.

The whole question of artillery was very much on the mind of the Academy. Three months earlier, Montalembert had read his paper on the Aix artillery trials that Fourcroy de Ramecourt had demanded so insistently.[21] Besides the controversy over the Gribeauval system, the memory of the failure of the French attack on Gibraltar in 1781 with d'Arçon's new floating batteries was still very much in the public's mind.[22] Whatever diffidence Le Roy had in dealing with a subject he felt belonged to military experts, the Academy was being dragged into a heated debate on the merits of rival technical systems for artillery and fortification, the two "armes savantes." It was in this context that Montalembert, in volume 5, had attacked military engineers and military engineering techniques.

On 26 February 1785, "in my absence," according to Montalembert, Fourcroy de Ramecourt read a memoir to the Academy in the name of senior officers of the corps of military engineering outlining the idea of moments of fortification, which he attributed to Vauban and presented as a scientific tool for comparing the merits of different fortifications.[23]

A week later the Academy reported on Fourcroy de Ramecourt's paper:

. . . it seems therefore to us that this method, whose author [Vauban] is one of the members of the Academy whose memory is respectable and dear to it deserves to be known, that its publication can be useful to the progress of an important art, that the method in itself is founded absolutely on experience and observation and that consequently it deserves to be published under the auspices [sous le privilège] of the Academy.[24]

The report finished, however, with a cautious remark:

. . . we believe that we should also note that the authors of the paper having only given general principles of the method and having published nothing on the data that enable it to be applied to particular cases, the Academy cannot judge the degree of exactitude and precision of the results one can attain in the present state of the art of constructing, defending, and attacking fortresses.[25]

Once again the Academy had drawn back from giving a wholehearted endorsement to one side in a very controversial military debate.

Fourcroy de Ramecourt had only been an associé libre of the Academy for a short time, although he had been a corresponding member for some years.[26] Fourcroy de Ramecourt had just about as many—or as few—claims to the honor as did his colleague Montalembert. Fourcroy de Ramecourt had contributed an article on the art of the lime burner to Duhamel de Monceau's *Description des Arts et Métiers*, a work sponsored by the Academy, and later he had presented a curious paper on the point at which birds become invisible to observers on the ground as they ascend higher in the sky, but his activity in the Academy was, if anything, less than that of Montalembert.[27] It was undoubtedly not for his scientific or technical talent that Fourcroy de Ramecourt was elected. The only military engineers with genuine scientific credentials, Coulomb and Jean-Baptiste Meusnier, had already been corresponding members of the Academy since 1774 and 1776 respectively and full members since 1781 and 1784 respectively. Charles-Marie-Thérèse de Tinseau was also a corresponding member.[28] Other engineers had been at least as active in presenting papers to the Academy.[29] The death of the Nicolas-Christian de Thy, comte de Milly, on 19 September 1784, had opened the first vacancy among the associés libres since Fourcroy had become the unofficial head of the military engineering corps. His election to the Academy was a routine honor for a senior technical servant of the state. The chief engineer of the Corps royal des ponts et chaussées, Perronet, had been a member since 30 June 1765. Fourcroy de Ramecourt's election may also have been both a sign that the Academy wanted to establish a balance with Montalembert, who was using it as a platform for his ideas, or—perhaps more likely—a signal of official governmental approval for its technical establishment.

Less than a fortnight after Fourcroy de Ramecourt's paper at the end of February, Montalembert requested commissioners of the Academy to examine volume 6 of *La Fortification perpendiculaire*, which contained a detailed refutation of a memorandum submitted to the Minister of War in February 1780 by Major Grenier of the Military Engineering Corps criticizing the first volume of Montalembert's work.[30] Along with the refutation, was a structural analysis of arches to refute Grenier's contention that Montalembert's masonry structures could not stand up to artillery fire. Resorting to calculations of an unnamed engineer from the Corps des Ponts et Chaussées, Montalembert argued that Grenier's use of Bélidor's methods for calculating strength of arches was outdated and resulted in grossly inflated dimensions.[31] This unnamed engineer was, in fact, the eminent civil engineer Prony.[32] The polemic was being deliberately dragged from the confidentiality of ministerial memoranda and thrust into the full glare of publicity. The commissioners Montalembert requested were named by a vote, a rather uncharacteristic practice of the Academy of Sciences that indicates either deep divisions or a unwillingness to deal with the matter. Buffon, Borda, and Bory were elected according to the minutes, but Le Roy again wrote the report, which appears to indicate that he was also elected. It took more than a year for the report (a crisp "deserves to be printed like the preceding [works] under the auspices of the Academy," with a comment that in view of the previous description of the earlier volumes "it would be useless to return to what has already been said about them") to come down.[33]

Montalembert was active not only in the Academy of Sciences but also outside it. He had launched a spirited and spectacular attack on Vauban, symbol and shield of the doctrines of the military engineering corps, both personally and by proxy.

The occasion for the first proxy attack on Vauban was an éloge of the Marshal written by Lazare Carnot, his Burgundian compatriot and military engineer, which had been awarded a prize by the Academy of Dijon on 2 August 1784.[34] In this prerevolutionary period, the éloge made much of Vauban's humanity, his civic virtue, and his efforts for fiscal reform. On more technical questions, Carnot implied that Vauban's great science, creative and dynamic in its origins, had degenerated to a set of rigid and unimaginative rules. Carnot was no blind adherent to the ideas of Fourcroy de Ramecourt or Cormontaigne: he went so far as to attack "a certain pretentious common fellow" [un certain vulgaire à prétention] and "ignorant plagiarists." He also attacked "the rudeness certain book scribblers have thought acceptable to attribute to military engineering officers." Subsequent events show that Carnot was thinking more of run-of-the-mill armchair fortifiers, but Montalembert perceived these words as a direct attack on himself. Reaction came soon, in the form of an anonymous reprint of Carnot's éloge "enriched with observations by an amateur." The "amateur" was none other than Montalembert.[35] He subjected Carnot to a brisk dressing down while praising perpendicular fortification as a system far superior to Vauban's.

Between the time Carnot received his prize and the time Montalembert responded, the Académie française had come forward with its own prize competition for eloquence. In this competition too the subject was a eulogy of Vauban.[36] This time, Montalembert got his staunch ally Choderlos de Laclos to respond. Although Montalembert's name never appeared, Laclos showed little originality in the arguments he used: they were just a repetition of the arguments of Montalembert. But he wielded them with superb literary skill to create a succès de scandale with the publication of his "Lettre à Messieurs de l'Académie française sur l'Éloge de M. le Maréchal Vauban, proposé pour sujet du Prix d'Éloquence de l'année 1787."[37]

Laclos patronizingly dismissed the choice of Vauban as the subject of a formal eulogy. Though understandable for a provincial academy proud of a native son, it was improper for the Académie française, a national institution whose opinion was supposed to reflect the true voice of the people honoring its greatest men. Vauban was not a "great man" as he was almost invariably called, nor should Frenchmen feel any sense of debt toward someone who, in building his fortresses, had "buried millions." Vauban had perfected the art of the attack, but he had not done anything new in defense, which had remained in thrall to old and defective methods of fortification.

There was an enormous outcry against Laclos' letter. Although the publication of *Les liaisons dangereuses*, with its strong whiff of immorality, had caused some grumbling among his military superiors, he had not suffered for it. Now the reaction was swift and brutal. Ségur, the Minister of War, yanked Laclos away from the woman he had just married and sent him to the bleak garrison at Metz, stolidly ignoring all appeals. Laclos' military career was stalled, and there was a flurry of defenses of Vauban by military engineers.[38] Besides one by Carnot, perhaps the most notable was by Le Michaud d'Arçon. D'Arçon lambasted the "new Erostratus" and rushed to the defense of Vauban's reputation ("consecrated by the voice of an enlightened century, reaffirmed by the judgment of a philosophical century").[39] He considered the attack on Vauban profoundly subversive because it was aimed at the engineers and the "grand edifice" they had constructed for the security and tranquility of the state. Dismissing systems of fortification in general and Montalembert's in particular, he subtly reminded his readers of Laclos' scandalous reputation while subtly mocking the support Montalembert's system had received from influential women as a result of boudoir politics: "One must nevertheless agree that these systems [of Montalembert] sometimes enjoy the advantage of being put forward by a sex that, barring the danger of liaisons, must know all about the art of defense."[40] D'Arçon had no doubt that Laclos' inspiration: he was the "harmonious trumpet of Montalembert, who wants to knock down the walls of Jericho."[41]

For Fourcroy de Ramecourt too there was no doubt whence the attacks on Vauban had originated. Rather than merely defending Vauban, he now publicly attacked Montalembert. On 16 June 1786 he presented to the Academy of Science a book that had just been published: *Mémoires sur la Fortification perpendiculaire par plusieurs officiers du Corps royal du Génie.*[42]

As Montalembert was to point out, there is something underhanded in Fourcroy de Ramecourt's book. The first part consists of the paper Fourcroy de Ramecourt read to the Academy back in February 1785 along with the Academy's report on it. But this forms only a small part of the book. No doubt unsatisfied with the less-than-categorical approval of the Academy, Fourcroy de Ramecourt had tacked on a "Suite du Mémoire; & Recherches sur l'utilité dont peut être au Service du Roi le Livre intitulé de *la Fortification perpendiculaire.*" Here the method of measuring the strength of fortifications discussed in his paper was applied to Montalembert's fortresses. The artillery trials at Aix and Filley's system of mezalectres were also discussed. In short, it was a broadside attack on Montalembert and all his works that enveloped people like Filley, whose work Montalembert had used to support his attacks against the corps of military engineering. Even though the paper read to the Academy was clearly responding to Montalembert, it maintained, albeit with some difficulty, a detached tone appropriate to an academic setting. The book, however, was quite openly polemical.

The book closed with a "Second Mémoire ou Observations sur le livre intitulé, *La Fortification perpendiculaire.* Par M. G***, Major au Corps royal du génie." Although the approval of the Academy had been printed at the proper place—at the end of the memoir that was actually read to the Academy—there was certainly room for confusion on the part of the reader: one could easily and mistakenly conclude that the entire book, including the very personal and bitter attack on Montalembert, had the Academy's approval.

Also confusing was the precise authorship of the work. It was ostensibly anonymous ("several officers of military engineering"), and Fourcroy de Ramecourt's name did not appear. The first memoir, including the part read to the Academy, was attributed to "several general officers of the Royal Corps of Military Engineering." The last part of the book was written by Major Grenier. It was an expanded and revised form of the memorandum Montalembert had rebutted in volume 6 of *La Fortification perpendiculaire*. Clearly there was an attempt to impart a double legitimacy to the book by implying the approval of both the Royal Academy of Sciences and the Royal Corps of Military Engineering. Both insinuations were dubious. The qualified and tepid approval of the Academy was limited to a very small part of the book, and the implication that the book represented the official opinion of the corps of military engineering was given no independent confirmation. Indeed, the veil of mystery covering the other putative authors has not been pierced to this day. It is almost certain that Fourcroy de Ramecourt was the primary author, assisted by a small number of relatively junior officers whom he had enlisted—or maybe even conscripted—in the undertaking.[43]

Fourcroy de Ramecourt was moving out of the arena of the Academy of Sciences into a wider one, that of public opinion, hoping to use the authority of the Academy and the Corps of Military Engineering to gain victory over Montalembert. This was necessary, he wrote, because their "officers' oath" obliged engineers to speak out for the safety of the kingdom: he was worried that ministers of the Crown and "men whose approbation has a character that necessarily sweeps along many other soldiers" would be misled by the credit Montalembert had succeeded in obtaining for a grossly defective system of fortification. Initially, engineers had kept silent because of their respect for military secrets and their belief that in the "exact sciences" ideas are reduced to their true worth once they are submitted to experience and calculation. "It is indeed the Authors' affair if it happens that their new ideas, analyzed in the crucible of *experience* and calculation evaporate in smoke."[44] Therefore it was now time to turn the guns of experience and science on systems of fortification that had obtained undeserved and dangerous publicity. He wished to present a method, based on experience and reason, that would permit schemes of fortification to be evaluated and compared in an objective manner. The yardstick that Fourcroy de Ramecourt proposed was in his opinion simple yet scientific. Unwittingly following Montalembert, who believed that specific gravity alone determined the strength of an artillery piece, Fourcroy de Ramecourt felt that one could calculate a number for any fortress that placed it on a linear scale of defensibility. The science of fortification could be reduced to a kind of elementary assaying. "Specific gravity is sufficient for our metals, without any other apparatus, to distinguish them from gold or platinum."[45]

The simple numerical criterion of excellence proposed by Fourcroy de Ramecourt was the "moment of fortification," a word he says he had borrowed from the "language of mechanics."

> It is the search for the *moments* of fortification that we call their *analysis*. . . . It is the use of this touchstone in our hands that places our Art in the rank of the positive sciences. [It] distinguishes and separates that which is military, useful, and conclusively proven from arbitrary ideas, chimerical speculations of the Study, conjectural properties and vain promises of all the Authors of good will who deal with this matter [fortification] without understanding it.[46]

The moment of fortification was the quotient of the time taken to force a surrender of a fortress divided by the cost of a front of fortification. (A front of fortification was that part of the fortification contained between the tip of one salient angle of a bastion to the tip of the salient angle of the adjacent bastion. In a perfectly regular fortress it was the module that was repeated to construct a fortress: for example, an octagonal fortress had eight fronts of fortification.) However, the simplicity of this number was complicated by the calculation of its constituent elements. Although construction costs could be estimated reasonably objectively, estimating the length of a siege was another matter. It involved use of the fictitious siege journals that Montalembert had rejected when discussing Cormontaigne's criticism of Bélidor's proposed system of fortification for Metz.[47] Fourcroy de Ramecourt claimed that this struck at the very root of the legitimacy of fortification as a science. It implied that "there could be nothing that was constant, uniform and well known in the progression of attacks. . . ."[48] But fortification was a true science:

> . . . an Art founded on a certain and evident theory, drawn from Nature itself, from *experience*, from reasoning, from Mathematics. It is its theory that distinguishes it from the mechanical Arts as well as from those of fantasy and which makes it a science. It is *experience* which makes it subject for all time to necessary changes, according to the different arms employed in War or the different manners of using them.[49]

It was this theory that made French military engineering superior and its successes so difficult to copy for those who read the pirated manuals of Vauban, the man who had reduced to a "true science the practice of attacking fortresses." Vauban's method of attack "*is really scientific, that is to say, proved by principles to be necessary and the best known to bring about most promptly the proposed effects.*"[50] Though others could copy his procedures, they could not copy his theory, which the military engineering corps carefully preserved as a military secret.

There is thus a major ambiguity in Fourcroy de Ramecourt's argument: while fortification is scientific ("a Physico-mathematical science") and can be subjected to objective discourse composed of mathematical terms, its theory is not public but restricted to a group of initiates.[51] This runs counter to the whole idea of a public science that was one of the major legacies of the Scientific Revolution of the seventeenth century, an idea solidly established in the Enlightenment and embodied in the very academy to which Fourcroy de Ramecourt was addressing his arguments. Fourcroy de Ramecourt wants to have it both ways: fortification is a public science but it is also a craft for initiates. Military secrecy justifies the need for keeping the "theory" of fortification—the very theory that is the mark of a mature and genuine science—hidden from the view of the profane.

Cormontaigne is superior even to Vauban in putting fortification in the ranks of the "sciences positives" because he does not keep it mysterious; he presents a transmissible theory.[52] Indeed, the role of Cormontaigne in fortification is explicitly identified with that of Newton in physics.[53] But at the same time there is a coyness in presenting only "some of the principles that guided and justified the works of M. de Vauban."[54] Perhaps part of the reason for this ambiguity lies in the status of experience in Fourcroy de Ramecourt's thinking. The epigraph for his book is "Artem experientia fecit" ("Experience creates art"), and the importance of experience is emphasized relentlessly and heavily throughout the book. Every time the word 'experience' appears, it is italicized. Cormontaigne had composed volumes "on the principles of *experience* to calculate the best manner of making fortifications."[55] On the other hand, the ideas of Filley "are seductive for amateurs and they have even been contagious for some modern officers of Military Engineering whose talents, instead of being improved by the study of war and the practice drawn from the *experience* proposed to us in the scientific productions of Vaubans and Cormontaignes, have become dull and absorbed in the harmful minutiae of Professors of the square and compass and in the trifles of their own inventions."[56] Here experience seems to be contrasted with theory. In another passage, Fourcroy de Ramecourt says "If to this *experience* we wish to join reasoning, although it is not very necessary. . . ."[57] This implies that experience is not only different from but also superior to reason. It is experience that tells engineers that earthworks 20 or 21 feet thick are adequate protection against 24-pound cannonballs, that casemates in revetments of earthworks are unhealthy for the troops manning them and do not permit adequate ventilation of smoke, that construction costs will double if earth has to be carted from a distance of as little as 500 toises, and that "a hundred *experiences*" show that it only takes 36 hours to open a breach in masonry that is 5–6 feet thick.

It is also experience of past sieges that enables military engineers to draft their fictitious siege journals and to take into account the "moral" factors that are so important and unpredictable not just in siegecraft but in war generally. Although in warfare nothing can be foretold with certainty, past experience of sieges makes it possible to calculate the "absolute" strength of fortresses, and it permits comparisons between different kinds of fortresses that include these moral factors.

Montalembert, too, had presented himself as an experienced military officer who was a veteran of several campaigns, including nine sieges. Fourcroy de Ramecourt retorted that the engineers had followed a hundred sieges, not only as participants but also through "our faithful tradition of journals." Here he was referring not to the fictitious sieges journals but to the actual siege journals that were kept for every siege and deposited in the archives of the military engineering corps.[58] In addition to personal experience, there was the vicarious experience of carefully documented history.

Thus, the foundation of fortification was experience, not merely of the living but of the dead, a cumulative and growing tradition that was invisible to the outsider either because it was the tacit knowledge of its practitioners or because it was hidden from the profane for reasons of national security. Properly distilled in the alchemist's cave of the secret military archive, it provided the justification, the data, and the basic principles for the "physico-mathematical science" of fortification and distinguished it from the science of the "minutiae of professors" and of "the square and the compass." In ways not specified, Experience had an absorptive and filtering function. It could integrate new discoveries and potential weapons, such as the balloons that had appeared 2 years before Fourcroy de Ramecourt's paper.[59] There was a constant dialectic between defense and offense in siegecraft that forced one to adjust to the other, and therefore fortification was by no means a static science. Yet Experience had laid down the framework in which future discoveries and changes could be infallibly analyzed and this framework had become fixed and unchangeable thanks to the work of Vauban and Cormontaigne. In practice this framework implied that the basic bastioned trace was the only possible fortification.

Thus, there is a mistrust of novelty that forms an unexpected and powerful counterpoint to the theme of fortification as a continuously progressing modern science. Novelty is perceived to be disruptive of basic principles that have been established scientifically for all time; it is a potentially regressive force that threatens the true basis for a science that should no longer be contested. Fourcroy de Ramecourt was no doubt aware that the Academy would remember Mesmer and Marat, whose claims to have overturned the basis of existing physical science were causing unpleasant publicity for the Academy in the early 1780s, when he wrote the following:

> We can apply to fortification a reflection of the *Histoire de l'Académie des Sciences* (1775, p. 64) concerning the research done even today by certain minds on the squaring of the circle, perpetual motion, etc., and which made the Academy resolve never to examine any solution to these problems.
>
> Seventy years of experience has shown the Officers of the Royal Corps of Military Engineering that none of those who, since M. de Vauban, have published new systems of fortification, that is, who have treated this Art as if it should be the product of speculations of the study knew either the principles or the usages of fortification. None of the methods they have followed have led to the goal they set themselves. This long experience has sufficed for these Officers to convince them of the feeble utility that would result for the Service of the King in examining these Fronts [of fortification that] claim to be better than the ground-plans of M. de Vauban. It has even convinced them that a new System of Fortification is today one of the characteristics of ignorance in this Art.[60]

These unfortunate final words implying that anything new is automatically wrong provided ammunition that was eagerly seized—and not just on one occasion—by Montalembert and Carnot to argue that the Corps of Military Engineering (or at least its leadership) was stuck in the rut of routine. It was a powerful argument at a time when 'routine' was a pejorative word. And Fourcroy de Ramecourt had placed himself solidly on the side of conservatism when he said that he would remain deaf to "all the vain clamor of the adepts, even more common and much more importunate in Fortification than in Chemistry."[61]

Moments of fortification were not the sole bone of contention between Montalembert and Fourcroy de Ramecourt. Fourcroy de Ramecourt reproached Montalembert for not having given a single detailed and dimensioned profile of his fortifications.[62] Defilading was a major element in the design of fortifications. Yet plan views of fortifications were always more interesting for Montalembert than profiles. He did not even reply to the criticism of inadequate attention to defilading until he brought up the subject in his attacks on the École Polytechnique a decade later.

Fourcroy de Ramecourt also rejected Montalembert's contention that siege batteries could only be effective at 20–30 toises. Montalembert himself had mentioned that walls had been breached at 250 toises at the British siege of Cartagena in 1741 and at 120 toises at the French siege of Tournai in 1745.[63] Fourcroy de Ramecourt, who had himself participated in the latter siege, said that the breaching distance was even greater—175 toises. Moreover, 24-pounder cannon could easily breach walls at 300 toises and destroy Montalembert's masonry gun towers at 375 toises.[64]

He also pointed out that the prescribed slope of approximately 55° for the earthworks for Louisville, Montalembert's hypothetical fortress, was much too steep to be self-supporting.[65] For earthworks that would not crumble, a slope of 33° was necessary, and this would increase the diameter of the fortified area of Louisville by about 12 toises. Furthermore, Montalembert's emphasis on the difficulties of breaching earthworks with cannon was irrelevant: earthworks with his prescribed slopes could be easily climbed by an assault party and would not even have to be breached.[66] Montalembert would allow this criticism to pass in silence. However, he would vigorously reject the cost estimates of the engineers, claiming that they had used sleight of hand in calculating masonry and earth required for his fortification. He claimed not only that his schemes were cheaper, but that they would be cheaper still if unspecified "abuses" in the payment of contractors were eliminated. According to Fourcroy de Ramecourt, however, the bastioned trace was considerably cheaper, and Montalembert had forgotten that any cartage of earth from a distance exceeding 500 toises from the work site doubled the cost of earthwork construction.[67] Grenier argued that Montalembert's estimate for the time to construct the Fort de Condé should be revised upward from 26 days to 40 days.[68] Montalembert was also accused of being ignorant of the capacities of workmen building fortresses, something that has the ring of plausibility in view of his experiences at Aix.

Another major point of contention was the efficacy of casemates. Fourcroy de Ramecourt did not mention that the artillery trials at Aix had been conducted to test the stability of the structure. Instead, he focused on the problem of smoke in casemates. Casemates were nothing new, and he was not impressed by Montalembert's claim that they worked well at Olmütz in the Austrian dominions or by the apparently satisfactory tests at Aix. The latter were not real casemates; they were gun decks, because there were no partitions between the guns. Casemates were described as a reversion to antiquated methods of fortification rejected in the seventeenth century. They could be used in special cases, such as the Château de Taureau in Brittany built by Vauban, but even he, Fourcroy de Ramecourt claimed without any documentary proof, had been reconsidering the tower bastions containing casemates that he had built at Neuf-Brisach. The question of casemates for the fortress at Landau had been discussed at length in the corps in 1740 but had been rejected.[69] He marshaled evidence from research by Duhamel de Monceau on air in mines and hospitals and even mentioned successful research by engineers at Marseille on evacuating smoke from casemates.[70] Joseph

Priestley's view that gases from gunpowder explosions were heavier than air and would not disperse easily was invoked. Furthermore, these gases were in large measure toxic "mephitic" gas.[71] Although satisfactory ventilation was possible in carefully controlled conditions, it was not practical in combat.[72]

Fourcroy de Ramecourt apparently felt that casemates were a central part of the argument: all the forts proposed by Montalembert would be useless if his casemates were not effective. Though prepared to admit on the basis of the experience he valued so highly that there might be doubts on the matter, he was confident that his assertion that casemates were dangerous and ineffective would ultimately be vindicated. Moreover, experience was more than just direct experience written or passed down; it was a sixth sense that was gradually acquired through long years of practice:

This author [Montalembert] has for himself only his own opinion. He lacks as we do the direct *experience* that we demand as necessary to destroy the objection about smoke. But it seems to us that meanwhile there exist enough data on this subject to decide on which side the most probable outcome can be found.[73]

Fourcroy de Ramecourt closed his portion of the *Mémoires sur la fortification perpendiculaire* with a final blast at Montalembert vindicating the military engineering corps' right to control the building of fortification:

By means of our method and all the military and geometrical talents acknowledged pretty generally in the officers who possess them, fortification must continue to improve itself both in its effects in war and economy of construction.

On the contrary, the old routine, that of Authors of good will, Professors and Amateurs of this kind, is only good for retarding the progress of the Art— as the four systems we have just analyzed prove—and even to plunge it back into all the vices of its childhood whence it was so luckily extracted by MM. de Vauban and Cormontaigne.[74]

Montalembert wasted little time in replying to Fourcroy de Ramecourt's onslaught. On 8 July 1786, less than a month after Fourcroy de Ramecourt's book was presented to the Academy, he asked it for a commission to examine his *Réponse au Mémoire sur la Fortification Perpendiculaire par plusieurs Officiers du Corps Royal du Génie, Présenté à l'Académie Royale des Sciences*, which was to become volume 7 of *La Fortification perpendiculaire*.[75]

The number of Montalembert's works continued to grow. He asked that the commission of the Academy examining this volume be composed of the academicians who had usually examined his books and those who had examined Fourcroy de Ramecourt's. This wish was partially satisfied: the commissioners were Condorcet and Laplace, who had reviewed Fourcroy de Ramecourt's book, and Le Roy and Borda, who had reviewed Montalembert's earlier books.[76] But the commission took almost 2 years to judge the book and then came forward with a cautious and diplomatic appraisal:

> One will conceive without difficulty that to support the reasons he opposes to the Gentlemen Officers of Military Engineering, M. de Montalembert has been obliged to go into depth on the various subjects of this discussion. This is also what has happened. Because of that it would be impossible to follow him in all that he says on this important subject without going into explanations and details [that are] too long and too difficult to grasp.
>
> Thus we will content ourselves with observing that M. de Montalembert in his answer has seemed to us to discuss in detail several important points of the art of fortification, and particularly those regarding the manner of judging the absolute strength and the real value of military strongholds; that in order to reply better to the Gentleman Officers of Military Engineering he has developed further several articles on which he had not gone into enough depth in his treatise on *Perpendicular Fortification*; [and that] finally this reply can only cast light on the important dispute that has for some time arisen on the inferiority of the art of defense of fortresses in relation to that of the attack.
>
> Consequently we believe that the reply of M. le Marquis de Montalembert, which can also form a continuation of his [previous] Work, deserves to be published under the auspices of the Academy without, it must be added, taking any position on the substance of the question.[77]

The original manuscript report in the archives of the Academy of Science shows hesitation among the commissioners. The original phrase "approved by the Academy as this [previous] work has been and to be" after the words "deserves to be" had been struck out, and the phrase "without, it must be added, taking any position on the substance of the question" had been added as an afterthought in lighter ink.[78] Again the Academy's commission apparently wanted to avoid controversy: admitting the interest of the questions raised and judging Montalembert's work to be a contribution to the debate, it avoided a detailed discussion of the work and taking a position on it. The commissioners may have been uncomfortable with the tone of the debate, which was getting to be highly personal. The Minister of War had ordered Montalembert to remove a direct reference to Fourcroy de Ramecourt from the work.[79] One is entitled to wonder whether the debate had not led to Condorcet's question to the Academy on 23 December 1786:

> I asked whether in case the Academy were consulted on the publication of a paper against one of its members it should examine this paper and comment on the basis of this examination. It was unanimously decided that it should not get involved [s'en mêler] in any way.[80]

The first point of interest in Montalembert's book is a letter from Lazare Carnot. Published in his preface, it marks a surprising development in the heated argument over Carnot's *Éloge de Vauban*—an argument that had led to Choderlos de Laclos' scandalous letter to the Académie française and to Montalembert's bitter attack on Carnot.[81] Like his many brother engineers who had sprung to the defense of Vauban and attacked Laclos, Carnot had also responded to the *Lettre à MM. de l'Académie française sur l'Éloge de M. le maréchal de Vauban proposé pour sujet du prix d'éloquence de 1787*.[82] As Marcel Reinhard pointed out, Carnot's approval of Vauban was not uncritical: his colleague d'Arçon felt that Carnot had not gone far enough in his reply to Laclos.[83] Carnot denied that unquestioning unanimity reigned on the subject of Vauban's ideas in the military engineering corps, as Laclos had claimed. Indeed, he displayed reserve not only toward the systems of Vauban but toward any system that claimed impregnability as a quality. Laclos had implied, agreeing with Montalembert, that such systems could be developed. Carnot retorted:

> True fortification, you say, must make up for both the number and even the quality of troops, as well as for the genius of commanders.
> Such fortification would be, sir, a beautiful secret to teach us. When you will have found it, I promise to show you cannon so well made that they will know how to put themselves into a battery and fire by themselves.[84]

Yet two years later, Carnot wrote a fulsome letter of support to Montalembert, the man who no doubt had egged Laclos on to attack Vauban (and incidentally Carnot) and who himself had bitterly attacked Carnot anonymously. Carnot denied that he had Montalembert in mind when he excoriated ignorant plagiarists in his *Éloge de Vauban*. Indeed,

> I did not want to denigrate your Work at all, General. I only knew it by name at that time. True, it appeared under the name of a general officer and a member of the Academy of Sciences, but ... these titles ordinarily barely announce more real merit in a Man of Quality than those of Duke or Cordon Bleu. ... I was ignorant then that you were such an honorable exception for yourself and the celebrated bodies of which you are a member.
> I have since read this most estimable work [that is] so full of genius. I have rendered it justice at all times when the occasion presented itself, and I will never cease rendering [justice] to it at all times when [the occasion] will present itself [again], holding nothing so dear as knowing truth and seeing its light shine, no matter whence it comes.
> Now that your casemates are known and tested, Fortification will take on a new face and become a completely new Art. It will not be permitted to fortify without [casemates] nor to employ the revenues of the State to make the mediocre when one can make the good. To you alone will go the honor of this, it is impossible that anyone can deny it to you. It is established that you have resolved the most important problem of military architecture, [whose solution] was sought for so long and so fruitlessly—procuring large numbers of easy to use protected guns [feux couverts].[85]

Here was a letter that Montalembert could and did flaunt in the face of the Royal Corps of Military Engineers. Here was the first military engineer who publicly approved Montalembert and hailed him as the genius who had resolved the most important problems of fortification of his time. For Fourcroy de Ramecourt, the blow must have been doubly painful. A pre-Revolutionary tone that enveloped the Academy of Sciences and general officers in the same off-handed disrespect was becoming common. Not only did Carnot publicly support Montalembert; he considered his casemates "tested" and consequently the basis of a "completely new Art." Not only was Carnot a brother officer, but he had recently been defended by his fellow officers against attacks from the man he was now praising in such an unrestrained manner. It is now known that Carnot's later imprisonment because of a scandal over a lady's honor on the eve of the Revolution was the result of strings being pulled by an influential relative at Versailles and not a result of a vendetta against him by his superiors in the corps of military engineering. But it is likely that his poor relations with his superior officers contributed to the ugliness of the incident.[86]

Carnot's letter was not the only weapon that Montalembert used against Fourcroy de Ramecourt. He pointed out that Fourcroy de Ramecourt had attempted to portray the approbation of the Academy for a paper of 35 pages as applicable to the entire book of 290 pages.[87] Moreover, a favorable review of Fourcroy de Ramecourt's book in the February 1787 issue of the *Journal des Savants* by the chevalier de Kéralio, the journal's editor, had truncated the Academy's report to make it appear that the method of moments of fortification had received its complete approval. Kéralio's opinion carried weight since he came from an old military family and was the brother of the military tutor to the Bourbon Prince of Parma.[88] Montalembert alleged that Kéralio had been duped into submitting a piece written for him by the engineers.[89]

Along with flattery about the Academy's omniscience, intellect, and rectitude, dangling the glory of association with a new and superior technology, Montalembert claimed that its previous approval of his work committed it to his defense. This argument evokes parallels with the one he had made to the prince de Montbarey during the Aix episode. He felt that the Academy's report on his own book vindicated him in spite of the caveat that the Academy "takes no position on the substance of the question." The Academy, wrote Montalembert, should not have added this phrase for, after all, it judged his work worthy of publication under its authority [privilège]. That alone was the guarantee of the truth of assertions contained in the book, and he invited his readers to draw the appropriate conclusions about his book and Fourcroy de Ramecourt's.[90]

The motto of Fourcroy de Ramecourt's book had been "Experience creates art." Montalembert's epigraph was "Artis perfectionem demonstrabit experientia" ("Experience will demonstrate the perfection of an art"). Whereas Fourcroy de Ramecourt had argued that existing techniques were solidly founded on a long and documented experience, Montalembert retorted that future experience would confirm the superiority of his new art. Montalembert disdained to answer all of his opponents' arguments: he ignored Grenier on the grounds that he had already dealt with him in volume 6 of *La Fortification perpendiculaire*. Indeed, he only answered selected points in other parts of Fourcroy de Ramecourt's *Mémoire*, contending that this was enough to destroy the case against him. The main thread drawing together his counter-arguments was the contention that yardsticks of quality for past technological practice could not be used for new technologies. Fourcroy de Ramecourt's Whiggish view of technology was that of a cumulative structure, built on a solidly established and unambiguous base, that could incorporate changes as radical as the use of balloons in warfare. Montalembert rejected this view in favor of a discontinuous view of technological advance: Vauban's ricochet shot and the increasing use of artillery called for a radical change in fortification. Not only existing techniques but also existing standards of evaluating those techniques must be rejected. Any criteria of measurement or evaluation from the old system were inapplicable to his new system, and therefore Fourcroy de Ramecourt's criticisms were beside the point.

Montalembert felt that the radical novelty of his fortifications made it impossible to construct any common scale of comparison between the resistance of the traditional bastioned trace and his own perpendicular fortifications. Rejecting the engineers' "method of sieges done in their studies," he argued that it was only applicable to bastioned fortifications.[91] He also claimed that the method was not that of Vauban or even that of Cormontaigne, but was ad hoc cookery designed to discredit his systems.

Quoting Vauban at some length, Montalembert argued that the great man, when estimating the length of sieges, had done so "more to serve as instruction than to propose A MOST CERTAIN RULE, because all fortresses being different one from another, one must conduct oneself in them according to the greater or smaller [number of artillery] pieces that they can oppose to the enemy and their more or less convenient access points [avenues]."[92]

Moreover, Cormontaigne, in his use of fictitious siege journals, had only computed the number of days the siege would last; he had not calculated the "moments of fortification" that Fourcroy de Ramecourt was presenting as the touchstone of the military engineer's art.[93] Montalembert is in fact correct in arguing that the concept of "moments of fortification," as opposed to days of resistance in a fictitious siege, was not a well-established calculus in the art of fortification but something concocted by Fourcroy de Ramecourt. No evidence has been found in the military engineering archives that it was ever used by anyone else before the writing of *Mémoires sur la Fortification perpendiculaire*.

Montalembert was also upset that the construction costs, which could vary from place to place depending on the price of materials and labor, were calculated for a *front* of fortification rather than for an entire fortress. A dodecagonal fortress with a cheaper front of fortification than a heptagonal fortress, say, could nevertheless be much more expensive overall. Indeed, he felt more generally that considerations of expense were extraneous and illegitimate in planning for national security. No matter what the cost, certain strategic points had to be made impregnable. The political and economic consequences of defeat would be incalculably greater than the savings predicted by a narrow and mechanical concentration on the cost of fronts of fortification.

Nevertheless, Montalembert did concede the possibility of using numerical indexes to quantify the strength of a fortress, but only when there was experience of sieges.[94] Such experience was, however, useless for fortifications of a radically different nature, as he claimed his were. Fourcroy de Ramecourt had argued that his scale was nothing more than a quantified and rational approach to uncertainty and that rejecting such an approach was tantamount to rejecting the possibility of a science of fortification. The moments of fortification were more sophisticated estimates of the kind a commanding officer would make on the basis of experience with known obstacles in calculating the time it would take a regiment to march from Paris to St. Petersburg.[95] Unwittingly anticipating Tennyson, Montalembert had replied that such an estimate would become meaningless if the road to St. Petersburg were lined with enemy cannon or if one were forced to march from Lyon to Geneva over glaciers and Mont Blanc instead of taking the road.[96]

The metaphor of the road is apt when contrasting the views of the science of fortification espoused by Montalembert and Fourcroy de Ramecourt. Both believe in the existence of such a road. But Fourcroy de Ramecourt's road follows what appears to him to be the natural lie of the land and becomes increasingly solid with the accumulated experience of building earlier parts of the route. Montalembert's road cuts across country, tunnels through hills, and bridges rivers; it uses hitherto unknown or neglected techniques that are inapplicable to Fourcroy de Ramecourt's road. For Montalembert, Fourcroy de Ramecourt's road is perpetually menaced by landslides and flooding, never reaching the Promised Land of Impregnability, while his own destroys the obstacles in its path and follows the shortest line to its destination. For Fourcroy de Ramecourt, Montalembert's road ignores reality and willfully plunges into unknown danger, while his own conforms to nature and sober economy.

Fourcroy de Ramecourt had not been the first to use a numerical index of strength for fortifications. Montalembert had done so in volume 5 of *La Fortification perpendiculaire*: the resistance of a fortress was a function of the product of the number of defenders' guns and the number of enceintes the attacker was obliged to breach to reach the interior of the fortress.[97] Fourcroy de Ramecourt had reproached him for inconsistency in attacking the engineers' attempts to quantify siegecraft with their number of days of resistance and moments of fortification while using his own magic number.[98] Montalembert's reply was that his number, though it could not be used to predict the exact number of days a fortress could resist, could be used to compare the strengths of different fortresses.[99] The components of his index were numbers that could be objectively determined by simple counting, whereas the number of days of resistance of a fortress—a number incorporated into Fourcroy de Ramecourt's moment of fortification—could be obtained only by relying on experience and making assumptions about the "moral" qualities of attackers and defenders.

Although there is naiveté and gross mathematical immaturity in Fourcroy de Ramecourt's "touchstone" of the military engineers' art, Montalembert's numerical criterion is not much better. Indeed, displaying a monomania that was always present in his work, he eventually identified the strength of his fortresses almost exclusively with the number of guns protected in casemates. He used this feature to deny the importance of experience on which Fourcroy de Ramecourt and the military engineers always harped:

... what need is there for experience to judge by the axiom above, that the best system is that which can oppose [attackers' guns] in a smaller space with many more guns [that are] better protected.

Do not deviate from this incontestable principle ... false reasonings will be pure loss for those who will dare to make them.[100]

To add insult to injury, Montalembert quoted d'Arçon, the unsuccessful engineering hero of Gibraltar, who had partly justified his defeat by pointing to the lopsided English superiority in firepower.[101] Military engineers had invoked their experience to demand deference for their expertise in general and their rejection of perpendicular fortification in particular. Montalembert rejected this experience both as a justification of expertise and as a tool for judging his system of fortification:

It has never been claimed that only sailors could decide that a vessel of a hundred [guns] is stronger than one of fifty. Similarly, one does not have to be a military man to judge two forts on land if one wants to consider them relative to the quantity of artillery that one can concentrate on the same point and the security of their emplacements. It is thus this essential property that makes their *absolute strength* or *their real value for war* and one must strive to find the most advantageous form for this great feature [qualité].[102]

Montalembert believed as fervently as Fourcroy de Ramecourt that fortification was a science both noble and certain. He believed even more than Fourcroy de Ramecourt that the certainty was quantifiable in a simple manner. But both men's yardsticks for a fortress' resistance were simplistic and flawed. The calculated lengths of mock sieges were not nearly so easy to determine as Montalembert's number of guns or number of enceintes. Experience and even intuition were necessary. Kéralio had said that the engineers' method was founded on "the doctrine of probability, on experience, on observation, on practice and the knowledge of siege warfare."[103] In his confused manner, Fourcroy de Ramecourt was groping toward some way of mathematically taming uncertainty—an intrinsic element of war that many military writers sought to identify and neutralize. His efforts were undoubtedly primitive and unsuccessful, but military engineers displayed a greater sensitivity for unknown factors than Fourcroy de Ramecourt's moments of fortification seem to imply. Montalembert's estimates were cruder and, if anything, more arbitrary than Fourcroy de Ramecourt's. Grenier and later d'Arçon argued that not only the number of guns but also their placement and positioning had to be taken into account. At 200 toises, a battery of five or six cannon behind earthworks provided only a small and difficult target for Montalembert's caponiers with almost ten times as many cannon, but the caponier would almost inevitably be hit and destroyed by the besieging battery.[104]

Another major theme of Montalembert's book is the accusation that the military engineers, posing as faithful heirs and defenders of the tradition of Vauban, are in fact betraying that tradition by falsifying it and by moving away from it while all the time condemning others who were guilty of that sin for being ignorant and incompetent in regard to fortification. The major falsification was attributing to Vauban the method of estimating the lengths of resistance of fortresses from the fictitious siege journals in which Fourcroy de Ramecourt put so much faith. Vauban had never in fact proposed such a method, and, as Montalembert had pointed out, even his estimates of the length of a siege were intended not as "a certain rule" but as instruction. In fact, for Vauban estimating the lengths of sieges seems to have been at least as much for the purpose of estimating provisions and munitions as for the purpose of estimating the strength of a fortress. Here Montalembert was on solid ground, for he was able to point to a small but significant change in Grenier's text. In his initial manuscript memorandum to the Minister of War, in February 1780, Grenier had apparently written this: "Cormontagne, an engineer of great merit, has consigned to widely diffused manuscript memoranda the principles of fortification that appear to be generally adopted today by the officers of the Royal Corps of Military Engineers." The printed version of the memorandum, included as the second part of the *Mémoires sur la Fortification perpendiculaire*, had inserted "of Vauban" in parentheses after "principles."[105] With a few strokes of a pen, Cormontaigne's methods had been transformed into Vauban's.

Furthermore, the military engineers had changed Vauban's original ground plans. First, they had abandoned his third system, with its bastioned towers and their casemated flanks, and reverted to his first system, which was essentially that of Pagan. Fourcroy de Ramecourt wrote that Vauban would no doubt have scrapped his idea of bastioned towers if he had lived longer.[106] He went on to claim that Cormontaigne was superior in some respects to Vauban because he had circulated his methods to his fellow engineers, thus raising fortification to the rank of a "positive science."[107] But the engineers, beginning with Cormontaigne, had deviated even more from the ground plans of Vauban's systems. Indeed, by putting their demi-lunes farther out from the fortress perimeter and making the angles between the faces of the demi-lunes and the faces of the bastions less and less oblique, they were insensibly approaching the basic tenaille trace of Montalembert. Later Montalembert would claim that the engineers were plagiarizing his ideas while attacking him.

The engineers had themselves accused Montalembert of being a plagiarist by referring to a work by Jean-Antoine d'Herbort published in Augsburg in 1735.[108] Montalembert indignantly rejected the charge, seeing it as proof that the engineers secretly admitted the value of his ideas since they were so eager to rob him of the credit. The accusation of plagiarism obviously pained him: it seems to have been Carnot's passing remark about "ignorant plagiarists" that jolted him into initially attacking Carnot. Montalembert was extremely proud of his originality and the novelty of his system; it was a major element of his argument that he was attempting to drag the art of fortification out of the rut of a dull, unthinking, and ultimately dangerous routine. "My system," he wrote, "is mine and it has incontestable characteristics of novelty since it renders the defense superior to the attack."[109]

In the final pages of his book, Montalembert, like Fourcroy de Ramecourt before him, had appealed to the Academy as a tribunal to judge the case between him and the military engineers:

> The assurance of the authors of this method [Fourcroy de Ramecourt's moments of fortification] in presenting it to an Academy where as much enlightenment reigns as wisdom, leads one to presume that they count on obtaining from it a blind confidence on a matter that they believe must be unfamiliar to it. They have rendered themselves prolix and diffuse, hoping that their obscurity would be attributed only to the nature of the question being decided and not to their style. But they do not know the Academy. No art's principles are unknown to it, all the sciences will find professors there, and military science in all its parts will not lack instructed judges. Besides, their paper's assertions are so little admissible that, with the soundness of

judgment and the habit of analysis so generally distributed among all [the Academy's] members, there is not a single one who could take the errors scattered with such profusion as truths. It is therefore to its tribunal that I appeal decisions pronounced against a work I published ten years ago with its approbation. This imaginary scale, this touchstone, this sensitive balance of the authors to evaluate all the fronts of fortification present and past, no matter what dissimilarity there may be among them, can only be regarded as errors capable of perpetuating the childhood of the art, while in uniting to bring it to its perfection, [the Academy] will be sure of sharing its glory.

But this important art will not fail to find here, like all the other arts, the protection and the help which have always been the immediate cause of their prosperity.[110]

We have already seen that Montalembert, like Fourcroy de Ramecourt, paid a great deal of attention to the official judgments of the Academy, reprinting them in full in his book, analyzing them at length, and attempting to show that the Academy had given him the seal of approval that it withheld from Fourcroy de Ramecourt. Throughout his reply to the engineers he expressed nothing but respect for the Academy, placing it in the position of supreme arbitrator of his contest with Fourcroy de Ramecourt. But even before the Academy had brought down the judgment he ostensibly valued so highly, and even before the book for which he was soliciting the Academy's approval was published, Montalembert appears to have despaired of its support and begun a preemptive and surreptitious attack on it. Almost immediately after the appearance of Kéralio's review in the February 1787 edition of the *Journal de Savants*, there appeared a "Lettre de M. le Baron de Montalembert à M. de Keralio, en Réponse au Compte qu'il a rendu dans le Journal des Savans, du Mémoire sur la Fortification perpendiculaire, publié sous le nom de plusieurs Officers du Corps Royal du Génie, en 1786." The letter, which is in fact a pamphlet of some 70 pages, has a title page indicating (possibly falsely) a London publisher and is dated 28 February 1787. Its later publication in Berlin indicates it was intended for wide diffusion.[111] Jean-Charles, baron de Montalembert (1757–1810), was a relative of the Marquis de Montalembert both by blood and by marriage (they had married sisters).[112] The baron unstintingly expresses his admiration for his better-known older relative and ostensibly wrote to defend him from Kéralio's attacks.

The public letter was almost certainly inspired, if not actually written, by the Marquis de Montalembert. The Baron de Montalembert "spent [his] life following [the marquis], step by step, in the never ending work to which he devotes himself in his study" and had access to his relative's work: he refers to having read the manuscript of the *Réponse au Mémoire sur la Fortification perpendiculaire* while it was still being considered by the Academy's commission.[113] Obviously it had the full approval of the marquis, who later included it in volume 8 of *La Fortification perpendiculaire* published in 1790. But this was after the Revolution had broken out and various hitherto sacrosanct institutions such as the Academy could be attacked with impunity thanks to removal of restraints on the press.

In 1787, however, the Academy was still deliberating on the reply to the engineers: a frontal attack was therefore not only exceedingly discourteous and inappropriate from a member of the Academy but also unwise. An attack by someone else could always be disavowed. When the Academy's commission had brought down its judgment and the marquis finally published his book, he had also added a rebuttal to Kéralio's review that was milder toward Kéralio personally than the baron's pamphlet.

However, the baron's arguments are essentially those of the marquis and the strident style is suspiciously familiar. The military engineers were accused of having "discouraged, denigrated, and insulted" all those who wanted to do research in fortification and "they were sure to stop the progress of an art on which they wanted neither to work nor permit [others] to work."[114] The jabs against the engineers' claim to be safeguarding the secrets of Vauban from prying profane eyes were even stronger, implying that the veil of secrecy was a shield for incompetence and cowardice:

Mystery was always the distinctive sign of CHARLATANISM and gallant soldiers should not show themselves under a wrapping so little worthy of them.[115]

If the engineers claimed that they revealed their secrets only to their military superiors, how was it that Montalembert's superiors had never heard of them when he was constructing fortifications on the island of Oléron in 1761?[116] The baron retorted to the charge that the marquis was a mere "isolated person" [homme isolé] as opposed to a knowledgeable and expert "entire corps" [corps entier] with a full list of the Marquis de Montalembert's military services and titles: He was a general officer, a member of the Academy of Sciences, had participated in numerous military campaigns, been a liaison officer with foreign courts and armies in the Seven Years' War, commanded the construction of fortifications, and as further support for all this the baron gave precise page references in *La Fortification perpendiculaire* to testimonial letters by Choiseul and Montalembert's superiors.[117] He asserted that a number of Fourcroy de Ramecourt's Directors of Fortification had refused to sign the *Mémoires sur la Fortification perpendiculaire* when it had been presented to them.[118]

But the most shocking aspect of the pamphlet was an attack on the Academy of Sciences and on its secrétaire perpetuel, Condorcet. The stage for the attack was set by praise for the Marquis de Montalembert's good breeding and elevation of spirit. He had refrained from reacting as would have been normal to an attack that had been tolerated and even welcomed by the Academy on one of its own members whose works it had approved. But, wrote the baron,

> I owe to truth to hide nothing of all that is reprehensible in this report. The conduct of M. le Marquis de MONTALEMBERT, in speaking about it in the introduction of his seventh volume cannot serve as my guide. He acted as a generous academician who seeks to justify confreres, as much as it is possible, whatever grounds he might have to complain of their injustice....
>
> But as for me, I do not have to consider his adversaries nor him. It is error that I attack and unveil. I would not have written if I had not had to demonstrate it, no matter from which direction it came.
>
> ... the wrongs which they [the Academicians] have inflicted on this occasion are all the more inexcusable in that they have damaged the public weal in approving that which they should have condemned and in condemning that which they should have approved.[119]

Thus the baron presented himself as a champion of Truth, moved by public spirit to point out the inappropriate and even dangerous conduct of an Academy that was supposed to be the guardian and herald of this Truth and was therefore guilty of negligence at best in its duties. The defenders of Mesmer and Marat had attacked the Academy as a bastion of privilege; the baron was implying that it was dangerous to the public. These were loaded words at a time when discontent with the Academy in particular and governmental institutions in general was running high. The attack on Montalembert had been presented by

> Mr. *Fourcroy*, Director of fortifications, favored by the Minister and residing with him. He had to be made happy in order to please his protector. The members of this Academy have therefore shown themselves on this occasion (and it is not the only one) better courtiers than good philosophers. What could come out of such dispositions? A judgment, where the partiality of the judges has made up for their lack of knowledge, it being evident that the Gentlemen Commissioners had none relative to that which they had to decide upon. This company, learned in so many other domains, is instructed in nothing concerning the military art. M. *De la Place*, examiner of the Artillery Schools, was the only one of the three commissioners who was in a condition to have an opinion based on his knowledge. But he had only one vote against two. It is known that all he could do was mitigate the report....[120]

The Academy, then, was servile and incompetent. It was no longer the impartial tribunal to which both Fourcroy de Ramecourt and Montalembert had appealed. Not only did it not have the necessary knowledge to judge in military matters, but it shared an epistemological bias with Fourcroy de Ramecourt that rendered it even more unfit for this role.

> EXPERIENCE can enlighten in Physics, and the Gentlemen Academicians, used to judging according to its indications, have made a false application of it. The experience of the length of resistance of a fortress concludes nothing for another siege....[121]

The baron did not hesitate to assail the personal qualities and working habits of the academicians and of the most publicly visible academician of all, its secrétaire perpétual:

It often happens that most of these gentlemen chat or write their letters during reports [of the Academy] yet they are nevertheless of the opinion of him who spoke, especially if it is one of the *fixers* [*faiseurs*] of the Academy. Mr. *de Condorcet,* who is known to have been the principal author of the report, was not accustomed to finding any contradictors there.[122]

At this point the pamphlet has degenerated into a bald attack on the Academy and smacks of Marat's *Les Charlatans modernes*.[123] More than just incompetent on epistemological grounds or the irrelevant specialties of its members, the Academy is pictured as bumbling, petty, and servile. It is a microcosm of all the inequity, selfishness, and arbitrary rule of privileged individuals that were seen as the hallmarks of a regime already under general attack. Although the Baron de Montalembert signs as "Colonel and sub-lieutenant in the Light Horse of the King's Bodyguard," there is an unmistakable and no doubt conscious whiff of Revolution in his pamphlet. In the years ahead, the baron, unlike the marquis, was to emigrate and fight against the Revolution. The Baron de Montalembert was not particularly an enemy of privilege; he was merely an enemy of the royal government and particularly its administration, as was most of the country, including a good part of the nobility. He (or more likely his better-known relative) was using a rhetoric that had become so common that its implications had become invisible. Thus, his pamphlet is less interesting for its rhetoric than for the Marquis de Montalembert's decision to resort to that rhetoric: while attacking the military engineers in the forum of the Academy of Sciences, he was simultaneously attacking the trustworthiness of that forum. Did the simultaneous pursuit of his goal along a high road and a low road merely show that he was adroit in political siegecraft, or was Montalembert losing hope that the Academy would come down on his side?

One can only infer Condorcet's relations with Montalembert; however, since it was no doubt Condorcet who had weakened the approval of the Academy for Montalembert's reply to the engineers, it is probable that they were not warm. Condorcet appears to have supported the conservative side in the debates over the new system of artillery proposed by Gribeauval in the 1760s and the 1770s.[124] As the official spokesman and one of the most influential members of the Academy (one of the faiseurs, as the Baron de Montalembert would have it), he was keenly aware of the threat posed to its prestige by aggrieved outsiders, circle squarers, and frustrated inventors. The bruising episode of the Mesmer controversy was behind him but still fresh in everyone's memory. Condorcet had rejected Mesmer on the grounds that, although many distinguished people had accepted Mesmer's ideas, only the opinion of "competent judges" should be taken into account. These had to be "physicists" with a "well-established reputation in France and even in Europe," not merely people with degrees who had published books. And this meant believing in authority, something Condorcet admitted was "hard for human reason [to accept]."[125] Clearly the authority was the Academy of Sciences, but such an attitude would have made him sympathetic to the arguments of a Fourcroy de Ramecourt, who also spoke on behalf of a corps of professionals and specialists against an "homme isolé." Even Fourcroy de Ramecourt's blundering attempts to mathematize uncertainty may have struck a chord with a man who dreamed of using probabilities and founding a social science.[126] After Fourcroy de Ramecourt died (12 January 1791), Condorcet gave him an academic eulogy that did not mention the debate with Montalembert.[127] There are hints of rigidity in Fourcroy de Ramecourt's character and narrow-mindedness in his interests, and he is described as a devout partisan of Vauban. But he also claims that Fourcroy de Ramecourt was the unanimous choice of his fellow Directors to be the Director at Versailles, gives a favorable evaluation of the corps of military engineers under his command, and approves the calculation of the length of resistance of fortresses:

It is thus that all the great divisions [of the subject matter] of war are linked to the science of the engineer and that he can submit its accidents to the certitude of an art that is not limited to constructing an isolated fortress following known rules.[128]

Montalembert's prolix, repetitive, and tiresome arguments, larded with insults and technicalities, in reply to arguments of the same sort by Fourcroy de Ramecourt, appeared at a time when France and its armed forces were in a state of pre-revolutionary effervescence. A first Assembly of Notables had met in January 1787, with little result; a second had been summoned for October 1788; the reforming minister Necker had been dismissed, then recalled for the first time in August 1788, and in that same month the États-Généraux had been summoned to meet in May of the following year—the year of the Revolution and the Bastille. In the army, debates raged in the Conseil de Guerre not only about new tactics and strategy, including the future importance of fortifications, but about rules of promotion and access to officer rank. The "aristocratic reaction" that preceded the Revolution had made entrance to and promotion in the army much more difficult from 1781 on; now it excited indignation within and without the army. The military engineering corps found itself fighting against those who, like Guibert, wished to reduce it in size and importance. Within the corps itself there were demands to reorganize it and to replace the single officer

(Fourcroy de Ramecourt) who was its effective commander with a committee of officers. Such a committee had been created in the ordinance of Saint-Germain of 1776, but it had been rather ineffective. Proposals were now made to give it more power and to expand and democratize its membership. Most military engineers seem to have been concerned with promotion policies, strategic doctrines, and the relations of the Royal Corps of Military Engineers with other corps of the army, such as the artillery and the ingénieurs-géographes, as well as with the civilian technical services of the government, such as the Corps des Ponts et Chaussées.[129]

With the exception of Lazare Carnot, perpendicular fortification seems to have had very few supporters. The stridency of Montalembert's attacks on the Academy through his mouthpiece indicates both the loosening of constraints on the eve of the Revolution and a sense of bitterness at what was evidently perceived as a rebuff by the most prestigious scientific institution in the country with which he had taken such great pains to associate himself and his system. On the surface, things seemed hopeless, and all that was left was venting one's frustration. But the Revolution was to change the situation radically, not just for France but also for Montalembert. Two weeks after the fall of the Bastille, Montalembert was presenting yet another paper to the Academy of Sciences.

IV
Engineers and Revolution

The cataclysmic upheaval of the Revolution did not interrupt or even slow the exchange of pamphlets by Montalembert and his enemies. Indeed, the controversy became more strident, and both Montalembert and the engineers showed themselves to be adept in adjusting to the rhetoric of new regimes. Shedding their particles and adopting a new lexicon of abuse, the combatants relentlessly persisted in their quarrel until the death of the two chief antagonists—Montalembert and d'Arçon—a few months apart in 1800. The persistence of the polemic into the Revolution, much longer than was the case for the debates on artillery and battle formations in the army, provides additional insight into the ideologies of the military engineering corps. The crucible of the Revolution forced many to confront, intensify, and occasionally to reject the tenets they upheld during the Old Regime. Radical revolutionaries were hostile to some of the ways in which military engineers thought about their profession and their society during this period. The engineers' reaction to this hostility, which was also exploited by the indefatigable Montalembert, shows that they were profoundly attached to ways of making war and to some aspects of the Old Regime that went back to the dawn of the Enlightenment and even before. If there was a "revolutionary" they admired, it was Vauban. If they were revolutionary, it was only as custodians of Vauban's tradition and as builders on the solid base that he had established.

The military engineering corps did change in significant ways, however. Even as it became still more military, its role and autonomy as a combat arm declined and it became more of a service corps. Only in the days of Vauban, and perhaps occasionally thereafter, had military engineers come close to achieving their dream of controlling the entire progress of a siege in all its aspects. Other military officers had never accepted their pretensions in this regard, and it did not take a Montalembert to incite a more educated sort of general officer to want to be more involved with sieges. Another major change was the attribution of specialized troops to the engineering corps, which vastly increased its numbers and changed the corps from being a corps of officers with no soldiers into a unit that resembled other branches of the army. Engineers were now drawn more tightly into the daily life of the army and the mundane duties of officering soldiers in peace and war.

Chapter 13 looks at the engineering corps and its ongoing battle with Montalembert during the Revolution. Closure of sorts was reached, but not to the taste of Montalembert or to some military engineers. Yet the corps of military engineering had finally acquired a stable institutional form—one that persists to this day.

Chapter 14 is an attempt to draw together the numerous threads of the story, to present some ideas on the legacy of French military engineering of the Enlightenment, and to try to integrate those ideas within a more general vision of engineering as it has developed to the present. This legacy was essentially somewhat pessimistic and conservative, not in the sense of reactionary and opposed to change, but in the sense of respect for the constraints of the material and social worlds.

13

Montalembert and the Engineers during the Revolution

The most incredible circumstances have led to revolutions of all kinds being not only possible but easy. One must take advantage of these circumstances because the destiny of our nation, at least for a long time to come, is only being able to do even wise things foolishly.
—Mirabeau, letter to Montalembert, 1789[1]

While battling in an Academy that showed distinct signs of coolness to his ideas, Montalembert never slackened in cultivating supporters in the government. After the episode at Aix and the death of Maurepas, this was becoming increasingly futile, but the momentous political changes of 1789 were to give him new opportunities and new hopes. On 29 July 1789 Montalembert was presenting a paper to the Academy of Sciences on the new military port and forts being constructed at Cherbourg. Once again, he claimed that the engineers were stealing his ideas and doing so incompetently: eviscerating them to hide their debt to him while at the same time making them ineffective. The forts at Cherbourg proposed by the military engineers are indeed remarkably similar to the artillery beehives of Montalembert, and the gun carriages used there, some of which were brought from the fort at Aix, were in fact modifications of Montalembert's gun carriages.

The Cherbourg Forts

However, Montalembert was addressing himself not only to an academic audience, as the correspondence with Minister of War La Tour du Pin included in his paper makes clear. The Cherbourg project, the largest military engineering construction project in France in the reign of Louis XVI, was highly publicized and controversial. It had attracted the interest of the whole country, including the king himself, who had visited the site in June 1786, the only time aside from his coronation that he had ever traveled away from Paris and Versailles.[2] The king came to inspect the works and to witness the ingénieur des ponts et chaussées Louis-Alexandre de Cessart sinking one of his great wooden cones to form part of a huge breakwater to protect the open port from the winds and waves of the English Channel. Strategically, the military port was a major challenge to domination of the Channel by Britain's Royal Navy, and had provoked considerable alarm in England. Interestingly, this new threat to England may have come from the expedition that had attempted to attack Rochefort and captured the Ile d'Aix in 1757. The project of Pierre Filley in 1762 to fortify Aix properly had been judged to be too expensive, and Choiseul had decided to fortify a harbor on the Channel instead.[3]

Like Aix, Cherbourg had been devastated in an English naval raid during the Seven Years' War in April 1758. Additional work on the fortifications of Cherbourg was ordered on February 1778; the following year, the project became more ambitious and additional forts were ordered.[4] The enormously expensive project was clouded with accusations of financial fraud and waste, controversy over its necessity and suitability of the site, and bitter infighting among the various participating parties—the military engineers, the engineers of the Corps des ponts et chaussées, the artillery, and the Navy. It was, in short, Aix writ large. Montalembert himself had argued for a military port at La Hogue rather than Cherbourg, and his proposals in volume 3 of *La Fortification perpendiculaire* for the fortification of Cherbourg, which included a series of ring forts, undoubtedly figured in an attempt to solicit a role in this project.

Montalembert's interest in the goings on at Cherbourg was probably heightened by the tests in 1787 that had compared his gun carriage and a new one developed by Meusnier. As Gillispie has pointed out, in spite of his exceptional talent for mathematics and physics, Meusnier was by temperament and inclination an engineer rather than a scientist.[5] The same year Montalembert had begun his work at Aix, Meusnier had been assigned to Cherbourg, where he distinguished himself both as an engineer and as a leader of young engineers who exposed corruption and faulty design in the early Cherbourg forts. He developed a system for vacuum distillation of sea water to produce fresh water on the waterless Ile Pelée, where the new Fort Royal was situated; methods of quickly heating cannonballs to red heat, which made them particularly dangerous to enemy ships; new casemates and embrasures at Fort Royal; and a new gun carriage to replace both the standard coastal battery gun carriages in use and the one Montalembert had described in volume 2 of *La Fortification perpendiculaire* and had built at the Ile d'Aix.[6]

In April 1787, Meusnier undertook a series of trials to compare the speed of firing of his new gun carriage and a number of other gun carriages, including Montalembert's.[7] The result was a primitive time-and-motion study over several hours that showed that the rate of fire for aimed red-hot cannonballs from Montalembert's gun carriages was around 1 minute 45 seconds between shots, whereas for Meusnier's gun carriage it was only 33½ seconds. Furthermore, Meusnier's gun carriage was serviced by a four-man gun crew, whereas every other gun carriage needed five men. At a slower rate of fire, Meusnier's gun carriage could be operated by a single soldier. Clearly, there was a significant difference in rates of fire.

The superiority of Meusnier's gun carriages was attributable not only to a better overall design but also to his attention to details. He was obviously well aware of the problems of friction and also corrosion from the salt air in a maritime fortress, and he took steps in his design to overcome these problems, which might seem initially insignificant but could cause grave and even crippling problems in the long-term operation of the cannon. Thus, although he used much sturdier iron instead of wooden axles, he inserted them into copper bearings that could be lubricated.

There is a significant similarity between the two gun carriages: both use an inclined plane to control recoil. The fact that Montalembert's gun carriages were brought from Aix to Cherbourg does indicate that there was interest in them, and it does appear that this gun carriage was the starting point for Meusnier's own research on an improved model. This and the engineers' use of casemates in the Cherbourg forts would certainly have been enough to lead Montalembert to accuse the engineers of copying his ideas. He began his July 1789 paper to the Academy of Sciences by claiming that the engineers had changed their views on the fortification of maritime sites. Now the engineers at Cherbourg were using casemates. Could one any longer refuse to believe, asked Montalembert, that his work and his fort at Aix had brought about this change of heart?[8] Yet at the same time, there is a curious rapprochement with the engineers even while he is pillorying them in his discussion of casemates. Casemates, he says, were an old invention and they had many defects, which he proceeds to list. Most of them happen to be the very ones the engineers had mentioned in their attacks on his fort at Aix: they were cramped and smoky, could not hold many cannon, had embrasures that were too large, and would be bested in any artillery duel with cannon in the open air that were free to change position. It was not surprising that they had been abandoned in military architecture. But he had changed all this. His casemates were no longer the same thing as the old defective ones. And structurally they were much stronger because the external wall no longer served as the buttress for the pier of the casemate vault, thus weakening the wall subject to battering by enemy shot, but the axis of the vault was now perpendicular to the external wall. Its piers were now acting as buttresses for the external wall and supporting, instead of being supported by it.

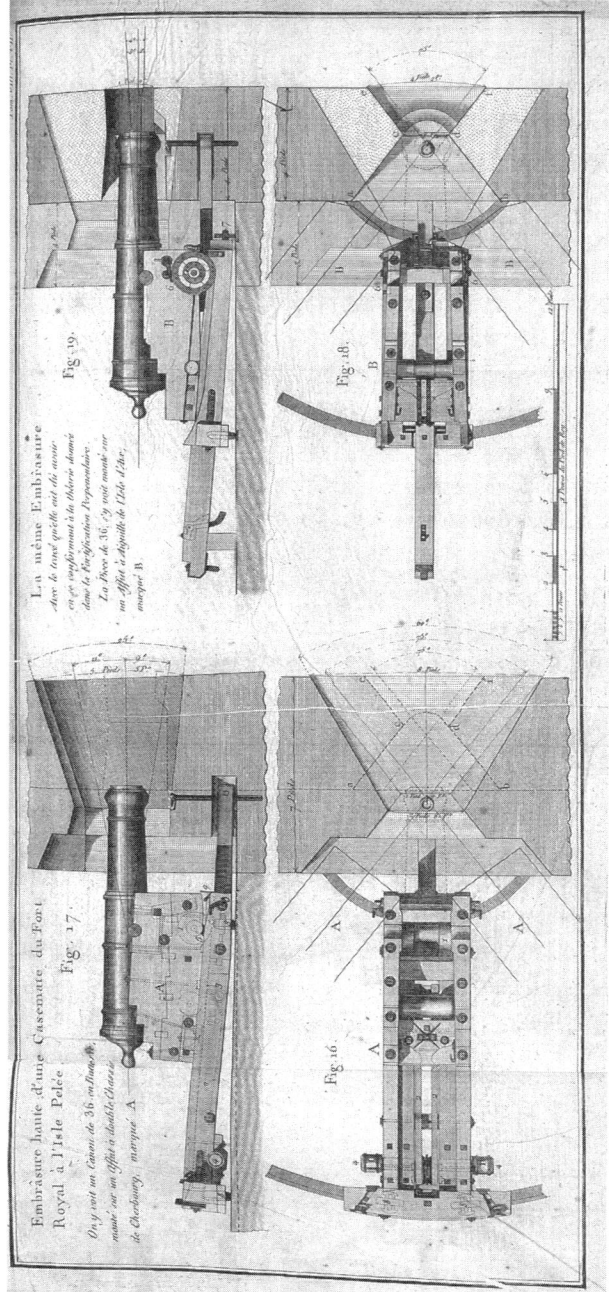

Figure 13.1
Montalembert's (right) and Meusnier's (left) gun carriages. Source: Montalembert, *La Fortification perpendiculaire* (volume 8, plate 7). Courtesy of Massey Library, Royal Military College of Canada.

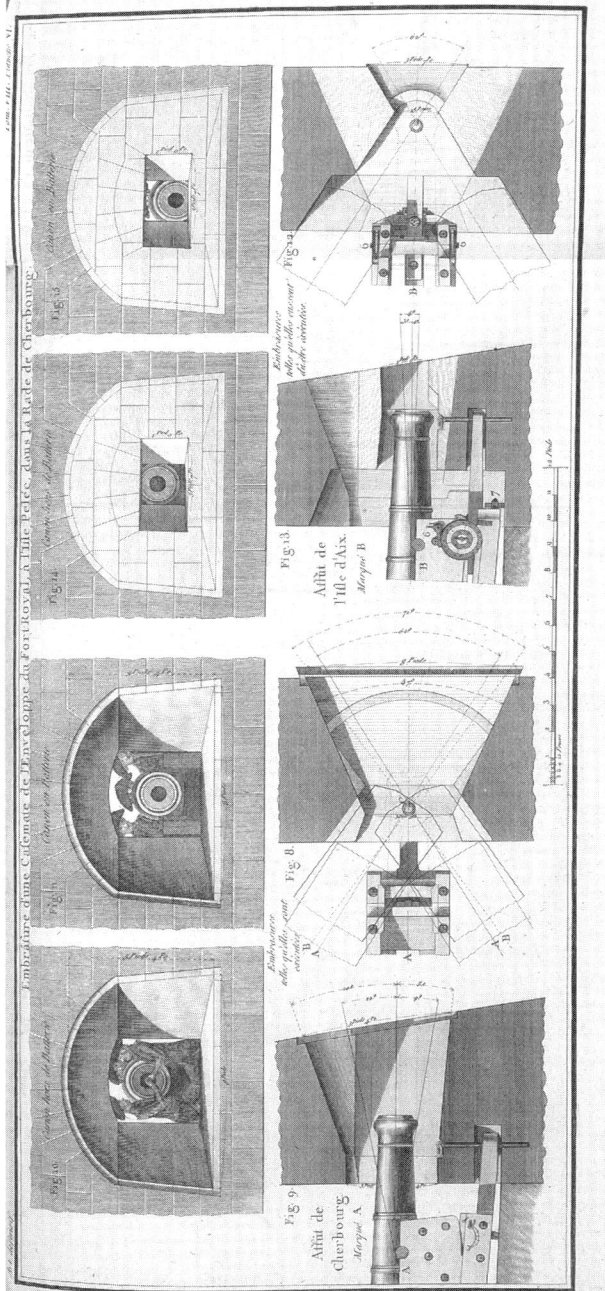

Figure 13.2
Montalembert's (right) and Meusnier's (left) embrasures. Source: Montalembert 1776–1797. Courtesy of Massey Library, Royal Military College of Canada.

Similar casemates have now been constructed at Cherbourg and Montalembert has no doubt what brought this about.[9] But the engineers have copied his ideas unintelligently. They have not been guided by any "constant principles," something which is clear to Montalembert from the bewildering variety of dimensions in different parts of Fort Royal. There are different thicknesses of walls, vaults, piers, different lengths and widths of casemates, although everywhere he sees an excessive amount of masonry.[10] The excessively large embrasures of the casemates will permit smoke to drift back into the casemate: he claims that Meusnier's casemates are so bad that they are even less safe to the soldiers in them than exposed positions on the ramparts.[11] And the engineers have failed to use as many cannon as they could have in the fort: Aix, with a floor area of about 600 square toises had 66 cannon in the interior fort, the engineers only have 45 cannon for a floor area of 1,300 square toises, and Montalembert predictably proposes a fort that would have 351 cannon.[12]

There is another reason to question whether the engineers had not borrowed at least some of Montalembert's ideas. The first embrasures built at the Fort du Homet were the regular funnel shaped embrasures with the wide end on the outside wall, but their shape changed in subsequent construction to give embrasures that looked in plan like two funnels with their wider ends on the interior and exterior walls respectively, coming together at a narrow neck between the walls. Provided that the mouth of the cannon protruded beyond the narrow neck, this enabled a greater field of fire. The field of fire had increased from 28° at the Fort du Homet to 47° at the lower battery of Fort Royal and 60° for its upper batteries.[13] Montalembert had suggested similar embrasures back in volume 5 of *La Fortification perpendiculaire*. And he argued that even wider fields of fire of more than 80° could be obtained. His own suggested embrasures gave a field of fire of 75° and he achieved this by making his walls considerably thinner than those of the engineers—4 feet versus 8 or more. On the other hand the engineers had a much wider vertical field of fire—approximately 25°—while Montalembert's was only 4°. As a result, Montalembert's embrasure openings toward the outside were only 4 feet wide and 1 foot, 9 inches high, while for the engineers they were 8 feet wide and 5 feet, 5 inches high.[14] The result was that each embrasure of the engineers presented an area of more than 40 square feet to enemy gunfire while Montalembert's smaller embrasures presented only 6 square feet.[15] While making embrasures that were safer for the men serving the cannon, he was making thinner and cheaper walls, as well as packing more cannon into his proposed forts. It is clear from an examination of the two proposals for Fort Royal (figure 13.3) that the engineers were not only more conservative in design but that

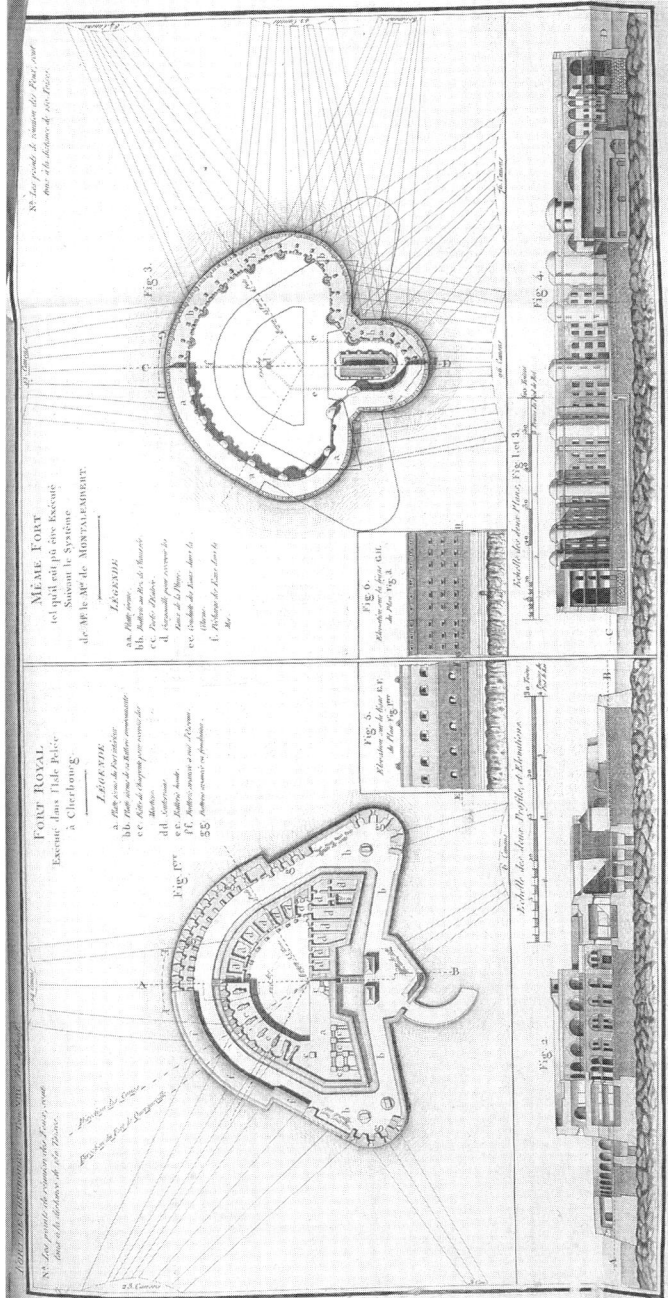

Figure 13.3
Forts proposed for Cherbourg by Montalembert and by the Military Engineers. Source: Montalembert 1776–1797. Courtesy of Massey Library, Royal Military College of Canada.

more detailed structural considerations and problems of operation and maintenance were not forgotten.

The Response of a Military Engineer at Cherbourg

No evidence has yet been found to indicate that Meusnier ever replied directly to Montalembert. In the following year, however, one of his colleagues at Cherbourg, Alexandre-Magnus Dobenheim (1754–1840), who was later to become the first professor of fortification at the École Polytechnique, took up the pen to reply to Montalembert. His unpublished paper is interesting because it gives the view of an engineer relatively junior in rank like Meusnier, but without the latter's scientific reputation. He not only dealt with Montalembert's criticisms of the work at Cherbourg, but he also more generally responded to the challenge to the military engineers that Montalembert had thrown down in the epigraph of his most recent book: "Do as I do, Sirs, or do better than me."[16] Although Dobenheim began with "We have not done as M. de Montalembert because it seemed to us that one could do less badly than him," his tone was actually measured if compared with Fourcroy de Ramecourt's book and d'Arçon's blasts at Montalembert that same year.[17] He is certainly prepared to admit that engineers make mistakes, even at Cherbourg, and he joined Montalembert in disapproving Fourcroy de Ramecourt's statement in his *Mémoires sur la Fortification perpendiculaire* that "every novelty proposed in fortification is certain proof of the ignorance of its author, because everything has been discovered in this great art." But he adds that everything has not yet been discovered in the great art of fortification, "in spite of the 120 well-executed engravings of M. de Montalembert."[18]

Mocking Montalembert's pride in his illustrations (and his belief in their demonstrative value) was common among his opponents in the engineering corps and by joining them Dobenheim clearly showed where his real sympathies lay. For in spite of certain drawbacks of Fourcroy de Ramecourt's book, Dobenheim agreed with much of it. Although he nowhere mentions the "moments of fortification" that for Fourcroy de Ramecourt were the "touchstone" of the engineer's art, Dobenheim devotes the first few pages of his paper to listing fifty-odd references to passages in Fourcroy de Ramecourt's book that he considered valid arguments against Montalembert's work. Passing on to the work at Cherbourg in particular, Dobenheim speaks as one who had been working there for seven of the nine years that Fort Royal on Ile Pelée was being built. Montalembert's walls were much too thin: experience of storms at Cherbourg had shown that 6-foot-thick buttressed walls reinforced with iron bars had been washed away. The thin walls of Montalembert's "castle of cards" would have trouble standing up to the elements and Cormontaigne's experiments had shown that cannonballs could penetrate up to 3 feet into masonry with an "infinite" amount of backing. The possibility therefore existed that shot could completely pass through 4-foot walls.

Rather than raising costs by ineptitude and even fraud as Montalembert had implied, the engineers had fought hard and successfully to have the first contract for the fort at Querqueville canceled and renegotiated on better terms for the Crown. The engineers were also not responsible for other features of the port or its siting. They had been forced to defer to their rivals, the ingénieurs des ponts et chaussées. There was no way of comparing costs of Fort Royal and Montalembert's proposed alternative fort because he had not given sufficient information to do so. What information he had given about the engineers' fort was often unreliable, such as the figure of 45 guns for Fort Royal, when it actually had 60. He had not mentioned the mortars that the engineers had in the fort and which Montalembert did not have in his, nor had he mentioned what the engineers considered to be a cardinal innovation that made their batteries much more fearsome than ordinary ones—facilities for heating red-hot shot. And when it came to expense, Montalembert's extra cannon (380 vs. 60) entailed enormous additional costs: iron cannon would need continual maintenance and replacement and bronze cannon would cost as much as the fort itself. The garrison to service them would also have to be raised from 500 to at least 1,900 men.[19]

The structural solidity of Montalembert's "castle of cards" was also questioned. Fort Royal at Cherbourg consisted of an almost circular single story of casemates behind which was a ditch and two more stories. Two small flanks, which had attracted Montalembert's scorn, were included on the interior perimeter of the land side of the fort. Montalembert, on the other hand, proposed a single perimeter consisting of three stories of casemates. Unlike the engineers, for whom each casemate was separately vaulted, Montalembert placed two and even three stories of his casemates under the same arch. And instead of solid masonry floors resting on the arches below them, his floors consisted of planks. A damaged arch could fall down on three successive casemates—the lower ones having little protection from flimsy plank floors—and even ricocheting shot entering an embrasure might smash the floors and send guns crashing down on those below. The relatively light piers could also be smashed by artillery fire and the loss of one could bring down two arches with three stories each at a time.

The separate arches for each of the casemates of the engineers made them quite different from those proposed by Montalembert. But even the idea of the axis of the casemate vault being perpendicular to the external wall was not original: such casemates were used in Spain at Saint Sebastian, Giron, Denia, and Fontarbia. They had also been proposed in the French Military engineering corps as well by Jean-François Garavague (1728–1800) in 1768, a decade before Montalembert published his first book on fortification.[20]

As for the embrasures and gun carriages, Dobenheim contented himself with some general remarks, leaving the detailed reply to Meusnier, who was in Paris at the time. Mistakes had been made in designing the first embrasures and the younger engineers had had to overcome resistance to better designs of embrasures by Etienne Poisson des Londes (1729–1787), who was Director of the district including Cherbourg. Yet Dobenheim felt that "the most defective of our embrasures is perhaps better than those M. de Montalembert is proposing to us."[21] Besides the smaller vertical fields of fire of Montalembert's embrasures, their wider field of horizontal fire and better masking of the artillerymen by narrower necks was purchased at the price of thinner walls and cheeks that could be more easily shattered by incoming shot. Furthermore, the more than four times as many embrasures in Montalembert's fort compensated to a significant degree the much larger external area of the embrasures in the engineers' designs. Unlike Montalembert's embrasures and casemates, those at Fort Royal were designed, Dobenheim claimed, to accommodate different types of gun carriages more efficiently, an important consideration if the custom-designed gun carriages were not available in adequate quantities and one had to rely on standard naval or coastal battery gun carriages.

To Montalembert's criticism that there is no uniformity in the forts, Dobeheim replies that the "true principles of fortification do not always lead to regular or similar forms."[22] The implication is clear. Rather than adhering rigidly and blindly to a fixed pattern, the engineers had designed to take into account particular circumstances of the site and the military situations that could be encountered. Furthermore, casemates themselves were no universal solution to the problems of fortification: ". . . we only admit casemates against ships and . . . we absolutely reject all those that can be exposed to enemy cannon from the landward side. . . ."[23]

Within the fortnight after delivering his paper to the Academy of Sciences, Montalembert had sent a summary to the Minister of War La Tour du Pin. Not only did he repeat that the engineers had mutilated his ideas to design something that was overly costly and downright dangerous to the troops garrisoning it, but he asked that the plans for the next fort to be constructed (at Querqueville) be sent to him so that he could prevent the same errors from being repeated. The Minister had sent his letter to Pierre-Jean de Caux (1720–1792), Director of Fortifications of Lower Normandy, whose reply, rejecting Montalembert's allegations, was transmitted by the Minister to Montalembert, who promptly fired off another letter which went without reply. This did not reduce Montalembert to silence at a time when the army, as much as the whole country was in a state of revolutionary crisis, and officers could safely ignore the prohibition of publishing their views on military matters without first clearing them with the Minister. He dashed off another pamphlet, where he outlined his basic arguments in the more technical paper delivered to the Academy of Sciences, revealed his correspondence with the Minister, and described his relations with de Caux over the years.[24] The main theme throughout, however, was a very public blast at the Royal Engineering Corps. The political revolution that had swept France committed it to a pacific and defensive policy, thus making fortifications ever more central in its strategic planning. Unfortunately, however, "the esprit de corps perpetuates even popular prejudices, it is the natural enemy of useful discoveries because the whole Corps values equally its acquired knowledge and its errors."[25] And institutionalized incompetence, engendered by a few senior officers thinking only of their own positions, who profited from the "insouciance" of ignorant ministers, had resulted in French fortifications being in a parlous state. The grave situation had been hidden from the public by a veil of secrecy by engineers who, like ministers of state, were automatically taken at their word. The principle of secrecy, "also the principle of the Ministers of Louis XIV, religiously observed by their successors up to our own day," was merely a pretext to maintain the existing dangerous state of affairs.[26] These were words bound to strike a chord in the heated political atmosphere of the first heady year of the Revolution. Invidiously lumping together military engineers and ministers, making the latter the accomplices and not the victims of the former, as he had done in earlier writings, was a tactic that Montalembert would not have used before 1789. There was now a new center of power in French political life—the National Assembly—and public opinion had become even more powerful than before.

In his *Réponse au Mémoire sur la Fortification perpendiculaire*, Montalembert had already condemned the engineers for insisting on their expertise and hiding their principles behind a veil of secrecy from their military superiors as being contrary to the proper subordination that they owed to their commanders. Here he continued to emphasize this idea, in expanding the circle of legitimate judges and creators of new fortifications to any officer who had read any of the many treatises on fortification. The reason was that the mystified science of the engineers was in fact extremely simple.

It was common to all the engineers of Europe, indeed all soldiers with interest in the subject; it could be understood and applied—and in fact was applied in practice—by ordinary surveyors and building contractors. There was only one principle, which Montalembert enunciated in capital letters:

FUNDAMENTAL PRINCIPLE
FORTIFICATION THAT WILL BE ABLE TO CONCENTRATE ON EVERY POINT OF ITS CIRCUMFERENCE THE LARGEST AND BEST COVERED QUANTITY OF CANNON IS ALWAYS THE ONE TO BE PREFERRED, & this principle, whose truth cannot be questioned, excludes all bastioned systems [when] compared with casemated angular ones, since the artillery is impossible to protect with the first and impossible to lose with the second [kind of systems].[27]

Boiling down the science of fortification to a single principle would clearly have been something that Dobenheim would have rejected. He was willing and even eager to submit technical questions to the arbitration of the Academy of Sciences and questions of reorganization of the Corps to the National Assembly, but he hoped that the commissioners of the latter body, who were at that moment debating such a new reorganization, would consider it a duty to reply

to the comparison M. de Montalembert makes of the officers of the Military Engineering Corps to surveyors . . . , they will be far from proposing that the construction of a building, once it is laid out be abandoned to the care of a contractor, they will feel the price, for example, of this constant supervision by officers of the Military Engineering Corps who themselves measure and trace the shape of each stone used, because they know that a badly cut stone, defective in one of its parts or badly placed can shorten the life of the parts on which it depends by several centuries, they will feel the necessity of this supervision that destroys every moment of the day the tendency that almost all contractors in the world have to do negligently or badly everything that adds nothing to their estimates, and finally the price of all the arranging without which more or less work to be executed every construction season would depend solely on the intelligence, the will or the considerations of the interest of a contractor.[28]

This passage shows the traditional mistrust of state engineers for private contractors, a mistrust that was strong among military engineers and closely related to their delicate role of responsibility for large disbursements, where suspicions of fraud always abounded and had plagued even the great Vauban himself. But it also displays a view of the role of engineer on the work site: intimate knowledge and involvement with the details of construction. The military engineer was valuable not only as a theoretician and designer but as a practical man capable of supervising workmen. In this regard there was a clear opposition between Dobenheim and Montalembert, who saw the essence of engineering residing in theoretical conception of fortifications, with the responsibility for details left to surveyors and contractors. A third point which it reveals about military engineers, and perhaps all French engineers in state service, was the belief that the engineer built for eternity, even if it went against economic sense. Rather than economic optimization of construction taking into account amortization and costs of renovation and replacement, military engineers aimed for perfection and durability. The classicism of the cultural ideals of the Sun King, the static zero-sum game economy of pre-industrial France, the abnormal demands of building to withstand the extreme conditions of war, and the conception of fortification as the ultimate guarantee of the state and the country all contributed to these views. Picon has argued that they were beginning to change with civil engineers who were beginning to perceive the exigencies of design in more dynamic and ephemeral terms as they were brought into contact with the civilian economy to a more intimate degree and in more varied situations.[29] Even so, these views died hard with state engineers even in the nineteenth century.[30] In the eighteenth century, as Dobenheim's remark indicates, they were even stronger.

Dobenheim was typical of a breed of younger engineers in the Corps, whose opinions had become visible now that the Revolution had stripped away the mask of corporate unity in the army. There was no automatic respect for the senior leadership of the Corps. Montalembert was correct in saying that Fourcroy de Ramecourt by no means represented all engineers. Those he did represent were probably the older generation of engineers, many of whom had not gone through the military engineering school at Mézières, and even here, one could not speak of unanimity of opinion. Yet the widespread malaise that Marcel Reinhard detected in the Corps of Military Engineering at the beginning of the Revolution was mainly over matters of pay, promotion, and administration.[31] There was undeniably impatience and even disrespect for an aging leadership increasingly incapable and also unwilling to entertain new approaches to the problems of construction and fortification, but this by no means meant that there was widespread support for Montalembert's new schemes of fortification either. The ultimate message of Dobenheim's reply is mistrust and rejection of Montalembert's ideas, belief in the role of the corps as the sole competent institution to tackle the problems of fortification, and faith in its capacity to respond to new developments, especially in light of the reorganization that the Revolution promised. Fourcroy de Ramecourt's criticisms are not rejected completely, as they were by Carnot; his points of detail are mentioned with respect and often accepted. Only the sweeping nature of his condemnation and his theoretical dogmatism elicits objections—not his criticism of perpendicular fortification.

A New Ally for Montalembert

There are therefore good grounds for believing that Lazare Carnot was a minority among his fellow engineers in his support of Montalembert. His boast that he was the only who dared to support Montalembert may in fact need to be tempered with the observation that he was the only one who wanted to support Montalembert. Yet he did this with increasing energy, which appears to have matched the increasing hostility of his superiors and to some extent increasing isolation from his colleagues. In 1787, the Conseil de la Guerre, consisting of the senior leadership of the army and including in its membership Guibert, had been created to propose fundamental reforms for the army.[32] Among other things, it was considering the dismantling of a substantial number of fortresses under the pressure of Guibert's arguments against fortification and financial constraints. Fourcroy de Ramecourt, who was a member of this Council, appears to have succumbed to pressure and agreed to cuts, and this attracted an attack from a number of engineers, including d'Arçon and Carnot.

In his memorandum to the Council, Carnot argued eloquently for the type of civilized warfare that was desired by both the military engineers and Montalembert.[33] In "this century of philosophy, where everything is submitted to the most rigorous discussion, where the antiquity of a custom cannot serve it as an authority, if it is not otherwise founded on reason," Carnot repeated the argument that French interests dictated a policy of peaceful defense of national frontiers rather than wars of expansion. France had more to lose than to gain in wars, it should do everything to discourage them, and fortification was the "conservative art" par excellence. He argued that rather than being a drain on the economy and the military budget that could potentially release funds to double the size of the army and enhance security, fortifications increased security because field armies could always be defeated and these defeats could be disastrous if they could not fall back on fortresses.

Claiming that the entire cost of fortifications over the centuries was the same as that of the cavalry alone for the 26 years since the Seven Years' War had ended, he argued that 100,000 men in fortresses were more effective than 200,000 without them. A hundred thousand men could thus be liberated for useful production and the wealth they generated could be used to raise an additional 100,000 men in wartime. He proposed a form of military service whereby a standing army of 100,000 men would be supplemented by trained reserves of 200,000 that could be raised in emergencies. Even apparently useless fortresses should not be demolished, but merely decommissioned. In case of an unforeseen need, they could be reactivated. If destroyed, however, they were gone forever.

Replying to arguments that it was good to have impregnable fortresses but fortresses simply were not impregnable, Carnot departed from the more widespread view among military engineers and said that if fortresses were not impregnable, they could be made impregnable.

> Indeed it is mathematically demonstrated that a place fortified according to modern principles, must hold out at least seven to eight months by the simplest possible defense and it is certain that, defended with bravery and intelligence, it would hold out more than a year.[34]

It is clear that although Montalembert is not mentioned by name, Carnot is thinking of his systems here. He was much more direct in his next paper that was written after the outbreak of the Revolution. Just before that time he had been imprisoned as the result of a personal scandal. He had disrupted the marriage of a young woman with whom he was romantically involved and outraged relatives had successfully pulled strings at Versailles to obtain a lettre de cachet. Carnot felt that his recent controversial stand in favor of Montalembert had led to a vendetta by Fourcroy de Ramecourt, who had merely used the incident as an excuse to imprison and harass him. His brief imprisonment ended 29 May 1789, just 6 weeks before the fall of the Bastille and the end of all lettres de cachet. Obviously bitter, Carnot drafted a memorandum to the National Assembly on proposed reforms in the Corps of Military Engineering, which he submitted for discussion to his fellow military engineering officers at Lille.[35] Among them was the complaint regarding a "single chief" for the corps and the proposal that he be replaced with a committee. There is a radical edge to the proposal in the suggestion that besides a given number of Directors, members of the committee should be elected by their colleagues of all ranks.

In their comments on the proposal, the officers at Lille had merely contented themselves with agreeing with Carnot's complaints about the centralization of command in a single officer: they did not propose democratic elections for a committee and left to the Assembly the responsibility for the precise details of a solution. It is noteworthy that Fourcroy de Ramecourt was attacked as much for his inability to stand up to pressure on the Corps and the erosion of its responsibilities as for autocratic rule.[36]

But this did not imply any endorsement of Montalembert. When Carnot asked for a formal disavowal by the Corps of Fourcroy de Ramecourt's *Mémoires sur la Fortification perpendiculaire*, his colleagues opined that such a disavowal could only come from a thorough study of the treatise of Montalembert and they left this to the Committee that Carnot had proposed and which they supported. If Carnot's radicalism in this memorandum is tempered by his colleagues, this is not true for the work that he eventually addressed to the National Assembly two months later. Carnot attacks the "oppressive regime" of the Corps of Military engineering with the inflated prose that would become ever more popular during the Revolution: "despotism is overthrown, you have broken your chains, France has nothing more to fear from its internal enemies."[37] But to protect this idyllic state from jealous powers, unassailable barriers to aggression must be erected and France must put its faith in fortresses and not large, expensive, and unreliable field armies. The National Assembly is called upon to prevent any further degradation or dismantling of fortresses as the Conseil de la Guerre had recently suggested and to rebuild them according to Montalembert's new principles. Carnot extolled the wonderful properties of casemates, which were supposed to increase the defensive power of fortresses by several orders of magnitude—they gave a fortress using them the same superiority over standard bastioned fortresses as a multi-decked warship had over a single decked raft.

Why had Montalembert's new and necessary ideas not been accepted? Blame was fixed squarely on Fourcroy de Ramecourt, who was attacked by name, whose intelligence was mocked, and who was described as the éteignoir (extinguisher) of novel and useful ideas in the corps. Carnot went on to describe his own difficulties after he had begun to defend Montalembert and pictured himself as a victim of "tyranny," of a "plot," and a veritable "inquisition." He again called for a committee of elected officers to examine new projects and concluded as follows:

It is then that a multitude of very interesting memoranda will emerge from their [the engineers'] studies, which the despotism of their organization [régime] forces them to lock up and, together, one will be able to form a coherent system of principles by which the very important art of the defense of frontiers will rise, so to speak, in an instant, so much above its present state as gunpowder is by its effects above the missile weapons of the ancients.[38]

There was little doubt that the perfect theoretical system of defense that would emerge once the "despotism" of Fourcroy de Ramecourt's leadership had been swept away and talent allowed free rein, subjected only to review and refinement by peers, would be that of Montalembert. Carnot had unequivocally made himself the defender of Montalembert and his system. His tone was now as intemperate and categorical as that of his mentor, who had chosen him as his official champion in the Military Engineering Corps and the National Assembly. At the end of 1789 he fired off a violent letter to the Minister of War:

Your prejudices against this system, Sir, are known, but that is yet another reason for you to observe the greatest impartiality in the choice of the members of the delegation charged to examine it. In the name of the fatherland that you serve, I dare to say to you that, now warned, you would be making yourself guilty of a crime if you would neglect the appropriate means for the defense of the kingdom.[39]

Even for the mounting tide of revolutionary rhetoric, the tone and style were judged "astonishing" and a passage against Fourcroy de Ramecourt was described as a "project of vengeance." The reason for this forwardness was that Montalembert, who had lost all influence he ever had with royal ministers, had found a new and powerful protector with a leader of the new revolutionary elite in the person of none other than Honoré-Gabriel-Victor Riquetti de Mirabeau (1749–1791).

Mirabeau, before he had come to assume his prominent and even predominant position in the Estates-General and the National Assembly, had acted as a spy for the French government at the court of Prussia, where he had become intimately acquainted with the military establishment of Frederick the Great. He developed a keen interest in military affairs and volume 7 of his *Monarchie prussienne* was devoted to this subject.[40] Mirabeau contrasted the elite French military engineering corps with the Prussian military engineers: they were considered to be the worst arm of the Prussian military, attracted the worst people, had the lowest prestige and pay, and were badly organized. Although he conceded that the Prussians were good in field fortifications and cartographic work, he claimed that Frederick himself did not know or care much about military engineering and consequently had always been the "dupe of charlatans or the victim of ignoramuses."[41]

Mirabeau's own ideas on military engineering, a profession he valued highly, ran counter to Vauban's in many ways. In opposition to Vauban and the whole trend of administrative developments in the French military engineering corps throughout the eighteenth century, he felt that the corps should be split up into two separate divisions: siege engineers and constructors of fortresses. Here was a reversion to the division between the ingénieurs des tranchées and the ingénieurs des places that Vauban, and all the successive administrations of the corps of military engineering, had struggled to eliminate. Indeed, he felt that a third division of field engineers for field fortifications should be added to the military engineering establishment. His ideas on the role of the military engineer suggest a romantic streak that could very well be compatible with Montalembert's ideas on the necessity of innovative genius in the ideal engineer:

. . . an engineer must not be a man of routine, but of invention. He most adjust his system of attack to the circumstances, follow the rules, and be sparing of his men when he can, but to depart from that and force obstacles, even at the price of blood, when necessary.[42]

Perhaps he saw such an engineer in Montalembert, "this officer who, no matter what has been insinuated by the charlatanism of the practitioners of the craft, has come up with very ingenious and very novel ideas in fortification. . . ."[43] At any rate Mirabeau was elected to the Military Committee of the National Assembly where he was in an excellent position to influence the major reforms the Assembly was planning for the armed forces. Already favorably inclined to Montalembert, Mirabeau may have had his opinions strengthened by Montalembert's paper to the Academy.[44] On 11 August 1789, loudly affirming his opposition to the "brigandage and the charlatanism of rival corps," he wrote to Montalembert:

> The National Assembly will be creating a Military Committee. Not only must you be on it, but it is also necessary that it names you Inspector General of the Fortifications of the kingdom, as well as reporter on the changes to be made there by means of the implementation of your system. The Artillery and Military Engineering must be joined together and amalgamated by you in common schools.[45]

Here finally was the moment Montalembert must have been waiting for. Montalembert hastened to present his works to the National Assembly, was acclaimed, and offered a seat in the gallery.[46] Supported by the most influential politician in France at a time when there was a mood for radical change in all aspects of the social and political system, including the army, he would be placed in a position to direct these changes and impose them on the Corps of Military Engineers. His tenacious struggle was about to be rewarded with success. Mirabeau summed up the opportunity admirably:

> The most incredible circumstances have led to revolutions of all kinds being not only possible but easy. One must take advantage of these circumstances because the destiny of our nation, at least for a long time to come, is only being able to do even wise things foolishly.[47]

But Montalembert was already old: at the age of 75 he needed a younger and more vigorous assistant. He turned for help to Carnot, his only ally in the Corps of Military Engineering, and together they began to make plans. In October 1789, Carnot wrote:

> I accept very gratefully, general, your offer to construct permanent lines [of fortification], provided that, as you know, it would be as chief [and] not under the orders of a superior, who would perhaps mutilate the projects that I have arranged with you.[48]

This mood of triumph perhaps explains the intemperate letters of Carnot to the Minister of War and Montalembert's own peremptory request to the Minister (in a letter of 8 November 1789) that de Caux and Meusnier, the principal architects of the Cherbourg works be summoned with their plans to be confronted with Montalembert's "demonstrations" of the inadequacy of their work.[49] But once again Montalembert was to be frustrated in his dream of reorganizing French fortifications. Mirabeau was drawn more and more into attempting to control the increasingly turbulent political situation in the country and the National Assembly. On 15 October 1789, after the royal family had been forcibly brought from Versailles to Paris by a mob, he began to correspond secretly with the king. High politics were now of greater moment for him than military reforms. Although a member of the Military Committee of the National Assembly that had begun its activity that same October, he rarely attended: the most influential member of the Committee that was heavily weighted with elderly, aristocratic military men not inclined to radical reforms was Edmond-Louis-Alexis Dubois-Crancé (1747–1814), one of the only two bourgeois members of the committee, who was also the only one to go on to become a member of the more radical National Convention.[50] On the face of it there might have been hope in the fact that Dubois-Crancé matched Carnot in political radicalism and commitment to profound military reforms. But he also had a brother and brother-in-law who were members of the Corps royal du génie.[51] Instead of turning to Montalembert and Carnot for advice on reorganizing the corps, he turned to Le Michaud d'Arçon, who had become the leader of the opposition in the corps to Montalembert and his ideas. Moreover, in December 1789 Carnot's classmate at Mézières, Jean-Xavier Bureaux de Puzy (1750–1806), deputy for the nobility to the Estates-General, was elected to the Committee. Although the aristocratic military engineer Bureaux de Puzy and the bourgeois soldier Dubois-Crancé apparently had little use for each other, they did seem to agree in not soliciting the advice of Montalembert and Carnot when it came to legislate on engineering matters.

This apparent allergy to Montalembert and his representative Carnot is yet another indication that whatever the discontent within the military engineering corps, it was hardly motivated by Fourcroy de Ramecourt's rejection of Carnot. If Fourcroy de Ramecourt had claimed at the time of the Aix episode that he had the support of 35 senior engineers in his rejection of the new system, the younger graduates of Mézières do not seem to have been any more enthusiastic for Montalembert's ideas than their elders. Unhappy with many things, including Fourcroy de Ramecourt's leadership, there was no one besides Carnot who seems to have shared his enthusiasm for Montalembert. Even a supporter of Carnot's memorandum of 2 August 1789 from Lille, expressed his reservations on making the proposed Committee of Fortifications elective lest "chance or intrigue" should lead to "a Montalembert or a runaway from the Conseil de la Guerre."[52] Napoleon was right to describe Carnot, in his defense of Montalembert, as an "original" among his comrades.[53] Carnot's active agitation on behalf of his new patron irritated and alienated rather than converted many of his colleagues, who were aware of his reputation and his ideas. Thus, along with d'Arçon and Bureaux de Puzy, Dubois-Crancé called on another classmate of Carnot, Bénézech, and Carnot was steadily held at arm's length. The battles within the Military Committee and later within the National Assembly as a whole led to a series of laws in 1790 and 1791. They resulted in an effective reduction in the size of the Corps of

Military Engineering by about a fifth, but the number of fortresses was held constant and none were ordered demolished. The Committee of Fortifications was reorganized, but none of its appointed members were partisans of Montalembert.[54] Of the four new positions of Inspector General of Fortifications, two were former Directors of Fortifications: Jean-Baptiste de Caux de Blacquetot (1723–1796) was the brother of the Pierre-Jean de Caux who was in charge of the work at Cherbourg and had attracted Montalembert's ire; Paul-Louis-Antoine de Rozières (1723–1794), as Director of Fortifications at Cambrai, had vainly attempted to prevent Carnot and his friends from holding a meeting at Arras to gain support for the petition he had drafted at Lille in August 1789.[55] Of the other two, both sons of military engineers who had reached senior rank, Louis-Thomas Lemercier de Senneton de Chermont (1717–1793) probably signed the tract written by d'Arçon against Montalembert in 1790, along with the two preceding inspectors general.[56] Little is known about Pierre-Joseph Légier (1730–1796?) except that he was apparently no radical, at least not in his politics: he resigned from the corps the week after the king was deposed and kept out of public view throughout the Terror. Although the law had specified only four Inspectors General of Fortifications, a fifth was added when the newly reorganized Committee of Fortifications met for the first time in January 1792: he was Montalembert's old nemesis Le Michaud d'Arçon.[57]

To make matters worse, Mirabeau died in April 1791 and in May Carnot was writing to Montalembert in a most depressed tone:

More than a year has passed since I have written you anything. The hope of finishing our military organization has continually kept me in uncertainty. Finally, here it is, but few advantages will come of it for the diffusion of your theory—new proof that it is easier to transform the constitution of a kingdom from top to bottom than to vanquish the sectarian spirit of a handful of individuals. As far as I am concerned, I cannot change my conviction, because it is founded on reason. I have, it is true, attracted a crowd of enemies, which would distress me a lot if I thought that I were in error, or if I had not renounced all ambitious hopes a long time ago.[58]

Carnot had not renounced ambition, however; on 1 October 1791, he would be elected to the new Legislative Assembly and, after the overthrow of the monarchy in August 1792, to the National Convention and the Committee of Public Safety, where he was responsible for military planning and acquired at least as much power as Mirabeau ever had. But in the interim, opposition to the ideas of Montalembert continued and even hardened. It was no longer Fourcroy de Ramecourt who led this opposition: he seems to have effaced himself after the Revolution and retired from the corps just a few days before his death on 12 January 1791. Le Michaud d'Arçon, more eloquent, more competent, and more respected than the former head of the corps, had already confronted Montalembert in closed military committees back in the 1770s. Now he too attacked in public. Montalembert's own public attacks on the engineers in his reply to Fourcroy de Ramecourt's book, his attack on the work at Cherbourg before the Academy of Sciences in 1789, and perhaps above all the prospect of his becoming the new chief of the military engineering corps galvanized d'Arçon into a vigorous and eloquent attack in 1790 with his *Réponse aux Mémoires de M. de Montalembert*.[59]

D'Arçon and Montalembert

D'Arçon's private papers reveal that not only was he active in advising the Military Committee of the National Assembly in its deliberations on the reorganization of the Military Engineering Corps, but that he was also coordinating resistance to Montalembert's attempts to impose his system on the corps. The reply he wrote to Montalembert's eighth volume of his works in 1790 was apparently circulated to a number of engineers for their approval. (This eighth volume was in fact a collection of papers: along with his papers on the Cherbourg forts, it included the Baron de Montalembert's response to Kéralio discussed earlier).

Among those commenting was David-Alexis Tholozé (1736–1802), who later became an inspector general of fortifications and member of the Comité central des fortifications. Tholozé's reply is of interest because it is an example of how revolutionary rhetoric could be mobilized in the defense of long-standing demands by the corps of military engineering that go back to the previous regime and had then been couched in an altogether different language.[60] The high-flown rhetoric of the first sentence sets the tone for the whole paper:

The sciences and the free arts untrammeled in their products cannot be affected by a despotic exclusion that would chain thought, limit their utility, and circumscribe their salutary influence. Truth is one, it must reign unbounded over thinking beings.[61]

Although this may sound like an endorsement of free spirits like Montalembert, there is no renunciation of the right, "justly acquired," of the corps of military engineering to examine and pronounce on new projects of fortification, a right which it has gained by "constant application" and the "lessons of experience." If the military engineering corps is a "constitutional corps of the state," if it is an integral part of the army, then one has to agree, writes Tholozé, that "[the right] of reporting on novelties relative to its art belongs to it alone. Must one, in excluding it from its acquired and natural rights deliver the security and the finances of the kingdom to guile, to surprises, and the versatile spirit of spinners of theories? If its functions were thus to be violated, if it merits this disgrace by its ineptitude, it must be disbanded."[62] Tholozé's words here, which with slight modification would have fitted in well with a rhetoric of corporatism in the Old Regime, fit in just as nicely with a rhetoric of the rational bureaucratic state. No longer part of the organic body politic of the Old Regime, the Corps of Military Engineering was now a constitutional part of the rational machine of the modern state where it fulfills a well-defined function. Implicit in both views is the idea of a corporate wisdom, either distilled with time and experience or formed by the mechanical forces of competing expert opinion refereed by reason, in its bureaucratic and statist forms. The military engineering corps is to have primacy over individual inventors. If there is a concession, it is that the state can set things right if errors are made, but even the process of correction flows along the proper and natural channels of a properly constituted state.

However, there was no arid and purely mechanical mode of reasoning in the military engineer's work. His art was not that of paper systems created by "ruler and compass"; it manifested itself when those systems had to be built on existing terrain with all its irregularities and disadvantages. It was then that the engineer had to improvise, to adapt, to break old rules, and to create new ones to obtain the uniform and constant result always desired in fortification: effective defense. He was even willing to consider "the systems of M. de Montalembert, in whole or in part."[63]

All in all, Tholozé rejects or minimizes most of Montalembert's ideas. But he accepts the need for more casemates—in the flanks of bastions—and the presence of outside officers in inspection committees, provided that they do not meddle with accounting and the actual construction of fortresses. He attempts to give an impression of even-handedness and high-mindedness in ignoring the abuse of Montalembert for the corps, even writing that "the systems of M. de Montalembert might be sprinkled with some sparks of genius that the state could advantageously collect."[64] Whether sly or sincere in his handling of Montalembert in this paper, Tholozé added his name to the list of the supporters of d'Arçon's considerably more violent and combative tract.

His rejection of Montalembert's proposal that military engineers be subjected to a watchdog committee composed of six or eight Inspectors of Fortifications drawn from officers outside the military engineering corps shows that Tholozé also knew how to draw on the well-stocked arsenal of revolutionary rhetoric. He depicted the dramatic and even tragic effects of forcing engineers to submit to the control of outsiders incapable of understanding the details of fortification: an outsider "would combine power with blind action, irregular and no doubt arbitrary; would destroy the rule of Truth, the rule of Liberty whose holy fire develops talents and animates the Will; never would despotism show itself with such frightening features, never would Slavery have a more humiliating character. What? Thought, reflection, and the productions of the Mind would be then enchained and branded by haughty ignorance!"[65]

Better the corps should be abolished than to suffer this tragic fate. But Tholozé was prepared to concede that joint committees of engineering officers and officers from outside the corps should examine major new projects and select the sites of new fortifications. Indeed, he wanted to see more regular contacts between engineers and the generals commanding the territorial divisions. This would, in fact, be a good thing, for it would bring together soldiers who would have to cooperate in wartime; the habit of close relations would make liaison easier in the future.

Another unidentified author in the d'Arçon papers compared Montalembert to certain other personages with bees in their bonnets and his friends are advised to dampen his zeal and his expenses, which will bring nothing good either to him or to the state. As for the engineers, they are advised to ignore Montalembert:

Engineers must not respond to M. de Montalembert. They have responded to him badly, he as responded to them worse still. It must be seen to that this miserable quarrel stops here.[66]

D'Arçon did not take this advice. Thirty years earlier, Fourcroy de Ramecourt had attempted to overcome Montalembert by ignoring him and it had not worked. The dispute was out in the open and Montalembert was more dangerous than ever with his appeals to the National Assembly and the support of Mirabeau at a time when the military engineering corps was being reorganized. D'Arçon and other engineers were apparently extremely effective behind the scenes in drafting new policy and thwarting Montalembert and Carnot, but much depended on the opinion of legislators who might be seduced by Montalembert's arguments. For d'Arçon also, this would be no gentlemanly fight, real or feigned. Montalembert was "a very insidious adversary": he had counted on the ignorance and laziness of the public, who were bored by technical discussions and did not bother to read the details of the argument. It was for this reason that the refutations of Montalembert by Fourcroy de Ramecourt and Grenier had failed. D'Arçon changed his tactics: He would deal with Montalembert's general principles. Reason had to defend itself but it would be difficult to do this against someone who promised *impregnable* fortifications—an idea that d'Arçon rejected a priori.

Moreover, he was not above using abuse and innuendo and exploiting revolutionary rhetoric in his own attack. The "old baby" [viel enfant], and the "former light cavalryman of the [Royal] guards" [ci-devant chevau-léger de la garde] was accused of hypocrisy in attacking ministerial whims and despotism. He himself had succeeded in getting the project at Aix approved by Maurepas, who had been attracted to the plays in Montalembert's personal theatre starring the pretty marquise de Montalembert, and was amused by the "singularity" of the Aix fort. Fourcroy de Ramecourt was reproached for being too complaisant to authority in not stopping the Aix fort, and his concentration on the question of smoke, which had led to the spilling of lots of ink and giving Montalembert another apparent proof of his system, had been unwise. At any rate, the fact that the fort did not collapse because of its "fibrous and elastic wood" also meant that the wood would quickly rot and the fort was merely a costly, provisional monstrosity.

D'Arçon considered as specious Montalembert's principle that a fortress with superior artillery to the attacker could not be taken. Mere numbers of artillery did not automatically give superiority because the attacker, unlike the defender huddled in Montalembert's casemates, could maneuver in the terrain around the fortress and could be adequately protected with earthen ramparts. At the same time the masonry protection for the fortress artillery was vulnerable: experience showed that revetted earthworks with walls eight feet thick could be breached at 300 toises. Montalembert's walls were only four feet thick with fewer buttresses than the regular walls built by the military engineering corps. By subordinating his military architecture exclusively to the needs of a numerous artillery, he had prevented the possibilities of sorties and an active defense by the garrison. And "the more M. de Montalembert would accumulate riches in artillery in a fortress, the more would be captured from him."[67] The military spirit of the troops would be undermined by Montalembert's obsession with artillery: coining a neologism, d'Arçon argued that one serious negative result would be to "cowardize" [apoltronir] soldiers.[68] The cost of fortifications was also increased with increasing artillery. While not wishing to begin a "trial of figures" with Montalembert, d'Arçon claimed as a rough rule that four times as much artillery meant four times as much expense. As dangerous as the principle itself was, the implication that "it suffices to keep this principle alone in mind to be an excellent judge" in fortification and that one can acquire the necessary knowledge in this subject within a few months was even more dangerous.[69]

D'Arçon mocked Montalembert's insistence on casemates: this was just another attempt, like using the term "fortification perpendiculaire," to use a catch phrase to get support for his ideas. Why not call the new version "fortification casanière" (hideaway fortifications)? Not that French engineers were against casemates; indeed, they approved of them and would like to have more of them but the lack of funds had not permitted their construction. They had already experimented with clearing smoke from casemates. More generally, d'Arçon bemoaned the lack of opportunity to build new fortresses and the confinement of engineers to routine maintenance. Montalembert was not original in the use of casemates, and here was another destructive effect of his antics. His ignorance of past work led him to attract excessive interest in technologies that had been known, tested, and modified or abandoned before he had taken them up. The result was confusion and disruption. Addressing Montalembert and other self-proclaimed innovators, he wrote:

Well, sirs! Consider then that you are plowing fields that are already plowed, cultivated, seeded, and in which only weeding is necessary. Don't you see that in overturning already germinating seeds that you lose all the fruit that a wise cultivation would have allowed you to harvest?[70]

D'Arçon also dealt with some of the specific items that Montalembert raised and here his responses were predictable and familiar from the writings of Dobenheim and Tholozé on the same subject. But they are harsher and more intransigent. He is against external inspection of the corps and particularly against this being one of the responsibilities of the embryonic general staff [état-major]. Perceived in the army as a fast lane for promotion, that institution was rank with favoritism, hostile to the military engineering corps, and contained many "turncoats" from the engineering corps, who had left to avoid the hard work, demanding standards, and poor opportunities for promotion in the corps. The casemates at Cherbourg were better than those proposed by Montalembert and the red-hot shot would make them even more formidable. He rejected Montalembert's insinuations of incompetence or financial corruption in the project at Cherbourg: the engineers did not have the necessary control in choosing the site or even placement of the works. And their rivals working on the harbor works, the engineers of the Corps des ponts et chaussées, were motivated by "hatred of the military spirit." Meusnier's gun carriages were also better, and Montalembert's scheme for fortified lines was "a wall of China." Already accused by Dobenheim of having a closed mind on anything that issued from the Corps of Military Engineering, Montalembert was accused by d'Arçon of wanting to contradict others to such an extent that he would not shrink from being self-contradictory to do so.[71] D'Arçon added the familiar jibe about Montalembert's "perfectly engraved" 130 plates.

Thus, unlike the papers written by Tholozé and Dobenheim, who had attempted to keep at least an appearance of objectivity and even conceded some value to some of Montalembert's ideas, d'Arçon's work is relentlessly and categorically hostile to Montalembert. However, d'Arçon was by no means extreme or atypical in his views. His pamphlet was signed by 47 military engineers who had read and supported the ideas in it.[72] And d'Arçon mentioned that others would have signed it but could not do so before the work went to press. Tholozé's name, for example, does not appear in the list. Nor does Fourcroy de Ramecourt's. Among the names which do appear, one can find nine out of the thirteen directors, six deputies or suppléants (substitute deputies) to the Constituent Assembly, Bénézech de Saint-Honoré, who had graduated first in Lazare Carnot's class, and Henri-Jean-Baptiste de Bousmard (1749–1807), future author of a major text on fortification. There were also Louis-Marie-Antoine d'Estouf-Milet de Mureau (1751–1825), later a Minister of War under the Directory and prefect for Napoleon, and Michel Vandebergue des Hautschamps (1732–1806), future interim director of the École polytechnique. Michel-François Beaudet de Morlet de Boisset (1750–1835) was heavily involved in the organization of the new school of military engineering at Metz to replace the Mézières school in 1797 and the Duvignau was most likely Antoine-Nicolas-Bernard du Vignau (1716–1795), former second-in-command at Mézières and "discoverer" of Gaspard Monge in the 1760s.

They were men who were either elderly or middle-aged with an average of 30 years of service in the corps. The two youngest engineers nevertheless had 13 years of experience to their credit, having entered the corps before the Ségur law restricting access to the officer corps exclusively to well-established nobility. There were at least six future émigrés, and five out of the six people elected to the Constituent Assembly had been elected as nobles. There is no reason to doubt that d'Arçon's "apology" for his corps had widespread support in the senior ranks of the corps. The absence of younger engineers from the list of signatures does not necessarily indicate a generational rupture in the attitude toward Montalembert. D'Arçon mentions that more names could have added to the list, and it is likely that he had restricted his initial choice to senior officers and colleagues of his own generation, with their greater experience, to give added weight to his tract. Furthermore, there were a proportionally smaller number of junior officers because the last years before the Revolution had seen a cutback in the number of cadets accepted at Mézières in order to bring down the total strength of the corps to that specified in the law. As mentioned earlier, no matter how discontented younger officers were with their elders and superiors over issues such as slow promotion, there is no trace of any widespread sympathy—indeed, much sympathy at all—for the ideas of Montalembert.

True to form, Montalembert replied to d'Arçon's attack, taking care to hurl an insult in the very title (*Réponse au colonel d'Arçon, auteur des batteries flottantes, sur son apologie des principes observés dans le corps du génie*) and epigraph of his work[73]:

Do good principles
Need apology?

Almost a decade after the unsuccessful Franco-Spanish siege of Gibraltar, whose culminating point was the disastrous assault by the "insubmersible and incombustible floating batteries" designed by d'Arçon that had turned out to be all too combustible and all too submersible under the fire of the defenders, d'Arçon was still smarting from this blot on an otherwise brilliant career. It is not really possible to determine with certainty whether the cause of the failure of the attack on Gibraltar was a result of d'Arçon's flawed designs for his floating batteries, as his detractors claimed, or whether it was interference by ignorant and envious superiors with his operational and design plans, as he himself asserted in a number of writings after the fiasco.[74] What is certain is that the Gibraltar siege was followed with intense interest by all of Europe and all its major participants, including d'Arçon, had become well known not just in military circles but among the public for their successes and failures.

As far back as his rebuttal in 1787 of Fourcroy de Ramecourt's memoir on perpendicular fortification Montalembert had quoted d'Arçon's remark, in d'Arçon's own apologia for the debacle at Gibraltar, that more guns will always overwhelm fewer guns in an opposing battery.[75] Montalembert now repeated this as well as many of the usual arguments he had made before. There were too many errors in d'Arçon's work to rebut in detail, the engineers counted on no one reading their works and everyone accepting their authority as engineers. They themselves invoked authorities to give weight to their arguments and he charged the engineers to refute him "not by vain speeches ... but by geometric propositions."[76] After reproaching d'Arçon for having used the name of Chambre on his pamphlet without the latter's permission, he accused him of publishing in an underhanded manner by not including the name of his publisher on his title page.[77] Proudly affirming that he "had never established an opinion unless it was demonstrated by engraved illustrations," he also reproached d'Arçon for the absence of plans: "They are all vague arguments, devoid of all proof, and announced with a confidence of which an example would be difficult to find."[78]

More painful to d'Arçon perhaps was Montalembert's rejection of any right of the engineers to judge on matters concerning the artillery. The true experts in the field were the artillerists who "alone" should place siege batteries. And the engineers were guilty by association with Louis XIV and his disastrous wars.

Predictably, there was a riposte from d'Arçon.[79] Cut to the quick, the "author of the floating batteries" adopted a mock apologetic tone: How could one imagine that a simple man, known to the world for failures that were as spectacular as his achievements were obscure, dared to reject the attacks of "an Academician, both royal and imperial."[80] Accusing Montalembert of ignoring the context of his remarks on the need for artillery superiority, he also attacked him for turning away from the military engineers, whose judgment he had initially solicited, to the Navy and the Artillery as judges in the quarrel, knowing full well that the corps were sufficiently rivals that he might get support for his system. Rejecting Montalembert's assertion that military engineers blindly accepted all of their own productions, he argued that at Cherbourg they would have been more severe with the excessive construction in certain cases if "if amateurs in favor had not forced their hand."[81] After another blast at the "triple-storied chicken coops for cannon," he finished "regretting" that Montalembert had not come up with at least one or two good ideas so that he could have "sincerely applauded, at least once."[82]

Introducing an interesting political twist in a cycle that was already drearily familiar, Montalembert replied in a pamphlet that was also submitted to the Jacobin Club on 1 May 1791.[83] Taking the form of a letter to Corps of Artillery, Montalembert praised the corps, professed his faith in the importance of the artillery, and declared that it was the artillery which should be responsible for deciding on the types of fortifications that they would be required to defend. As for d'Arçon's reply to his previous pamphlet, he claimed that he knew of its existence but would not bother to read it because it probably contained the same old "indecent" remarks. Once again, he argued that the engineers should be amalgamated with the artillery, leaving no doubt about which was the better corps and should have the dominant role in such a union. To the artillery he passed on the torch of the battle he had waged alone for many years for better fortifications and improved security for the country.

D'Arçon was not prevented from writing yet another reply in spite of Montalembert's claim not to have read him. This occurred sometime after Montalembert published the ninth volume of his works in 1793.[84] By this time the political situation was rapidly changing. The king had been overthrown and executed and, although Montalembert still mentioned his academic affiliations with the academies of Paris and St. Petersburg (no longer "royal" nor "imperial") on the title page, he had shed his particle and title and was now simply "Marc-René Montalembert." If anything, and as incredible as it may sound, Montalembert had become even more intransigent. *La Fortification perpendiculaire* was henceforth an inadequate title, and he would no longer refer to his work as an "essai" since his principles were evidently the basis of the defensive art: his new title for volume 9 was now *L'Art défensif supérieur à l'Offensif, par une nouvelle manière d'employer l'artillerie.* . . . The role of the artillery was exalted even more than it had been, and it was now the artillery, rather than military engineering, which was the sole basis of the "conservative art." In a sense, this was stripping military engineering of even its name: the "conservative art" was always considered a proud synonym of the occupation of the corps. Worse, Vauban was again used against the engineers. He was described as a great artillerist! It was on this that his reputation deserved to be based rather than on his mediocre engineering that was slavishly and incompetently copied by his successors.[85]

Draping himself in the mantle of the patriotic and self-sacrificing citizen concerned for the security of his country, Montalembert filled the rest of the volume with attacks against the "systematic iniquity" of conventional engineering design. But he did not forget to attack engineers individually and collectively, including the aboyeur (barking dog) of the corps who was obviously d'Arçon, all the while professing great respect for individuals in the corps and claiming that his only enemy was a conservative esprit de corps.[86] It is in this volume that Montalembert published many of the details of the 20-year-old struggles in closed committees with d'Arçon and the military engineers and gave interesting information on the construction of the Aix fort. For example, he clearly states that the fort was ordered constructed over the objections of the Minister of War by Maurepas.[87] The praise of the royal minister and his ministerial despotism when it was no longer profitable or even wise to do so provides weighty evidence of Maurepas' role during the Aix affair. Montalembert also did not shrink from deploring nationalist prejudices regarding Vauban, who was much inferior to the Dutchman Coehoorn.[88]

At other points Montalembert almost sounds like an agitator in a revolutionary political club. In the context of the Republic of Virtue, the following passage is chilling:

> ... there should only exist that which is good in a nation, when it is wisely organized. Everything that does not serve this great aim must be considered as a defective member of the social body, which cannot be amputated too quickly.[89]

Here it is clearly implied that the military engineers are animated by the "mad passion to rule by means of prestige." In a savage attack on Bureaux de Puzy's proposals in the National Assembly on the reorganization of the military engineering corps and an expansion of its powers, Montalembert came out with a frank proposal for its outright abolition. Mocking those who think that every engineer is a Vauban, he wonders why no one has ever noticed or suffered from the losses of all these "Vaubans." In arguments that borrow the language of revolutionaries attacking the regular clergy and look forward to Saint-Simon's strictures on traditional social elites, he calls engineers "passive beings who produce nothing."[90] They are also a non-military corps where promotion is no longer based on campaign services. The pride and intrigues of these "military Jesuits" will lead to their downfall and the dissolution of their order.[91]

Since they do not want to perfect their art and stubbornly stick to their bastions, they should and can be replaced by "an architect to direct and a contractor to construct." If a "perfect" execution of a project is required, give the job to an engineer of the Corps des ponts et chaussées.[92] Artillery officers are at least the equals of the engineers in the superior scientific knowledge much vaunted by Bureaux de Puzy. Indeed, if Montalembert's methods would be adopted, the science of the artillery officers would be "everything" and the science of the engineers would be "nothing."[93] Montalembert's solution is not surprising: abolish the corps of military engineering and turn over its "theoretical" functions, which it refuses to perform by ignoring his system, to the artillery; and turn over its "mechanical" functions to the engineers of the Corps des ponts et chaussées. Dispersing the engineers throughout the army after their corps is abolished will lead to a general improvement in the level of the officers in the other arms who will be stimulated by the considerable talents of the former engineers to emulate them. The engineers themselves will benefit from being liberated from the yoke of technical conservatism enforced by powerful chiefs and be able to develop those talents more effectively.[94]

In d'Arçon's reply to Montalembert, he too, had shed his particle and he also demonstrated that he knew how to use revolutionary rhetoric:

The time for vanity is gone; the esprit de corps is overthrown, prejudices are vanquished, all the traces of charlatanism are effaced and the love of the fatherland is going to be reinforced with all the personal passions that it reproves and absorbs.

Perhaps it might appear that these generous renunciations have not at all penetrated in the retarded soul of Marc Montalembert. . . . He wants to reign tyrannically over the whole of fortifying Europe, not by the force of his lay-outs but only by the belittling of artisans [depression des artistes]. . . .[95]

He admitted, however, that the art of the military engineers was not perfect. Carried away by the authority of Vauban, they had perhaps tried to imitate him in too servile a fashion. This "great citizen" was, after all, only a man and could not have laid down the boundaries to "the indefinite progress of perfection." Maybe they had in fact neglected casemates too much, but at the same time they should be reproached for not having been severe enough in rejecting the (bad) casemates of Montalembert. One could not make any kind of compromise with Montalembert because he categorically rejected all bastioned systems and anything that was not his own idea.

D'Arçon also made what at first sight looked like a major retreat from the engineers' contention that there was no such thing as an impregnable fortress. This claim of Montalembert's had also been enshrined in revolutionary ideology and legislation with draconian penalties for commanders who surrendered their fortresses to the enemy. It was no longer wise or healthy to argue against impregnability in a Republic where all citizens were expected to be heroes. Impregnability was now possible in new conditions, admitted d'Arçon, because armies were motivated by the moral superiority of patriotism. It was morale, not structures and ground plans, then, that led to impregnability, and Montalembert's own fortresses possessed no guarantee of impregnability because they did not in themselves prevent treason and cowardice. One should not forget that all fortresses, no matter what their design, can only be saved in the long run by energetic and courageous relief armies.[96] Reversing Carnot's sobriquet of Fourcroy de Ramecourt as the éteignoir (candle snuffer) of ideas in the corps of military engineering, he called Montalembert the "éteignoir of French virtues" because he made no provision for sorties and vigorous counterattacks by the garrison, trusting entirely to massive firepower.[97] Revolutionary élan was constricted and debilitated in his fortresses.

Montalembert was also accused of variability and flux in his schemes. He "has been successively perpendicular, then angular; he as appeared in round forms, he became flat, then circular etc. but he has nonetheless been faithful to his cumulative principle in artillery."[98]

There were the now-routine cutting remarks about the "former officer of the light cavalry" and the bad old days when credit, titles, and intrigue could be a barrier to common sense and easily fool men in high positions who were unwilling or unable to listen to their technical experts.

Toward the end of his tract, d'Arçon comes closest to suggesting a civilized dialogue between the military engineers and Montalembert. If the latter would only diminish the height of his towers, replace cannons by muskets, thicken his walls, and make flanks, then one could talk.

Carnot at the Committee of Public Safety

The elevation of Carnot to the Committee of Public Safety in August 1793 must have appeared as yet another surprising turn of the wheel in Montalembert's fortunes. His earlier patrons, Maurepas and Mirabeau, had died at crucial and unfavorable times, but neither the royal minister nor the brilliant manipulator of revolutionary assemblies was ever as powerful as one of the group of twelve men who sat on the Committee of Public Safety and ruled France during the terrible Year Two of the Revolution. With Carnot on an all-powerful body that brooked no dissent, one would expect that Montalembert's ideas would have been proclaimed official revolutionary gospel. Yet nothing of the sort happened.

There were signs of attention from the Convention. Two weeks after Carnot's arrival at the Committee of Public Safety, Montalembert was granted a pension of 11,700 francs by the government.[99] (He had been fruitlessly and insistently soliciting a supplement to the pension he had been granted for his services at Aix from the royal government for a decade.) A month later the National Convention ordered that the copy of his works that he had donated to the government be placed in the National Archives and, a few weeks after the fall of Robespierre, the Committee of Public Safety invited him to continue his work on fortification.[100] Montalembert fulfilled this order by writing a scathing review of a treatise on fortification recently published by the military engineering officer Noizet-Saint-Paul.[101] Tests were performed on his gun carriages at Vincennes on 28 floréal year 2 [17 May 1794] and 12 brumaire year 3 [2 November 1794].[102]

But none of this seems to indicate anything more than perfunctory gestures. Given the disastrous financial situation of the government, it is doubtful that even his pension was ever paid in full. His promotion to general was followed by an order immediately retiring him.[103] He appears to have been reduced to requesting military rations and an undated note in his service record mentions his dining with Carnot and that the latter had supported his request. It would therefore seem that he maintained his contacts with Carnot and that these remained cordial. Yet Montalembert's biographer Gaudin mentions that Carnot, solicited by Montalembert to help pay his general's salary, intervened on his behalf publicly but sent a private note saying that the salary was superfluous.[104] It seems that not only were the attentions of the Convention purely perfunctory and formal, but even his ally Carnot had become cool to him. Entrusted with the overall military strategy of the Revolutionary armies, Carnot the engineer coordinated an aggressive strategy of field battles by large armies on all fronts and devoted little attention to fortifications. After the summer of 1794, with the French on the offensive in Europe, there was little need and few resources for fortifications. Military engineers, who gave no sign of being any less hostile to Montalembert, were used in attacks on enemy fortresses and could not be spared for constructing new ones in any case. Even on the principles of fortification, there are signs, considered in more detail below, that Carnot was no longer an unqualified supporter.

Was this an evolution in Carnot's own thinking? The taming effect of high responsibility on the formerly maverick junior officer? The continuing sullen resistance to Montalembert's ideas by an essential corps of officers? Or embarrassment with Montalembert's monotonous stridency and aggressiveness toward expertise at a politically and socially chaotic time when Carnot was striving to bring cohesion and acceptance of authority in an army and society torn by dissension and the blandishments of various self-appointed saviors both political and technical? Montalembert had also associated himself with the Lycée des Arts, a forum for disgruntled craftsmen and anti-academic forces.[105] It was the head of the founder of the Lycée des Arts, Gaulhard-Charles-Émile Desaudray (1740–1832), who read the only public eulogy of Montalembert after his death a few years later, a eulogy he never received from his former colleagues at the Academy of Sciences.[106] It is perhaps a sign of his relations both with Carnot and the revived Academy (now a section of the republican Institut National) that he failed to be elected to the place that had been vacated by Carnot's expulsion from that body after the coup of Fructidor in 1797. True, Montalembert had competed against Bonaparte, who was assiduously courting the scientists and intellectuals of the Institut National at that time, but he was not even runner up in the election.[107]

Montalembert and the École polytechnique

In 1795 Montalembert published the tenth volume of his works. It was a production that looked even more ragged that its predecessor, consisting of the reports on the trials of his gun carriages at Vincennes the previous year and his review of Noizet-Saint-Paul's book. A large part of the book consisted of an incomplete military encyclopedia by A. P. Julienne de Belair, a former artillery captain in the service of Holland who was an enthusiastic admirer of Montalembert. An almost plaintive note can be detected in his signature of his review against Noizet-Saint-Paul: "Marc-René Montalembert, former general officer, in his 81st year." Was Montalembert finally flagging in his crusade against the engineers and their bastions? One could almost think so from the page of "Citations of the principal places in the volumes of perpendicular fortification, where *officers of the Corps of military engineering* are spoken of favorably" he included in his book.

D'Arçon was also no longer young: His request for retirement from the army on grounds of health had been approved on 28 November 1793. But the Minister of War had expressed the hope that "les jeunes ingénieurs sans-culotte," deprived of his example would still benefit from "[his] lessons and [his] precepts so profoundly analyzed in [his] writings" and d'Arçon had not remained inactive.[108] There is some evidence that he advised Carnot on military affairs—yet another sign that there was some rapprochement between the former defender and most effective enemy of Montalembert. After an absence of a few months from Paris at the height of the Terror, he was recalled by the Committee of Fortifications as a consultant. One final clash between Montalembert and his nemesis was about to take place.

During the summer of 1794, overshadowed by the more dramatic events of the victories of the Revolutionary armies over their foreign enemies and the fall of Robespierre, a number of scientists, politicians, and engineers had been working on the foundation of the future École polytechnique, initially known as the École centrale des travaux publics. Along with advanced instruction in the mathematical and physical sciences, the school was to teach the elements of civil and military engineering. Indeed, the official pronouncements of the founders indicated that the school would replace all the old technical schools of the Old Regime, such as the school at Mézières, the École des Ponts et Chaussées, and the school of Mines, all of which were in a state of extreme disorganization or collapse.[109] An illustrious faculty had been chosen to teach the students. To "jump start" the school and to get the first graduates ready in a year, rather than the regular three years it would have taken under normal circumstances, "revolutionary" crash courses were instituted to provide a quick overview of the subjects to be taught. Thus, a summary of the chemistry to be taught in the first year was given in the first month. Similarly, a course for fortification, a subject to which most of the normal third year of the school was devoted, was allotted a month of "revolutionary courses" which began on 1 ventôse year 3 [19 February 1795].

The regular teacher for fortification at the school was Dobenheim, who has already been observed responding to Montalembert's criticisms of the Cherbourg project, and he was assisted by Jacques-David Martin de Campredon (1761–1837).[110] But the person chosen to teach the inaugural "revolutionary" course was d'Arçon.[111] It is almost certain that the Committee of Fortifications was consulted on staffing for the school, for all the first faculty members teaching fortification at the school were its members. The choice of d'Arçon to give the first course, which was a series of twelve lectures, indicates the esteem in which he was held by his colleagues and the importance attached to his views on fortifications. If the choice may also have partly been motivated by charitable considerations—d'Arçon was not being paid his pension and suffering severe financial distress—it seems certain that d'Arçon's reputation as the most eloquent and vigorous representative of the ideologies and attitudes of the military engineering corps was the main reason for the appointment. The course was immediately published as a book and widely distributed by the government and it was eventually translated into German (1801) and English (1815).[112]

The members of the Committee of fortifications named by the Commission des travaux publics in 1794 were the generals Bouchet, d'Arçon, Pierron, Favart, the chefs de brigade (colonels) Claude-Marie Carnot and Crublier, the chefs de bataillon (majors) Lapisse, Sanlot, Morlet, Foissac, Noizet-Saint-Paul and Milet-Mureau, and the captains Dobenheim, Boucher, Martin and Gobert.[113] Among these officers, Bouchet, Favart, Crublier, Sanlot, Morlet, et Milet-Mureau (more than a third) had signed d'Arçon's blast against Montalembert in 1790.[114] D'Arçon himself and Dobenheim must be added to these names as opponents of Montalembert. And it is difficult to believe that they were the only ones among the other officers whose opinions we do not know. If the presence of Lazare Carnot's younger brother may indicate some degree of balance in attitudes toward Montalembert, it is clear that the official institutions of the military engineering corps had had their way regarding nominations to the staff of the new school. Montalembert certainly thought so.[115]

D'Arçon's "revolutionary" course, as well as its summary published by Dobenheim in the first issue of the journal published by the École centrale des travaux publics, the résumé of the regular courses of fortification taught by Dobenheim and Martin de Campredon, and other articles on fortification by Dobenheim's successor, Horace Say, in subsequent issues of the journal elicited an almost predictable round of attacks from Montalembert, clearly showing that any apparent decline of his earlier vigor was illusory.[116] They appeared in the form of rebuttals in a sort of "shadow journal" titled *L'Ami de l'Art Défensif ou Observations sur le Journal Polytechnique de l'École Centrale des Travaux Publics* that was published from 1796 to 1797 and later bound into a volume that is sometimes classified as the eleventh and final volume of Montalembert's works.[117] These strange publications, perhaps unique for being a "journal" devoted to attacking an article printed in another journal, contain the by now familiar litany of complaints, abuse, and attacks in the also familiar format of a long-winded preliminary discourse and quotations from the writings of d'Arçon, Dobenheim, and Say, each followed by a long and hostile exegesis and commentary. "These Observations," the first page promised, "will only continue as long the Professors of the defensive art will continue to present their errors as truths."[118]

The tone was set. It was the engineers who were now being inflicted with the indignity of being assimilated to professors. For Montalembert, the new École Centrale des Travaux Publics with its vaunted polytechnical curriculum that was supposed to mark a new beginning in technical education was nothing more than the old school of Mézières resuscitated after having been dismembered and dismantled during the Revolution.[119] The Corps of Military Engineers was invited to repudiate Fourcroy de Ramecourt's *Mémoires sur la Fortification perpendiculaire* so that it could be rehabilitated in the public's opinion.[120] A new school dedicated to innovation by a revolutionary government publicly committed to change should encourage new ideas and methods in fortification. But the heavy hand of conservatism was still strong. This gave Montalembert occasion not only to begin a diatribe against his old and now dead enemy Fourcroy de Ramecourt, but also to attack Gribeauval as yet another conservative in the mold of Fourcroy de Ramecourt.[121] (From this it appears that his blandishments to the artillery had not been fruitful.)

True, good things were taught at the new school, but many of them were useless. It was not necessary that a military engineer know all the subjects taught to the civil engineers of the Corps des ponts et chaussées as specified by the initial syllabus of the school.[122] Furthermore, although all the other subjects were taught well, fortification was taught badly.[123] Chemists no longer taught alchemy, astronomers no longer taught Ptolemy, why did military engineers persist in teaching Vauban?[124] Concerning Vauban, Montalembert pointed out some discrepancies. The articles in the *Journal de l'École Polytechnique* contradicted Fourcroy de Ramecourt's earlier assertions that there was a system of Vauban. Was there then or wasn't there a system of Vauban?[125] Dobenheim had argued that, if Vauban had lived longer, he would have changed his ideas on fortification to take into account his own improvements in siegecraft such as the use of the ricochet shot. But if Vauban had been surprised by death, why had his loyal successors the military engineers not corrected the master's system?[126] He rejected the justification, also advanced by Dobenheim, that the military engineers had been kept in a state of idleness by a lack of building projects. This was untrue. They were and still were building and were continuing to build in the old wrong way.[127] Furthermore, demolition of the old system, provided that it was replaced with a good system of fortification like his own, was both practical and cost-effective.[128]

Montalembert was prepared to concede that the engineers had made some changes in their techniques that were often invisible to the layman. They were continuing to make changes that for the most part were plagiarisms of his work. D'Arçon in particular was accused of this crime—a crime that was characteristic of unimaginative mediocrities.[129] D'Arçon was accused of hiding his plagiarism by resorting to laconic descriptions of his methods.[130]

Turning to the specific themes that Dobenheim had raised in his articles, Montalembert was predictably critical. One did not need a lifetime of practice in fortification, as suggested by Dobenheim, to become truly competent in the field. Furthermore, not mathematics but genius was needed.[131] He also rejected the claim that the public should be knowledgeable in fortifications in order to be able to evaluate different kinds of fortification being presented to it. In spite of the apparent openness implied in this attitude, it was useless mystification and not necessary. The simple principle of superiority of firepower, the basis of Montalembert's systems, was enough to judge among different methods of fortification.[132]

Diffusion of the principles of fortification among engineers, other soldiers, and members of the public was not enough; it was crucial that the *right* principles—Montalembert's—should be disseminated. Otherwise error would merely become more widespread.[133] He was, for example, against the study of defilading, the subject of an entire article by Horace Say. The increasing use of mortars and howitzers with more curved trajectories made the science of defilading completely useless.[134] In a development that contradicted Carnot's views on the subject of fortifications, Montalembert argued that fortifications, rather than being built to facilitate sorties by the defenders, should make such sorties unnecessary. They were to be self-sufficient.[135]

On the whole, however, Montalembert was pessimistic that his ideas would be accepted in his own lifetime in his own country, in spite of the list of foreigners he gave who had adopted his ideas.[136] Yet, for "the glory of his nation," he had to continue his work, for someday someone would hear him.[137] He hoped that his old colleague at the Academy of Sciences, Charles Bossut, former examiner of entrants to Mézières and the Corps of Military Engineering would help him and use his influence to change the curriculum. The final number of his *Observations* contains his correspondence with Bossut, who gently responds that he is a mathematician with no competence in fortification and notes Montalembert's "animosity [which] is a little strong." "Truth should strive more coolly toward its aim," Bossut writes, and he begs Montalembert to pardon this "philosophical reflection."[138] A retort by Montalembert went unanswered.

This did not discourage him from continuing to fight for his ideas. The minutes of the Institut national record that on New Year's Day 1800, less than three months before his death, he read a paper on a new type of naval gun carriage, which was forwarded to a commission composed of Prony, Sané, and Bory.[139] A cautious report by Prony and Bory praised his gun carriage for land based cannon but was somewhat skeptical about its shipboard use, suggesting that trials be carried out at Brest.[140] It was probably about this session of the Institut that the astronomer Lalande wrote in his obituary of Montalembert:

A few months ago he came to read to the Institut a new paper on naval gun carriages. He was received with veneration and listened to with religious silence. Never had a man of 86 been seen to read with such a loud voice a paper so important to France.[141]

The man with the strong voice had displayed energy of another sort in a second marriage to Rosalie Cadet (niece of the academician and pharmacist Cadet-de-Vaux to whom the world is grateful for the invention of the marshmallow and the peppermint drop) at the age of 80 that resulted in the birth of his first and only child, a daughter, at the age of 82.[142] Another obituary claims that the pleasures of domesticity consoled Montalembert in the final years of his life, yet according to Lalande, they do not appear to have been enough to prevent him from frequent thoughts of suicide. Still ready to offer to test his naval gun carriages at Brest at his own expense, he was in great financial difficulties, having lost what was left of his fortune during the Revolution. The doors of the Institut remained closed, and the kinder treatment he received in the writings of military engineers such as Gayvernon, professor of fortifications at the École Polytechnique from 1797 on, was not translated into any major adoption of his ideas. A quatrain with a distinct whiff of paranoia and self-pity he composed a few days before his death leaves the impression that Montalembert had finally resigned himself to hoping only for posthumous recognition in his own country.

Persecuted all his life
Though worthy of a better fate
Time will dry up envy
Death was his final happiness![143]

14

The Conservative Art of Military Engineering in Old Regime France

Various causes of instability continually act on the institutions of men: the fluctuations that they produce sometimes appear dangerously in the science of arms, which contributes so powerfully to the stability of states. Among the means that wisdom produces to prevent this torrent of changes that in the long run undermines and destroys the improvement of our lot, let us dare to say, while awaiting proof, that fortresses alone are capable of stopping its course and to direct it in a way that contributes directly to the security of states.
—Jean-Claude Éléonor Le Michaud d'Arçon[1]

Montalembert did not bring about any major changes in French fortifications, with the possible exception of some coastal fortifications and some forts in mountainous areas.[2] And French military engineers could and did claim that those few examples came from a rich pre-existing tradition of designs for their particular sites and owed nothing to Montalembert. The military engineers, unlike their colleagues in the line army and the artillery, who eventually accepted the changes of Guibert and Gribeauval, won their fight against Montalembert and his new military architecture. They succeeded in this despite the fact that the controversy persisted under both the Old Regime and the Revolution and despite Montalembert's ability to mobilize such powerful allies as the royal minister Maurepas, the revolutionary tribune Mirabeau, and Lazare Carnot, "Organizer of Victory" of the Jacobin dictatorship.

Yet the military engineering corps was itself profoundly transformed during the Revolution. One can say that both Montalembert and the old royal corps died during the Revolution. Military engineers were not as favorable to the Revolution as Lafayette believed them to be. Of the 300 engineering officers who were in the corps that was reconstituted by the new government in 1791, only 153 were still in the corps in early 1793; by 1806, at the beginning of the Napoleonic Empire, that number had dropped to 66.[3] True, those who stayed were usually in the upper ranks of the corps. The traditions of a rigorous scientific education and hard work for new entrants had only been strengthened at the École polytechnique, which was in many ways the heir of Mézières. The new corps was no less hostile to Montalembert and owed much to the legacy of its predecessor, but in many important ways its role had changed.

The Revolution finally made what Saint-Germain had decreed in 1776 a reality: it made the corps a truly military corps, a long-standing aim of most military engineers from Vauban on. For Vauban, military engineering [génie] was truly a "craft . . . beyond our strength." However, for the old Marshal its core was military. For the man who thought that the ultimate test of bonne noblesse was military service and that the military calling was the highest of all secular vocations, the technical skills of engineers did not undermine their primary function of being military men. The Revolution had gone farther in another direction where there had been less unanimity among military engineers, for it made them a corps that was not merely a corps composed exclusively of officers but a corps that also had its own troops. In 1800 its strength had reached 8,261. Although there was a sharp increase in the number of officers (to 593 from the 300 specified in the reorganization law of 1791) because of the increased size of the French army that was fighting most of Europe, the bulk of this increase came from the thousands of sappers and miners who were now permanently and directly under the command of the engineers.[4] This 27-fold increase in numbers doomed forever any hopes that military engineers, as officiers du cabinet in the calm of their chamber, could devote themselves to contemplating the theoretical parts of war. For some this was a welcome development, but it was a sign of a radical rupture with the ways in which engineers had thought about and waged war in the past.

Along with the definitive achievement of full military status and the acquisition of engineering troops during the Revolution, came a related change in the military nature of the corps. No longer as visible a corps as it had been in the siege wars of Louis XIV or a disembodied brain directing the army in combat, as some dreamed it would become, the corps became a service corps. This implied a shift from the central position in warfare that military engineers had occupied in the glory days of siege warfare to an auxiliary role. Siegecraft and fortification remained important (indeed, their increasing involvement with field fortifications made military engineers as necessary as they ever had been), but they were no longer involved to the same extent as previously in combat, which was taking place more and more on battlefields and less and less at fortresses. As a service arm they were as necessary as before, perhaps even more so, but engineering wars of the kind in which technical skill had played a preponderant role were in a temporary decline. Military engineers no longer had to worry about their status as true soldiers—all modern armies realized their need for them—but their duties were now more precisely circumscribed. Hegemonic impulses that engineers like du Portail may have harbored for the corps as a potential brain of the army were put to rest along with any lingering doubts about their status as true soldiers.

This state of affairs came about rather quickly and was essentially complete in the Napoleonic army. The stability that came swiftly and decisively by administrative fiat during the Revolution may seem surprising in view of the instability that existed before the Revolution. The rapidity with which the institutional status and professional mission of the military engineering corps was stabilized during the Revolution can partly be explained by its being part of a military structure—orders come from above and they are obeyed. Under the Old Regime, however, orders were often evaded. More effective centralization, which was one of the hallmarks of the Revolution, reduced such evasion. There was less of the incoherence that continued to plague the centralizers and ministers of the Old Regime in spite of their best intentions and plans. On the other hand, the relative success of post-Revolutionary regimes in dealing with army reform and administration is also the final fruition of attempts to do so by the previous regime.

Preceding chapters have shown that, in spite of its formal institutionalization and legitimization in 1776, which gave it prestige and apparent viability, the functional domain of military engineering was in flux and by no means settled. This must be remembered when considering the Montalembert episode, which at once aggravated and exemplified its unsettled state. In the century between the creation of the Department of Fortifications in 1691 and the reorganization of the Royal Corps of Military Engineering during the Revolution in 1791, there were five major administrative reorganizations of military engineers. The many centrifugal forces acting on the corps of military engineering both from within and without were results not only of its failures but also of its successes. The fact that the military engineers were amalgamated with the artillery in 1755 for a brief period is further evidence that their institutional independence was contested.

Bureaucratically, it may have appeared that military engineers occupied a niche that was beyond the reach of rivals and those who denied its right to autonomous existence because it was under the jurisdiction of the Ministry of War, one of the primary organs of the state. Yet their colleagues in the administration of Roads and Bridges (the ingénieurs des ponts et chaussées) were under an even more important and influential ministry: the Contrôle Général des Finances. Increasingly throughout the eighteenth century, this was the ministry that counted, in every sense of the word, not only for its own departments but for every other department of the government as well, most of whose purse strings were ultimately controlled by it. The administration of the Corps des Ponts et Chausées, enjoying centralized, capable, and continuous leadership under Perronet and the Trudaines, was unceasingly active in binding the country together with a modern system of communication, which gave its engineers experience and opportunities not normally available to military engineers, who were locked into a domain defined by the needs of military security. This was a domain that military engineers anxiously guarded as an element of their legitimacy but which also imprisoned them in routine technical work that enervated their younger members and undermined their claim of indispensability and uniqueness. In some areas they were not successful in breaking through the often self-defined walls around their occupational domain, and toward the end of the eighteenth century they were in full retreat before an expanding Corps des Ponts et Chaussées, which had begun to encroach on them. At Cherbourg, the Corps des Ponts et Chaussées played a major role in the construction of the harbor proper, much to the chagrin of the military engineers. In the final years of the Old Regime, the military engineers were stripped of their right to build ports. (They fought back hard to regain this right, and port construction, both military and civil, was included as part of military engineering in the first draft of the program of instruction at the École Polytechnique, founded during the Revolution. However, it was transferred to civil engineering in the final organization of the school a few months later.[5]) Not only did this represent a loss of professional turf; it deprived the military engineers of the opportunity to develop the skills that had received such a fine theoretical basis at Mézières.

In other domains, it was the military engineers who were attempting to expand their range of activity. Less powerful groups of engineers, such as the ingénieurs géographes, were subordinated to the military engineers after 1776 (although they regained their autonomy after the Revolution).[6] It was the military engineers' successful encroachment on the functions of these engineers, who were the topographical arm of the army, which contributed to an expansion of their functions and led some of them to consider redefining the military engineer as superior kind of staff officer.

With the stability and consolidation of the military engineers' status within the army during the Revolution came a loss of leadership as the technical elite within the engineering profession. There was a definite lessening of the envy that had been a backhanded tribute to their elite status both within the army and with other kinds of engineers. With the foundation of the École Polytechnique, their educational institution (amalgamated after 1802 with the artillery and moved from Mézières to Metz) no longer had the salient role that it had before the Revolution. The civil engineering schools of the Corps des Ponts et Chaussées and the Corps des Mines attracted better students and were more active in research in engineering science. Military engineering took its place as merely one of a number of proliferating specialties of engineering.[7] No longer was military engineering the prime field of engineering, nor were military engineers the only ones who received a good formal education.

Closure came not only because of a new regime but also because of the new regime's commitment to the new kind of war that the French had learned from Frederick the Great and developed so brilliantly on their own.[8] Although there were many sieges in the Revolutionary and Imperial wars, those more mobile wars were rather different from the war that military engineers had learned to fight in the previous century. Both Montalembert and the military engineers agreed in calling the art of the military engineer "the conservative art." The conservative art was a feature of a conservative war—essentially a defensive and relatively static war.

Conservative War

Military engineers were conservative in a number of ways. They were conservative socially, in spite of the spectacular examples of the military engineers Lazare Carnot and Prieur de la Côte-d'Or on the Committee of Public Safety. The same Carnot, before the Revolution, had energetically but unsuccessfully solicited confirmation of a dubious title of nobility from the official court genealogist at Versailles.[9] (In 1797 he would be purged from the government of Republic and forced to flee into exile after being accused of crypto-royalism.) The proportion of nobles in the engineering corps rose steadily in the years before the Revolution, and, as Bien reminds us, revolution is not necessarily the culmination of reform. Reform and revolution can be distinct and even opposite rather than sequential and related. The military reformers in the army, like the engineers, were politically and socially conservative, and many of them were to fight against the Revolution. To judge from officers' loyalty to the Republic, the military engineers were the most loyal of the major military corps—even more so than the artillery officers, whom Alder has characterized as "revolutionary."[10] However, only one-third of the military engineers remained loyal. All signs show that, in spite of frustration and dissatisfaction on the eve of the Revolution, military engineers were thoroughly integrated into the army and shared the attitudes of its officer corps—a state that they had energetically struggled to achieve throughout the century.

In their military architecture, the military engineers were, of course, conservative in their adherence to bastioned fortification and in their mistrust of novelty. According to their most eloquent and articulate spokesman, d'Arçon, it took some courage to be an imitator. People had a natural admiration for the novel, and for d'Arçon this was a tendency to be checked.[11] In their technical thinking, engineers should mistrust grand ideas, unique solutions, systems, and radical changes. In their art, d'Arçon conceded, the "best could only be relative."[12] There were no "absolute solutions," only "principles, views, motives, means, . . . material that could be put to work."[13] There are always exceptions. "Grand ideas cost little; details kill."[14]

It is not difficult to see here a kind of Burkean conservatism that is wary of rationality and of intellectual elegance. It is a conservatism that is congenial to what Huntington calls the "conservative realism" that is central to the pessimistic moral vision that is at the core of the professional military ethic.[15] Wars are seen to arise from human weakness and are inevitable for this reason. There is an "acceptance of existing institutions, . . . limited goals, and . . . distrust of grand designs." "The military ethic emphasizes the permanence, irrationality, weakness, and the evil in human nature."[16] For the French military engineers this pessimism was only tempered to some extent by the belief that rich and civilized societies had progressed in civilization. Even so, it was a progress that was always vulnerable and fragile. At the level of tactics, there is also a pessimism regarding the possibilities of defense. Fortifications can always be overwhelmed, and the best the engineer can do is delay the inevitable. Impregnability is dismissed as a chimera.

But if civilization is vulnerable, it should not try to seek safety in imitating the "Tartars without a fatherland." Peaceful, wealthy, and civilized nations such as France should avail themselves of the fruits of their civilization, their science, and their industry. They should use more technology in war. It is in this manner that the hazards and unpredictability of war can be controlled to some degree.[17] Instead of resorting to the primitivism of a Guibert, instead of a war of movement and battles, the engineers preached a war of "industry" and "exhaustion." Risky battles could destroy, if successful, but they could never lead to possession. Moreover, one could not "undefeat" battles, but one could retake fortresses. The language of statics and immobility is visible everywhere in d'Arçon's discourse. Fortresses are part of the "general conservative dispositions" of a state; they are its columns; they are permanent and concrete; they provide "a kind of force of inertia favoring a noticeable reduction of the totality of moving forces, [which] must necessarily produce the most happy influence."[18] Moreover, technology and structures would have a restraining and calming influence that would be good for the tempestuous French, whose "disorganizing passions" would be calmed by the steadying spectacle of solid fortifications.

These kinds of wars were not the wars of Revolutionary France, and it is noteworthy that there was little trimming of d'Arçon's rhetoric to accommodate it to the official view when such trimming was prudent. D'Arçon admitted, when it was not entirely wise to do so, that "fortresses were not at all revolutionary."[19] The implication was that neither were engineers revolutionary. During the height of the Terror, d'Arçon had conceded that impregnability might be possible, but in his lectures at the École polytechnique he reverted to the more typical view of engineers. The marvelous military prodigies of the Revolutionary wars were a result of highly abnormal and, he hoped, unique conditions that would never recur. In 1795 France was paying a great cost to field its 14 victorious armies, and the disruption could not be tolerated forever. It would be necessary to strive for a future in which violence would be diminished and the effects of war would not be felt by the mass of the people. The ultimate aim of fortifications was static—an "equilibrium of forces" that one day might lead to the "mutual respect of nations." In this, the viewpoint of military engineers faithfully reflected the attitude of the monarchy in the last years of the Old Regime, when it generally hewed to a pacific policy and, after the American War, seemed to have lost all interest in military adventures.[20]

The Conservative Practice of Military Engineering in Eighteenth-Century France

The demands of administration and management in the life of the working engineer in the eighteenth century were further reasons for a tendency to conservatism. In the jargon of social constructionists of technology, the great art of the engineer was quite literally "closure," and this was achieved through the indispensable mediation of administration. Design, innovation, and many other things had roles to play, but they were ultimately subservient to the chief end of technological practice: to finish a job or construct an artifact. This was by no means an unexceptional task in the preindustrial era, when success was not routine and failure was more frequent than today. Yet it lacked drama or color. For those, both contemporaries and historians, who observed Montalembert's imaginative schemes in the first volumes of his works, it seemed clear that the dull plodding of military engineers was a trait marking inferiority rather than sensible practice. It was seen as a manifestation of "routine," that terrible and damning sin of the Enlightenment, which no one, including the engineers, was prepared to justify.

In his perceptive book on the aeronautical engineering in the twentieth century, Walter Vincenti modifies the definition of engineering given by the British engineer G. F. C. Rogers:

> Engineering refers to the practice of organizing the design, construction, and operation of any artifice which transforms the physical world around us to meet some recognized need.[21]

Vincenti's book is essentially about the epistemology of engineering as it pertains to design, the first and the most often studied of the three aspects of engineering mentioned in his definition. But he is careful not to forget the other aspects. Indeed, it is Vincenti who adds the third feature—operation, which subsumes maintenance—to Rogers' original definition. Operation, that humble and apparently routine activity of the engineer, is often forgotten by historians of engineering. Many consider it uninteresting and not posing any intellectual challenges. The documentation associated with it is at once overwhelming and ephemeral, rarely surviving to be analyzed by historians. For Popperian epistemologists and others for whom the central problem of epistemology is the growth of knowledge, design will certainly look more interesting than operation and maintenance.

Nathan Rosenberg, writing more particularly about the aeronautical industry, is an exception to this general trend. For him, knowledge can grow from the unprepossessing soil of mere operation. In his article on "learning by using," he argues that a technology, even after the development stage, can be improved as a result of the experience of users.[22] The improvements can take the form of "embodied knowledge" (feedback in the form of actual design changes and structural improvements) or "disembodied knowledge" (in the form of improved techniques of usage, such as cutting operation costs for airplanes because of increasing experience on the job and increasing confidence that original projections of required maintenance were excessive). Rosenberg argues that such "learning by using" is particularly relevant to technologies with large capital equipment costs and a high degree of systemic complexity. His remarks apply to most of today's advanced technologies, but fortification in the eighteenth century also fits these criteria. Concerns about vegetation growing on both masonry and earthworks, and the necessity for removing it, were mundane but everpresent operational problems that conditioned engineers' thinking about higher levels of design and were generally invisible to outsiders without daily and prolonged experience managing fortifications. In Vincenti's schema, this kind of operational problem has a feedback relationship with other levels of the design process, itself hierarchical and consisting of five stages: project definition, overall design, major component design, design of minor components of the preceding category, and design of further subdivisions.[23] This schema for aeronautical design lends itself to fortification design as well. Beginning with the choice of a site, which is conditioned by strategic, social, and economic considerations, the designer proceeds to the choice of the number of sides of a fortress and the kinds of major outworks, such as hornworks, crownworks, and demi-lunes, that seem necessary. Each bastion, outwork, or retaining wall requires detailed design, as do finer elements (casemates, posterns, powder magazines, barracks). Each of these levels of design generally has a subdivision of labor: a site may be chosen by a minister or by the king; a junior engineer or contractor will design a postern. At all levels of the design process there will be interaction and feedback with the corresponding levels of the construction and maintenance functions of engineering. There is not only interaction up and down the hierarchy of design functions but also lateral interaction with the similar hierarchical functions of construction and maintenance. This matrix of functions, spread out in time as well as in space, demands coordination and management, even when some of the functions are conflated or interact informally. This leads to the importance of "organizing" mentioned in the definition above.

Organizing and organizations are therefore fundamental features underlying engineering's shifting intra-professional and inter-professional boundaries. They have become more evident and more imperative as systems have become more complex and expensive, but they were also visible in the systems, such as fortification, that existed in pre-industrial societies. No matter how crucial the eighteenth century was for the development of scientific engineering and for its ultimate application to design, the central features of military engineering, masked by today's concentration on the centrality of design in engineering, were construction and maintenance, which ultimately conditioned and constrained design and absorbed the energies of most military engineers. This led to a certain neglect of engineering science, more honored and studied than practiced. Operation led engineers into the world of detail, administration, and management, which, along with their martial activity, was the primary focus of the work of the majority of engineers.

What their opponents saw as routine, military engineers saw as habitude (habit), the subject of much praise by Lazare Carnot as a young engineer.[24] Habitude, the triumph over natural unproductive penchants of human beings and a product of self-discipline, hard work, and careful attention to detail, was an engineering virtue that was both honorable and necessary to the success of the engineer's work. It is also a conservative virtue. The attention to detail required for efficient administration was important for routine maintenance, project management, accounting, and reporting. No matter how much engineers publicly denied their links with administration, that lowly craft of clerks, no matter how much they ridiculed it, it was the central part of their work. Even Bélidor, notwithstanding his focus on engineering science, agreed that drawing up a tender properly was the key element of the engineer's art.

Conservative Engineers versus a "Revolutionary" Marquis

Military engineers were men of detail; Montalembert, a man of great schemes and plans, certainly was not. Not for him the petty calculations of detailed costs and the work involved with detailed design. Rarely in the archives and in his publications are there detailed calculations of the costs that he claimed made his fortresses significantly cheaper than those built by the engineers. In the case of the projects for the successive forts at Aix, the dubious nature of his financial estimates is evident. Even then, Seuillet, one of the representatives of the Corps du génie at Aix, was probably right when he claimed that Montalembert's contractors and master masons were the ones who actually did his detailed design work.[25] To some extent this had no doubt been true for the engineers in earlier years of the century, but it was something that had changed under the administrative pressures of reporting, accounting, and control of construction projects. When Montalembert resorted to anonymous aid from Prony to respond to Grenier's criticisms (based on calculations taken from Bélidor's *Science des Ingénieurs*) of the strength of Montalembert's arches, it was not merely another example of appealing to specialists for justification (as long as they were hostile to military engineers); it was a sign of his incapacity to do the calculations himself. Although there is no direct evidence of this, it is highly likely that any detailed calculations he made as well as the drawings that were presented to the public were the work of his draftsman Mandar, a competent architect who later went on to publish an anthology of various systems of fortification and became a professor of architecture at the École des Ponts et Chausées after the Revolution.[26] Montalembert's consideration in not boring ministers with details was also consideration for himself.

No doubt Montalembert's inattention to detail was part of the reason for his fertility in producing new schemes of military architecture. By taking an overall view of things, ably assisted by draftsmen such as the Mandar, he had not shrunk from criticizing Vauban and proposing remedies for the increasing threat of a more powerful and numerous artillery. It focused his attention on the more prestigious—both then and now—aspect of design in fortification and military engineering and it loosened the hold of tradition on his thinking.

For the engineering corps, the collective experience of more than half a century of fortification and siegecraft played a primary role in the inspiration for problems and solutions, as well as the formation of attitudes. In the case of the military engineers, this collective experience was necessarily refracted through the prism of a formally institutionalized military corps. Hierarchical group effort was a natural and indispensable element of a military technology like fortification, which involved major construction projects. Edward Constant, modifying Kuhn, convincingly argues that technical tradition, based on experience, creates a "community of technological practitioners."[27] There is an additional reciprocal feature that must be considered here. For eighteenth-century French military engineers, the formally structured community defines technological experience as much as that experience defines the community of technological practitioners. Consensus is formed not only by successful practice and voluntary adherence, as in the case of most such communities, but also by a deference to authority enforced by military discipline.

Rather than see themselves as individual inventors, as Montalembert did, engineers saw themselves as a corps. This corps was even more cohesive than Constant's communities of practitioners. For this corps, social and institutional cohesion was more than a basic social necessity in the corporatist world of the Old Regime; it was the guarantee of retention, transmission, and exploitation of experience, itself seen as the fount of any worthwhile technical activity. Experience was, as Fourcroy de Ramecourt never tired of saying, the touchstone of their art, but the effectiveness of that experience depended on social and institutional factors. Thus, discipline and group cohesion, already valued by military engineers as soldiers, were also considered crucial for the preservation and diffusion of indispensable technical experience.

The fact that experience could be neatly integrated into the epistemology of Condillac, which was an essential feature of French scientific thinking at the time, gave added legitimacy to the engineers' approach and added a modern aspect to what could be seen as traditionalist routine. Indeed, experience was invoked to justify the distrust of novelty mentioned earlier. Montalembert and the engineers—men of the Enlightenment all—emphasized the "scientific" nature of their work. But most senior engineers still saw scientists as collectors of facts, continuing and developing a cumulative tradition of technological progress, while Montalembert saw the scientist as the inventor of genius who struck out on his own against a stifling tradition that he aimed to destroy.[28]

It is not surprising that Montalembert's situation on the theory side of the divide between theory and experience also locates him on the side of novelty and individualism in the confrontations between novelty and tradition and between individual creation and planning by a group. Montalembert's emphasis on novelty—something that fitted well into the ideology of progress that begins to be articulated after the middle of the eighteenth century—is probably what attracted Carnot to his ideas. Carnot was one of the engineers who had begun their professional careers just about the time when the polemic with Montalembert erupted. As Napoleon put it, Carnot had "never gone to war," and he is out of place in the schema of the typical military engineer presented here. Better trained in mathematics and science and less involved in military operations than the preceding generation, a pioneer in developing the scientific theory of work, thwarted in possibilities of promotion as the senior leadership aged but tenaciously hung on to their posts, and witness to the decline of the frequency and prestige of fortress warfare, it is not surprising that Carnot was open to some of the arguments of Montalembert. Carnot's adherence to Montalembert's views was in fact less thorough than might be surmised from his enthusiastic and even exaggerated public letter, but the two men drew together on the question of the importance of novelty.

Applying the terminology of Thomas Kuhn as developed by Constant, one could say that Fourcroy de Ramecourt, Montalembert, and Carnot all believed in the existence of "paradigms" for fortification.[29] Fortification, they all agreed, was truly a science; for Fourcroy de Ramecourt it was even a "positive" science. The Corps of Military Engineers, a community of practitioners clearly defined in bureaucratic terms, possessed a paradigm of praxis based on the work of Vauban and Cormontaigne. The work of the latter formed the "exemplar" essential to their technological paradigm. The difference between Fourcroy de Ramecourt and Montalembert lies in the former's belief that his technical paradigm (which he goes so far as to assimilate, for unclearly specified reasons, to the Newtonian paradigm) is henceforth immutable in general, although it can and should be improved in detail. It is a good paradigm of the Old Regime—an organic paradigm that does not admit of revolutions. For Carnot and Montalembert, however, technological revolutions are not only possible; they are indispensable, and inevitable in the case of fortification. While Fourcroy de Ramecourt and his military engineers are (in Constant's terminology) patient cultivators of "normal technology," Montalembert and Carnot dream of technological "revolutions" (a term that was becoming popular in science in the 1780s).[30] The virtues of the "normal" technologist are patience and discipline; those of the "revolutionary" technologist are creativity and novelty. And rather than the group, it is the individual genius who will play the decisive role in smashing old molds and bringing about technological revolutions.

Moreover, an emphasis on design almost inevitably leads to a search for novelty, whereas thinking about maintenance and functioning has the contrary effect of upholding established procedures.

Military engineers were trying to establish and protect a fragile exemplar of practice that was by no means the frozen body of doctrine described by Montalembert. Their attachment to Vauban the person took the form of attachment to his siegecraft and to his heroic image as symbol and patron spirit of their corps. Their attachment to any of his alleged systems was considerably weaker, even though they felt that the bastion was a naturally superior form that should be incorporated in all fortifications except certain special cases such as coastal fortifications. (Even Montalembert claimed, somewhat contradictorily, while accusing them of slavish adherence to a general schema of fortification, that they had not hesitated to modify Vauban.) Strict adherence to a standard ground plan was not the central concern of the military engineering corps, as Montalembert charged—they were too involved in building and maintaining fortifications in real situations to be bound by a single pattern. Moreover, it would have gone against their innate prejudices against the armchair fortifiers whom they associated with this genre of geometrical fantasy. More than anything else, perhaps, in their daily life and actions, military engineers had absorbed Vauban's advice that there was no fixed system of fortification, that terrain determined all, and that only in special circumstances, such as at Neuf-Brisach, could they indulge in delightful geometrical exercises and construct regularly shaped fortification.

Working both in the provincial *directions* of fortifications and at Versailles, and thus familiar with the daily workings of governmental bureaucracy, military engineers understood the constraints and advantages of working within large organizations. Their personal behavior was severely monitored and controlled, and their superiors routinely entered remarks on their morals and conduct in their personnel dossiers.[31] No strangers to intrigue, they were nevertheless farther from the world of favoritism and from the court circles in which Montalembert had influential allies. The fact that Montalembert was on the verge of overcoming the formidable and organized inertia of the engineering corps on two occasions—once under a royal regime and once under a Revolutionary one—is a tribute to his ability to persuade powerful and influential individuals. The fact that he ultimately failed once his protectors disappeared indicates that his achievement was very much an individual one and that he was isolated from bureaucratically entrenched technical administrations, which were becoming more important than individuals.

Systematic Conservatives and Conservative Systems

The divide between the individual inventor and the organization men reflects a final major difference between Montalembert and his opponents. Montalembert is deservedly known in the history of fortification as the author of a system of fortification. The word 'system' had a rather negative connotation during the Enlightenment. In his Preliminary Discourse to the *Encyclopédie*, d'Alembert associated "systems" with rigidity and dogmatism in thought that led to elegant theories accepted because of their coherence and plausibility rather than for their correspondence to the facts or fruitfulness for science. Certainly there is some justification for both parties to accuse each other of this sin. In fortification, however, although both Montalembert and the military engineers used the pejorative meaning of 'system' in attacking each other, it could have a neutral sense meaning a particular design of a fortification.

Ultimately, however, any weighing of the relative degree to which Montalembert and his opponents suffered from the defect of the esprit de système is less interesting than a comparison of their views on systems from another perspective—the revealing contrast between Montalembert as an author of systems of fortification (in the neutral use of the word) and the military engineers as designers and builders of technological systems in the sense in which 'system' has been used by historians of technology. Montalembert was, to a distinctly greater degree than his opponents, an author of systems, in the sense that he was primarily a designer and an architect of fortifications.

Montalembert concocted systems (as in système de fortification) whereas the military engineers as a corps built systems (as in Thomas Hughes' concept of technological systems). Hughes' concept of system, developed in his seminal study of electrical power generation and distribution systems at the end of the nineteenth century and the beginning of the twentieth, is also applicable to the system of fortification built in France from the time of Vauban on.[32] D'Arçon and many of his colleagues, following Vauban, thought in terms of a national, integrated, defensive system of fortifications composed of fortresses perceived as subsystems, instead of isolated punctual works whose role was defined purely in terms of local defense. The system was perceived as having an external environment beyond its control (the territory of foreign powers) over which it attempted to assert control and which at the same time posed a threat to the system's integrity. The system was dynamic in the sense that there was a tendency to expansion and rearrangement of territory to produce a more efficient defensive network and also in the sense that it could respond as a whole to demands made on particular points or areas of the system. There was interaction between the elements of the system, perceived to include not only the fortresses but also the army and navy which worked within it.

It was a system, like an electrical power system, that demanded centralized control and standardized procedures for efficient operation. It had a product and that product was security and power of the state. Even less tangible perhaps than the intangible electrical power produced and distributed by electric systems, it was perceived as being no less real. For a country that had been saved from military catastrophe during the War of the Spanish Succession by Vauban's frontière de fer, security was necessary and almost palpable. Although it may have been conceived and in its initial stages created by an individual or a small number of individuals, the system needed a coherent organization and management to continue to grow and even to survive.

Hughes sees several stages of technological systems: invention and development, diffusion, growth, increasing momentum and sometimes acceleration with qualitative change. In the case of electrical systems, the period of invention and development is associated with the person and work of Thomas Edison, whom Hughes cites as an example of an "inventor-entrepreneur" because of his keen eye for the market as well as his technical talent. In the case of classical artillery fortification, it is Vauban who would conform best to this role. (In Vauban's case, the word 'entrepreneur' is stripped of the commercial connotation it has in Edison's case. It hews more closely to one of its French meanings: someone who undertakes or takes in hand.) The "technological momentum" of the system, another concept proposed by Hughes, can be seen in the almost universal adoption of bastioned fortifications in France during and after Vauban's time.[33] One could also use the phrase "technological inertia" for "technological momentum." For mature systems, technological momentum increases; technological inertia leads to greater conservatism.

In the phase of qualitative change of electrical systems Hughes sees the increasing and eventually dominant role of financiers and consulting engineers. In the case of fortifications in France in the eighteenth century an analogous development can be found with the growing importance of the formally trained military engineer working within an increasingly articulated and regularly established bureaucracy. In both systems, however, even in the initial phases of the inventor-entrepreneur, organized management and an administrative bureaucracy are crucial. Without them, the system cannot survive or develop. Even the contributions of the most exceptional individuals must be mediated by bureaucratic and managerial support; indeed, it is usually a trait of the more important technical inventors involved in the creation and development of systems that they make equally great contributions to organization and control.

Technological systems, by virtue of their intrinsic complexity, require subdivision and coordination of labor and resources. Because of this complexity and the interconnectedness of their multiple elements, they also require acceptable functioning of all elements to prevent disorder and collapse of the system. The often routine nature of this coordination in no way detracts from its indispensability. Hence, good management and control are crucial at the upper hierarchical levels of the system, while attention to routine details is equally necessary at the lower levels of the system. In the case of the engineering that goes on in the system, there are inevitable parallels with the performance of design, construction, and operation. Planning for a new national electrical grid will also be dependent on the daily operational procedures of linesmen. Indeed, the parallels are so close that it is difficult to see where systems management ends and engineering begins. (The call for joint programs in business and engineering in the twentieth century is an illustration of this fusion of management and engineering, something that historians focusing exclusively on design have a tendency to overlook.) The example of Vauban shows that a concern with systems often extended into attention to the economy as a whole. After his death, Vauban was known almost as much for his revolutionary proposals on reforming taxation as for his fortifications.[34] During his travels across France, he was always careful to include economic and social data in his notes.

Any kind of working technological system is difficult to achieve in practice, and there is a constant struggle to maintain stability and growth. Harmonious operation of technological systems requires negotiation, compromise, and coordination. The French military engineering corps was attempting to achieve such harmony and coordination at a time when it was under attack in a pre-industrial society and rightly perceived such system stability to be indispensable for successful performance of its duties. The benefit of stability and harmony is that complex technologies, which would be impossible without systems, work more or less as planned. On the other hand, this can lead to the persistence of technical options that correspond less and less to the needs of a given society and adjust more slowly to perceived technical needs and problems. Powerful individuals, radically different ideas, and competitive systems can disrupt and transform well-entrenched technological systems, but such disruption and transformation gets more difficult as the system increases in size and scope. In the eighteenth century, with simpler technologies and fewer and more immature technological systems, it was easier. Well-connected and forceful individuals with radical technical ideas, like Montalembert, could get a hearing and could pose a threat to the managers of existing technological systems.

The refinement of Hughes' systems model by Rosalind Williams also has applicability to military engineering and fortification in eighteenth-century France.[35] Williams distinguishes between two kinds of systems and sees them as having roots in the eighteenth century. Williams' work has the benefit of reminding us that systems are not an exclusively industrial phenomenon that arrived with the second industrial revolution.[36] One kind of system is inspired to a great degree by Cartesian mechanistic models: a system, like a machine or a factory making machines, is composed of material elements that fit together harmoniously and perform some common function. Such a model could be extended to a Taylorized factory in which production flows are carefully related to one another and to the machines that create or regulate them according to some centralized plan. It is a system that is ultimately finite, limited, and material. The other kind of system, according to Williams, is more a physical network in space, such as a communication system. It permits the flow of information or goods and binds together a polity and a given society. The former is that of the mechanical engineer in, say, a machine; the latter is the system of the civil engineer in, say, a road network. The working of the second type of system is often less visible than that of the first type, but its influence can be much greater. It can privilege certain areas of space, direct development to some areas and not to others, and force conformity to itself over a much larger space and longer time. The system of the military engineers partook of both of Williams' aspects of systems. Each node of their network of security was a fortress constructed of many elements and constituting a whole. Montalembert was fixated on this perspective of fortification. But these fortresses controlled roads and movement, exerted an influence on patterns of economic growth and human settlement, and molded an entire nation according to the plan of central controllers. Military engineers were aware of these aspects and always thought of their nominal task of building or fighting over fortresses as merely a component of their ultimate aim. Hence the almost universal disdain for the problem of impregnability, perceived as a local obsession and ultimately irrelevant to the proper functioning of the overall system.

It was this kind of "systems engineering" that called for a collective effort to plan and to build systems that had to be integrated into a local physical and human environment. This collective effort was necessary both at the level of each individual fortress and for the defensive system as a whole. Hence the importance of the organization over the individual and the necessity of a bureaucracy in designing, building, and maintaining these systems. Much of the system builder's skill was in forging or imposing alliances between various groups and interests in large technical structures and other organizations linked to them. Montalembert was an inventor but not, in Hughes' parlance, an inventor-entrepreneur, who is the only kind of inventor relevant to the systems technologies so characteristic of modern technology. Hence his failure in creating a technical system of his own or combating effectively a system that, no matter how fragile and how undeveloped by comparison with more modern systems, had acquired the capability of repelling the attacks of a lone inventor, no matter how brilliant and no matter how well connected at court.

Montalembert was not, of course, functioning in the commercial context of a modern capitalist market, but in the milieu of a royal court and administrative bureaucracy. But this milieu was no less amenable to system building than more modern ones. Vauban's success provides the proof. He was a Hughesian inventor-entrepreneur; Montalembert was not. In the context of a systems technology, such as fortification, Montalembert was therefore a failure as an inventor.

Being a technical entrepreneur was not just a question of forging alliances, negotiating deals, or imposing procedures. It was also necessary to display competence at the mundane level of details, something Montalembert never did and does not appear to have wanted to do, whereas military engineers' involvement with the details of projects was increasing. Economic, corporate, and technical considerations had forced even those engineers who once had left details to civilian entrepreneurs to pay more attention to them. Engineers had to be more careful watching money in a profession with many temptations to dishonesty and a perennially impecunious governmental administration perpetually vigilant in guarding its coffers. They could not leave a vacuum where other corps and rivals were only too eager to penetrate. They also had to know more and more about what they were doing because of the evolution of technology (including the development of a more powerful artillery, which had attracted Montalembert's interest). In jealously guarding their corporatist turf, military engineers were necessarily forced to become more competent and more involved in the details of their work, which Montalembert could ignore in such a cavalier manner.

The inherently conservative tendencies of systems builders tend to assert themselves after the heroic phase of the inventor-entrepreneur. By the last quarter of the eighteenth century, the system builders of the French military engineering corps had a massive technical infrastructure in place. This conditioned further growth in a number of ways. Since the vast majority of existing constructions involved the bastioned trace, this became the classical model that was taught and assimilated at Mézières and on the job. Maintaining and repairing these fortifications—the bulk of the military engineer's duties—further imposed this trace as the canonical form. The military experience of decades of sieges again made these fortifications seem natural. Hardly any fortifications were built from scratch, and the opportunity for radically new kinds of fortification was virtually non-existent.

However, the conservatism of systems builders did not preclude a commitment to growth and development. Another aspect of systems is their tendency to expand and grow—a feature of technological momentum whose other face is technological inertia. Financial constraints and the relatively stable situation of French borders, rather than unwillingness, kept the Corps royal du génie from pursuing a more vigorous building program. These constraints worked to accentuate the conservatism in ideas and projects of military engineers. This never entirely stifled their tendency to develop improvements on their basic model of the bastioned trace. When they did have a chance to build in new situations that they felt precluded bastions by their situation, such as at Cherbourg, they built imaginatively, even provoking the charge of plagiarism from Montalembert. Graduates of Mézières and men of the Enlightenment, military engineers had no quarrel with progress. Indeed, they felt that their being members of a corps savant made them its natural ministers and defenders against the ill-advised encroachments of "charlatans." Their daily work, their integration into the army of the Old Regime, and the traditional conception of fortification as the defense of civilization, made them appreciate the virtues of conservatism. That same daily work, their modern education, and the spirit of the times made them, along with most of the educated elites of Old Regime France, believers in progress. They truly were progressive conservatives. The paradox of the oxymoron was the paradox of their profession, which had to continually build and preserve new things cautiously on a solid base.

It is worth repeating that there was no simple dichotomy between the brave genius inventor and the blinkered and lazy technical establishment or the charlatan versus the experts. The military engineers, however, represented a technology of the future in precisely those aspects that look the least modern—their corporatist mindset and their adherence to routine, which they would have called habitude. If they might be accused of sterility in innovation, they and their mode of operation were indispensable for the actual success and development of any technology in fortification. By his insistence on the future and his worship of novelty, the apparently revolutionary Montalembert was displaying the attitudes and the mindset of the individualistic aristocratic frondeur of the past, doomed to extinction in the new technological milieu, not the standard bearer of scientific progress as he would have liked to see himself. The engineers, ancestors of the Organization Man, apparently rooted in conservatism and corporatism, were portents of the technology of future.

Engineers and Artillerists

To conclude this discussion of the conservative art of military engineering in eighteenth-century France, it is useful to compare military engineers and artillerists. The contrast between the two corps savants of the Old Regime is as revealing as the contrast between the military engineers and the Corps des ponts et chausées. Scrutinizing the two military corps, whose members were brothers in arms yet often rivals, leads to mutual illumination of features that are not always evident when restricting attention to each of them taken alone. A caveat is necessary at the outset, however. Engineering is not now nor was it then a monolithic profession. Among military engineers, even when welded together by esprit de corps and military discipline, there were a number of divergent views on the nature of engineering and the role of engineers in the army and society. The views of d'Arçon and du Portail differed radically on their attitudes to the involvement of engineers in the details of day-to-day construction activity. Moreover, the corps of military engineers, although the most prestigious and influential of the various engineering corps, was only one of several. Civil engineers, topographers, and mining engineers had different functions, different loyalties, and different visions of the world in which they lived and worked. The artillery corps, too, was not monolithic, as the struggle over the Gribeauval reforms indicates.

Ken Alder's brilliant and stimulating book *Engineering the Revolution: Arms and Enlightenment in France, 1763–1815* provides a convenient vantage point of the French royal artillery corps and the profound changes it underwent after the Seven Years' War, notably the Gribeauval reforms. His examination of the artillery corps leads him to claim that the artillery officers were revolutionary in a way that went beyond mere politics; indeed, they succeeded in imposing many of their ideas of rationality, meritocracy, and service to the nation-state on the agenda of the Revolution. Indeed, he suggests that the French Revolution was a revolution of engineers. Is this true of military engineers? I will argue that it is not, and I will also argue that there is room for revision in Alder's argument. To begin, what were the similarities and differences between artillerists and engineers?

It is perhaps best to start with the similarities already mentioned. They were both members of military corps that had fought a long battle to get recognition as true soldiers and not merely as useful hangers-on to the army. For both, there was a long-standing distinction between warriors and technicians that had only gradually been effaced and whose signs were still visible at the end of the century. Indeed, Alder's designation of artillerists as "engineers" can be understood with this in mind. Artillery officers were, of course, not engineers. Neither then nor now were they called such. Indeed, the friction between the artillerists and the military engineers is a persistent theme in the history of these corps all throughout the eighteenth century, and Montalembert attempted to use it in his attacks on the military engineers. Nevertheless, both corps savants had well-developed systems of technical education that trained candidates for the officer's epaulettes after rigorous selection examinations that were heavily mathematical.[37] Laplace, one of the giants of mathematics and mathematical physics, was the official examiner of the artillery corps at the end of the Old Regime.[38] Moreover, a subset of artillery officers were involved with the production of weaponry and arms procurement as inspectors and managers of state arsenals. It is entirely fair to call these particular artillery officers engineers; today they might be called mechanical and metallurgical engineers, and in France they would be called ingénieurs de l'armement.

Both artillerist and military engineering officers were exemplars of the new kind of military professionalism described earlier. They both acquired commissions not by purchase but by on the basis of merit validated by competence in mathematics. They were given advanced technical training, and they were specialists in their respective military crafts. Both had a strong ideal of service to the state, and for both the conflation of the state with the monarch was increasingly leading to the effacement of the latter in their conception of military and civic loyalty.

Yet there were fundamental differences between artillerists and engineers. The number of artillerists involved in strictly technical tasks was much smaller than the corresponding number of military engineers. Both artillerists and military engineers were soldiers, and this was a cherished and central part of their self-image. However, artillerists, unlike military engineers, were playing an ever-increasing role in warfare, retaining their old functions in siege artillery and expanding them in the operations of the field artillery that were becoming more important in inverse proportion to the declining role of siegecraft as mobility and field battles became the rule rather than the exception. The artillerist Napoleon Bonaparte assiduously cultivated the image of a scientist with his well-publicized candidacy for election to the Institut National, in which he took the position vacated by the proscribed military engineer Lazare Carnot. There is no doubt, however, about who was the better scientist and engineer. Napoleon and his brother artillerists saw themselves primarily as people who fired guns rather than people who made them. They were a combat arm, and they were becoming an increasingly important and prestigious one, rather than a service arm, which is what the military engineering corps was becoming during the Revolution.[39]

In the early part of the eighteenth century, the artillerist Bélidor deserves the name of engineer as much as any member of the military engineering corps, but he was by no means a typical artillerist, always isolated and on occasion hounded by his superiors. His successors in the artillery were people like Jean-Louis Lombard and the du Teil brothers. The former's primary interests lay in the domain of ballistics, and he lacked the scope of Bélidor's ecumenical engineering interests; the latter were essentially tacticians.

I cannot go as far as Alder to claim that the values suffusing the regenerated post-Revolutionary French state were those of his artillerist-engineers, who were responsible for not only generating them but also establishing their pre-eminence during the Revolution, when many of them were in influential positions with real political power. For Alder, it was not provincial lawyers like Robespierre and Danton who made the Revolution what it was but technocratic engineers like Lazare Carnot and Prieur de la Côte-d'Or. Even Napoleon Bonaparte is seen as an engineer rather than as a soldier of the Revolution. The values of meritocracy, scientism, and service to the impersonal state were not peculiar to engineers, whether military engineers or artillerists. They were shared by wide sections of France's pre-Revolutionary elites, and they were inspired by the ideas of the Scientific Revolution, by the popularity of Cartesianism among educated Frenchmen, by the centralizing and rationalizing tendencies of Bourbon enlightened despotism, by classical models of taste in literature and the arts, and by the encadrement of schoolboys in Jesuit colleges and other educational institutions emerging from the Counter-Reformation, to mention a few examples that come to mind immediately. Untangling these diverse strands would lead beyond the scope of this book if it were in my power to do so, but it seems reasonable to assert that the specifically engineering ideals described by Alder, though they resonated with many of these attitudes and views, were by no means predominant or generative. That is not to say that those military engineers and artillerists who stayed loyal to the Revolution were not used to good effect by the Revolutionary authorities. Some of them, including Lazare Carnot and the lesser-known Prieur de la Côte d-Or, attained positions at the very pinnacle of power during the brief period of the Terror. Carnot was not, however, typical of his corps, and Napoleon was not the only one to notice the fact that Carnot was an "original."[40] At a time of war and national crisis, the republican authorities looked everywhere for loyal specialists, and few of them were to be found. For matters dealing with the crucial war effort it was natural that the National Convention would look to the handful of regicides in its midst with military and organizational skills. Prieur and Carnot were exceptional in their own corps, the solid majority of whose members had deserted or withdrawn from the army. Carnot served the Directory and then Napoleon's dictatorship but was always removed from power by his masters whenever he attempted to follow an independent line of action.

Carnot, perhaps the most visible engineer during the Revolution, was more a superb organizer and a technical specialist than a technocrat. It was only during the short and exceptional period of the Terror that Carnot and his associates approached the condition of technocrats—rulers with technical qualifications who could ignore political processes and impose their solutions. They were by no means the first technical specialists in government close to positions of power—Vauban comes immediately to mind. Like Vauban, they were responsible to political masters. It can be argued, although not without difficulty, that Louvois and Louis XIV did not have the clout of the National Convention and that Carnot could more readily ignore the Convention than Vauban could ignore Louvois. Yet if Carnot had ever wanted to do so, he did not have complete freedom to act without the support of his politician colleagues on the Committee of Public Safety and in the legislature. In fact there was close coincidence of views between the technical specialists and their political colleagues, and this was due to mutual agreement rather than imposition by "technocrats."

Governments have used technical specialists for a long time and continue to do so. This gradually became more systematic during the last decades of the Old Regime, and the French Revolution marked a new high in this activity, as Gillispie and Alder have shown so well.[41] The relations between government and science and technology continued to strengthen, especially during other periods of national emergency such as World War II, when they became permanent and institutionalized. Nevertheless, governments and engineers remained distinct groups, notwithstanding their increasing closeness. In spite of the technocratic dreams of experts and their admirers, governments used rather than served the experts to whom they accorded material support to serve the state more effectively. That is not to say that technologists and scientists are apolitical and devoid of political and social opinions, conscious or unconscious. However, only those whose views are compatible with the views of those of rulers, whether revolutionary or conservative, are allowed to participate to some controlled degree in formulating and executing policy.

Of course, influence may not necessarily be direct; it may be indirect at the level of ideologies and attitudes. It is true that the École polytechnique, the nursery of much of France's administrative elite as well as of its technical elite, was much influenced by the technical military schools of the Old Regime, particularly the military engineering school at Mézières. The École polytechnique and other educational institutions formed in its mold exemplified meritocracy, rationality, and government service—the values of France's state engineers. However, the influence of scientists interested in increasing the industrial power of the nation and promoting the advance of science was at least as great, in spite of the ostensible reasons for the founding of the school. The history of the first years of the school shows that Monge and other scientists managed to marginalize engineers in the school's early years.[42] Engineers were not the only ones active in the creation of the school. When Napoleon militarized the school and subordinated it more closely to the needs of the state, he did so not as an artillerist-engineer but as the supreme warlord of a country that was in an almost continuous state of conflict with its neighbors.

My argument about the effect of engineers on the political and social ethos of France is more limited and tentative than Alder's. I have suggested that military engineers informed a model of military professionalism among a discontented, numerous, and increasingly influential class of officers who were the real workhorses of the army and whose opposition to the court nobility monopolizing the army's upper ranks and wealthy commoners attempting to buy their way into the army was so intense as to take on almost revolutionary undertones. Yet what Alder calls the "technological life" (the "coherent social and ideological world which gives purpose and meaning to a set of material objects") of military engineers was profoundly conservative.[43] Like the reforming officers who were influenced by their example, military engineers were ultimately conservatives, in spite of their modern technical tasks and their unhappiness with many of the incoherent aspects of the regime they served. In military architecture they adhered to basic designs that went back to the Renaissance and had been cast in canonical form more than a century earlier. They were wary of innovation in spite of their adherence to the Enlightenment belief in progress, and they still defended notions of warfare that assigned central importance to sieges and fortifications.

The struggle over the Gribeauval reforms indicates that the influential artillerists were considerably more innovative than engineers. They had accepted the principles of more mobile warfare, they had rejected sieges as the primary sites of warfare, and they were willing and able (after a serious internal struggle) to modify their materiel and weaponry to take account of the new mode of warfare. Is it correct, however, to attribute this to a more revolutionary bent among artillerists? In the two decades before the Revolution, both artillerists and military engineers were achieving what they had always wanted: recognition as true soldiers. But the martial aspect of the artillery outweighed the technical aspect. Increasingly artillerists fought as a true combat arm rather than as a service arm, which is what the military engineers were becoming. They were also fighting more in new situations, on battlefields instead of before fortresses, as the nature of war changed, especially during and after the Seven Years' War. To repeat, they were more interested in firing guns than in making them. Napoleon, like the majority of his fellow officers, was a soldier first and an engineer-artillerist second. Even Gribeauval's attention to reforming the artillery is more plausibly explained by the need to catch up to Prussian and Austrian developments in light field artillery than by any revolutionary desire to break away from the present to a future to be created by engineers. Prussia and Austria were not revolutionary powers. Gribeauval had witnessed Prussian and Austrian developments, some of which dated back to the War of the Austrian Succession, when he fought the Prussians in the Austrian service during the Seven Years' War. Frederick the Great's stunning victories, which featured rapid marching and deployment on the battlefield as well as the extensive use of light field artillery, had taught the French a costly lesson that Gribeauval and others absorbed and perfected. This was not so much a revolutionary attempt to transform suddenly the nature of eighteenth-century war as an attempt to stay abreast of and to surpass potential enemies who had demonstrated the effectiveness of new technologies.

It is undeniable, however, that Gribeauval transformed French artillery radically, and that he attempted to transform the manufacture of small arms even more radically by trying (with limited success) to institute what is a recognizable precursor of methods of making interchangeable parts. This technical feat, traditionally attributed to the American small-arms industry in the first half of the nineteenth century, was to become the basis for the mass production of durable consumer goods such as automobiles. To succeed in this attempt in the context of Old Regime France required profound socioeconomic transformations. Alder has documented the ultimately unsuccessful attempt of the artillery corps to accomplish such transformations. This failure, in turn, condemned the artillerists' efforts to bring about interchangeable-parts manufacture. Alder is surely correct in his belief that "engineers' [i.e., artillerist-engineers'] technological life revolved around the management of large systems of workers, soldiers, and weapons."[44] Large systems of artifacts and people are the working domain of most engineers in traditional engineering, not just artillerist-engineers. Alder also argues that an inevitable consequence of this is involvement in the social domain and that French artillerist-engineers attempted to engineer that domain in a way that would conform to their own social ideals. The ultimate result was an "engineering of the French Revolution." Indeed, he generalizes,

> in the broadest sense, engineering is a revolutionary enterprise, perhaps the quintessential revolutionary activity. In principle, engineering operates on a simple, but radical assumption: that the present is nothing more than the raw material from which to construct a better future. In this process, no existing arrangement is to be considered sacrosanct, everything is to be examined in the light of present aspirations, and all practices refashioned according to the dictates of reason. The immediate purview of engineering is technological design; but as we have seen, engineers thereby seek to shape an entire techno-social world. In this sense, engineering denies history. It has no sympathy with the conflict, compromise, and happenstance that brought the world to its present state. It collapses *is* into *ought* to build (or promise to build) a world more consonant with human desires as it understands them.[45]

In the revolutionary situation of the French Revolution, untrammeled by the constraints usually imposed by society on the utopian tendencies Alder describes, engineers supposedly got their way and put an indelible stamp on modern France. Even in the case of his artillerist-engineers, however, Alder's view needs further examination, partly because of the difficulties in clearly identifying exactly who these supposed engineers were, partly because effects seem to be disproportionate to causes, and partly because of the example of other kinds of engineers, such as the military engineers of the Old Regime.

The creative aspect of design makes it both more interesting to the historian and central to a flattering self-image of the engineer. One implication is that there is a future-oriented and projective element in the ethos of the engineer that makes him a revolutionary. There is an important sense in which this is true. The modern engineer is a revolutionary in the domain of conceptualization because he often attempts to create something that did not exist before and because he uses a continually changing engineering science that is expanding because of his efforts. But one must not forget both the material and the socioeconomic context in which the engineer-designer works. The engineer does not create ex nihilo. Such considerations are all the more pertinent in less glamorous features of engineers' work: operation and maintenance. Good engineers know that the socioeconomic context is often crucial for a technology. As Alder points out, the artillerist-engineers who tried to transform the production of small arms in France during the Revolution were ultimately defeated by it. Unlike their nemesis Montalembert, whose cavalier mistreatment of his work force and inattention to accounting practices increased the purely technical difficulties in his construction at the Ile d'Aix, the military engineers, like their great model Vauban, were always attentive to the socioeconomic context. Indeed, one might suggest that their involvement with large ongoing projects like the construction and maintenance of fortresses, with no specifically dedicated support personnel of their own, gave them more respect for this context than the artillerists, who were more accustomed to imposing their authority on their permanent force of more tractable subordinate soldiers.

Alder rightly points out that neither "derived society" nor "applied science" is technology; technology is neither the result of unmediated social forces nor a formulary of recipes engendered by science.[46] Technologists, among whom engineers form a distinct and increasingly dominant breed, are technology's proximate authors. They are a social group occupying a certain place in a given society, and they have certain interests. Preserving and enhancing their particular group interests is as important for them as for other social groups. However, these interests, in spite of their particularisms, usually also reflect those of the wider social group. Engineers, whether military engineers, artillerist-engineers, or civil engineers, were members—perhaps in a modest way, but members nevertheless—of the social elite of the Old Regime. They undoubtedly moved up the social scale of the elite that was emerging from the Revolution, and many of their values were noticeably diffused within the broader elite. However, the extent to which this happened is not necessarily a function of the ephemeral positions of power of a handful of their number. Nor should their role in determining the ideology of the post-Revolutionary elite be overestimated. It was because their values resonated with that ideology that their agency can sometimes be perceived as being so formative. Their values and attitudes were indeed influential, but at a less general level and among a more restricted section of the elite: the armed forces, where the corps savants, particularly the military engineers, influenced reformers.

That the military engineers did not attempt anything quite so ambitious and forward looking as the institution of interchangeable-parts manufacture was attributable to their conservatism—a conservatism that was more inclined to respect the social environment than it was to depart in radical ways from an already well-beaten path. Aware of their socioeconomic context, they had long known that they had to work in it and sometimes against or around it. Long experience, not only as designers but often as maintainers and operators of the massive system of fortifications in France, had convinced them that this context was even more difficult to transform than the material world they were building. The possibilities of transforming the context in radical ways were limited, especially from within such conservative institutions as the army and the administration of the Old Regime. Military engineers and the engineers of the Corps des ponts et chaussées tried to engineer France, but there are few signs that they tried to engineer the French Revolution. Engineering France often brought them into conflict with a social milieu in which they had to use social tools to achieve their aims. The example of the notorious and unpopular corvée system of using conscripted labor to build and maintain roads shows the engineers of the Corps des ponts et chaussées using the social tools they had at their disposal to achieve their aims. Engineers were perceived as defenders of the corvée system. This put them squarely in the camp of the reactionaries when Turgot attempted to abolish it and replace it with a tax that fell more equitably on landowners.[47] However, the engineers' attitudes can be explained not only by their greater social affinity with the elites who rejected the abolition of the corvée but also by the engineers' pragmatic realization that their work could not be accomplished without the support of these elites, whose resistance was more redoubtable than that of a relatively powerless peasantry in retarding their projects. Here, too, military engineers displayed one of the characteristics that Huntington detected in the professional attitudes of military officers: a tragic and pragmatic sense of life that accepts evil and strife as inevitable features of the human condition that must be fought—indeed that is the officer's function—but which will never be eliminated. Engineers did try to transform their social environment, but only to a limited degree, and they were dependent for any radical changes on powers to whom they were subjected. The minority who stayed loyal to the Republic served it, sometimes regretting the new style of warfare that was a hallmark of the Revolutionary government; however, they did not try to control or mold it, nor could they alone do so. The socioeconomic milieu of the Old Regime was more rigid and resistant in many ways than the material world that they attempted to mold. On the other hand, the social flux of the Revolution was too intangible to capture and channel.

In addition to the powerful social forces that act to make the engineer conservative, ultimately the engineer is a technologist, and technology consists of materials, energy, tools, and technique.[48] There is strong and intimate interaction between technology and society, but they are not the same thing. Blurring the line between them is neither analytically fruitful nor ontologically accurate. Unquestionably, the socioeconomic context is always crucial, but it is the context of technology; it is not the technology itself. The material world is a solid and concrete thing that imposes respect on the engineer grappling with it. Much as the engineer may wish to transform or change it, he has to take it into account. It is a material world created by the past, which imposes its burden. The engineer does not, as Alder would have it, "deny history"; he cannot afford to do so. Neither does he accept it, but he is forced to account for both its material and social legacy. Alder is right when he says that for the engineer "the present is nothing more than the raw material from which to construct a better future."⁴⁹ But it is also nothing less. The present is precisely the result of the past, and this raw material, both social and material, is ignored at one's peril. The engineer is not a reactionary, and he does want to transform the present. But it is just because he does want to change it, and because he is a technologist entrapped by the toils of the material and social world, that he cannot afford the luxury of utopianism and remains a conservative and not a revolutionary.

The example of Frederick Taylor indicates that the inevitable contact engineers have with the social domain has tempted many to transform that domain, at least in the confines of the factory or workplace and sometimes beyond them. Taylor's attempt to apply engineering rationality to the work force, to make the engineer independent of the traditional and inarticulate artisanal skill of workers, and to increase the efficiency of production took place in late-nineteenth-century America precisely at a time when the technology of interchangeable parts was coming to full maturity. Like the identical material artifacts coming off the assembly line, the Taylorized worker was to be shaped into a predictable and measurable homogeneous working entity. Ideally this would give the engineer complete control of production. The fact that in this case the engineer did attempt to manipulate the social world and to integrate into his technologies of production appears to confirm that engineers were never quite as modest in their aims as I have implied above.⁵⁰ Moreover, French military engineers such as Coulomb were also attempting to measure the ergonomic capacities of slaves in Martinique, and Picon has suggested that the need to deal with a large and unskilled work force provided by the corvée led engineers to begin attempts at quantifying and rationalizing human labor not subjected to the traditional practices of skilled artisans.⁵¹ Nevertheless, these eighteenth-century studies were pioneering efforts that were rarely pursued. It was only in the context of American industrial capitalism of the late nineteenth century, when a relatively undifferentiated labor force whose links to the world of craft skills and organization had either been considerably eroded or never existed in the first place, that a Taylor was able to work and thrive.⁵² In the France of the eighteenth century, no matter how dominant the engineer was on the work site, he nevertheless had to tread warily and take account of a more rooted and established tradition that could not be ignored even when it was being circumvented or suppressed.

In the wake of the Second World War and the explosive growth in the size and complexity of mega-projects, we may be witnessing a transformation of engineers and engineering. In recent work, Thomas Hughes, one of the pioneers of the historiography of technical systems, has suggested the existence of new kinds of "postmodern" technical systems in which external socioeconomic context plays a larger role and requires more intimate involvement by the engineer than the traditional systems he described in earlier work.[53] He argues that the projects that led to the Atlas missile, the SAGE air defense system, and Boston's Central Artery/Tunnel have required new kinds of skills and procedures that have involved the engineer more intimately than ever in the politics of technology and politics tout court. Contrasting earlier systems of electrical power distribution with postmodern systems, Hughes proposes a contrasting list of characteristics for each.[54] It is noteworthy that among other contrasts, such as that between "production system" (for his "modern" systems) and "project" (for his postmodern systems) and that between "Taylorism" and "systems engineering," Hughes also contrasts "maintenance" versus "construction" and "unchanging" versus "continuous change." Whether this is really engineering or whether it is the typical kind of engineering practiced today needs further study. Rosalind Williams also sees an emerging distinction between "big E" and "little e" engineering in the postwar period.[55] The fragility of the engineering ethos that Williams describes in postwar America is real even though it is neither a contemporary phenomenon nor an exclusively American one. It is also plausible that the transformations that engineering is undergoing today are greater than any since the Enlightenment. The world that engineers face today is considerably more fluid than any they have ever faced. The purely material world and the technology that is characterized solely by technique, matter, energy, and tools appear to have weakened their hold on the modern engineer, who seems to live more and more in a virtual world.

But the world of the engineer of the French Enlightenment was being built carefully and slowly, and it was a world that he saw as better than the world that gone before. It was also a world worth conserving and maintaining, because—like the profession of engineering—it was tenuous and threatened by the forces of disorder and change. Unrelenting human industry was required to defend the acquisitions of a glorious past, and the engineer saw himself as one of the main defenders of those acquisitions and as the primary builder on the solid ground of that past.

Conclusion

Engineering is in a state of turmoil today. "All that is solid melts into air," "the center is not holding," and "dissolution" are words and phrases used by Rosalind Williams in her seminal article on the subject of engineering at the beginning of the twenty-first century.[1] Even students at MIT seem to be more and more unsure of what they will actually be doing once they become engineers.[2] This appears to be an unsettling and novel situation—a far cry from the days of Samuel Smiles and Rudyard Kipling, whose panegyrics about engineers described a robust and confident tamer of the earth and its natural forces for the benefit of man. As I have argued, however, engineering was in a state of turbulence and self-doubt during the French Enlightenment. Such a condition is perhaps inherent in engineering in spite of the appearance of stability its material creations imply. But during the French Enlightenment, a recognizably modern and familiar engineering was being born. Like all births and all beginnings, it was painful and it deserves special attention.

French military engineering was the template for various kinds of engineering that emerged in France and in some other parts of the world during the eighteenth century. During that time, in spite of apparently increasing institutional coherence, the Corps du génie was the scene of internal debates about its true nature. The delicate and dynamic balance among the heterogeneous elements of military engineering was changing as a result of changing warfare, the appearance and initial developments of an engineering science, a new form of engineering education, and the continual struggle for status and for a place in the cluttered corporate world of Old Regime France, itself subjected to the corrosive forces of the Enlightenment and the pressures of the absolutist monarchy. Moreover, toward the end of the century it came under an exceptionally vigorous and sustained external attack by an influential enemy—the marquis de Montalembert—who went so far as to suggest that the corps be abolished outright. On the eve of the Revolution, it was in a state of ferment and deliquescence.

That the corps triumphed over its enemies, survived, and was successfully transformed from an aspiring and frustrated combat arm into an integrated and respected service arm of the army was the result of a number of factors. The primary one was the fact that military engineers were a community of practitioners who built and managed a technological system: the system of fortifications in Old Regime France. Essentially created during the reign of Louis XIV, with the energetic and intelligent help of Vauban, it required sophisticated administration, proven competence of its designers, builders, and managers, and rigorous discipline. It had already become difficult for an individual, no matter now brilliant and well connected, to overcome such a community. Also, there was more cohesiveness in the military engineering corps than there was in the artillery corps, which was the scene of another major debate between those advocating novelty and those attempting to maintain a status quo. Although sieges never disappeared and military engineers were as useful as ever in them, eighteenth-century French engineers were witnessing a slow decline of fortress warfare. As a consequence, they felt threatened as a community and threatened in their status, and they drew together against outsiders. Even engineers who were bitter enemies, such as the radical junior officer Lazare Carnot and the elderly general Fourcroy de Ramecourt, agreed that the "conservative art" of fortification and "industrious war" were marks of a civilized and enlightened nation, and that consequently military engineers were a necessity because of their central role in fortification and fortress warfare.

It was as defenders of a system rather than as military architects that military engineers had an advantage in their struggle with the innovator Montalembert. In the artillery corps, Gribeauval, a successful innovator, was also helped by high patronage and the increasing role of artillery in warfare that was becoming more mobile and less a succession of sieges. But perhaps more important was the fact that Gribeauval presented a complete system, not just an artifact, and enjoyed the support of many of his fellow artillery officers. Within the artillery there was a clash of systems; within the military engineering corps there was a struggle of a single outsider against a system.

However, the Corps du génie did not emerge unchanged from its struggle, and it paid a price for its victory. The transformation into a service corps with command of troops was also a retreat from hegemonic aspirations. Most notably, it did not become the "brain" of the army, although it contributed greatly to the idea of a general staff. Nevertheless, it left important legacies to modern engineering, to engineering education, and to the army. To engineering it provided an exemplar of technical bureaucracy that flourished in France in the next century; to engineering education it provided a well-developed institution of formal engineering education with a strong emphasis on theoretical science and mathematics; to the army it furnished a handy example of technical professionalism that the army itself was beginning to develop. Moreover, a small number of its most brilliant members were laying the groundwork for what would bloom as engineering science during the nineteenth century.

In the final analysis, however, the cauldron of conflict that was French military engineering during the Enlightenment rested on a solid base, which enabled it to survive and leave its brilliant legacy to other engineers. Engineering is at the interface between the material and social worlds. The relatively narrow community of engineering practitioners was itself part of a broader society. Society writ large acted as a force on the engineering community, which shared the values of the society in which it was imbedded. Simultaneously, engineers attempted to manipulate matter for purposes defined by society—or at least the dominant elites of that society. To do this required not only technique but also the support and cooperation, willing or unwilling, of social actors who wielded political and economic power. In their different ways, both the social and the material worlds were tractable only to a limited degree. Laws that could be exploited but not ignored governed the natural world. In theory social and economic institutions were man-made and could be unmade, but in practice they could turn out to be the most recalcitrant forces in creating technologies or building technological systems. To be sure, engineers attempted to manipulate social forces to achieve their material aims as they attempted to manipulate natural forces. But these social forces (or actors, if one prefers to be more concrete) were never entirely malleable, particularly during the Old Regime in France. Indeed, because engineers were simultaneously members of their community of practitioners and members of the broader social community—and often relatively privileged members—they subscribed more willingly than not to the values of their society. Bound by the constraints of matter on the one hand and the constraints of society on the other, they had to respect both, and they worked around rather than against them to bring projects to a successful conclusion.

This led to a rather conservative ideology among engineers, an ideology that is eminently reasonable in view of the inevitably amphibious nature of engineering, which has one foot in the world of things and another in the world of people. André Maurois has said that the wisdom of the engineer is the wisdom of Candide, that other hero of the Enlightenment. Candide's wisdom is resigned and pessimistic but not inactive—he does not cease from cultivating his garden and improving his material world, even though he harbors no illusions about the goodness, the rationality, or the possibilities of perfection of the social world. Here we rejoin what Huntington calls the "conservative realism" that is the core of the pessimistic moral vision of the professional military ethic. The fact that engineers have served kings and revolutionaries equally well in roles crucial to the functioning of the state does not make them the creators of either sociopolitical system. Too much has been made of the ephemeral role of a handful of politically radical engineers such as Lazare Carnot during the Revolution in suggesting that engineers were inherently revolutionary. If some individual engineers were revolutionary in their politics or in their technology, engineers as communities of technical practitioners were not. The evidence that they, as a community, were the primary vehicle for imposing an ethos of meritocracy, hierarchical technocratic thinking, and devotion to state service is tenuous. Aside, perhaps, from the particular case of the army, they reflected such ideals; there is little evidence that they radiated them. To be sure, in order to succeed in their technical projects they were forced to attempt, within circumscribed limits, to manipulate and work around social forces and economic factors. Experience, that "touchstone of their art," according to Fourcroy de Ramecourt, had taught them not only that matter imposed constraints on design and construction, but also that the social world they lived in and on the whole accepted imposed constraints that were just as imperative in practice.

Engineers had succeeded remarkably well in engineering frontiers and landscapes in Old Regime France. In fact, this was a more effective way of affecting the social topography than in attempting to change it directly. Subject to the discipline of their corps (military or civil), to the social prejudices, mores, and institutions of their milieu, and to the pressures of an absolutist monarchy they had done much to strengthen, they worked within and with the structures of their time and their society, not against them. If they sought change, it was more in the material world that they transformed incrementally and cautiously than in the social world they accepted. Military engineers were the proud inheritors of the legacy and program of Vauban. They viewed it as progressive in the sense Condorcet would have given to that word. The material infrastructure and the technological system that Vauban had built and they maintained and carefully extended was also part of a society that they saw as worth conserving, not only from the ravages of war but from what they perceived as ignorant and unenlightened enthusiasm. The roads, bridges, canals, and fortifications they had created were the concrete and indispensable armature of a civilized, progressive, peaceful society.

It may not be unwarranted to speculate that engineers have always seen themselves primarily as tamers of matter. Of the two recalcitrant forces involved in their work, the social is not only more difficult to tame, though apparently more pliable than the material; it is the one they are less willing to confront, since it is part of them. Thus, the military engineers of the French Enlightenment worked within the Old Regime and the

Revolutionary regime. Engineers have never been rebels, in spite of Veblen's hopes for a revolutionary class of technicians.[3] Their discontent has been limited to the complaint that they have not been given full recognition for maintaining and conserving the working order of civilization.

After the Second World War, particularly in America, this may be changing. There are major signs of transformation in both the material and the socio-economic sphere. Information technology has led to increasing numbers of engineers moving beyond "hardware" into "software," an immaterial world of code and symbols. The unprecedented growth of the market and of capitalist enterprises with a global reach have made today's extremely dynamic, even volatile, market forces perhaps the dominant ones of the socio-economic framework in which engineers function. They are finding that their footing in both of the once-solid worlds in which they lived becoming unstable. Once again, as in the Enlightenment, there are tendencies toward expansion and diversification of their fields of action, even as their role is being questioned and they are urged to transform themselves. Perhaps engineers, hitherto made naturally conservative by the checks of the material world and wrapped in the toils of their professional communities and society, are on the threshold of truly revolutionary change, with all its challenges and opportunities. If so, perhaps engineering will see a profound transformation that will lead to a radically new profession and a radically new role in society. If so, could the engineer who was coming into being during the Enlightenment, with his commitment to stability and order, be transformed beyond recognition and disappear?

Time will tell. My own feeling is that there will still be life in the traditional engineer who came out of the Enlightenment. Respect for matter must always be a feature of engineers and indeed all technologists. Veblen may be right: the love of manipulating and transforming matter for human needs is a primeval instinct that is common to engineers. It may be enervated by the virtual world that engineers themselves did so much to create, but it cannot be destroyed. In the final analysis, matter must be accommodated, and thus engineers of the kind that emerged from the Enlightenment will still be necessary.

Notes

SHAT: Service historique des archives de l'armée de terre
BIG: Bibliothèque de l'Inspection du Génie
AIG: Archives de l'Inspection du Génie

These archives have been amalgamated under the general control of the Service historique des archives de l'armée de terre in Vincennes.

Chapter 1

1. Cited by Mallagh (1981, vol. 2, p. 13).

2. Perhaps the best general history of fortification is Q. Hughes 1974. Also useful are Hogg 1981 and Hogg 1975. For more particular periods before the age of gunpowder, see Arnold Walter Lawrence, *Greek Aims in Fortification* (Clarendon, 1979); Pierre Letriche and Henri Trézigny, *La fortification dans l'histoire du monde grec* (CNRS, 1987); Elliott Frederick Winter, *Greek Fortification* (University of Toronto Press, 1971); E. W. Marsden, *Greek and Roman Artillery* (Clarendon, 1969–1971); John R. Kenyon, *Medieval Fortifications* (St. Martin's, 1990); and the fascinating if not always accurate Eugène-Emmanuel Viollet-le-Duc, *Annals of a Fortress* (Boston: James R. Osgood, 1876) and *An Essay on Military Architecture of the Middle Ages* (Oxford and London: J. H. and J. Parker, 1860; reprint: Greenwood, 1977).

3. On the lengths of the most famous sieges that occurred on French territory, see Rocolle 1973.

4. The fortress of Monte San Giovanni in the Kingdom of Naples, according to Duffy (1979, p. 9).

5. John Rigby Hale, "The Early Development of the Bastion: An Italian Chronology c. 1450–c. 1534," reprinted in Hale 1983 (see pp. 12–16).

6. de la Croix 1972, p. 41; Pollak 1991.

7. Hale 1977, p. 41.

8. de la Croix 1972, p. 41.

9. Hale, "Bastion," p. 14.

10. John Rigby Hale, "To fortify or not to fortify? Machiavelli's contribution to a Renaissance debate," reprinted in Hale 1983.

11. On Francesco di Giorgio's originality and influence, see Hale, "Bastion," pp. 19–21.
12. Mallagh 1981, vol. 1, pp. 228–231.
13. Reti 1963.
14. Ferguson 1992, p. 115.
15. Mallagh 1981, vol. 1, p. 42.
16. Ettlinger 1977.
17. Ibid., p. 97.
18. Alberti 1988. The original version was published in Latin in 1486.
19. Wilkinson 1977, p. 131.
20. Ibid., p. 136.
21. Böhme 1978.
22. Wilkinson 1977, p. 134.
23. Vasari, cited by Mallagh (vol. 1, p. 42).
24. On Alberti and his work, see Gadol 1969.
25. Shelby 1972.
26. On Tartaglia, see the introduction to Drake and Drabkin 1969.
27. Vitruvius was relatively well known before this date, but it was his "rediscovery" by humanists at the beginning of the fifteenth century that made his work a central one for Renaissance artists and intellectuals. See Cervera Vera 1978.
28. Vitruvius 1914, book 1, chapters 1 and 2.
29. Wilkinson 1977, p. 141.
30. Hale 1977, p. 19. See also Pollak 1991, p. xiv ff.
31. In the course of an argument with Antonio da Sangallo the Younger, Michelangelo is supposed to have said "I don't know very much about painting and sculpture, but I have gained a great experience of fortifications, and I have already proved that I know more about them than do you and the whole tribe of the Sangallos." (quoted by Duffy (1979, p. 40), citing G. Clausse, *Les San Gallo*)
32. Kruft 1994, chapter 9, esp. p. 117.
33. There were in fact few books published on fortification in the first half of the sixteenth century, and mentions of fortification in published books were usually found in books dealing with military and political affairs or architecture. For a bibliography of writings (including some manuscripts) on fortification and siegecraft in the period 1502–1554 that can be used in conjunction with de la Croix 1963, see Bury 1985.
34. Tartaglia 1554. The shorter first edition had appeared in 1546.
35. According to J. Hughes (1974, p. 103), the covered way, of which the first recorded example occurs at Brescia in 1428 has been generally attributed to Tartaglia.
36. Mallagh 1981, vol. 1, p. 20.
37. Lamberini 1986.

38. Wilkinson 1977, p. 139.

39. On the concept of the military intellectual, see Barker 1975.

40. Galileo's treatises on fortification can be found in volume 2 of Favaro 1932.

41. This occupies a mere 58 pages in Favaro 1932. The other treatise—*Trattato di Fortificazione*, consisting of private lessons—takes up 67 pages.

42. Galilei 1978.

43. According to Lamberini (1986, p. 15), Galileo had rediscovered Belluzzi's manuscript treatise on the subject of earthwork fortifications and had inserted it into his own treatise. See also Campillo 1986, p. 251.

44. Marchi 1599. On the numerous reprints and bibliography of de' Marchi, see Bury 1985.

45. Campillo 1986, p. 225.

46. Mallagh 1981, esp. vol. 1, p. 75 ff.

47. Mallagh 1981, vol. 1, p. 26 (citing Castriotto).

48. See chapter 8 below.

49. Errard de Bar-le-Duc 1594, 1584.

50. Préface à la Noblesse françoise, Errard de Bar-le-Duc 1600.

51. Errard de Bar-le-Duc 1594, Dedication to the King.

52. As his basic module for determining other dimensions of the fortress he uses the length of the flank of the bastion (16 toises for the hexagon, 19 toises for the heptagon, and so on), a rather inconvenient dimension that was not used by his successors, who used the side of the polygon on which the fortress was being constructed or else the "line of defense," in his case 120 toises (720 French feet, the usual range of a musket). The most characteristic feature of his system was his practice of making the shoulder angle of the bastion 90°, which inevitably led to an acute angle between the flank and the curtain wall. This was contrary to Galileo's injunction that the flank of the bastion should meet the curtain wall at a right angle. Since most of this flank in Errard's system was taken up by an orillon, the flanking capabilities of his small bastions were rather poor, and this feature was quickly abandoned by his successors.

53. Errard de Bar-le-Duc 1600.

54. Ibid.

55. Ibid.

56. See the introduction to Drake and Drabkin 1969.

57. Ramelli 1976, pp. 46–53.

58. Errard Bar-le-Duc 1600, book IV, p. 102.

59. Blaise de Vigenère, *L'Art militaire de Onosender* (1605), cited in Favé 1846–1871, vol. 2, p. 264 (E. M. Lloyd's translation).

60. de la Croix 1963, p. 40.

61. Pepper and Adams 1986, pp. 181–182.

62. Duffy 1979, p. 40.

Chapter 2

1. Rochas d'Aiglun 1910, vol. 1, p. 119, citing Cosseron 1869.
2. Augoyat 1860–1864, vol. 1, pp. 6–7; Duffy 1979, p. 45.
3. Duffy 1979, p. 45.
4. Ibid., p. 54; Augoyat 1860–1864, vol. 1, p. 10. See also Shelby 1967.
5. Duffy 1979.
6. As distinct from engineers sent to other countries on missions by their own governments. Examples of this were French military engineers sent to Turkey and to the United States at the time of the American Revolutionary War.
7. Daniel Speckle (1546–1589) in Germany claimed a particular style, but according to Duffy (1979, pp. 55–57) he merely made sensible modifications of Italian methods. For more insight into the friction between foreign Italian engineers and native engineers see White 1978.
8. Augoyat 1860–1864, vol. 1, p. 25. See also Buisseret 1964.
9. Augoyat 1860–1864, vol. 1, p. 24.
10. According to Pernot (1987, p. 408), he was Pompeo Targone.
11. Blanchard (1979, p. 105, n. 194) quotes Cosseron de Villenoisy's denigrating remarks about the schoolmasters. Cosseron de Villenoisy mentions Claude Famuel as the most important teacher, while Chandler (1976, p. 273) mentions Georges Fournier as the most important and respected author of fortification texts before Vauban. Chandler also mentions Pagan as the true developer of classical fortifications.
12. Vérin 1993, pp. 300, 306.
13. Du Breuil 1665, p. 77.
14. See de Dainville 1734/1954. On education in various military and religious colleges in the eighteenth-century, see Taton 1964.
15. Marolois 1627. For more on Dutch fortification, see Duffy 1979, chapter 4.
16. Naudin 1695, p. 154.
17. Blondel 1683.
18. Naudin 1695, p. 157.
19. Guillaume Leblond (1704–1781), professor of mathematics to the "pages de la grande écurie du Roi," was the author of numerous works on mathematics and artillery and of the popular *Elémens de fortification*, published in six editions over a period of nearly 30 years. He taught mathematics to the royal children between 1751 and 1778.
20. Milliet Deschales 1677.
21. Ibid., p. 192.
22. According to Augoyat (1860–1864, vol. 1, p. 81).
23. Lalonde 1665.
24. Lalonde 1689.

25. Cohen (1982, p. 39) estimates that there were 400 people who could be considered "mathematically minded" in all of England at the end of the seventeenth century. These included the members of the Royal Society.

26. On the porvedditore di fortezze, see Lamberini 1987. There seem to be some differences in the role and status of this official in various Italian states. Lamberini had the Florentine provedditore in mind but the Venetian provedditori alle fortezze seem to have been a somewhat different group of people, according to Hale ("The first fifty years of a Venetian magistracy: The provedditore alle fortezze," in Hale 1983).

27. Guerlac 1943.

28. De Ville is quoted to this effect by Guillaume Leblond in his article on fortification in the *Encyclopédie* (vol. 7, p. 196).

29. White 1967, pp. 149–150.

30. Pernot 1987.

31. See Hall 1997.

32. Duffy 1979, p. 136.

33. Gille 1964.

34. Duffy 1985, p. 96.

35. Choderlos de Laclos 1951, pp. 566–589.

36. Saint-Simon 1988, vol. 2, p. 299.

37. Ibid., vol. 2, p. 299.

38. There are a number of letters in which Louvois transmits orders from the king either to Vauban directly or to his superiors forbidding his presence in the zone of fire and travel by sea. E.g. Rochas d'Aiglun 1910, vol. 2, pp. 114, 117, 150, 159, 179. Writing to the Maréchal de Humières on 17 November 1677 ordering him to besiege St. Ghislain, Louvois adds: "You know enough the displeasure His Majesty would have if some inconvenience were to happen to …Vauban for it to be unnecessary for me to recommend his conservation and to use your authority to prevent [it]. . . ." Vauban was to be kept out of the siege trenches and their management was to be left to his subordinate. On 3 November 1688, Louis XIV himself finishes a letter to Vauban by ordering him "absolutely to preserve yourself for the good of my service." (Rochas d'Aiglun 1910, vol. 2, p. 298; see also ibid., letters from Louvois dated 16 August 1674, 3 October 1674, 17 July 1677, 17 November 1677, and 15 February 1679)

39. See Alexander 1981; Cavanaugh 1967.

40. Vauban, "Mémoire sur le rappel des Huguenots adressé à feu M. de Louvois en décembre 1689" (with an addition dated 5 May 1693) and "Idée d'une excellente noblesse et des moyens de la distinguer par les générations," both reprinted in Rochas d'Aiglun 1910.

41. Vauban 1707. The view that Vauban attracted the ire of the king for the boldness of the ideas in *Dîme royale*, put forward so forcefully by Saint-Simon, has been recently debunked by M. Morineau (1985). Moreover, Louis XIV continued to show signs of esteem for his great engineer after the incident. Nevertheless the king was irritated that Vauban had dared to publish the work without authorization and it was officially condemned. Vauban's economic ideas have attracted the attention of economic historians on this subject. See Michel et al. 1891; Cadet 1970.

42. Saint-Simon, vol. 2, p. 880: "Patriote comme il était, il avait toute sa vie été touché de la misère du peuple, et de toutes les vexations qu'il souffrait." Note 3 on pp. 1595–1596 of volume 2 contains remarks on the word 'patriot'. The Académie française only accepted this word in 1762, and Saint-Simon has sometimes been given the credit for having been the first, or at least one of the very first, to use the word in its modern sense. Morineau (1985, p. 261) argues that in fact Saint-Simon borrowed the word from English Whigs, who used it to describe themselves in opposition to Walpole.

43. Fontenelle 1709/1969, vol. 1, pp. 197–221.

44. The Association Vauban is dedicated to maintaining the memory of its namesake, publishes the journal *Vauban,* and continues to flourish today. Biographies, both popular and scholarly, continue to be published, of which Blanchard 1996 is the best recent example of scholarly excellence.

45. Quoted by Duffy (1975, p. 91).

46. Lloyd 1887, p. 61. Duffy (1975, p. 17) claims that Prussian engineers, supposed to be extremely sensitive to national prestige and parsimonious in their praise of foreigners, made an exception for Vauban and continued to refer to him with respect up until the 1860s.

47. According to his own service record presented in March 1703 and reproduced in Rochas d'Aiglun 1910, vol. 1, p. 22 ff.

48. Augoyat 1860–1864, vol. 2, p. 595.

49. For an extended bibliographical note and inventory of Vauban's manuscripts see Rochas d'Aiglun 1910, vol. 1, pp. 81–96.

50. Mention (1884, pp. 9–11) claims that ministers had to contend with a multiplicity of treasurers, whose accounts were irregular and out of date at the best of times.

51. According to one of the better contemporary sources on the finances of the royal regime (Forbonnais 1758, vol. 3, p. 102).

52. Forbonnais 1758, vol. 3, p. 102; vol. 4, p. 291.

53. Rocolle 1973, p. 404.

54. Augoyat 1860–1864, vol. 1, p. 254, citing Forbonnais. Rocolle (1973, p. 404) cites almost exactly the same figure (4,059,360 livres) for the slightly shorter period 1682–1707.

55. Augoyat 1860–1864, vol. 1, p. 235; Peter 1994, p. 245.

56. Augoyat 1860–1864, vol. 2, p. 638.

57. Ibid., pp. 574, 577–578.

58. Ms. Fol. 206a, vol. XXI, "Comptes des Fortifications entre 1755 à 1757," Bibliothèque de l'Inspection du Génie, Service historique de l'armée de terre, Vincennes.

59. Rochas d'Aiglun 1910, vol. 1, p. 237 quoting from a manuscript treatise *L'Attaque des places* presented to the Dauphin, the Duc de Bourgogne in 1705.

60. "Abrégé des services" (Rochas d'Aiglun 1910, vol. 1), p. 19. Even later, at the siege of Luxembourg in 1684 under Vauban's command, four out of 60 engineers were killed and 31 wounded (Rochas d'Aiglun 1910, vol. 2, pp. 241–244).

61. The only one actually published in his lifetime was *Directeur General des fortifications* (The Hague: Van Bulderen, 1685).

62. Lloyd 1887, p. 88.

63. [Joseph Sauveur], ms. "Traité de fortification," ms. 41B, Bibliothèque de l'Inspection du Génie. Guillaume Leblond, author of the article "Fortification" in the *Encylopédie*, also refers to Vauban's "three systems."

64. Rochas d'Aiglun (1910, vol. 1, p. 127) cites Cosseron de Villenoisy (who put the whole thing rather forcefully and convincingly but has not been always followed in the more recent literature) as denying that Vauban had a system.

65. Blomfield (1971, p. 43) mentions that Fournier, earlier in the seventeenth century, specified 260 feet as the depth of defensive perimeter.

66. A toise was a French fathom, equal to 1.949 meter.

67. For additional details on the traces of various features of classical artillery fortifications see Hogg 1975, chapter 4.

68. Rochas d'Aiglun 1910, vol. 1, p. 119, citing Cosseron.

69. "Voulez-vous que j'enseigne qu'une courtine est entre deux bastions, qu'un bastion est compose d'un angle et deux faces, etc.? Cela n'est pas mon fait." (Rochas d'Aiglun 1910, vol. 1, p. 119, citing Cosseron)

70. The crucial differences between the second system and the third are an increase in the depth and size of the detached bastions and a recessed curtain that has embrasures for cannon on the ground level to provide flanking fire for the curtain. See figure 2.1.

71. Rochas d'Aiglun 1910, vol. 1, p. 129.

72. Guntowers of stacked casemates were popular both above and below the Alps in the fifteenth century. Casemates in bastions were also suggested by Francesco di Giorgio and they were also used in the papal fortifications of Rome at the Porta Ardeatina bastion built by Antonio da Sangallo the Younger in 1542. See Pepper and Adams 1986, pp. 22–27.

73. On the whole question of warfare, fortification, and national character, see Le Michaud d'Arçon 1795, pp. 70–77.

74. Casemates were used not only by Italian during the Renaissance but also by the French in sixteenth-century fortifications such as the Château du Taureau in Brittany and the gun tower at Camaret near Brest constructed by Vauban himself and now called the Tour Vauban. This tower bears some resemblance to the gun towers later proposed by Montalembert.

75. Pagan 1689. The complete title is *Les Fortifications du Cte. Pagan, Nouvelle editioin. Augmentée d'une Idée générale de la Fortification; où les termes de cet Art sont expliqués et rapportés à de nouvelles Figures: Avec des Notes sur le Texte. Et des Eclaircissemens qui contiennent la solution des principales difficultez qu'il y ait dans cette Science: et la Manière de Fortifier de Monsieur de Vauban. Par Mr. Hebert Professeur Royal de Mathematique.*

76. Cambray 1689.

77. Source of the 20 maxims: Cambray 1692, p. 3 ff.

78. The gorge is the interior side, or neck, of a bastion or other defensive work, not protected by a parapet.

79. Presumably the angles referred to are the internal angles of the polygon. This would thus be just be a necessary corollary and another form of maxim 5.

80. Naudin 1695, n. 16.

81. Du Fay 1707. The testimonial is dated 2 March 1691. This edition is billed as a "new and last edition" although it was in fact reprinted again. The first edition had appeared in 1681. An English translation, *The new Method of Fortification, as practised by . . . de Vauban . . . With an explication of all terms appertaining to that art. Made English [by A. Swall]. . . . Second edition. In which is added, exact draughts of Dunkirk, Maestricht, etc.*, was published in 1692 in Amsterdam.

82. Guerlac 1943.

83. Blanchard 1962, p. 57.

84. The French royal army increased sevenfold from the time Henri IV planned an invasion of Germany at the end of his reign to the height of the War of the Spanish Succession under Louis XIV, according to Lynn (1980, p. 578). During the same period, the number of engineers expanded from 6 to approximately 300 engineers with the numbers continuing to increase to a maximum of 390 in 1772. The earlier figure for engineers may be somewhat low since military officers and others occasionally served as engineers during the early part of the sixteenth century. (Buisseret 1964a; Blanchard 1979, p. 213)

85. Corvisier 1983, p. 490.

86. Truttmann 1975, p. 22. For Gaston Zeller's views see Zeller 1928. Another scholar who thought along similar lines was H. Chotard (1890).

87. Duffy 1985, p. 81.

88. Corvisier (1983, p. 437) claims that at least some of the aggressive program of military expansion traditionally attributed to Louvois was inspired by Vauban and his desire, driven by topographical considerations, to create his pré carré (discussed below).

89. For an example of such correspondence, see Chotard 1890. For another collection of letters see Favé 1847. The correpondence of Louvois with the Hüe de Caligny family of engineers that takes up most of the latter is analogous to the classic two-volume anthology of Vauban's correspondence and writings by A. de Rochas d'Aiglun.

90. In 1671, Louvois had intervened behind the scenes to save Vauban from potentially career-shattering charges of fraud (Blanchard 1996, p. 156).

91. Duffy 1985, p. 65. On Coehoorn, see ibid., chapter 2.

92. Simon Deodat Lefèbvre, *Oeuves Completes de M. Lefèbvre*, cited by Duffy (1985, p. 128).

93. King 1949.

94. Richet 1973, p. 91.

95. Jean Racine, "Relation de ce qui s'est passé au siège de Namur," in Racine 1952, vol. 1, p. 65 ff. Further indications of the royal interest are in Racine's "Notes et fragments" dated 1678 and his "Précis historique des campagnes de Louis XIV depuis 1672 jusqu'en 1678."

96. Racine to Boileau, letter from Namur dated 3 June [1692] (ibid., p. 509). In another letter to Boileau a few days later [15 June] Racine comments on Vauban restraining the impetuosity of the troops ("mes enfants") so that they do not get killed in vain.

97. These instructions, probably written by the engineer Tarade, Director of Fortifications of Alsace, are quoted by Bélidor (1729, chapter 10, pp. 394–395).

98. Westfall 2001, p. 327. See also Solon 1981.

99. Pelletier 1990.

100. See d'Arçon, quoted on p. 141 of Pelletier 1990.

101. Vauban to Hüe de Caligny from Paris, 9 March 1698 (Rochas d'Aiglun 1910, vol. 1, pp. 590–595).

102. Ibid., p. 594.

103. Cited by Lazard (1934, p. 540).

104. Perhaps the most famous of Vauban's reports is his description of the area around Vézelay (Vauban 1986). The report ends with a table of 50 columns giving such data as the names of parishes, names of landlords, numbers of houses (both standing and ruined), males and females subdivided by marital status, males and females above and below 14 and 12 years of age respectively, detailed breakdowns of land usage, numbers of various kinds of animals, numbers of plows, and notes on the soil quality of each area. For his own domain of Bazoches, which happened to be in the Vézelay area, he wrote that the soil was very mediocre, producing "only [enough] wheat and wine to feed its inhabitants with no commerce. . . ."

105. See Daire 1843; Cadet 1970; Michel 1891. Morineau (1985) claims that Saint-Simon was responsible for creating the myth of Vauban's disgrace in the year of his death, and has convincingly demonstrated that Vauban did not fall into disgrace as a result of this book, even though the book was condemned for not having passed through the regular censorship.

106. Westfall 2001, p. 321.

107. Desportes and Picon 1997, p. 25.

108. Hoog et al. 1982.

109. Desportes and Picon 1997, p. 11. This is my rather loose translation of *géographie volontaire*. I use the term "voluntarist" in a sense similar to Joravsky's (1970). Picon uses "géographie volontaire" to describe modern French regional planning (aménagement) that includes the creation of infrastructure as well as the distribution of economic activity,

110. Mariage 1990.

111. Vérin 1991.

112. Mesqui 1985, p. 89.

113. Vauban, "Memoire sur le Canal de Languedoc" (Rochas d'Aiglun 1910, vol. 1, pp. 551–576).

114. Vauban, "Projet de Navigation pour les Pays Bas" (1706), cited in Mesqui 1985.

115. Vauban to Louvois from Guise, 20 January 1673 (Rochas d'Aiglun 1910, vol. 2, p. 89).

116. Vauban to Louvois from Valenciennes, 23 November 1678, cited by Rocolle (1985, p. 52).

117. Vauban, "Mémoire des places frontières de Flandres qu'il faudrait fortifier pour la sureté des pays de l'obéissance du Roy," November 1678 (Rochas d'Aiglun 1910, vol. 1, pp. 189–191).

118. Duffy 1985, p. 44.

119. Sautai-Dossin et al. 1994.

120. Vauban, "Projet de Navigation."

121. Corvisier 1983, p. 437.

122. Smith 1969, pp. 139–155.

123. Livet 1985.

124. Vauban, "L'Importance dont Paris est à la France; et le soin que l'on doit prendre de sa conservation" [1689?] (Rochas d'Aiglun 1910, vol. 1, pp. 188).

125. Bornecque 1985.

126. Zeller 1933.

127. Vauban to Racine from Paris, 13 September 1696 (Rochas d'Aiglun 1910, vol. 2, pp. 445–447).

128. Vauban to Le Peletier from Antibes, 17 February 1693 (Rochas d'Aiglun 1910, vol. 2, p. 376).

129. Vauban, "Intérêt present des États de la Chrétienté" (Rochas d'Aiglun 1910, vol. 1, pp. 491–496). Rochas d'Aiglun believed that the memorandum was written around 1700, when Philip V was offered the throne of Spain.

130. Ibid., p. 492.

131. Vauban, "Projet de Paix assez raisonnable pour que tous les intéresséz a la guerre présente, en deussent être contens, s'il avoit lieu et qu'il plut a Dieu d'y donner sa bénédiction," 2 February 1706 (Rochas d'Aiglun 1910, vol. 2, pp. 496–532, quoted passage on p. 504).

132. Ibid., p. 510.

133. Ibid., p. 508.

134. Desportes and Picon 1997, p. 219.

135. Lepetit 1988, pp. 21–32. This definition of a network [réseau] appears on p. 21: "Réseau territorial [est] un ensemble de voies de transport dont la hiérarchisation et l'articulation spatiale font système, et dont les qualités de fonctionnement dépendent précisément des modalités de cette articulation et de cette hiérarchisation."

136. Guillerme 1988. Guillerme bases his analysis on Cormontaigne's *Architecture militaire ou l'art de fortifier* (The Hague, 1741; second edition 1746).

137. See, for example, his "Memoire sur le Canal de Languedoc" (Rochas d'Aiglun 1910, vol. 1, pp. 551–576).

138. In the reorganization of 1691 the numbers of engineers in the departments of Colbert and Louvois were 112 and 164, respectively—a total of 276 (Blanchard 1979, p. 82).

139. Louvois to Vauban, 20 October 1667, cited on p. 132 of Lazard 1934.

140. Blanchard 1979, pp. 84 and 95.

141. Ibid., p. 95.

142. Ibid., p. 111.

143. Vauban to Le Peletier, 17 February 1693, Rochas d'Aiglun 1910, vol. 2, pp. 379–380). Blanchard (1979, p. 104) points out that Vauban was using the word 'génie' for "military engineering" before this became common usage in French. Antoine Furetière's dictionary of 1690 does not attribute to 'génie' the meaning Vauban gave it, but by 1778 the dictionary of the Académie française gives "l'art de fortifier, d'attaquer, de defender une place, un camp, un poste" as one of the definitions of 'génie'.

144. Blanchard 1979, pp. 105–106.

145. Rochas d'Aiglun 1910, vol. 2, p. 380.

146. According to Truttmann, Louis XIV used to devote Monday evenings, and later Sunday afternoons, to discussions of fortification.

147. Blanchard 1979, p. 116.

148. Vauban 1685, p. 73. According to Augoyat, this work was written around 1680. It was reprinted a number of times in collections of other works on fortifications. Extracts were reprinted in Bélidor's *La Science des ingénieurs* (1729).

149. Vauban 1685, pp. 79–80.

150. Blanchard 1979, p. 123. Blanchard's statistics are not complete, and she has only partial and variable samples of the social and professional backgrounds of her engineers. Thus, for the batch entering between 1692 and 1715, only 65 percent of backgrounds are known with certainty. However, this is probably as complete a collection of data as it will ever be possible to get and, all things considered, Blanchard's work is exceptional in the precision of its information.

151. Augoyat 1860–1864, vol. 1, p. 175.

152. Blanchard 1979, p. 123.

153. Ibid., p. 213.

154. Augoyat 1860–1864, vol. 1, p. 451. For an example of the state of incompleteness of the fortifications of a major military port—Toulon—see ibid., p. 320.

Chapter 3

1. Guignard, Le Chevalier Pierre-Claude de, *L'École de Mars*, cited by Duffy (1985, p. 96).

2. Baxter 1976.

3. Blanchard 1979, p. 158.

4. Ibid., p. 159. This scarlet uniform was later changed of reasons of safety to a gray uniform. In 1776 engineers received a new blue uniform with red piping.

5. Augoyat 1860–1864, vol. 2, p. 390.

6. Blanchard 1979, p. 170–171. Blanchard was able to identify the socio-professional origins of 91 percent of the engineers between 1716 and 1747.

7. Blanchard 1979, p. 171. Two out of three candidates who were sons of engineers were successful in their examinations while this was true for only one out of four who had no relations among engineers.

8. Bluche 1986.

9. Truttmann 1976, p. 51.

10. Blanchard 1979, pp. 191, 294–296.

11. The annual rate of engineers dying in or as a result of combat was 5.5 between 1701 and 1714. It was 9.6 between 1744 and 1748.

12. Marc-Pierre de Voyer, Comte d'Argenson (1696–1764), is not to be confused with his brother, René-Louis de Voyer, Marquis d'Argenson (1694–1757), who also served Louis XV as a minister (of External Affairs) from 1744 to 1747. The marquis was the author of the eight-volume *Journal et mémoires du marquis d'Argenson* (Paris: Société d'histoire de France, 1859–1867), where one can get an interesting view of the regime of Louis XV and a clear indication of the d'Argenson clan's dislike for d'Asfelfd (se e.g. vol. 1, p. 210 and vol. 3, p. 55).

13. On the Regemorte family, see Petot 1958, pp. 128–129. See also Augoyat 1860–1864, vol. 2, pp. 56–57; Blanchard 1981; Tiersonnier 1913, pp. 220–231. There seems to be some confusion on the dates of birth and death of various members of the family.

14. Blanchard 1979, p. 200. Tiersonnier extends the date of Noël de Regemorte's stay at the bureau of fortifications until 1757, when he is supposed to have left as a result of the fall of his patron from grace and his departure from the Ministry.

15. Le Sancquer, in addition to his influential post of Premier Commis de la Guerre (effectively the head of the administration of the ministry, often with long terms of service who stayed to advise ministers as they came and went) since 1771, also took over at the subsidiary functions of the offices in charge of the Coast Guard, the Artillery, and Military Engineering. I thank Patrice Bret for his comments and information from the archives of the Service historique l'armée de terre.

16. Dossier Cremilles, SHAT, L.G. 843, 1ère série.

17. Picon 1992, chapter 5.

18. There are cases of attempts to intercede with examiners, such as for the military engineer and one of the future founders of the École polytechnique, Prieur de la Cote d'Or, but it is highly doubtful whether they were effective. During the last decade before the Revolution there are some signs of loosening the strict entrance requirements in favor of birth. However, they were energetically resisted by the leadership of the school. See Taton 1964, pp. 600–601.

19. Augoyat 1860–1864, vol. 2, pp. 284–287.

20. Ibid., p. 289, citing Cormontaigne's *Mémorial pour l'attaque des places*.

21. See chapter 12 below.

22. On this short-lived union, see Augoyat 1860–1864, vol. 2, pp. 455–460; Nardin 1982, pp. 103–104.

23. Blanchard 1979, p. 206 and p. 209.

24. On Mézières, see Taton 1964.

25. Besides Taton 1964, see Picon 1992, Artz 1966, and Léon 1961.

26. On artillery schools, see Alder 1997; Hahn 1964; Le Puillon de Boblaye 1858.

27. For more on the contributions of Chastillon, scion of a distinguished dynasty engineers, see Belhoste 1990 and Taton 1964b.

28. There seems to be relatively little published on Bourcet, who is almost universally considered one of the great military thinkers and cartographers of the eighteenth century. He has been called a precursor of the idea of a general staff and an inspirer of Napoleon Bonaparte. See Pallière 1979.

29. Augoyat 1860–1864, vol. 2, p. 441.

30. Aside from articles in biographical dictionaries, there is no biography of Bossut. There are several biographies of Monge; the best is Taton 1951.

31. See Bledstein 1976 for the American case.

32. The best study on the school is Picon 1992.

33. Picon 1992, p. 214; Blanchard 1979, p. 221.

34. For example, du Portail, whose ideas are described in more detail in chapter 8, was placed under arrest for a year because of his role in leading the harassment of a fellow student not considered of sufficiently noble birth.

35. Picon 1992, p. 92–93; Dartein 1906, p. 71.

36. Reinhard 1956, p. 23.

37. Examiner Camus to Fourcroy, 16 December 1751, cited in Taton 1964b, p. 575.

38. The inventory of the military engineering archives contains a reference to a memorandum of Cormontaigne on Mézières, but the document has been misplaced or lost.

39. The marquis d'Argenson, in his memoirs (vol. 1, p. 212), cites a contemptuous reference to someone as a "faiseur de mémoires." It seems that writers of memoranda were a distinct and identifiable species for Louis XV's ministers.

40. Cosseron 1869, p. 218.

41. Cormontaigne had submitted a manuscript copy to d'Asfeld in 1718 (Cosseron 1869, p. 223).

42. "Memoire sur les contestations survenus dans les deniers sièges des pays bas, entre l'artillerie et le génie à l'occasion de l'emplacement des batteries destinées à ruiner les deffenses des Ouvrages opposés aux attaques," 19 July 1748, BIG/Fo83.

43. d'Argenson to Cormontaigne, 24 July 1746, in BIG/Fo87.

44. Vauban 1685. According to Augoyat, this work was written around 1680.

45. For an inventory of one such portfolio see Blanchard 1979, pp. 511–516.

46. See Cosseron 1869, p. 233.

47. Bousmard 1797–1799; Cormontaigne 1803.

48. Cosseron 1869, pp. 224–226.

49. There were a number of trials in Marseille on the problem of dissipating smoke in casemates and attempts to develop techniques to do this (Archives du Génie, SHAT, article 21, section 1, §1, carton 6, pp. 16, 17). Further trials were conducted in October 1799 at Neuf Brisach; the results were published and circulated to encourage further trials and "perfect this part of the art" (ibid., carton 9, p. 8). The results of the trials seemed to favor the use of casemates to cover heavy artillery (cannon) but there was still a widespread belief among military engineers that smoke would be suffocating and impede the action of the artillerists.

50. In the context of the turf wars between the artillery and military engineering, d'Arçon stressed the importance of mines and countermines. He saw the latter as a means of resisting artillery, which even the military engineers agreed was becoming increasingly powerful and effective. (Le Michaud d'Arçon, Jean-Claude-Éléonor, "Mémoire du Corps du Génie 1776," ms. 491G, Bibliothèque municipale de Besançon)

Chapter 4

1. The petard is the ancestor of the limpet mine, which was often used in the early sixteenth century in surprise assaults on fortified posts or houses. It consisted of a conical metal container, packed with gunpowder whose base was attached to some part of a wall, almost always a door, and exploded to blow an opening or blow down the door. It was never really of much use against more substantial fortifications and largely went out of use as the century advanced.

2. Translated into English as Vauban 1968. The full original title—*Mémoire pour servir d'instruction dans la conduite des sièges et dans la défense des places, dressé par M. le Maréchal de Vauban et présenté au roi Louis XIV en 1704*—is misleading. The work is in fact based on a manuscript of Vauban that was probably written between 1667 and 1672 and is an earlier and less polished work than one based on a later manuscript that was published earlier, in 1737. It is the 1737 book, not the 1740 book, which was in fact based on a manuscript presented to the king around 1704. On the complex bibliography of Vauban, see Rochas d'Aiglun 1910, vol. 1, chapter 2.

3. Duffy 1985, p. 10.

4. Cited ibid.

5. Lazard 1934, p. 31. Duffy (1985, p. 221) also mentions the possibility that Vauban got his ideas for siege parallels and *pierriers* from French veterans of Candia. The siege seems to have attracted famous German and Dutch fortifiers like Rimpler and Coehoorn as well.

6. Charles Goulon, one of Vauban's most talented Huguenot officers and the chief of his miners, who after the Revocation of the Edict of Nantes defected to England and then to Holland, had served at Candia. See Hebbert 1997.

7. Lazard 1934, p. 452.

8. Vauban 1968, p. 169

9. Ibid., p. 142.

10. Ibid., pp. 35, 51.

11. Ibid., pp. 157–161.

12. Ibid., p. 157.

13. Ibid., p. 61.

14. Ibid., p. 112.

15. Parts of Vauban's list of necessary supplies in a siege is reproduced on p. 244 of Chandler 1976.

16. Bélidor 1729/1830, p. 223.

17. Vauban 1968, p. 34.

18. Ibid., p. 151.

19. Ibid., p. 30.

20. Ibid., p. 140.

21. It was eventually published in The Hague as *De l'attaque et dela defense des places par M. de Vauban, Maréchal de France et directeur général des fortifications du Royaume* (Pierre de Hondt, 1737).

22. Rochas d'Aiglun 1910, vol. 1, p. 243.

23. Jean Racine, "Relation de ce qui s'est passé au siège de Namur," in Racine 1952, vol. 2, p. 265 ff. Vauban is mentioned only once in this account, but Racine mentions him as the driving force of the siege and as extremely solicitous of the safety of his men in letters to Boileau from the siege (pp. 509, 513). See also Sterne 1759.

24. On the impact of fortifications on literary figures of the eighteenth century, see Dejean 1984.

25. Vauban 1968, p. 96–97. Vauban's advice had merit. The course of the Great Northern War may have been changed by the death of Sweden's warrior king Charles XII in the siege trenches before Fredriksten in 1718.

26. See Fourcroy de Ramecourt 1786, discussed below in chapter 12.

27. Senneton de Chermont, "Mémoire de ce qui s'est passé de plus considérable à l'attaque de Mons pendant tout le temps du siège," ms., AIG, article 15, section 3, §1, Mons.

28. Ibid. A fascine is a bundle of brushwood tied together. Numbers of them can be piled up and placed to give protection or thrown down in marshy and wet places to permit passage of infantry.

29. Ibid.

30. Lynn (1991, p. 319) says there were 2,300 casualties among the attackers. Duffy (1985, p. 41) says there were 1,450 and considers this a relatively small number in view of the Allies' incompetence in siegecraft.

31. Augoyat 1860–1864, vol. 2, p. 377–380.

32. Blanchard 1981.

33. "Journal of the siege of Bergen-op-Zoom," ms., AIG, article 15, section 3, §1, Berg-op-Zoom.

34. A claye (modern French: claie) is a wattled hurdle (rectangular frame). According to the OED, in fortification they were laid down "on marshy ground or across a ditch to provide a firm passage, etc., or, often covered with earth, to stop up a breach, to strengthen a battery, or to protect a position from an enemy's fire."

35. A siege journal for the siege of Münster during the Seven Years' War by Duvignau, the second-in-command of the Mézières school, mentioned by Augoyat (1860–1864, vol. 2, pp. 495–497) as an exceptionally good one, seems jejune and uninformative compared to the one for Mons in 1691. The possibility that important parts of the journal have been lost, however, cannot excluded.

36. [Jean de Camelin du Revest], "Service des ingenieurs a la suite des armées" [1745], ms., AIG, article 3, section 9, carton 1, no. 18 bis.

37. Ibid.

38. Ibid. The functions of the "Major et de l'aide major" are included in article 15: "to procure and furnish by his efforts, according to the rank of each, bread and forage, and to busy himself enough with all this [de s'intriguer assés sur tout cela] so that no one suffers."

39. Cited by C. R. Fourcroy de Ramecourt in "Essai sur la méthode des officers du génie pour examiner les projets et systèmes de fortification, et pour connoitre la force des places de guerre," ms., 25 December 1768, AIG no. 21, section 1, §1, carton 6, no. 1.

40. Charles-René de Fourcroy de Ramecourt, "Extrait des Observations sur la manière de reconnoître les Places de Guerre et de les visiter," ms. Condé, 3 May 1761. AIG, article 3, section 9, carton 1, no. 46.

41. Ibid.

42. Ibid., p. 6.

43. Fourcroy de Ramecourt, "Extrait des observations . . . ," AIG, article 3, section 9, carton 1, no. 46.

44. Fourcroy de Ramecourt, Extrait "Extrait des observations . . . ," AIG, article 3, section 9, carton 1, no. 46.

45. Parker 1988; Adams 1990, pp. 28–52. A number of refutations and modifications of Parker's view can be found in Lynn 1990a.

46. Van Creveld 1977, p. 28.

47. Duffy 1985, pp. 291–292.

48. Dejean 1984, p. 65.

49. Faucherre 1996, pp. 64–87.

50. Richet 1973, p. 81, citing R. Mousnier et al.

51. According to Corvisier (1983, p. 363).

52. For example, the Corps des Ponts et Chaussées, whose founding is usually dated 1716, had similar kinds of territorial divisions and hierarchies as the military engineering corps. See chapter 3 above for a discussion of this civil engineering corps. See chapter 8 below for Puységur's reference to the military engineers as his model for developing a more systematic and rational kind of warfare.

53. Duffy 1985, p. 6.

54. Corvisier 1983, p. 188. The chevalier de Fourilles, appointed the following year to the position of Inspector General of the Cavalry, appears not to have left such a mark on history or the language. Was it because he was less successful with an arm that was traditionally more noble and recalcitrant to the imposition of order from above?

55. On Le Notre's gardens at Versailles, see Mariage 1990.

56. Duffy 1985, pp. 11–12, quoting the Dutch Council of State and the Earl of Orrey.

57. Cited by Léonard (1958, p. 10).

58. According to Corvisier (1983, p. 363).

59. Le Michaud d'Arçon (1795, p. 10) talks about "industrie militaire," of which fortification is a branch. For him, fortification must use both science and industry, the latter in the sense of hard work and workmanship that reminds one of Thorstein Veblen. Although d'Arçon thinks that the necessary "bellicose genius" would be lost by too much reliance on "industry" (p. 151), it nevertheless is an important element of the military engineer's trade. In his earlier *Considérations sur l'influence du génie de Vauban dans la balance des forces de l'État* (Strasbourg, 1786, p. 70), he argued that defense must inevitably succumb to the attack since the latter became *industrieuse*, i.e. organized, workmanlike, systematic, relying not on physical bravery but technology and construction.

60. Rousset 1864–1879, vol. 1, p. 396.

61. Hüe de Caligny 1846, pp. 167–187.
62. Duffy 1985, p. 152.
63. Pepper and Adams 1986, p. 168.
64. Duffy 1979, p. 23.
65. Marlborough, cited by Duffy (1985, p. 39).
66. Duffy 1975, p. 90.
67. Ibid., p. 91.
68. Lauerma 1956, p. 140.
69. Camon 1914. The five types of fortification are places de barrage, places de dépôt, centres d'opérations, barrières fortifiés, and camps retranchés.
70. Prost 1991.
71. "Note dictée par l'Empereur sur un projet de fort en terre avec réduit de maçonnerie," ms., AIG, article 21 section 1, §1, carton 9, p. 23.
72. See Duffy 1985, p. 291.
73. Arbelot et al. 1987, p. 15.
74. Duffy 1985, p. 154.
75. Lynn 1980, p. 578. Lynn calls this growth "revolutionary."
76. That number jumped to 1,200 in 1794.
77. The ordinance of 10 March 1759 prevented engineers from serving on the staff. It was during Saint Germain's ministry that this prohibition was lifted (Mention 1884, p. 184).

Chapter 5

1. Charles-Augustin Coulomb, then Captain in the Royal Corps of Military Engineers, in a memorandum addressed to the Comte de Saint-Germain, Minister of War, 1 September 1776, AIG, article 3, section 10, carton 2, no. 5a.
2. Mousnier 1974–1980, pp. 546–558.
3. Mumford 1964, Noble 1977, T. Hughes 1983, and T. Hughes 1998 are only a few of the books that discuss the importance of the managerial functions of engineers.
4. Picon 1992, pp. 47–53.
5. Lundgreen 1990, p. 41.
6. On Sully's organization of military engineering, see Augoyat 1860–1864, vol. 1, pp. 25–27.
7. Vauban 1685. According to Augoyat (1860–1864, vol. 1, p. 63), this work was written around 1680. It was reprinted a number of times in collections of other works on fortifications (including Bernard Forest de Bélidor's *La Science des ingénieurs*, which reprinted extracts of this work).
8. Vauban 1685, introduction.
9. *Instruction pour les Ingenieurs Directeurs des Fortifications des Places du Roy*, n.d. [ca. 1704]. AIG, article 3, section 9, carton 1, no. 3.
10. Vauban 1685, p. 41.

11. Ibid., p. 44.

12. See chapter 2 above.

13. Vauban 1685, p. 45. Presumably these more frequent reports were to be kept only when new work was being done, although there is no explicit indication of this.

14. This particular illustration is one of the interesting drawings of the engineer Claude Masse (1651–1737). For a color reproduction see Hebbert 1997.

15. Vauban 1685, p. 131; Vauban, "Instruction générale sur la fortification de Longvuic," reproduced on p. 21 of Garcin 1972.

16. Vauban 1685, p. 94.

17. Ibid., p. 73.

18. See chapter 9 below.

19. See Gillispie 1957, 1980.

20. Vauban 1685, pp. 15–16.

21. Ibid., p. 61.

22. Ibid., p. 70.

23. See chapter 9.

24. Lamberini 1987.

25. John Rigby Hale, "The first fifty years of a Venetian magistracy: The Provveditori alle Fortezze," in Hale 1983.

26. Lamberini 1987, p. 11. The marrioli, so called because of the type of mattocks they used, were press-ganged peasants who were employed for excavations.

27. Vauban 1685, p. 76.

28. [Le Peletier], "Instruction pour les Ingenieurs en chef," ms., AIG, article 3, section 9, carton 1, no. 5. The solicited replies from Directors Tarade and Valory are dated 11 December 1709.

29. "Memoire des principales choses qu'un Ingenieur en chef doit observer," undated ms., AIG, article 3, section 9, carton 1, no. 7 bis.

30. [Le Peletier], "Instruction pour les ingenieurs subalternes," [1709], ms., AIG, article 3, section 9, carton 1, no. 6a.

31. Lamberini 1987, p. 13.

32. "Memoire des principales choses qu'un Ingenieur en chef doit observer," undated ms., AIG, article 3, section 9, carton 1, no. 7 bis.

33. In addition to conventions on indicating different materials, such as bricks, masonry, earthworks, by different colors, there were conventions on shading introduced by the head of the Mézières school, Chastillon. The principles underlying shading were intended to make it easy for ministers to consult and to have a better idea of elevations, always of central concern to the working military engineer, not the attractive plan views that delighted amateurs. The sun was always considered south of the structure and its rays hit it an angle of 45°. See an analysis of this in Belhoste 1990. Engineers were quick to mock competitors like Montalembert, who did not follow these conventions. They claimed that the plethora of his illustrations were merely intended to impress. See e.g. Le Michaud d'Arçon 1790, p. 114.

34. "Memoire des principales choses qu'un Ingenieur en chef doit observer," undated ms., AIG, article 3, section 9, carton 1, no. 7 bis.

35. "Détails concernant le service de l'Ingenieur en Chef d'une place ou lon fait connoitre 1o la maniere dont il doit rendre compte au Directeur et au Ministe avec l'indication des tems et des circonstances qui luy obligent 2o L'arrangement qu'il peut mettre dans ses papiers pour les avoir sous la main et sans confusion 3o Les differentes formules des etats et mémoires par lesquels il rend compte des ouvrages faits, à faire ou proposées" (ms., [Claude-Antoine-Pierre de Rault de Ramsault], [1738], AIG, article 3, section 9, carton 1, no. 11).

36. "Copies des différentes lettres et décisions des Ministres de la Guerre concernant le service des Officiers du Corps royal du génie dans les places en Languedoc . . . ," ms., AIG, article 3, section 9, carton 1, no. 58.

37. "Copies des différentes lettres . . . ," letter dated 13 May 1763.

38. "Copies des différentes lettres . . . ," letter dated 10 July 1755.

39. "Copies des différentes lettres . . . ," letters of 18 and 26 July 1757.

40. "Ordonnance du Roy pour la conservation des fortifications des places du 30 juillet 1695" transcribed in another ms. "Recueil des lettres du Ministre," AIG, article 3, section 9, carton 1, no. 32.

41. "Copies des différentes lettres . . . ," letter dated 10 October 1754 and 13 September 1753.

42. "Copies des différentes lettres . . . ," letters of 28 March and 17 August 1730 and 21 February 1731.

43. "Copies des différentes lettres . . . ," letter dated 13 May 1763.

44. "Copies des différentes lettres . . . ," letter dated 17 April 1765.

45. "Copies des différentes lettres . . . ," letter dated 2 October 1738.

46. Garcin 1972.

47. For an English translation and commentary on another project of Vauban's, see Hebbert 1997.

48. Cited on p. 31 of Garcin 1972.

49. [Ramsault], ms. "Détails concernant le service . . . ," [1738], AIG, article 3, section 9, carton 1, no. 11.

50. Garcin 1972, pp. 71–84.

51. The service des places also included the actual design and planning of fortresses, according to Louis Le Bègue du Portail, ms. "Mémoire sur le Corps du génie," 22 décembre 1773, Archives de l'Inspection du Génie, article 3, section 10, carton 1, no. 73, p. 66.

52. Le Michaud d'Arçon, Jean-Claude-Éléonor, "Mémoire du Corps du Génie 1776," ms. 491G, Bibliothèque municipale de Besançon.

53. Ibid.

54. Bien 1979, p. 79.

55. Le Michaud d'Arçon, "Mémoire du Corps du Génie 1776," ms. 491G, Bibliothèque municipale de Besançon, p. 213v.

56. Ibid.

57. On du Portail or Duportail, see Kite 1933. For more on du Portail's role and the role of other French military engineers during the American Revolution, see Walker 1981. For his service record and other details, see Blanchard 1981.
58. Louis Le Bègue Du Portail, "Mémoire sur le corps du génie" (22 décembre 1773), ms., AIG, article 3, section 10, carton 1, no. 73, p. 66.
59. Ibid., p. 70.
60. See Vauban 1686.

Chapter 6

1. Williams 2000, p. 658.
2. This article, written after the 1776 reorganization of the Corps of Military Engineering, calls for a spectrum of skills, both intellectual and manual, that is truly Vitruvian. In addition to being competent in the sciences and possessing "genius," the engineer must directly supervise the tasks of construction: "L'exécution exige de lui toutes les connoissances qui y ont rapport, il ne peut en mépriser aucune sans que ses travaux n'en souffrent; maçon, tailleur de pierre, charpentier, forgeron, serrurier; tous ces métiers deviennent les siens tout à tour puisqu'il doit en employer les ouvriers, les éclairer, souvent même les conduire comme architecte militaire ou civil." (*Encyclopédie; ou, Dictionnaire raisonné des sciences, des arts et des métiers, par une société de gens de lettres. Mis en ordre & publié par M. Diderot, & quant à la partie mathématique, par M. d'Alembert*, Lausanne: Les Sociétés typographiques, 1779–1982, vol. 18, p. 713)
3. Vérin 1993.
4. Memorial University (St. John's, Newfoundland) introduced a program in "software engineering" in 1996, which provoked a lawsuit by the Canadian Council of Professional Engineers that was discontinued in 1999 after the parties agreed to set up an independent panel to offer recommendations on the use of the term.
5. This definition of technology is from p. 83 of Alex Roland, "Theories and models of technological change: Semantics and substance," *Science, Technology, and Human Values* 17 (1992): 79–100.
6. On engineering science, see Layton 1971. Vincenti 1990 and Vérin 1993 are also useful.
7. Alder 1997, p. 60 ff. See also Picon 1989, 1987.
8. Huntington 1972, p. 8.
9. Vérin 1993, p. 18 ff.
10. See chapter 9.
11. Williams 2000.
12. T. Hughes 1998.
13. Meadows et al. 1972.
14. Davidson and Cox 1983, p. 73.
15. See Donkin 1936–37; Buchanan 1983, 1989. The romantic vision of Rudyard Kipling of the robust tamer of nature and creator of material civilization, who was not at all a gentleman, was also popular among engineers—see Harvie 1977.

16. Abbott 1988, p. 20.

17. Perrucci and Gerstl 1969.

18. Fourcroy 1794. The revolutionary chemist Antoine-François Fourcroy was a distant relative of Charles-René Fourcroy de Ramecourt, head of the Royal Corps of Military Engineering.

19. Buchanan 1983; White 1999. See also Raymond H. Merritt, *Engineering in American Society, 1850–1875* (University of Kentucky Press, 1969).

20. Layton 1971a.

21. Noble 1977.

22. Smith 1990.

23. Elias 1950.

24. Vérin 1998.

25. On shipbuilding as another major focus of the creation of the modern engineer, see Vérin 1993. Vauban, usually thought of as a fortifier and a siege engineer, was also involved in the construction of ports and arsenals, e.g. at Dunkerque and Toulon. See Peter 1994.

Chapter 7

1. Augoyat 1860–1864, vol. 2, p. 303.

2. Schalk 1986.

3. Corvisier 1978, p. 355.

4. Chaussinand-Nogaret 1976.

5. Jardin and Tudesqu 1973.

6. Doyle 1972.

7. Bien 1974, 1979.

8. Corvisier 1978, p. 353.

9. Arcq 1756.

10. Notably by the Abbé Coyer in *La Noblesse commerçante* (Paris: Duchesne, 1756).

11. Bien 1974, p. 30.

12. Rochas d'Aiglun 1910, vol. 1, pp. 632–641. See also pp. 642–646.

13. Bien 1979 p. 95. Of the 22 generals who were members of the committee advising the Minister of War on military reform that functioned between 1780 and 1784 and who lived into the Revolution, 15 emigrated. Five stayed in France to defend the king and were imprisoned or killed.

14. Corvisier 1978, p. 353. Corvisier does not give a source for this assertion. On the basis of what appears to be a more thorough analysis, Bodinier (1983) claims that military nobles were as liberal as the rest of the nobility, which did have a significant liberal faction at the beginning of the Revolution.

15. Corvisier 1970, p. 78.

16. Bien 1979, p. 80.

17. Corvisier 1970.
18. Langins 1990a, pp. 97–98.
19. Augoyat 1860–1864, vol. 2, p. 303.
20. According to Corvisier (1959), 17.3–39 percent of infantry officers were promoted, vs. 1.5–3.4 percent of engineering officers.
21. For a detailed look at the greatest French collège of all see Palmer 1975. On the ideal of the honnête homme and engineers, see chapter 2 of Weiss 1982a.
22. Bien 1979, pp. 85–86.
23. Vigny 1835. Vigny only spent some time in the army during the Restoration, but his vignettes of the military life should not be ignored.
24. Huntington 1972.
25. For a stimulating lexicographical analysis of the word 'engineer' see chapter 1 of Vérin 1993.
26. Huntington 1972, p. 10.
27. Ibid., p. 16.
28. Vérin 1998.
29. Huntington 1972, p. 19.
30. Guibert 1773, vol. _, chapter 16.
31. An ordinance of 1759 explicitly banning military engineers from serving on the general staff was bitterly attacked by both by d'Arçon and du Portail.
32. Elias 1950, p. 308.
33. Lavisse 1911–1926, vol. 9, p. 53.
34. According to Rosen (1981, pp. 202–203), Saint-Germain did not have a particular commitment in favor of one or the competing "systems" of artillery (Gribeauval's and Vallière's, described in chapter 7 of this volume); he was simply interested in restoring harmony to the artillery corps, and by this time Gribeauval's ideas were winning the battle against Vallière's.
35. On Saint-Germain, see Mention 1884 and Léonard 1958.
36. *Ordonnance du Roi, concernant le Corps du Génie, du 31 décembre 1776* (Imprimerie Royale, 1777), preamble.
37. Ordonnance du 4 décembre 1762, cited by Augoyat (1860–1864, vol. 2, pp. 551–552).
38. Blanchard 1979, p. 213.
39. Memorandum to the government at the end of 1762 cited by Augoyat (1860–1864, vol. 2, pp. 550–551).
40. Blanchard 1979, p. 220. The age of the retirees ranged from 72 to 88. One 48-year-old engineer was being retired for frequent absences from his post.
41. Bureaux de Pusy 1790a, p. 13.
42. Augoyat 1860–1864, vol. 2, p. 595.
43. Augoyat 1836a, pp. 648–666.
44. Augoyat 1860–1864, vol. 2, p. 589.

45. Information compiled from Blanchard 1981 and from volume 2 of Augoyat 1860–1864.

46. Letter from Dajot at Niort to Fourcroy de Ramecourt dated 25 April 1778. AIG, article 3 section 8, carton 1, p. 4.

47. Cited in Louis-Joseph-François-Julien de Robien, "Constitution et service du Corps Royal du Génie," ms. [1781?], AIG, article 3 section 16, carton 2, p. 29.

48. Augoyat 1860–1864, vol. 2, pp. 649–650.

49. Robien, "Mémoire," ms. [1783], writing in the name of a number of unidentified Directors (AIG, article 3, section 10, carton 2, p. 29a).

50. Robien, "Constitution."

51. Just before the ordinance of 1776, de Castries, a lieutenant general in Flanders and a future Minister of the Navy, complained that the engineer at Maubeuge had refused to allow him to see plans of the fortress without specific permission from the Minister of War (AIG, article 3, section 9, carton 1, p. 57a). He was still complaining of this and the lack of coordination of the engineers with the army as a whole after the ordinance (SHAT/A13703, Correspondance Officielle du génie, October–December 1777).

52. Picon 1992, pp. 211–212, 226–227.

53. Reinhard 1950–52, vol. 1, chapter 9. For more on this council see Bien 1974.

Chapter 8

1. *The Foundations of the Science of War* (London, 1925), p. 19.

2. In writing this chapter I relied heavily on Gat 1989. For details of tactics I used Quimby 1957.

3. Gat 1989, pp. 13–24.

4. See Stoye 1994 and Lund 1999.

5. Gat 1989, p. 29.

6. Folard 1726.

7. Quimby 1957, pp. 26–41.

8. Léonard 1958, p. 126.

9. Castel de Saint-Pierre 1713.

10. Puységur 1748, "Avertissement de l'Auteur."

11. Ibid., "Avertissement de l'éditeur."

12. Guibert 1773, vol. 1, p. 4.

13. Based on data on p. 25 of Gat 1989. Gat cites bibliographic work by Pöhler as a source.

14. Order, 26 August 1786 (Augoyat 1860–1864, vol. 2, p. 633).

15. Lynn 1984; Scott 1978.

16. Quimby 1957, p. 4.

17. J. F. C. Fuller, *A Military History of the Western World*, cited by Rosen (1981, p. 14). According to Quimby (1957, p. 9), pikes, after being reduced to one-fifth of infantry formations of the French army in 1688, were abolished in 1703.

18. Rosen 1981, p. 65.

19. Hahn 1964/1986, pp. 513–545.

20. Nardin 1982.

21. Gribeauval's success in reforming the French artillery has led to a rather francocentric view of his achievement and the importance of the Austrian background needs more emphasis. An important exception is Lund 1996.

22. *Ordonnance*, 23 August 1772, cited by Rosen (1981, p. 175).

23. Jean-Paul Charnay, "Sociologie et technique de l'Ordre militaire selon Guibert," in Charnay 1981, p. 101.

24. Dupuget 1772.

25. The colors refer to the colors of the breeches of the old uniform (red) and the new uniform (blue) of the artillery (Chalmin 1968).

26. Besides the work of Rosen and Nardin on Gribeauval, see Favé 1846–1871 on the history of French artillery generally.

27. Lauerma 1956.

28. Nardin 1982, pp. 339–347. On the French effort to develop interchangeability of parts in small arms, see Alder 1997. On the American case, see Hounshell 1984.

29. On "command technology," see McNeill 1982.

30. See Charles Tronson Du Coudray, *L'Artillerie nouvelle* (Amsterdam, 1772); Jean du Teil, *De l'Usage de l'artillerie nouvelle dans la guerre de campagne: connaissance nécessaire aux officiers destinés à commander toutes les armées* (Metz: Marchal, 1778). Jean du Teil's brother Joseph was the commander of the artillery school at Auxonne, where Napoleon served as a young artillery lieutenant.

31. Quimby (1957) argues that Sir Charles Oman's traditional English view is based on the experience of the Peninsular War where much was made of the contrast between the "thin red line" and the French columns. In fact, says Quimby, the French maneuvered in columns but usually the final attack was also a line. Wellington's genius was skillful use of the terrain to catch them with the firepower of his lines while they were maneuvering.

32. Mesnil-Durand was for a time a military engineer, being one of the last of those who could become one without having to pass through the school at Mézières.

33. Quimby 1957, p. 6.

34. Ibid., p. 119, no. 25. Hublot (1987, p. 296) cites Dumouriez, the French commander at the epochal battle of Valmy in 1792, reporting that 20,000 cannon shots had been fired to inflict about 100 Prussian casualties, i.e. about 100 shots per dead Prussian. Assuming the average field cannon to be an 8-pounder, this would result in about 800 pounds of cast iron per dead Prussian, about 5 times their weight in metal.

35. Quimby 1957, chapter 7.

36. According to Léonard (1958, p. 285), this council took over much of the administration of the Ministry of War, and the Minister became much less important.

37. Le Michaud d'Arçon 1774.

38. Guibert 1773, vol. 2, p. 78.
39. Clairac 1749.
40. Reinhard 1950–1952, vol. 2, chapters 5–10.
41. Carnot 1812. The first edition, published in 1810, remained in the depot of the military engineering corps without being distributed to the students at the Metz school of artillery and military engineering.
42. Louis XIV's remark—a reprimand addressed to a tenacious officer defending Béthune in 1710—was reported by Saint-Simon and cited by Léonard (1958, pp. 29–30).
43. See the remarks of Chamlay quoted in chapter 4 above.
44. Puységur 1748, p. 2.
45. Clausewitz 1832/1976, p. 133.
46. Saxe 1757.
47. Cited on p. 30 of Gat 1989.
48. On Saxe, see Quimby 1957, chapter 2; Gat 1989, pp. 29–33.
49. Clausewitz 1832/1976, p. 133.
50. Gat 1989, pp. 67–78.
51. I have found no confirmation of Gat's claim that Lloyd was with the French army at Bergen-op-Zoom in French sources. Lloyd may have been one of the ingénieurs volontaires, still widely used before 1750, and may not have made it into the records.
52. Konvitz 1987, p. 16.
53. Gat 1989, p. 74.
54. Ibid., pp. 79–94.
55. Ibid., p. 90.
56. Gat (1989, pp. 91–93) shows that the geometrical reasoning is in fact wrong and contradictory.
57. Cited by Gat (ibid., p. 82).
58. Guibert 1773, vol. 1, p. xxi.
59. Guibert 1773., vol. 2, p. 88.
60. Guibert 1773., vol. 2, p. 2.
61. Guibert's "petite tactique" and "grande tactique" would roughly correspond to today's "tactics" and "strategy" respectively. "Petite tactique" was "élémentaire et bornée"; "grande tactique" was "composée et sublime," the science of generals.
62. Guibert 1773, vol. 1, p. xx.
63. Ibid., p. vii.
64. Ibid., p. xxxvi.
65. Ibid., p. xliii.
66. Guibert 1773., vol. 2, p. 85.
67. Cited by Reinhard (1950–1952, vol. 1, p. 127).

68. Jean-Paul Charnay, "Sociologie et technique de l'Ordre militaire selon Guibert," in Charnay 1981, p. 105.

69. Guibert 1773, vol. 1, p. xxx.

70. Guibert 1773, vol. 1, p. 83.

71. Ibid.

72. Guibert 1773, vol. 2, p. 78. For more on the etymology of 'génie' and 'ingénieur' see Vérin 1993, chapter 1.

73. Guibert 1773, vol. 2, p. 122.

74. Camon 1914. In spite of his reputation as an apostle of a war of movement, Napoleon derided those who thought that fortifications were of no use whatever. Indeed, they were essential for offensive warfare.

75. Guibert 1773, vol. 2, p. 90.

76. Ibid., p. 98.

77. Louis Le Bègue du Portail, "Mémoire sur le Corps du génie," ms., 22 décembre 1773, Archives de l'Inspection du Génie, article 3, section 10, carton 1, no. 73. Augoyat (1860–1864, vol. 2, p. 585) considers this the "most remarkable" of the memoranda (including those of Coulomb and d'Arçon) addressed to Saint-Germain by military engineers before the reorganization of 1776.

78. Du Portail, "Mémoire sur le Corps du génie," pp. 100–101.

79. Ibid., p. 9.

80. Guibert, a second edition of whose *Essai général de tactique* had appeared in the same year as du Portail's "Mémoire sur le Corps du génie," is mentioned and attacked by du Portail, who mentions having read the *Essai* but claims that his own work is not intended to be a refutation.

81. Du Portail, "Mémoire sur le Corps du génie," p. 10.

82. Ibid., p. 12.

83. Ibid., p. 52.

84. Ibid., p. 19.

85. Ibid., p. 21.

86. Ibid., p. 13.

87. Vauban 1968, p. 30.

88. *Ordonnance du Roi Concernant le Corps du Génie & les Compagnies de Sapeurs & de Mineurs du 10 Mars 1759*, art. XXIX, cited by Augoyat (1860–1864, vol. 2, p. 489).

89. Du Portail, "Mémoire sur le Corps du génie," p. 28.

90. The ordonnance of 31 December 1776 fixed the number of military engineers at 329 (all officers) while the full strength of the seven regiments of the corps of artillery was 12,000 (Gay de Vernon 1805, vol. 2, p. 22). During the Revolution, Gay de Vernon dropped his particle and was simply Citizen Gayvernon.

91. Du Portail, "Mémoire sur le Corps du génie," p. 43.

92. Picon 1992, chapter 5.

93. Du Portail, "Mémoire sur le Corps du génie," p. 65.
94. Ibid., pp. 65–66.
95. Ibid., p. 66.
96. Ibid., p. 81.
97. Ibid., p. 70.
98. Ibid., p. 72.

Chapter 9

1. From pp. 22–23 of the 1830 edition of Bélidor 1729 (published in Paris by Didot with notes by Louis-Marie-Henri Navier).

2. The first two of the four "days" of Galileo's classic *Discourses and Mathematical Demonstrations concerning Two New Sciences* (1638) are on the strengths of materials.

3. Galilei 1974, "Second Day."

4. Timoshenko 1953. See also Heyman 1972.

5. Bélidor 1729, p. 512 ff. The sixth and final book of Bélidor's work ("Manière de faire les devis") is on making estimates and specifications.

6. Bélidor (1729, p. 229) explicitly acknowledges reprinting the pertinent parts of Vauban's *Directeur Général des fortifications*.

7. Grandjean de Fouchy 1761.

8. The *Index Biographique de l'Académie des Sciences* (Gauthier-Villars, 1979) indicates that Bélidor was first named a correspondent of Saulmon (3 May 1722), then Pitot (24 March 1734), and finally Bouguer (sometime before 1748) before becoming an associé libre surnuméraire on 30 March 1756 and a regular associé libre on 26 October of the same year after the death of La Galissonnière. When Bélidor died, his place was taken by the Inspector of the Artillery, Vallière.

9. Another claim to fame during his lifetime was his correct insistence, for which he suffered some harassment in the artillery, that the charge of powder in guns could be reduced by a third. Yet another was his idea, contradicting Vauban, that there was no limit to the effectiveness of mines.

10. The last French edition mentioned in the Bibliothèque nationale catalogue appeared in 1757. The Russian edition was published in Saint Petersburg in 1766–1769.

11. See e.g. Straub 1964.

12. Bélidor 1725, "Préface."

13. Ibid.

14. *La Science des ingénieurs* was to be followed by books on hydraulics and the attack and defense of fortresses. Bélidor did keep his promise regarding a book on hydraulics by publishing *Architecture hydraulique*, but seems not to have published a formal treatise on fortifications. He was, however, active in this field as well and was involved in acrimonious debates with the military engineering corps on this subject. Although his ideas circulated in manuscript—the fortifier Montalembert referred to them at length—it may be that his military superiors ordered him not to publish for reasons of national security.

15. Bélidor says architecture rather than engineering, but he clearly means what today would be called structural engineering.

16. Bélidor 1830, p. 159. This 1830 reprint of the 1813 edition contains notes by Navier.

17. Ibid., p. 302.

18. Ibid., p. 62.

19. A copy of one of these can be found in AIG, article 21, section 7, §4, carton 1, no. 5. It contains a marginal note by Fourcroy de Ramecourt asserting that Vauban had these profiles distributed to all fortresses (followed by a comment by the nineteenth-century archivist colonel Augoyat that there is no proof for this assertion). The date of the original drafting has been given as 1684 or 1687. It is highly probable, however, that these profiles were considered standard by the corps of military engineering.

20. Bélidor 1830, p. 83.

21. Bélidor 1830, p. 91.

22. Bélidor 1729, p. 15.

23. Bélidor 1830, p. 309.

24. Ibid., p. 301.

25. Ibid., p. 87.

26. Ibid., p. 27.

27. Ibid., p. 208.

28. Ibid., p. 248.

29. Ibid., p. 159.

30. Ibid., p. 160.

31. Ibid., plate 9. Plate 10 gives a table for the dimensions of revetments not topped by parapets, something that was useful for more civilian projects such as dikes, elevated roadways, and civilian structures not likely to come under artillery attack.

32. Bélidor 1830, p. 55.

33. Bélidor 1729, book 1, p. 43. Navier did not include this table in his 1830 edition. The formula that is used to calculate the table is rather straightforward.

34. Blanchard (1979, p. 179) mentions that d'Asfeld recommended Bélidor to all his engineers.

35. Blanchard (1981) states that Buchotte (1673?–1757) entered the military engineering corps in 1706.

36. This work seems to have remained in manuscript (Bibliothèque de l'Inspection du Génie, Vincennes, France, ms. 96, "L'ingénieur de place, ou la manière de servir dans les places"). The undated manuscript of the preface is in the military engineering archives at Vincennes, AIG, article 21, section 7, paragraph 4, carton 1, no. 9. Buchotte probably was the author of a published work on drawing dedicated to d'Asfeld: *Les Règles du dessin et du lavis pour les plans particuliers des ouvrages et des bâtimens. . . par M. Buchotte* (Paris: C. Jombert, 1722).

37. AIG, article 21 section 7, paragraph 4, carton 1, no. 9.

38. AIG, article 21 section 1, paragraph 1, carton 3, no. 17.

39. Augoyat (1860–1864, vol. 2, p. 178) quotes a circular letter dated 13 May 1737 from the marquis d'Asfeld to the Directors of Fortification [District Engineers] asking them to get all engineers to subscribe to the first part of Bélidor's forthcoming *Architecture hydraulique*. D'Asfeld's support for the *La Science des Ingénieurs* was mentioned above.

40. See Coulomb, Charles Augustin, "Essai sur une application des règles *de maximis & minimis* à quelques problèmes de statique, relatifs à l'architecture," *Mémoires de Mathématique & de Physique, présentés à l'Académie Royale des Sciences par Divers Savans, & lûs dans ses Assemblés* 7 (1773), pp. 343–82. This classic memoir is reproduced with a translation and an excellent historical commentary in Heyman 1972.

41. Navier, in his notes to the 1830 reprint of *La Science des Ingénieurs*, mentions that Coulomb's ideas were little known and taken up by Prony in 1794 (Bélidor 1830, p. 92).

42. Augoyat (1860–1864, vol. 2, p. 183) refers to a "pont-levis à bielles pendantes" invented by Frézier.

43. Frézier 1737–1739.

44. Frézier 1760, vol. 1, p. iii.

45. Ibid., p. vi.

46. Ibid.

47. Ibid.

48. This is no doubt Jacques-François Blondel (1705–1774), a well-known teacher of architecture, who was also taught architecture to many students at the École des ponts et chaussées.

49. Frézier 1760, vol. 1, p. 2.

50. On the importance of visualization in engineering, see Ferguson 1992. Graphical methods in statics and the strength of materials go back to Varignon and even before. The tradition remained vital among engineers in the nineteenth century with contributions by Poncelet, Clapeyron, Lamé, Carl Culmann, and Otto Mohr. See Timoshenko 1953.

51. The best biography of Monge is Taton 1951.

52. For further bibliographic references to Meusnier's life and work see Gillispie 1980, p. 533 ff.

53. For a family tree of the Chastillons, see Blanchard 1979, p. 197.

54. Belhoste 1990.

55. [Chastillon, Nicolas-François-Antoine de], ms. "Traité des ombres dans le dessin géométral de Chastillon," [1763] located at the Archives du Génie, Vincennes, and cited by Belhoste (1990 p. 115).

56. Cited on p. 112 of Belhoste 1990.

57. Originally there was a clear distinction and little connection between work done in a class, essentially mathematics, and practical exercises like stonecutting and drafting done in workshops and drafting rooms. According to Belhoste, an increasingly theoretical stereotomy taught to the students in conjunction with their practical work blurred the distinction between the salles de théorie and the salles de pratique to some extent. The influence of Bossut and Monge in raising the level of mathematical instruction, however, revived the distinction between theory and practice on a new and higher plane (Belhoste et al. 1990, pp. 73ff.).

58. Augoyat 1860–1864, vol. 2, pp. 550–552.

59. Belhoste 1990, p. 117.

60. Ibid., pp. 120–124.

61. Ibid., pp. 131–134. There are more general histories of engineering drawing such as Booker 1963 and Baynes and Pugh 1981.

62. Sakarovitch 1989.

63. Taton 1951, p. 13.

64. Hall 1983, pp. 13–15.

65. Gayvernon 1802, 28e Leçon.

66. An excellent description of various techniques of defilading, which provides the basis for much of this discussion, can be found in Belhoste 1795. Sakarovitch 1989 is valuable for its description of the techniques of defilading and their influence on descriptive geometry.

67. [Chastillon, Nicolas-François-Antoine de], ms. "Traité de Relief, Commandement, et Defilement de la fortification à l'usage des écoles du Génie à Mézières," n.p., n.d., Bibliothèque de l'Inspection du Génie, Vincennes, folio 116. Augoyat dates, and Belhoste confirms, that this work was written in 1764. There are a number of manuscript copies in the Archives du Génie at Vincennes as well as in the library of the École nationale des ponts et chaussées in Paris. Students at Mézières copied this and other similar works in extenso as part of their course of studies. It is sometimes included with the manuscript copies of Duvignau's treatise mentioned below and catalogued under the latter's name. On the Mézières school, see Taton 1964b.

68. Du Vignau, Antoine-Nicolas-Bernard, ms. "Exercice sur l'art du Tracé, du Relief, de la construction, de l'attaque et de la defense des fortifications composé en 1768 pour l'instruction fondamentale des jeunes officiers Eleves de l'École du génie à Mézières . . . ," Bibliothèque de l'Inspection du Génie, ms. 118a. That it had attentive readers and a long life as a manual is suggested by its eventual publication in the nineteenth century more than 50 years after it was written. See Duvignau 1830. The text has a valuable preface on the history of defilading by the editor, Captain Hanus.

69. Say 1796.

70. Meusnier, Jean-Baptiste, ms. "Mémoire sur la Détermination du Plan de Site," [1777], Archives of the École Nationale des Ponts et Chaussées, ms. 1894. Copies can also be found in the SHAT/AIG, e.g., article 21, section 1, §1, carton 6, p. 12.

71. Augoyat 1860–1864, vol. 2, pp. 568–569. Augoyat also mentions Cruquius, who put depth contours on his Dutch maps of the mouth of the river Maas (Meuse) in 1733 as a precursor, albeit probably unknown, of Buache. Konvitz (1981, p. 98) mentions Halley having used contour lines to represent the magnetic field in 1701, Cruquis and Buache having represented depths in maritime maps in 1720 and 1737 respectively, and Ducarla and Dupain Triel having done the same for terrestrial maps in 1782 and 1790.

72. Belhoste (1992, pp. 545–546) thinks that Monge may have also used contour lines and it is not necessary to assume any influence from Ducarla.

73. Bret 1991.

74. Noizet 1823. On the use of the first contour lines in topography, see Prost 1991, pp. 121–129.

75. Gaspard Monge, "Leçons de Monge: Cinquième Leçon (11 ventóse an III [1 March 1795])," in Dhombres 1992, p. 349.

76. Gaspard Monge, "Mémoire sur la théorie des déblais et des remblais," *Mémoires de l'Académie Royale des Sciences, 1781*, pp. 686–704, Paris, 1784. Monge read the paper to the Academy on 27 January and 7 February 1776 and reread it after his admission as a regular member of the Academy.

77. Taton 1951, pp. 193–204.

78. Noizet 1823, p. 9.

79. Ms. "Exercice sur la fortification," Archives du Génie, Vincennes, article 21 section 1, §1, carton 7 (1775–1797), p. 8. The work is anonymous and bears no date but internal evidence suggests that it was written shortly after the fall of the monarchy in August 1792.

80. I have never seen nor read about a full analysis of defilading as a school exercise or in the design of any fortress.

81. According to Rocolle (1973, vol. 1, p. 484), defilading started to be taken seriously after the development of rifled artillery.

82. Augoyat 1860–1864, vol. 3, p. 582.

83. Ibid., pp. 597–598.

84. Lesage, ms. "Cours et projet de fortification permanente à l'École impériale de l'artillerie et du génie," 1810–1811, Bibliothèque de l'Inspection du Génie, Vincennes, folio 241.

85. Bousmard 1797–1799; Noizet 1792.

86. Bret 1991; Berthaud 1898–1902.

87. Ms. "Exercice sur la fortification," Archives du Génie, Vincennes, article 21 section 1, §1, carton 7 (1775–1797), p. 8, fol. 16.

88. Dorn 1970.

89. Bélidor 1729 includes a testimonial by senior officers of the military engineering corps and by the director of the Academy of Architecture.

90. Bélidor 1830, pp. 22–23.

91. Rouse and Simon 1957.

92. Picon 1992, p. 228.

93. Petot 1958, p. 321.

94. Augoyat 1860–1864, vol. 2, pp. 14, 754, and 638.

95. For a beautiful sampling of contemporary illustrations of the work of the engineers of the ponts et chausses, see Picon and Yvon 1989.

96. See Arthur Young, *Travels in France and Italy during the Years 1787, 1788, and 1789* (1792; reprint: Dent, 1915). See also the results of their work on the eve of the Revolution in Arbelot et al. 1987.

97. Kuhn 1977.

98. Ibid., p. 63.

99. Gillmor 1971.

100. See chapter 12 below.

101. Bensaude-Vincent and Stengers 1992.

Chapter 10

1. Argenson 1859–1867, vol. 9, p. 156. The entry is for 22 December 1755.

2. The most substantial biography is Gaudin 1938. Biographical information can also be obtained from biographical dictionaries such as Arnault 1820 and Michaud and Michaud 1811, but it is not always reliable. I have mostly relied on Gaudin, on Montalembert's service record at the Service historique de l'armée de terre, and on Montalembert himself. See also Martin 1988; Lalande 1800; Desaudray 1800; Delisle de Sales and Sulpice Imbert de Laplatière 1801. There was no éloge of Montalembert at the Institut National (successor of the old Royal Academy of Sciences) since Montalembert, unlike most of the members of the latter institution was not reelected to its successor body.

3. On the Prince de Conti, see Capon 1907.

4. Arnault 1820, "Montalembert, Marc-René."

5. Capon and Yves-Plessis 1907, p. 52.

6. A seat in the Academy became vacant in April 1747 when the king's surgeon, La Peyronie, died. On 13 May, Montalembert read a paper on saltworks to the Academy. Two days later, the king's Minister of the Royal Household with responsibility for the Academy, Maurepas, wrote a flattering letter to Conti about Montalembert's performance. He was elected a member on 8 July.

7. This did not include the needs for cannon in the colonies. See Pritchard 1987a, p. 146.

8. During the first half of the eighteenth century, the sole right to supply cast-iron guns was in the hands of the Landouilette de Logivière family whose two forges (Rancogne and Planchemenier) were located only a few miles from Angoulême (Pritchard 1987a, p. 145).

9. Favé (1846–1871, vol. 4, p. 50) gives the weight of a 36-pounder cast-iron cannon as 7,300 livres (3,573 kilograms, or about 3.6 tonnes).

10. Pritchard 1987a, p. 145. The total number of guns in the ships of the line of the French Navy in 1746 was about 3,800; this gives some idea of the importance of Montalembert's contract. This figure is estimated from p. 39 of Pritchard 1995.

11. On the French metallurgical industry, besides Pritchard 1987a, see Woronoff 1984 and Gille 1960.

12. Decker 1984, p. 146.

13. Notes by Antoine-Marie Augoyat written in the 19th century giving the chronology of Montalembert's dealings with the government regarding his ordnance contracts (AIG/article 21 section 15, §1, carton 1), p. 27.

14. Pritchard 1987a, p. 144. The contract was actually let on the basis of weight of iron. Thus, Montalembert's contract in November 1750 specified 46,680 quintals. (A quintal was 100 livres, or 107.92 English pounds avoirdupois.) An estimate of about 20 million livres for the French naval budget in 1750 is not unreasonable. Government budgets are extremely difficult to esti-

mate in Old Regime France, in large measure because of archaic accounting practices. According to Mention (1884, pp. 9–11), the complexity and incoherence of accounting practices ensure that it is as difficult for modern historians to get accurate figures for budgets as it was for French ministers of finance, such as Necker and Brienne. Thus, it was possible for the Minister of Navy Sartine to appropriate 25 million livres for his department under the nose of as sharp-eyed a minister of finance as Necker. See H. Carré, E. Lavisse, and P. Sagnac, *Louis XVI (1774–1789)*, in Lavisse 1909, vol. 9, p. 65. It was during Sartine's tenure as Minister of the Navy (1774–1780) that naval budgets, generally low relative to the army, ballooned from 25 million to 169 million livres. The best treatment in English of French naval finances during the Old Regime can be found in Pritchard 1987a. A copy of Montalembert's agreement with Brassac to assume his contract with the government for 200 cannon, dated 4 November 1750, can be found in Archives Nationales, Paris, Minutier Central, Etude CXVI (Lenoir) liasse 403, 26 January 1761.

15. Gaudin gives the text of such an official protest.

16. Pritchard 1987a, p. 148.

17. On the involvement of the court nobility in business enterprises, an activity from which they were barred in theory, see Chaussinand-Nogaret 1976, p. 146. Woronoff also speaks of a kind of "entrepreneurial nobility."

18. There seems to be little published information on this prominent family in the history of French metallurgy except some entries in biographical dictionaries on Jean Maritz (1711–1790); see Hoefer 1852–1866. The best treatment in English so far seems to be that on pp. 96 ff. of Rosen 1981. See also Jackson and de Beer 1973; Conturie 1951.

19. The standard blast furnace of the eighteenth century could produce barely enough metal for the 18-pounder cannon: between 2.1 and 2.3 tonnes (Decker 1984, pp. 142–143).

20. Pritchard 1987a, pp. 147–148.

21. Archives Nationales, Paris, Minutier Central Etude CXVI (Lenoir) liasse 366, 30 March 1752.

22. SHAT, Montalembert military service record; Argenson 1859–1867, vol. 9, pp. 149–150, 156, entries for 11 December and 22 December 1755.

23. AIG/article 21 section 15, §1, carton 1, p. 27 (Augoyat, ms. notes).

24. Pritchard 1987a, p. 150; Augoyat, ms. notes.

25. Pritchard 1987a, p. 150.

26. Ibid.

27. Lacombe 1933; Argenson 1859–1867, vol. 9, p. 164, entry for 2 January 1756.

28. Montalembert, ms. "Observation sur la Reponse de Mrs. les officiers d'artillerie de Rochefort au Mémoire de M. le Mis. de Montalembert." AIG/article 21, section 15, §1, carton 1, p. 4.

29. Private communication from M. Michel Decker, 4 July 1992.

30. Ms. reply to artillery officers, AIG/article 21, section 15, §1, carton 1, p. 4.

31. Ms. reply to artillery officers, AIG/article 21, section 15, §1, carton 1, p. 4.

32. Pritchard 1987a, p. 150.

33. Augoyat, ms. notes.

34. Ibid.

35. It is likely that Montalembert did not invent this machine himself but that it was the product of one of his employees, Robert Xavier Ansart (private communication from Decker, 4 July 1992). In a study on Ansart's son, who came to the United States during the Revolution and produced cannon for the America rebels, David B. Ingram, US Artillery (ret.) indicates that Ansart had married Montalembert's sister (Colonel Louis de Maresquelle: The Years before Dracut, typescript).

36. Augouyat, ms. notes; Pritchard 1987a, p. 150.

37. Bigot de Morogues, 1 July 1754, Archives Nationales, Marine B3521.

38. Pritchard 1987a, p. 149.

39. Montalembert from Angoulême 14 December 1753, AIG/article 21, section 15, §1, carton 1, p. 27.

40. Augoyat, ms. notes.

41. Ibid.; Pritchard 1987a, p. 152.

42. Conturie 1951.

43. Augoyat, ms. notes.

44. Ibid.

45. Archives Nationales, Paris, Minutier Central Etude CXVI (Lenoir) liasse 381, 26 October 1755.

46. Gaudin 1938.

47. Ibid. Because of this court case, Montalembert's daughter, Gasparine-Rosalie de Montalembert, gets a separate mention in the biographical dictionary of Arnault, who dates the beginning of the case in 1784.

48. Argenson 1859–1867, vol. 9, pp. 149–150, entry for 11 December 1755.

49. Argenson 1859–1867, vol. 9, p. 156, entry for 22 December 1755. Mauchault had also been Minister of the Navy (replacing Rouillé) and, according to Conturie, was much less favorable to Montalembert than his predecessor.

50. Argenson 1859–1867.

51. Montalembert 1768.

52. Pritchard 1987a, p. 152. This is confirmed by Gaudin, by Conturie, and by my own work.

53. Montalembert 1768.

54. Ibid.

55. Pritchard 1987a, p. 158.

56. Bigot de Morogues, commissaire général for the port of Toulon on an inspection trip to Montalembert's foundries, to the Minister from Angoulême, 1 June 1754 (AN/Marine B3521, fol. 105–106).

57. Villars de la Brosse, ms. "Observations sur les forges de M. le Mis. de Montalembert, Rochefort 2 février 1754, AIG/article 21, section 15, §1 carton 1, p. 27.

58. Conturie 1951.

59. Ibid., pp. 137–138.

60. Pajol from Angoulême 28 November 1758, Archives Nationales, Paris, Marine B3542, fol. 263–264.

61. Fleury de Saint-Charles 1903; Montalembert 1777.

62. Etienne François, duc de Choiseul-Stainville (1719–1785), was one of Louis XV's most powerful and reforming ministers in the domain of the military. Protégé of Mme. de Pompadour, he was also the patron of Gribeauval and had served as the French ambassador to Vienna.

63. Montalembert 1777, vol. 2, p. 140.

64. Grigorian 1968, p. 87. He was elected a member on 22 December 1760, the fourteenth Frenchman to receive this honor.

65. Woodbridge 1985.

66. Hebbert 1990.

67. For his correspondence with various ministers and officials and the field fortifications he proposed around the Château d'Oléron, see volume 3 of *La Fortification perpendiculaire*.

68. This Old Regime rank corresponds to the modern brigadier-general.

Chapter 11

1. Cosseron 1869, p. 282.

2. Montalembert, Marc-René, marquis de, *La fortification perpendiculaire, ou Essai sur plusieurs manières de fortifier la ligne droite, le triangle, le quarré, et tous les polygônes, de quelqu'étendue qu'en soient les côtés, en donnant à leur défense une direction perpendiculaire. Où l'on trouve des méthodes d'améliorer les places déjà construites, et de les rendre beaucoup plus fortes. On y trouve aussi des Redoutes, des Forts et de Retranchemens de campagne, d'une construction nouvelle,* Paris: Philippe Denys-Pierres, 1776. The first volumes were published by Philippe Denys-Pierres, Printer of the King's Great Council and the Collège de France in sumptuous editions—volume 1 in 1776, volume 2 in 1777, and volumes 3 and 4 in 1778. Originally intended as a single volume, the work had overflowed into four. These volumes form the most coherent and systematic part of his work. Volume 5, published in 1784 and originally billed as the "first part" was printed by Alexandre Jombert (le jeune), who also published the "supplement" to the fifth volume, which is now catalogued as volume 6, in 1786. Philippe Denys-Pierres again published volume 7, which appeared as a separate volume with a separate title *Réponse au Mémoire sur la Fortification perpendiculaire, par Plusieurs Officiers du Corps Royal du Génie* in 1787 (see chapter 12). This was Montalembert's contribution in the polemic that moved to the Royal Academy of Sciences. Volume 8, published in 1793 by F. Didot was a collection of pamphlets and papers that had been published separately to which was added a preface. Beginning with volume 8, the title page read *L'Art Défensif supérieur à l'Offensif ou la Fortification perpendiculaire...*" or some variant thereof and all were essentially collections of published works with a new introduction. In 1793, his fortune rapidly shrinking, he also reprinted his earlier volumes with his new title in a considerably less ornate edition. Volume 9 was published in the same year (1793) by a consortium of publishers who included the Cercle Social publishing house, Firmin Didot, Magimel, Volland, and L'Esclapart. Volume 10 was published

by Magimel alone in 1794. Some libraries list as an eleventh volume the various numbers of a journal called *L'Ami de l'Art Défensif, ou Observations sur le Journal Polytechnique* . . .and a pamphlet appearing between 1796 and 1797 (see chapter 13). Subsequent references to this collection of Montalembert's works will take the form Montalembert, *Fort. Perp.*

3. "Préface de l'auteur" and "Discours préliminaire," *Fortification perpendiculaire*, 1761 (printed prospectus, AIG, article 21, section 1, §1, carton 5, p. 10).

4. Montalembert, *Fort. perp.*, vol. 2, chapter 1.

5. Montalembert, *Fort. perp.*, vol. 1, p. xl.

6. Ibid., p. xxxii.

7. Engineers had followed at a "pas de tortue." Montalembert, *Fort. perp.*, vol. 9, p. 84.

8. Montalembert, *Fort. perp.*, vol. 1, p. 69.

9. The average musket had a range of 200–250 meters and the heavier muskets used in the defense of fortresses (the *fusil de rempart*) a range of 250–300 meters in Montalembert's time. The 24-pounder and 16-pounder cannon had average ranges of 1,000–1,200 meters, although their maximum ranges were between 3,000 and 4,000 meters and their point-blank ranges between 400 and 600 meters. These figures are from Gayvernon 1802.

10. Montalembert, *Fort. perp.*, vol. 1, p. 55.

11. *Essai sur plusieurs manières de fortifier la ligne droite, le triangle, le quarré, et tous les polygónes, de quelqu'étendue qu'en soient les côtés, en donnant à leur défense une direction perpendiculaire. Où l'on trouve des méthodes d'améliorer les places déjà construites, et de les rendre beaucoup plus fortes. On y trouve aussi des Redoutes, des Forts et de Retranchemens de campagne, d'une construction nouvelle*, Paris: Philippe Denys-Pierres, 1776.

12. Montalembert, *Fort. perp.*, vol. 1, p. 66.

13. See Pepper and Adams 1986.

14. See figure 2.2.

15. De Ville and Milliet Deschales were among many who regarded casemates as smoky death traps.

16. In a letter to Le Peletier dated 26 February 1711, the engineer Valory says that the experience of casemates at Neuf-Brisach, Vauban's masterwork completed just a few years ago, showed that smoke caused problems (AIG, article 21 section 1, §1, carton 1, p. 15a). Tests on smoke in the casemates of Metz, Perpignan, and Besançon were ordered by the government during the Revolution and the printed results indicate that smoke, while it was a nuisance, did not impede the defenders efforts (*Comité des Fortifications: Procès-Verbaux des Expériences faites sur les effets de la fumée de la poudre, dans les reduits de sûreté* . . . , Paris: Imprimerie nationale, Germinal an III [April 1795] in AIG, article 21 section 1, §1, carton 7 p. 12). It is not impossible that the more tolerant attitudes to casemates and various bunkers may have been encouraged by pressure from above and the participation of artillery officers and the other arms in the trials. On the other hand, there always was some interest in casemates by military engineers themselves in spite of the reigning opinion that they were useless and even dangerous (see chapter 12 below).

17. Montalembert, *Fort. perp.*, chapter 5.

18. As a rough guide, Vauban had suggested 60 cannon and 30 mortars for a hexagonal fortress.

19. Blomfield (1971, p. 43) gives the depth of fortifications as 702 feet at Neuf-Brisach, compared to 260 feet for Fournier earlier in the 17th century. 702 feet is about 110 toises.

20. Montalembert, *Fort. perp.*, vol. 1, p. 143.

21. Duffy (1985, pp. 182–197) gives illustrations and examples of Dahlberg's towers in his discussion of his work.

22. Montalembert, *Fort. perp.*, vol. 2, plates 6 and 7.

23. Ibid., chapter 7.

24. Arguing that expensive bronze cannon costing 200–250 livres/quintal could be replaced by cheaper cast-iron cannon costing between 30 livres/quintal (using current methods) or 15 livres/quintal using his new methods of casting could allow on average a twelvefold increase in the number of cannon for the same cost, Montalembert (*Fort. perp.*, vol. 2, pp. 240–242) postulated that a traditional twelve-sided Vauban fortress with 120 guns would be replaced by 1,440 guns at Montalembert's Louisville. He seems to have been not at all inhibited by the thought that this amounted to more than a third of the firepower of the entire French navy at mid-century.

25. Cosseron 1869, p. 282.

26. Montalembert, *Fort. perp.*, vol. 2, chapter 4, plates 14, 15, 16. See figure 11.8.

27. Montalembert, *Fort. perp.*, vol. 2, p. 304 ff.

28. A senior artillery officer (one of the prominent anti-Gribeauvalists), the marquis de Saint-Auban, rejected Montalembert's calculations for Fort Royal as gross underestimates of the numbers of troops he would need. Saint-Auban claimed that at least three soldiers would be necessary to operate each cannon and that one needed a reserve of 400 as replacements for killed and wounded, not to mention soldiers needed to relieve cannoneers who at times would be in action night and day. On top of this, space would have to be made for the enormous quantities of supplies and munitions for the garrison. (Marquis de Saint-Auban, "Extrait du Journal Militaire du 15 octobre [1779] in *Mémoire sur les nouveaux systèmes d'artillerie*, Paris, n.p., n.d., vol. 2)

29. Bombardment was a relatively imprecise barrage of mortar shells or incendiaries intended to set a city or an arsenal on fire. Battering was a more focused fire with heavy caliber artillery shot to breach walls or fortifications. The former was held in some disrepute because it could be used as a terror weapon to drive civilian populations to put pressure on the garrison to surrender, and was often resorted to by inept besiegers who could not break through the fortifications and penetrate the defensive perimeter of the fortress.

30. Cosseron 1869, pp. 279–280.

31. Montalembert, *Fort. perp.*, vol. 6, pp. 2–3.

32. Montalembert (*Fort. perp.*, vol. 9, p. 171) mentions the date of October 1777. However, according to another reference, he was ordered to do this in October 1778 (Ms. "Note concernant la date des différents projets de fortification . . ." by Montalembert, Archives nationals, Marine, C7 215, dossier Marc-René Montalembert). But this appears to be a slip and the date of October 1777 appears to be the right one.

33. According to Montalembert's reply to Sartine from Maumont dated 23 January 1778. AIG, article 8, section 1, île d'Aix, carton 1, no. 57/1.

34. Ms. "Mémoire sur la nécessité et sur les moyens de fortifier l'Isle d'Aix dans le Rade de Rochefort" by Montalembert, January 1778, AIG, article 8, section 1, île d'Aix, carton 1, no. 57/2.

35. Montalembert, *Fort. perp.*, vol. 3, p. 73 ff.

36. AIG, article 8, section 1, île d'Aix, carton 1, no. 57/2. (In a later memorandum (14 and 15 February 1779) Montalembert estimated the cost of the masonry fort as 900,000–1,000,000 livres (AIG, article 8, section 1, île d'Aix, carton 2, no. 4/1). Fourcroy, in his initial evaluation of Montalembert's masonry fort, had estimated the cost at 1,800,000 livres. (AIG, article 8, section 1, île d'Aix, carton 1, no. 64/2).

37. AIG, article 8, section 1, île d'Aix, carton 1, no. 57/1.

38. Ibid.

39. Sartine to Montbarey, 12 June 1778, AIG, article 8, section 1, île d'Aix, carton 1, no. 57/3

40. Ms. "Observations" by Gribeauval, n.d. [1778], AIG, article 8, section 1, île d'Aix, carton 1, no. 58.

41. Picciola 1999.

42. Montalembert to Monsieur le Comte [de Maurepas?], 1 August 1778, AIG article 21, section 1, §1, carton 7, no. 24.

43. Ibid.

44. Ibid.

45. Montalembert to Monsieur le Comte [de Maurepas?], 1 August 1778, AIG article 21, section 1, §1, carton 7, no. 24.

46. For jibes about Maurepas' being an eager spectator of the performances of Montalembert's plays in his private theatre, where his pretty wife often took the main roles, see Le Michaud d'Arçon 1790.

47. Ms. "Note concernant la date des différents projets de fortification . . ." by Montalembert, Archives nationales, C7 215, dossier Marc-René Montalembert.

48. Ms. "Mémoire sur la Défense de l'Isle Daix," 14 January 1779, AIG, article 8, section 1, île d'Aix, carton 1, no. 59. In the same portion of his work where he had discussed coastal defenses and his first (masonry) fort, he had also criticized Vauban's wooden forts at Dunkerque as less solid and just as expensive as masonry (Montalembert, *Fort. perp.*, vol. 3, p. 92).

49. Montalembert to Ségur, 9 December 1780, AIG, article 8, section 1, île d'Aix, carton 2, no. 1/1.

50. Ms. "Mémoire sur la Défense de l'Isle Daix," 14 January 1779, AIG, article 8, section 1, île d'Aix, carton 1, no. 59.

51. Ms. "Mémoire sur la Défense de l'île d'Aix," 15 February 1779, AIG, article 8, section 1, île d'Aix, carton 1, no. 60/1.

52. Indeed, Montalembert refers to warships as "forteresses mobiles," which are invincible unless confronted by his more powerful fortresses on land (Ms. "Mémoire sur la nécessité et sur les moyens de fortifier l'Isle d'Aix dans le Rade de Rochefort" by Montalembert, January 1778, AIG, article 8, section 1, île d'Aix, carton 1, no. 57/2).

53. According to Fourcroy de Ramecourt (Ms. "Mémoire sur les ouvrages de fortification proposés à faire pour la défense de l'isle daix et de sa Rade," 15 November 1779, AIG, article 8, section 1, île d'Aix, carton 1, no. 64/2), Montalembert had requested 140 cannon for the fort itself and an additional 55 cannon and 8 mortars for a covering battery. This would be the normal complement of a large fortress.

54. There was 3 feet 7 inches (French measure) of earth on the roof of the fort and an additional 4 feet and 7½ inches for the rampart (Seuillet report, AIG, article 8, section 1, île d'Aix, carton 2, no. 2.)

55. A report of 1 messidor year 9 [20 June 1801] suggests that there were 24 cannon on the landward side, considerably fewer than on the seaward side. AIG, article 8, section 1, île d'Aix, carton 2, no. 62/1.

56. Montalembert to Montbarey, 21 May 1779, AIG, article 8, section 1, île d'Aix, carton 1, no. 62/1.

57. Ms. "Mémoire sur les forts en bois," 25 May 1779, AIG, article 21, section 1, §1, carton 7, no. 23.

58. Coulomb to Fourcroy, 28 August 1779, AIG, article 8, section 1, île d'Aix, carton 1, no. 63.

59. Ms. "Note particulière relative à l'Estimation de quatre à 500 mille Livres de dépense pour La construction du fort provisionel," 1779, AIG, article 8, section 1, île d'Aix, carton 1, no. 61/4.

60. Fourcroy claimed that he had not obtained a detailed plan of the Aix fort when asked for an opinion on it at some unspecified date (but presumably well after work had begun) by Sancquer, executive assistant dealing with engineering matters to the Minister of War. Fourcroy to Sancquer on 15 September 1779, AIG, article 8, section 1, île d'Aix, carton 1, no. 64/1.

61. Ms. "Observations" by Gribeauval, n.d. [1778], AIG, article 8, section 1, île d'Aix, carton 1, no. 58.

62. AIG, article 8, section 1, île d'Aix, carton 1 p. 64/1.

63. Ibid..

64. Fourcroy, ms. "Memoire sur les ouvrages de fortification entrepris à l'Isle d'Aix relatif à l'Etat ou se trouvent ces ouvrages au mois de Xbre. 1779," 31 December 1779. AIG, article 8, section 1, île d'Aix, carton 1, p. 67/1.

65. Montalembert, *Fort. perp.*, vol. 9, "Avant-Propos," pp. x–xiii.

66. On Coulomb, see Gillmor 1971.

67. Montalembert, *Fort. perp.*, vol. 9, "Avant-Propos," p. xi.

68. Coulomb to Fourcroy Aix, 28 August 1779. AIG, article 8, section 1, île d'Aix, carton 1, p. 63/2.

69. Coulomb to Fourcroy Aix, 20 October 1779. AIG, article 8, section 1, île d'Aix, carton 1, p. 63/3.

70. Coulomb to Fourcroy Aix, 28 August 1779. AIG, article 8, section 1, île d'Aix, carton 1, p. 63/2.

71. Coulomb to Fourcroy Aix, 20 October 1779. AIG, article 8, section 1, île d'Aix, carton 1, p. 63/3.

72. Montalembert, *Fort. perp.*, vol. 9, "Avant-Propos," p. xiv.

73. Ibid.

74. Ibid., p. xv. Although Coulomb was only a corresponding member of the Academy in 1779, he was elected a full member on 12 December 1781.

75. Coulomb to Fourcroy Aix, 20 October 1779. AIG, article 8, section 1, île d'Aix, carton 1, p. 63/3. Documentation that Carpilhet supplied Fourcroy with information has not been found in the military engineering archives. There is clear evidence, however, that Coulomb did.

76. [Coulomb], ms. "Description du fort de l'Isle d'Aix," AIG, article 8, section 1, île d'Aix, carton 1 p. 66 and the letters mentioned earlier (ibid., p. 63).

77. Coulomb to Fourcroy Aix, 20 October 1779. AIG, article 8, section 1, île d'Aix, carton 1, p. 63/3.

78. Fourcroy, ms. "Memoire sur les ouvrages de fortification entrepris à l'Isle d'Aix relatif à l'Etat ou se trouvent ces ouvrages au mois de Xbre. 1779," 31 December 1779, AIG, article 8, section 1, île d'Aix, carton 1 p. 67/1.

79. Ibid.

80. Ibid.

81. Montalembert to Montbarey, 29 February 1780, AIG, article 3, section 8, §1, carton 1, no. 5/22.

82. Ibid.

83. Montalembert, "Observations sur un memoire relatif aux ouvrages de fortification entrepris à l'Isle d'Aix," 10 February 1780, AIG/Art.8, section 1, île d'Aix, carton 1, no. 69/2, p. 5.

84. A number of Montalembert's plays were published: *La Statue, comédie en 2 actes, en prose, mêlée d'ariettes, par M. le marquis de Montalembert, musique de M. de Cambini, représentée, pour la première fois, sur le théâtre de l'hôtel de Montalembert, au mois d'août 1784*, n.p., 1786; *La Bohémienne supposée, comédie en 2 actes, mêlée d'ariettes, par M. le marquis de Montalembert,... musique de M. Thoméony, représentée, pour la première fois, sur le théâtre de l'hôtel de Montalembert, le 7 mars 1786*, n.p., 1786; and *La Bergère de qualité, comédie en 3 actes, mêlée d'ariettes, par M. le marquis de Montalembert,... musique de M. de Cambinij, représentée, pour la première fois, sur le théâtre de l'hôtel de Montalembert, le 24 janvier 1786*, n.p., 1786. Besides plays, he authored the quasi-philosophical work *Essai sur l'intérêt des nations en général et de l'homme en particulier...*, n.p., 1749. His wife, Marie-Joséphine de Comarieu, marquise de Montalembert, was also the author of *Horace, ou le Château des ombres* (Paris: Maradan, 1823).

85. On Choderlos de Laclos, along with the classic biography by Dard 1905, Poisson 1985 can be consulted with profit. It has been suggested that Choderlos de Laclos was writing his famous novel in his leisure moments at l'île d'Aix, but Poisson (p. 117) thinks that the time at l'île d'Aix may have been an interruption in the writing.

86. Choderlos de Laclos, ms. "Mémoire en réponse pour la partie de l'artillerie à celui présenté au Ministre sur les travaux de l'isle d'Aix par M. de Fourcroy," n.d. [1780], [p. 10]. AIG, article 8, section 1, île d'Aix, carton 1, p. 70.

87. Laclos, "Mémoire," pp. 1–2.
88. Ibid., p. 1.
89. Ibid., p. 5.
90. AIG, article 8, section 1, île d'Aix, carton 2, no. 4/1.
91. Montalembert, *Fort. perp.*, vol. 9, "Discours préliminaire," p. vii.
92. Montalembert, "Mémoire sur l'effet du canon dans les casemates, avec le Procès-verbal de l'épreuve faite du Fort en bois de l'île d'Aix, le 7 octobre 1781, lu à l'Académie Royale des Sciences le 29 Novembre 1783."
93. Montalembert, *Fort. perp.*, vol. 5. The more extended paper to the Academy, which included the complete procès-verbal, was published in volume 8.
94. Ms. "Mémoire [on de Voyer's letter of 7 october 1781]," 15 October 1781. AIG, article 8, section 1, île d'Aix, carton 2, p. 6/2. This memorandum was probably written by Fourcroy.
95. de Voyer to Ségur, 13 February 1782, AIG, article 8, section 1, île d'Aix, carton 2, no. 8/3.
96. Ibid.
97. Meulan d'Ablois, 16 September 1780, AIG, article 8, section 1, île d'Aix, carton 1, no. 75/4.
98. Montalembert to Montbarey, 3 October 1780, AIG, article 8, section 1, île d'Aix, carton 1, no. 75/6.
99. Copy of letter from Choderlos de Laclos to Latour Dupin, 1 January 1783, AIG, article 8, section 1, île d'Aix, carton 2, no. 16.
100. Alexander (1981) argued that Vauban's defense of royal absolutism was somewhat unique in that he felt that the justification for absolutism was not traditional deference to authority, but for the possibilities royal absolutism held for improving the lot of the king's subjects.
101. Ms. "Memoire sur l'Isle d'Aix par M. le Mis. de Montalembert," [more precise date] 1782, AIG, article 8, section 1, île d'Aix, carton 2, no. 12/2.
102. de Voyer to Ségur, 25 July 1782, AIG, article 8, section 1, île d'Aix, carton 2, no. 11/2.
103. Ms. "Mémoire sur l'Isle d'Aix" by Latour Dupin, 25 February 1783, AIG, article 8, section 1, île d'Aix, carton 2, no. 23/1.
104. Seuillet to Fourcroy, 6 January 1783, AIG, article 8, section 1, île d'Aix, carton 2, no. 18. Coulomb, who was transferred to Lille for 1780–81, probably left Rochefort in late 1779.
105. Ms. "Memoire," 16 February 1783, AIG, article 8, section 1, île d'Aix, carton 2, no. 19/2.
106. Ms. "Memoire sur l'isle d'Aix," 8 February 1783, AIG, article 8, section 1, île d'Aix, carton 2, no. 20/1. (Seuillet had mentioned leaks almost a year earlier.)
107. Ibid.
108. Montalembert to the Minister, 29 September 1783, AIG, article 3, section 1, §1, carton 1, no. 5/1.
109. Jean-Baptiste de Pinsun (1743–1799), Director of Fortification for Aunis, 26 December 1783, AIG, article 8, section 1, île d'Aix, no. 30/2.

110. Letter and memorandum of Montalembert to Minister of War, 7 October 1787, AIG, article 8, section 1, île d'Aix, carton 2, nos. 40/1 and 40/2.

111. Letter and memorandum from Montalembert, n.d., SHAT ms. 1740(37).

112. Ms. reports dated 10 July 1793, AIG, article 8, section 1, île d'Aix, carton 2, nos. 46/1–46/3.

113. This seems to have occurred in the first decade of the nineteenth century, although the fort had been renovated and modified during the Revolution. Napoleon ordered new fortifications for the Rochefort roadstead at this time. They included a new fort on the northern tip of the island as well as a fort on the sandbank of Boyard, which was only completed under the regime of Louis-Philippe. The Boyard fort does indeed look like a moored stone ship, very much a Montalembert fortress.

114. Ms. "Memoire sur l'Isle d'Aix, 1782, AIG, article 8, section 1, île d'Aix, carton 2, no. 12/2.

Chapter 12

1. Le Michaud d'Arçon 1786, p. 9.

2. Montalembert, *Fort. perp.*, vol. 5, p. 325.

3. Ibid., p. 325.

4. Montalembert, *Fort. perp.*, vol. 1, "Avant Propos," p. xx.

5. Montalembert, *Fort. perp.*, vol. 10.

6. Montalembert to Monsieur le Comte [de Maurepas?], 1 August 1778, from Bourbonne, AIG article 21, section 1, §1, carton 7, no. 24.

7. He referred specifically to the engineers Chambre and Sauvagère and claimed that the latter had been hounded from the corps for his support of Montalembert. "Discours préliminaire," Montalembert, *Fort. perp.*, vol. 9, p. lxxii ff. and p. lxi ff. Although Sauvagère is identified as a chief engineer at Oléron in 1761, he is not included in the authoritative *Dictionnaire des Ingénieurs militaires 1691–1791* (Blanchard 1981).

8. E.g. "Avant-Propos," Montalembert, *Fort. perp.*, vols. 5–7.

9. "Avant-Propos," Montalembert, *Fort. perp.*, vol. 5, pp. xxvii–xviii.

10. "Discours préliminaire," Montalembert, *Fort. perp.*, vol. 5, p. iv.

11. Dossier Montalembert, Archives of the Academy of Sciences, Paris. Praise of a sample of paper from Montbron near Angoulême (a mill that was probably owned by Montalembert). Académie des Sciences, procès-verbaux, 30 July 1780.

12. Montalembert 1766, 1755. His solution to spinning was proper padding of the cannonball when inserted in the bore. On a foray by Montalembert into icthyology, see Romero 1999.

13. Dossier Montalembert, Archives of the Academy of Sciences, Paris, item dated 13 March 1776.

14. On Mesmer, see Gillispie 1980 and Darnton 1968. On Marat, see Gillispie 1980.

15. Dossier Montalembert, Archives of the Academy of Sciences, Paris, item dated 13 August 1777.

16. Montalembert had submitted volume 5 on 30 August 1783, the report is dated 2 April 1784 (Dossier Montalembert, Archives of the Academy of Sciences, Paris). The book was published in 1784.

17. Le Roy report, 2 April 1784, Dossier Montalembert, Archives of the Academy of Sciences, Paris.

18. On this controversy, see Alder 1997 and Nardin 1982.

19. Montalembert, *Fort. perp.*, vol. 5, chapter 2. He also rejected the gun carriages of the Gribeauvalists for field artillery ("Avant-Propos," ibid., p. xiv).

20. Marquis de Saint-Auban, *Mémoire sur les nouveaux systèmes d'artillerie*, 2 vol., Paris, s.d. The second volume of this work contains a reprint of an article by the author on Montalembert in the *Journal Militaire et Politique* of 15 October 1779.

21. Montalembert, "Mémoire sur l'effet des canons dans les casemates, avec le procès-verbal de l'épreuve faite de celles du fort en bois de l'isle d'Aix, le 7 octobre 1781, lu à l'Académie royale des sciences, le 29 novembre 1783," par M. le marquis de Montalembert (Paris: A. Jombert jeune, 1784); Montalembert, *Fort. perp.*, vol. 9. This publication was reprinted from volume 8. The minutes of the Académie des Sciences indicate that the paper was actually read on 29 September 1783 rather than on 29 November.

22. Russell 1965. The epic siege of Gibraltar, which made it a symbol of impregnability, was followed by all of Europe, and d'Arçon's "floating batteries" attracted a great deal of attention. See the apologia in the controversy over the failed attack with his new invention in Le Michaud d'Arçon 1783. The siege even provided fodder for one of the stories of Baron Münchausen.

23. Académie des Sciences, procès-verbaux. For Montalembert's remark about his absence, see Montalembert, *Fort. perp.*, vol. 7, "Avant Propos," p. 7.

24. Academy report on Fourcroy's memoir, Académie des Sciences, procès-verbaux, 16 March 1785; text reproduced by Montalembert in *Fort. perp.*, vol. 7, "Avant Propos, p. 14.

25. Ibid.

26. According to the *Index Biographique des Membres et Correspondants de l'Académie des Sciences du 22 décembre 1666 au 15 novembre 1954* (Gauthier-Villars, 1954), Fourcroy de Ramecourt was elected on 28 November 1784. He had already been a correspondent of the abbé Nollet since 1767. Just a month after he read his paper, on 23 April 1785, he was reconfirmed as an associé libre of the Academy after its administrative reorganization by Lavoisier, according to *Tableau Chronologique de l'Académie Royale des Sciences de Paris, depuis son établissement en 1666, jusqu'en 1774* (copy with handwritten additions in archives of Académie des Sciences). It is possible that the later date may be the official confirmation of the election.

27. Fourcroy de Ramecourt, *L'art du chaufournier*, 1766.

28. *Index biographique* des membres et correspondants de l'Académie des Sciences du 22 décembre 1666 au 15 décembre 1967 (Gauthier-Villars, 1968).

29. According to the minutes of the Academy for 1786, papers were also presented by the engineers Flachon de la Jomarrière and de Rozières, in addition to those by engineer members of the Academy.

30. Commissioners named "au scrutin" for examination of "la seconde volume partie de son cinquième volume" at Montalembert's request. Académie des Sciences, procès-verbaux, 9 March 1785.

31. Montalembert, *Fort. perp.*, vol. 6, p. 339 ff.

32. [Prony, Gaspard-Clair-François-Marie Riche de], ms. "Observations sur les objections faites contre la solidité de la Caponniere casematée decrite dans la fortification perpendiculaire de M. De Montalembert, Tome 1 page 139 et suivantes," Archives of the *École nationale des Ponts et Chaussées*, ms. 1738. Although the work is not signed, the catalogue indicates that this was written by Prony. That Montalembert had contacts with Prony, a member of the archrival Corps des Ponts et Chaussées, is indicated by another letter by Montalembert to Prony from Paris, dated 21 August 1791, where Montalembert solicits advice on a hydraulic machine he wants to send to Santo Domingo that would have the motive force of two mules of average strength. Ibid., ms. 977(811).

33. Académie des Sciences, procès-verbaux, 7 April 1786.

34. Carnot 1784. For a detailed analysis of the eulogy, see Duthuron 1940.

35. The complete reprint of Carnot is interspersed with Montalembert's lengthy "observations." [Montalembert, M.-R. de], *Éloge de Sébastien Le Prestre, chevalier, seigneur de Vauban . . . Ouvrage enrichi d'observations par un amateur* (La Haye, 1786).

36. The prize was offered on 25 June 1785, according to the editor of Choderlos de Laclos, *Oeuvres Complètes* (1951).

37. The work, dated 21 March 1786, is reprinted in Choderlos de Laclos 1951.

38. Dard (1905, pp. 8, 117) and Reinhard (1950, vol. 1, p. 111) refer to additional unpublished manuscripts on the same subject in the military engineering archives.

39. Le Michaud d'Arçon 1786, p. 3.

40. Ibid., p. 9.

41. Ibid., p. 58.

42. This was the date Fourcroy's book was presented by Condorcet to the Academy (Académie des Sciences, procès-verbaux).

43. One of them was Firmin-Paul-François de Bosquillon de Frescheville (1747–1807). His calculations on the cut and fill necessary for Montalembert's fortresses that Fourcroy de Ramecourt used can be found in AIG/article 21 section 1, §1, carton 7, p. 4.

44. Fourcroy 1786, p. 37.

45. Ibid., p. 9.

46. Ibid., p. 23.

47. Montalembert (*Réponse au Mémoire sur la Fortification Perpendiculaire par plusieurs Officiers du Corps Royal du Génie, Présenté à l'Académie Royale des Sciences*, Paris: Philippe-Denys Pierres, 1787, p. 277) says that he had done this in an earlier volume of his work (Montalembert, *Fort. perp.*, vol. 3, p. 136.). The *Réponse* is catalogued as volume 7 of *La Fortification perpendiculaire*.

48. Fourcroy 1786, p. 101.

49. Ibid., p. 12.

50. Ibid., p. 7.
51. Ibid., p. 118.
52. Ibid., p. 109.
53. Ibid., p. 108.
54. Ibid., p. 6.
55. Ibid., p. 108.
56. Ibid., p. 112.
57. Ibid., p. 99.
58. See chapter 4 above. He indicated the existence of 130 such siege journals in the military engineering archives, where most of them can still be consulted today (ibid., p. 19).
59. Ibid., p. 17.
60. Ibid., p. 38.
61. Ibid., p. 118.
62. Ibid., pp. 41 and 71.
63. Montalembert, *Fort. Perp.*, vol. 3, p. 115.
64. Fourcroy 1786, pp. 73 and 159. Fourcroy's assertions are also confirmed by Wellington's commander of engineers, General Jones, in his descriptions of sieges during the Peninsular War.
65. Fourcroy 1786, p. 66. Fourcroy refers to *Mémoires de l'Académie Royale des Sciences*, 1769, p. 235 ff. to support his contention.
66. Fourcroy 1786, p. 144.
67. Ibid., p. 66.
68. Ibid., p. 178.
69. Ibid., p. 93.
70. These trials are no doubt those conducted by the engineer Elie-Marie Pierron de la Chambre (1732–1800). Pierron's reports on the trials, which are generally optimistic about evacuating smoke from casemates by means of special chimneys with thermally induced drafts can be found in the military engineering archives. (Ms. "Mémoire et projet pour expulser la fumée de la poudre dan s les casemates voûtés," AIG article 21, section 1, §1, carton 6, no. 16). Another report by Pierron, Ibid., no. 17, described trials on his proposal, and yet another (Ibid., no. 18) was a proposed design for vaulted bomb-proof batteries. Fourcroy de Ramecourt was supportive of the scheme for eliminating smoke but called for more trials. The matter does not seem to have been followed up.
71. Fourcroy 1786, p. 87.
72. Ibid., p. 88.
73. Ibid., p. 91.
74. Ibid., p. 118.
75. The title page of Montalembert's reply to Fourcroy bears the date 1787. Montalembert appears to have published an earlier edition of the book before the Academy's judgment. Condorcet's verification of the report of the Academy's commissioners dated 14 April 1788 is

printed at the end of the book, and a separately paginated "Avertissement," "Avant-Propos," and "Discours préliminaire" also contain material dating from 1788: a letter from Carnot dated 21 August 1788 and a note by Montalembert asking the reader to delete an angry jab at Carnot, who had become his supporter, on p. 10 of the text.

76. La Rochefoucauld is also mentioned as a member of the commission in the index of reports in the archives of the Academy of Sciences, but he does not appear to have signed the report.

77. Montalembert, *Réponse* . . . , "Avant-Propos," pp. 19–20.

78. Dossier Montalembert. Archives de l'Académie des Sciences.

79. Montalembert, *Réponse* . . . , "Discours préliminaire," p. xxiii. The offending passage followed a sentence lambasting the incoherent ideas of the *Mémoire sur la fortification perpendiculaire* and casting doubt on their origins: "On ne peut alors les attribuer avec certitude qu'à celui (M. de Fourcroy) qui s'est chargé de les présenter à l'Académie." Montalembert restored this passage to the text in the second edition of the work during the Revolution.

80. Académie des Sciences, procès-verbaux.

81. The letter that Carnot wrote to Montalembert (discussed below and published in Montalembert's reply to Fourcroy's book) says that it was sent after Carnot had read Montalembert's reply. Carnot to Montalembert, no date, no place, cited in Montalembert, *Réponse* . . . , "Avertissement," pp. 11–12. (Montalembert later reprinted this letter in volume 9 of his works and indicated that it was written in Bethune on 21 August 1788.)

82. Carnot 1786.

83. Reinhard 1950, vol. 1, p. 112.

84. Carnot, *Observations*, cited by Reinhard (1950, vol. 1, p. 112).

85. Carnot to Montalembert, 21 August 1788, cited in Montalembert, *Réponse* . . . , "Avertissement," pp. 11–12.

86. Reinhard 1950, vol. 1, chapter 10.

87. Montalembert, *Réponse* . . . , "Avant-Propos," pp. 16–17.

88. Louis Félix Guinement de Kéralio (1731–1793) was editor of the journal. His brother Auguste Guinement de Kéralio (1715–1805) was the military tutor of the Prince of Parma. I am grateful to Annie Geffroy for this clarification.

89. Montalembert, *Réponse* . . . , "Avant-Propos," p. 8 ff. Kéralio's article is in *Journal des Sçavans*, Février 1787, pp. 75–82. A more balanced review, which did not take a position on the polemic and merely asserted that it was a useful exchange of ideas, appeared in *Le Mercure* (31, 5 août 1786, p. 32).

90. Montalembert, *Réponse* . . . , "Avant-Propos," pp. 21–23.

91. Montalembert, *Réponse*, 1787, p. 268; "Avant-Propos," pp. 11–12.

92. Vauban, cited in Montalembert, *Réponse* . . . , pp. 22–23.

93. Montalembert, *Réponse* . . . , p. 297.

94. Ibid., p. 280.

95. Fourcroy 1786, p. 101.

96. Montalembert, *Réponse* . . . , pp. 71–72, 281.

97. Montalembert, *Fort. perp.*, vol. 5, 1784 (reprint: Magimel, 1793), p. 261.

98. Fourcroy 1786, p. 105.

99. Montalembert, *Réponse . . .* , p. 278.

100. Ibid., "Discours préliminaire," pp. ii–iii.

101. Ibid., p. vi.

102. Ibid., pp. iii–iv.

103. Le Chevalier de Kéralio, "Mémoires sur la Fortification perpendiculaire, par plusieurs Officiers du Corps Royal du Génie," *Journal des Sçavans*, Février 1787, pp. 75–82.

104. Fourcroy 1786, pp. 159–160.

105. Montalembert, *Réponse . . .* , pp. 106–107.

106. Fourcroy 1786, p. 25.

107. Ibid., p. 109.

108. Ibid., p. 82.

109. Montalembert, *Réponse . . .* , "Avertissement," p. 8.

110. Montalembert, *Réponse . . .* , pp. 310–311.

111. According to Montalembert's admirer and disciple A.-P. Julienne de Belair (1787, p. xviii).

112. Jean-Charles de Montalembert, a member of the cadet branch of the Montalembert family, became the head of its senior branch when Marc-René de Montalembert died. The baron was an ardent royalist who emigrated during the Revolution and fought with the English army (attaining the rank of general) and died in Trinidad. He was the translator of Carnot's works on fortification and the father of the eminent liberal monarchist and Catholic Marc-René-Anne-Marie, comte de Montalembert (1777–1831), who served as an ambassador under the Restoration and is the progenitor of the present-day main branch of the Montalembert family.

113. Le Baron de Montalembert, *Lettre de M. le Baron de Montalembert à M. de Keralio, en Réponse au Compte qu'il a rendu dans le Journal des Savans, du Mémoire sur la Fortification perpendiculaire, publié sous le nom de plusieurs Officers du Corps Royal du Génie, en 1786*, London: T. Spilsbury et Fils, n.d., p. 2, in Marc-René de Montalembert, *Fortification perpendiculaire*, vol. 9, Paris, 1790.

114. Baron de Montalembert, *Lettre . . .* , p. 7.

115. Ibid., p. 13.

116. Ibid., p. 62.

117. Ibid., pp. 61–62.

118. Ibid., p. 17.

119. Ibid., pp. 15–16.

120. Ibid., p. 24.

121. Ibid., pp. 25–26.

122. Ibid., p. 28. For an uncannily similar evaluation of the working habits of the Academy in session, see the judgment of Franz Anton Mesmer, cited in Ernest Maindron, *L'Académie des Sciences* (Paris: F. Alcan, 1888), pp. 58–59.

123. Marat 1791.
124. Rosen 1981, p. 181.
125. Darnton 1968, appendix 6.
126. Baker 1975.
127. Condorcet 1847.
128. Ibid., p. 445.
129. Reinhard 1950, vol. 1, chapter 11.

Chapter 13

1. Mirabeau to Montalembert, Paris, 11 August 1789, quoted in Montalembert, *Fort. perp.*, vol. 9, "Avant-Propos," pp. xxxi-xxxii.

2. On Cherbourg, see Chantereyne 1900; Demangeon 1978; Lepelley 1990.

3. Report dated Compiègne 18 August 1763 and cited on p. 56 of Jean-Charles de Montalembert n.d.

4. Reinhard 1950, vol. 1, pp. 54–57. Lepelley (1990) mentions that orders to construct Homet and Pelée were issued on 3 July 1779.

5. Gillispie 1980, p. 533 ff.

6. On his experiments with red hot cannonballs, see ms. "Détails des Expériences faites à Cherbourg par ordre de M. le Duc de Harcourt sur l'effet des Boulets Rouges, et sur la Meilleure manière de les Servir Pour la deffense des Postes Maritimes Par M.M. de St. Paul Directeur du corps royal d'artillerie, et Meunier Lieutenant en Premier au corps royal du Génie, de l'académie royale des Sciences," MR 1740(37), SHAT, Vincennes.

7. Meusnier, Jean-Baptiste, ms. "Procès verbal des Epreuves faites suivant les ordres de Monseiur le duc de Harcourt, et de Monsieur le Duc de Beuvron, d'un nouvel affut de casematte, destiné à servir de modèle pour l'armement de la batterie basse du fort Royal et des autres batteries casemattées des forts de Cherbourg," MR 1740(37), SHAT, Vincennes.

8. Montalembert, *Mémoire sur les casemates exécuteés à Cherbourg, sur celles exécutées au Fort de l'île d'Aix, & sur les Affûts de Canons; avec le Projet d'un Port à la Hogue; Où l'on démontre sur plusieurs Planches les défauts des Méthodes qui y ont été employées, les avantages de celles qu'on eût dû suivre. Lu à l'Académie Royale des Sciences le 29 Juillet 1789*, Paris: Philippe-Denys Pierres, 1790, p. vol. This appears in volume 8 (published in 1793) of Montalembert's works.

9. Ibid., p. 28.
10. Ibid., p. 34.
11. Ibid., p. 60.
12. Ibid., p. 29.
13. Ibid., p. 26. On p. 60, this last value is given as 62°.
14. Ibid., pp. 26, 60–61.
15. Ibid., pp. 60–61. In fact, the product of 4 and 1.75 is 7 and not 6.

16. "Epitre à Messieurs les Officiers du Corps Royal du Génie," Montalembert, *Fort. perp.*, vol. 8.

17. Dobenheim, A.-M., ms. "Différentes observations sur les derniers memoires de M. de Montalembert, à la tête desquels se trouve une epitre à MM. Les officers du Corps Royal du Genie," Querquerville [near Cherbourg], 30 July 1790. AIG, article 21 section 1, §1, carton 7 p. 7.

18. Ibid., article 2.

19. Ibid., article 15.

20. Garavague's letter to that effect can be found in the archives, along with a note by Fourcroy de Ramecourt urging its publication. See AIG, article 21 section 1, §1, carton 7, p. 6.

21. Dobenheim, "Différentes observations...," article 36.

22. Ibid., article 50.

23. Ibid.

24. Montalembert 1790.

25. Ibid., p. 1.

26. Ibid., p. 9.

27. Ibid., p. 25.

28. Dobenheim, "Différentes observations...," article 50. The page reference ("23") is to Montalembert's *Observations sur les nouveaux forts*.

29. Picon 1992, p. 226 ff.

30. Smith 1990. See Shallat 1994 for the impact on American engineers. Shallat (1990) detects a split in attitudes and even in design policies between American engineers in the civilian and the military sectors, but this view is challenged by Angevine (2001). For the argument that French state engineers were considerably *more* attuned to a liberal economy than they have been hitherto credited, see Geiger 1994.

31. Reinhard 1950, vol. 1, esp. chapters 11 and 12.

32. On the Conseil de Guerre, see Bien 1974; Reinhard 1950, vol. 1, chapters 9 and 11.

33. Carnot 1789. An abridged reprint of this work can be found in Charnay 1984 (vol. 1, pp. 411–422). According to Charnay (ibid., p. 85), this work was written in August 1788 and published in January 1789.

34. Carnot, *Mémoire... [sur] les places fortes*, in Charnay 1984 (vol. 1, p. 419).

35. Ms. "Délibération de Messieurs les Officiers du Génie de la Garnison de Lille. Mémoire de Mr. de Carnot ainé," 2 August 1789. AIG, article 3, section 10, §1, carton 3, p. 6.

36. Ibid., fol. 2–3.

37. Lazare Carnot, *Réclamation adressée à l'Assemblée nationale, contre le régime oppressif sous lequel est gouverné le Corps royal du Génie, en ce qu'il s'oppose aux progrès de l'art et au bien qu'il serait possible de faire (28 septembre 1789)*, Paris: Impr. de Vve Delaguette, 1789, p. 1. This work is reprinted in volume 1 of Charnay 1984.

38. Ibid., p. 430.

39. Cited by Reinhard (1950, vol. 1, p. 157).
40. Published separately as *Tactique prussienne* (Mirabeau 1789).
41. Ibid., p. 169.
42. Ibid., p. 177.
43. Ibid., p. 187.
44. Mirabeau's library contained a number of Montalembert's works. They included the first five volumes of *La Fortification perpendiculaire* ("whose illustrations were hand colored with the greatest care"), a duplicate set of that work, *Réponse au Mémoire sur la Fortification perpendiculaire*, and *Observations sur les nouveaux forts . . . de Cherbourg* (items 1845–1848, *Catalogue des Livres de la Bibliothèque de Feu M. Mirabeau l'Ainé*, Paris, 1791, p. 269). The great revolutionary was not the first to benefit from hand-tinted plates; the copy presented to the king also had them.
45. Mirabeau to Montalembert, Paris, 11 August 1789, quoted in Montalembert, *Fort. perp.*, vol. 9, "Avant-Propos," pp. xxxi–xxxii. Also quoted by Reinhard (1950, vol. 1, p. 162). (Montalembert gives the full text).
46. On 22 September 1789, according to *Le Moniteur*, 21–22 September 1789.
47. Mirabeau to Montalembert, Paris, 11 August 1789, quoted in Montalembert, *Fort. perp.*, vol. 9, "Avant-Propos," pp. xxxi–xxxii.
48. Quoted by Reinhard (1950, vol. 1, p. 162).
49. Montalembert 1790, pp. 16–17.
50. Of the other members, at least four are known to have later emigrated. Another later led the federalist forces at Normandy in 1793.
51. The sister of François-Ignace Ervoil d'Oyré (1739–1799), a military engineer and the son of a former Director of Fortifications, was the wife of one of Dubois-Crancé's brothers. Another brother, Jean-Baptiste Dubois de Crancé (1740–1793) was a graduate of Mézières and served in the Corps of Military Engineering until his death in battle.
52. Reinhard 1950, vol. 1, 154.
53. Charnay (1984, vol. 1, p. 32), citing Las Casas, *Mémorial de Sainte-Hélène*, 9 June 1816.
54. Blanchard 1979, p. 221.
55. Reinhard 1950, vol. 1, p. 153.
56. Le Michaud d'Arçon 1790. This tract contains the names of alleged signatories who supported the views d'Arçon put forward in this work. Because there were two engineers with the name Lemercier de Senneton de Chermont, there is some doubt as to which of them signed this work.
57. Augoyat 1860–1864, vol. 3, p. 570.
58. Carnot to Montalembert, 1 May 1791. Text given in Montalembert, *Fort. perp.*, vol. 9, "Avant-Propos," pp. xxxiii–xxxiv. Partial quote in Reinhard 1950 (vol. 1, pp. 164–165).
59. Le Michaud d'Arçon 1790.
60. Tholozé, David-Alexis, "Observations sur les mémoires de Mr. de Montalembert relativement à la fortification dite perpendiculaire; la Rénovation des cazemates etc. et le Projet

d'Enceindre les Limites du Royaume par des Lignes impénétrables. Ainsy que sur la réponse du Colonel d'Arçon pour servir d'apologie aux principes observés dans le corps du Génie," Bouchain, 19 September 1790. D'Arçon Papers, Bibliothèque Municipale de Besançon, ms. 491G. A copy of this work in another hand has been attributed to Joseph-Gabriel Monnier de Courtois (1745–1818) by the nineteenth-century historian of French military engineering Augoyat (AIG, article 21, section 1, §1, carton 7 p. 10a). Monnier de Courtois, also an engineer at Bouchain and had organized a military engineering school for the Ottoman Empire between 1784 and 1788. The usually reliable Augoyat seems to have erred, unless one assumes that the document was drafted by Monnier and Tholozé and perhaps other officers in the Bouchain garrison as a collective opinion.

61. Tholozé, ms. "Observations" 1790, fol. 1.

62. Ibid.

63. Ibid., fol. 2.

64. Ibid., fol. 11.

65. Ibid., fol. 9.

66. Ms. "Mémoire sur la discution(?) entre Mr. de Montalembert et les Offrs de Génie," n.d. D'Arçon Papers, Bibliothèque Municipale de Besançon, ms. 491G.

67. Le Michaud d'Arçon 1790, p. 18.

68. Ibid., p. 110. This word seems a neologism coined by d'Arçon. It does not appear in standard French dictionaries.

69. Ibid., p. 57.

70. Ibid., p. 59.

71. Dobenheim, "Différentes observations," article 49. Dobenheim pointed out that Montalembert was ready to correct the defects of the fort at Querqueville without even having seen the plans of the fort.

72. Montalembert claimed that one of those listed (Chambre) had not in fact read it or signed it, but there is no reason to believe that the other people on the list did not subscribe to the views expressed by d'Arçon.

73. Paris: Didot, 1790. The work is dated October 1790.

74. There is some support for the view that d'Arçon was a victim of meddling by the duc de Crillon, commander of the besieging army. See Russell 1965.

75. Montalembert, *Fort. perp.*, vol. 7, p. 184, quoting from d'Arçon 1783.

76. Montalembert 1790b, p. 5.

77. There is no confirmation in Blanchard 1981 of Montalembert's allegation that Chambre was discharged from the Corps of Military Engineering in 1790 because he agreed with the principles of perpendicular fortification.

78. Montalembert 1790b, p. 16.

79. Le Michaud d'Arçon, Jean-Claude-Éléonor, *Suite des Réponses du Colonel d'Arçon, pour éclairer les Répliques de M. de Montalembert*, n.p., n.d.

80. Ibid., pp. 1–2.

81. Ibid., p. 23.

82. Ibid., p. 38.

83. *Lettre de Marc-René de Montalembert, Maréchal de camp, au Corps Royal de l'Artillerie*, n.p., n.d. According to the catalogue of the Bibliothèque nationale, this 11-page piece is dated 29 March 1791.

84. Montalembert 1793. This was volume 9 of *La Fortification perpendiculaire*. It was not, however, to be the last one, as promised.

85. Montalembert, *Fort. perp.*, vol. 9, "Avant-Propos," p. i.

86. Ibid., "Discours Préliminaire," p. xi.

87. Ibid., p. v.

88. Ibid., "Avant-Propos," pp. xxi–xxii.

89. Ibid., pp. xx.

90. Ibid., "Discours Préliminaire," p. xvii.

91. Ibid., p. xiv.

92. Ibid., p. xv.

93. Ibid., p. xxvii.

94. Ibid., pp. lxxvii–lxxix.

95. [Le Michaud d'Arçon, Jean-Claude-Éléonor,] *Des Fortifications et des relations générales de la guerre de siège, pour servir de Réponse au dernier ouvrage de Marc-René Montalembert, par le citoyen Michaud, Inspecteur des Fortifications*, Paris: Magimel, an 2 [1794], pp. 1–2.

96. Ibid., p. 16.

97. Ibid., p. 64.

98. Ibid., p. 24.

99. SHAT, dossier personnel Marc-René de Montalembert, decision of Conseil Exécutif provisoire, 28 August 1793.

100. The decree of the Convention is dated 26 September 1793 and the decree of the Committee of Public Safety is dated 22 thermidor year 2 [9 August 1794], according to Montalembert, *Fort. perp.*, vol. 10. But *Le Moniteur* for 28 September 1793 gives the date as 27 September. On that occasion, according to *Le Moniteur*, Lakanal praised Montalembert as "notre meilleur officier du génie" and his system as even better than Vauban's.

101. Montalembert, *Fort. perp.*, vol. 10.

102. An official description of the tests is published in volume 10 of *Fort. perp.*

103. Extrait des registres du Directoire exécutif du 4 ventôse an 4 [23 February 1796]. SHAT, Dossier personnel Marc-René de Montalembert.

104. Gaudin 1938, p. 115.

105. On the Lycée des Arts and the similarly named Lycée and its various precursors and heirs, see Smeaton 1955; Hahn 1971, esp. p. 217 ff.

106. Desaudray 1800.

107. The top candidates were Bonaparte (411 votes), Dillon (371), Montalembert (367), Lamblardie (348), and Molard (303) (*Procès-verbaux de l'Institut National*, vol. 1, p. 296, 26 brumaire an 6 [16 November 1797]). Acording to Lalande (1800, p. 123), Montalembert had written a letter renouncing his candidacy to Carnot's vacated seat in the Institut, but there is no other evidence for this.

108. Rochas d'Aiglun 1867, pp. 110–111.

109. Fourcroy 1794.

110. They were assisted for a short time by the military engineers François-Henri-César Catoire and Horace Say (1771–1799). The latter was a brother of the economist Jean-Baptiste Say (1767–1832).

111. On "revolutionary" courses during the French Revolution, see Langins 1987; see also Janis Langins, "Words and institutions during the French Revolution: The case of 'revolutionary' scientific and technical education," in *The Social History of Language*, ed. P. Burke and R. Porter (Cambridge University Press, 1987).

112. Le Michaud d'Arçon 1795. A German translation was published in 1801.

113. Augoyat 1860–1864, vol. 3, pp. 576–577.

114. Le Michaud d'Arçon 1790.

115. Montalembert 1795–96, p. i.

116. Articles on fortifications in *Journal de l'École Polytechnique* (published under different names in its early years): A.-M. Dobenheim, "Fortification," *Journal de l'École Polytechnique* 1, cahier 1 (Germinal an 3) [1795: 38–77; Horace Say, "Fortification. Cours Préliminaire," ibid. 1, cahier 12 (Floréal et Prairial an 3) [1795]: 77–99; H. Say, "Fortification," ibid. 1, cahier 3 (Messidor, Thermidor et Fructidor an 3) [1795], pp. 346–348; H. Say, "Fortification," ibid. 1, cahier 4 (Vendémiaire, Brumaire et Frimaire an 4) [1795], pp. 617–618; H. Say, "Mémoire sur le défilement des fortifications," ibid., 588–616. Montalembert also responded to the official syllabus of the fortifications course published in *Programmes de l'enseignement polytechnique de l'École centrale des travaux publics*, Paris: Imprimerie Nationale, pluviôse an III [1795] (reprinted in Langins 1987, Annexe G). All these writings are from 1795.

117. Numbers 1–3 of the *Observations*, dated Year 4 of the Republic, were published by Louvet in Paris. Number 4, dated Year 5, was published by Magimel.

118. "Avertissement," *L'Ami de l'Art défensif, ou Observations sur le Journal Polytechnique de l'École Centrale des Travaux Publics, par le Général Montalembert. Mois Germinal. Article Fortification. No. Ier.*, Paris: Louvet, an 4 [1796].

119. The school at Mézières was abolished by the republican government on 24 pluviôse an II [12 February 1795]. Those parts of its curriculum dealing with siegecraft were to be taught in a school to be set up at Metz for that purpose, and the parts of curriculum dealing with "theory" were be taught at the École des ponts et chaussées in Paris, where the library and models were to be transferred. This was indeed the germ of the future École polytechnique, created a month later.

120. Montalembert 1795–1797, no. 1, p. iv.

121. Ibid., no. 2, p. vii.

122. Ibid., no. 1, p. 4.
123. Ibid., no. 3, p. xxvi.
124. Ibid., no. 3, p. xxi.
125. Ibid., no. 3, p. iv.
126. Ibid., no. 3, p. 8.
127. Ibid., no. 3, p. 8.
128. Ibid., no. 3, p. 4.
129. Ibid., p. vii.
130. Ibid., no. 2, "Discours préliminaire," p. i.
131. Ibid., no. 1, p. 14.
132. Ibid., no. 3, p. 17.
133. Ibid., no. 1, p. 18; no. 3, pp. 12–13.
134. Ibid., no. 1, p. 9 ; no. 3, p. xxiv.
135. Ibid., no. 2, p. 33.
136. The most eminent was Karl Friedrich von Lindenau, former adjutant of Frederick the Great, for whom he probably translated Montalembert, and military mentor to the Austrian Archduke Charles, opponent of Napoleon.
137. Montalembert 1795–1797a, no. 1, pp. ii-iii.
138. Bossut to Montalembert, Paris 6 nivôse year 5 [26 December 1796], ibid., no. 4, pp. 7–8.
139. Procès-verbaux de l'Institut National, 11 nivôse an 8 [1 January 1800], vol. 2, p. 76.
140. Procès-verbaux de l'Institut National, 16 nivôse an 8 [6 January 1800], vol. 2, pp. 79–80.
141. Lalande 1800.
142. His marriage to Rosalie Cadet occurred on 4 nivôse year 3 [25 December 1794], according to his service record, Dossier "Montalembert, Marc-René," SHAT, Vincennes. He had divorced his first wife, who had decided not to return to Paris from England, a few months earlier, on 26 August 1794.
143. Quoted by Desaudray (1800).

Chapter 14

1. Le Michaud d'Arçon 1795, p. 1.
2. There is some evidence that Montalembert's ideas enjoyed more favor during the Napoleonic period, but it appears that this was mostly the case with mountain forts, which by their location could not have defenses that hugged the ground. The evidence is slight that the idea of towers in coastal fortifications, such as the Martello towers that proliferated on the coasts of England and its colonies, owed anything to Montalembert (Prost 1991).
3. Blanchard 1970, p. 105.
4. Gay de Vernon 1805, vol. 1, p. 42.

5. Langins 1987.
6. Berthaud 1898–1902.
7. Belhoste and Picon 1996.
8. Lynn 1984.
9. Reinhard 1950, vol. 1, p. 72.
10. Bodinier 1983 (table I, p. 68) gives a figure of 19–21 percent of artillery officers of the old royal army remaining under the new colors of the Republic at the end of April 1794 versus 35 percent of engineers.
11. Le Michaud d'Arçon 1786, p. 11.
12. Le Michaud d'Arçon 1795, p. 237.
13. Ibid, p. 148.
14. Le Michaud d'Arçon 1786, p. 15.
15. Huntington 1972, pp. 62–79.
16. Ibid., p. 79.
17. Le Michaud d'Arçon 1795, p. 13.
18. Le Michaud d'Arçon 1786, p. 4.
19. Le Michaud d'Arçon 1795, p. 79.
20. For example, it had drawn back from a confrontation in Holland when Prussia intervened to suppress the revolt of the pro-French Patriot party against the House of Orange.
21. Vincenti 1990, p. 6.
22. Rosenberg 1982, pp. 120–140.
23. Vincenti 1990, p. 9.
24. Reinhard 1950, vol. 1, p. 96 ff.
25. Seuillet at Fouras to Fourcroy de Ramecourt, 6 January 1783, AIG, article 8, section 1, Ile d'Aix, carton 2, no. 18.
26. On Mandar, see Picon 1995 and Mandar 1801.
27. Constant 1980.
28. Langins 1990.
29. On the applicability of models of scientific change to technological situations, including those of Kuhn, see Laudan 1984. Also see other articles in *The Nature of Technological Knowledge*, ed. Laudan.
30. Particularly in the case of chemistry (Cohen 1985; Levin 1984).
31. The tradition continued into the nineteenth century, when engineers of the Corps des Ponts et Chaussées were expected to integrate into respectable and influential local society to make the work of the administration easier. See Ringrose 1995.
32. T. Hughes 1983. For more on systems, see T. Hughes 1987, 1998, 2001. On French electrical systems and the context in which they were created, see Frost 1991.
33. T. Hughes 1969.

34. See chapter 2.
35. Williams 1993.
36. In her more recent work (see chapter 6 above), Williams seems to stress the discontinuity rather than the continuity of today's engineering and engineering before the Second World War.
37. See chapter 3.
38. Duveen and Hahn 1957.
39. Gillispie makes this point in his critique of Alder 1997 (Gillispie and Alder 1998).
40. Napoleon made his remarks at St. Helena specifically referring to Carnot's lonely defense of Montalembert in his corps (Reinhard 1950, vol. 1, p. 115). In his eulogy of Carnot at the Academy of Sciences, Arago painted an uplifting picture of the studious Carnot being a great contrast to his fellow officers frequenting cabarets and cafés while the serious Carnot spent his time in libraries. Arago (1854, vol. 1, pp. 520–521) claimed—rather unjustly, I think—that 'original' was synonymous with 'philosophe', something considered outlandish in the military engineering corps.
41. See also the earlier and more traditional treatment of the military industrial effort of the Revolution in Richard 1922.
42. On the early history of École polytechnique, see Belhoste 1989; Langins 1991.
43. Alder 1997, p. xii.
44. Ibid., p. xii.
45. Ibid., p. 15.
46. Ibid., p. 87 ff.
47. Picon 1992, pp. 47–52.
48. See the definition of technology in Roland 1992.
49. Alder 1997, p. 15.
50. On Taylor, see Kanigel 1997; Aitken 1960; Haber 1971.
51. Picon 1992, p. 51; Gillmor 1971, pp. 32, 77–78.
52. Even in the United States, Taylor was frustrated by organized and skilled workers who could enroll the support of their political representatives (Aitken 1960).
53. T. Hughes 1998, p. 305.
54. Ibid., p. 305.
55. Williams 2000, p. 656.

Conclusion

1. Williams 2000.
2. Ibid., p. 656.
3. Layton 1973.

Bibliography

Anonymous and Collective Works

Association Vauban. 1992a. *Vauban et ses successeurs dans le Territoire de Belfort.* Paris: Association Vauban.

Association Vauban 1992b. *Vauban et ses successeurs dans les Alpes-de-Haute-Provence.* Colmars-les-Alpes/Paris: Amis des Forts Vauban de Colmars; Association Vauban.

Association Vauban. 1994. *Vauban et ses successeurs en Hainaut et d'Entre-Sambre-et-Meuse.* Paris: Association Vauban.

Association Vauban. 1995. *Vauban et ses successeurs en Briançonnais.* Paris: Association Vauban.

Encyclopédie, ou Dictionnaire raisonné des sciences, des arts et des métiers, par une société de gens de lettres, mis en ordre et publié par M. Diderot et quant à la partie mathématique, par M. d'Alembert. Paris: Briasson, 1751–1765.

Thèses de Mathématique, de Géométrie, de Trigonométrie rectiligne, et de Fortifications, qui seront soutenues au Collège de Louis le Grand, le Vendredy 25 Juin 1751, à trois heures de l'après midi. 1751. Paris: Thibout, Imprimeur du Roi.

Ordonnance du roi concernant le corps du génie du 31 décembre 1776. 1777. Paris: Imprimerie Royale.

Mémorial de l'Officier du Génie, ou Recueil de mémoires, d'expériences, observations et procédés généraux propres à perfectionner la Fortification et les Constructions militaires. 1803.

Textbook of Fortification and Military Engineering, for use at the Royal Military Academy, Woolwich. 1877. HMSO.

La Science Classique XVIe-XVIIIe siècle: Dictionnaire Critique. 1998. Flammarion.

Other Works

Abbott, Andrew. 1988. *The System of Professions: An Essay on the Division of Expert Labor.* University of Chicago Press.

Acerra, Martine. 1996. "Villes-ports, Villes-arsenaux." In *La ville et la guerre*, ed. A. Picon. Besançon: Editions de l'Imprimeur.

Adams, Simon. 1990. "Tactics or politics? 'The Military Revolution' and the Hapsburg hegemony, 1525–1648." In *Tools of War*, ed. J. Lynn University of Illinois Press.

Aitken, Hugh G. J. 1960. *Taylorism at the Watertown Arsenal: Scientific Management in action, 1908–1915.* Harvard University Press.

Alberti, Leon Battista. 1988. *On the Art of Building in Ten Books.* MIT Press.

Alder, Ken. 1997. *Engineering the Revolution: Arms and Enlightenment in France, 1763–1815.* Princeton University Press.

Alexander, R. 1981. Vauban and Absolutism. M.A. thesis, University of Toronto.

Allen, Michael Thad, and Gabrielle Hecht, eds. 2001. *Technologies of Power: Essays in Honor of Thomas Parks Hughes and Agatha Chipley Hughes.* MIT Press.

Allent, Pierre-Alexandre-Joseph. 1805. *Histoire du corps impérial du génie.* Paris: Magimel

André, Louis. 1942. *Michel Le Tellier et Louvois.* Paris.

Angevine, Robert G. 2001. "Individuals, organizations, and engineering: U.S. Army officers and the American railroads, 1827–1838." *Technology and Culture* 42: 292–320.

Antoine, Michel. 1989. *Louis XV.* Fayard.

Arago, François. 1854–1859. *Oeuvres complètes.* Paris: Gide et Baudry.

Arbelot, Guy, Bernard Lepetit, and Jacques Bertrand. 1987. *Atlas de la Révolution française: Routes et communications.* Paris: Éditions de l'École des Hautes Études en Sciences Sociales.

Arcq, Philippe Auguste de Sainte-Foix Chevalier d'. 1756. *La Noblesse Militaire, ou le Patriote français.*

Argenson, René-Louis marquis d'. 1859–1867. *Journal et mémoires du marquis d'Argenson.* Société d'histoire de France.

Arnault, A. V. 1820. *Biographie nouvelle des contemporains; ou, Dictionnaire historique et raisonné de tous les hommes qui, depuis la Révolution française, ont acquis de la célébrité par leurs actions, leurs écrits, leurs erreurs ou leurs crimes, soit en France, soit dans les pays étrangers.* Paris: Librairie Historique.

Aron, Raymond. 1976. *Penser la guerre, Clausewitz.* Gallimard.

Artz, Frederick B. 1966. *The Development of Technical Education in France 1500–1800.* MIT Press.

Augoyat, Antoine-Marie. 1825. "Mémoire sur la pénétration et l'effet des projectiles." *Mémorial de l'Officier du Génie, ou Recueil de mémoires, expériences, observations et procédés généraux propres à perfectionner la Fortification et les Constructions militaires* 7: 86–160.

Augoyat, Antoine-Marie. 1836a. "Notice historique de M. de Fourcroy, maréchal de camp du corps du génie." *Le Spéctateur militaire* 21: 648–666.

Augoyat, Antoine-Marie. 1836b. "Notice historique du lieutenant-général Filley de la Côte du Corps du Génie." *Le Spéctateur militaire* 21: 534–552.

Augoyat, Antoine-Marie. 1860–1864. *Aperçu historique sur les Fortifications, les Ingénieurs, et sur le Corps du Génie en France.* Paris: C. Tanera, J. Dumaine.

Bachaumont, Louis Petit de. 1777–1789. *Mémoires secrets pour servir à l'histoire de la république des lettres en France, depuis MDCCLXII jusqu'à nos jours; ou Journal d'un observateur, contenant les analyses des pieces de théâtre qui ont paru durant cet intervalle ; les relations des assemblées littéraires ; les notices des livres nouveaux*. Londres: John Adamsohn.

Baker, Keith Michael. 1975. *Condorcet: From Natural Philosophy to Social Mathematics*. University of Chicago Press.

Baker, Keith Michael. 1987. "Scientism at the end of the Old Regime: Reflections on a theme of Professor Charles Gillispie." *Minerva* 25: 21–34.

Baldet, Marcel. 1964. *La vie quotidienne dans les armées de Napoléon*. Hachette.

Barker, Thomas Mack. 1975. *The Military Intellectual and Battle: Raimondo Montecuccoli and the Thirty Years War*. State University of New York Press.

Baxter, Douglas Clark. 1976. *Servants of the Sword: French Intendants of the Army 1630–1670*. University of Illinois Press.

Baynes, Ken, and Francis Pugh. 1981. *The Art of the Engineer*. Lutterworth.

Belair, Alexandre-Pierre-Julienne de. 1787. *Nouvelle science des ingénieurs*. Berlin and St. Petersbourg: Société du Pôle Arctique.

Belair, Alexandre-Pierre-Julienne de. 1792. *Eléments de fortification . . . suivis d'un dictionnaire militaire*. Paris: F. Didot.

Belhoste, Bruno. 1989. "Les origines de l'École polytechnique. Des anciennes écoles d'ingénieurs a l'École centrale des travaux publics." *Histoire de l'Education* 42: 13–54.

Belhoste, Bruno. 1990. "Du dessin d'ingénieur à la géométrie descriptive: L'enseignement de Chastillon à l'École royale du génie de Mézières." *In Extenso: Recherches à l'École d'Architecture Paris-Villemin* 13: 103–136.

Belhoste, Bruno. 1992. "Les problèmes de défilement." In *L'École Normale de l'An III: Leçons de Mathématiques*. Dunod.

Belhoste, Bruno. 1994. "De l'École des Ponts et Chaussées à l'École Centrale des Travaux Publics." *SABIX: Bulletin de la Société des Amis de la Bibliothèque de l'École Polytechnique* 11: 1–69.

Belhoste, Bruno, and Antoine Picon, eds. 1996. *L'École d'application de l'artillerie et du génie de Metz (1802–1870): Enseignement et recherches (Actes de la journée d'étude du 2 novembre 1995)*. Paris: Ministère de la Culture Direction du Patrimoine Musée des Plans-Reliefs.

Belhoste, Bruno, Antoine Picon, and Joël Sakarovitch. 1990. "Les exercices dans les écoles d'ingénieurs sous l'Ancien Régime et la Révolution." *Histoire de l'éducation* 46, May: 53–109.

Belhoste, Bruno, Amy Dahan Dalmedico, and Antoine Picon, eds. 1994. *La Formation Polytechnicienne 1794–1994*. Dunod.

Bélidor, Bernard Forest de. 1725. *Nouveau cours de mathématiques, à l'usage de l'artillerie et du génie où l'on applique les parties les plus utiles de cette science à la Théorie & à la Pratique des differens sujets qui peuvent avoir rapport à la guerre*. Paris: Nyon.

Bélidor, Bernard Forest de. 1729. *La Science des ingénieurs dans la conduite des Travaux de Fortification et d'Architecture civile*. Paris: Claude Jombert.

Bélidor, Bernard Forest de. 1734. *Le Bombardier françois, ou nouvelle methode de jetter les bombes avec precision*. Amsterdam: Aux depens de la compagnie.

Bélidor, Bernard Forest de. 1737. *Architecture hydraulique ou l'Art de conduire, d'Elever, et de Menager les eaux pour les différens besoins de la Vie*. Paris: Charles-Antoine Jombert.

Bélidor, Bernard Forest de. 1755. *Dictionnaire portatif de l'ingénieur*. Paris: Charles-Antoine Jombert.

Bélidor, Bernard Forest de. 1830. *La science des ingenieurs, dans la conduite des travaux de fortification et d'architecture civile*, ed. C. L. M. H. Navier. Paris: F. Didot.

Bennett, Jim, and Stephen Johnston. 1996. *The Geometry of War 1500–1750*. Oxford: Museum of the History of Science.

Bensaude-Vincent, Bernadette, and Isabelle Stengers. 1992. *Histoire de la Chimie*. Découverte.

Berniard, Pierre-Antoine. 1988. *Histoire de l'Ile d'Aix*. Ottawa: Beauregard.

Berthaud, Col. 1898–1902. *Les ingénieurs géographes militaires 1624–1831: Etude historique*. Paris.

Best, Geoffrey, ed. 1982. *War and Society in Revolutionary Europe, 1770–1870*. St. Martin's Press.

Bien, David D. 1969. "Military education in 18c France: Technical and non-technical determinants." In *Science, Technology and Warfare*, ed. M. Wright and L. Vaszek. US Government Printing Office.

Bien, David D. 1974. "La réaction aristocratique avant 1789: L'exemple de l'armée." *Annales Economies Sociétés Civilisations* 29: 23–48, 505–534.

Bien, David D. 1979. "The Army in the French Enlightenment: Reform, reaction and revolution." *Past and Present* 85: 65–98.

Blanchard, Anne. 1962. "'Ingénieurs de Roi' en Languedoc au 18e siècle." *Revue d'histoire moderne et contemporaine* 9: 161–170.

Blanchard, Anne. 1970. "Les ci-devant ingénieurs du roi." *Revue internationale d'histoire militaire* 30: 97–107.

Blanchard, Anne. 1973. "Ingénieurs de Sa Majesté Très Chretienne à l'Etranger ou l'école française de fortifications." *Revue d'histoire moderne et contemporaine* 20: 25–36.

Blanchard, Anne. 1979. *Les Ingénieurs du "Roy" de Louis XIV à Louis XVI: Étude du Corps de Fortifications*. Centre d'histoire militaire et d'études de défense nationale, Université Paul Valéry (Montpellier III).

Blanchard, Anne. 1981. *Dictionnaire des Ingénieurs militaires 1691–1791*. Centre d'histoire militaire et d'études de défense nationale, Université Paul Valéry (Montpellier III).

Blanchard, Anne. 1996. *Vauban*. Fayard.

Bledstein, Burton J. 1976. *The Culture of Professionalism: The Middle Class and the Development of Higher Education in America*. Norton.

Blois, E. de. 1865. *De la fortification en présence de l'artillerie nouvelle*. Paris: Dumaine.

Blomfield, Reginald. 1971. *Sebastien le Prestre de Vauban, 1633–1707*. Barnes & Noble.

Blondel, François. 1683. *Nouvelle manière de fortifier les places*. Paris: F. Blondel, N. Langlois.

Bluche, François. 1980. *La vie quotidienne au temps de Louis XVI*. Hachette.

Bluche, François. 1986. *Les magistrats du Parlement de Paris au XVIIIe siècle*. Economica.

Bodinier, G. 1983. "Les officers de l'armée royale et la Révolution." *Révue internationale d'histoire militaire* 55: 49–69.

Böhme, Gernot, Wolfgang van den Daele, and Wolfgang Krohn. 1978. "The 'scientification' of technology." In *The Dynamics of Science and Technology*, ed. W. Krohn et al. Reidel.

Boltanski, Luc. 1982. *Les Cadres: La formation d'un groupe social*. Paris: Minuit.

Bony de la Vergne, Ferdinand-Ernest-Alexandre comte de. 1842. *Anecdotes, Bons Mots, Saillies, Balourdises, Eccentricités Evénements singuliers avec quelques souvenirs de l'École de Mézières, à l'usage des rieurs de bon aloi*. Metz.

Booker, P. J. 1963. *A History of Engineering Drawing*. Chatto and Windus.

Bornecque, Pierre. 1984. *La France de Vauban*. Paris: Arthaud.

Bornecque, Robert. 1985. "Une paradoxe: la ville citadelle de Montdauphin." In *Vauban réformateur*, ed. C. Brisac et al. Association Vauban.

Bornecque, Robert. 1995. *Vauban et les Alpes*. Saint-Léger-Vauban: Association des amis de la Maison Vauban.

Bousmard, Henri-Jean-Baptiste de. 1797–1799. *Essai général de fortification, et d'Attaque et Défense des Places; Dans lequel les Deux sciences sont expliquées et mises l'une et l'autre à la portée de Tout le Monde*. Berlin: Decker.

Bradley, Margaret. 1992. "Engineers as military spies? French engineers come to Britain, 1780–1790." *Annals of Science* 49: 137–161.

Bragard, Philippe. 1994. "La Reconstruction des forteresses belges aprés 1815: Une application des théories françaises de la fin du XVIIIe et du début du XIXe siècle à propos de Namur, Huy et Dinant." In *Vauban et ses successeurs en Hainaut et d'Entre-Sambre-et-Meuse*. Paris: Association Vauban.

Braudel, Fernand. 1979. *Civilisation matérielle, économie et capitalisme XVe-XVIIIe siècle*. Armand Colin.

Braun, Georg, and Frans Hogenberg. 1572–1618. *Civitatis Orbis Terrarum*. Cologne.

Bret, Patrice. 1989. "Les Oubliés de Polytechnique en Egypte: Les artistes mécaniciens de la Commission des sciences et des arts." In *Scientifiques et Sociétés pendant la Révolution et l'Empire*, Actes du 114e Congrès national des Sociétés savantes.

Bret, Patrice. 1991. "Le Dépôt général de la Guerre et la formation scientifique des ingénieurs-géographes militaires en France (1789–1830)." *Annals of Science* 48: 113–157.

Brisac, Catherine, Nicolas Faucherre, and Johel Coutura, eds. 1985. *Vauban réformateur*. Association Vauban.

Brown, John K. 2000. "Design plans, working drawings, national styles: Engineering practice in Great Britain and the United States, 1775–1945." *Technology and Culture* 41: 195–238.

Brunot, A., and R. Coquand. 1982. *Le Corps des Ponts et Chaussées*. Paris: CNRS.

Buchanan, Robert Angus. 1983. "Gentlemen engineers: The making of a profession." *Victorian Studies* 26: 407–427.

Buchanan, Robert Angus. 1985. "The rise of scientific engineering in Britain." *British Journal for the History of Science* 18: 218–233.

Buchanan, Robert Angus. 1989. *The Engineers: A History of the Engineering Profession in Britain, 1750–1914.* Kingsley.

Buchwald, Jed Z., and I. Bernard Cohen. 2001. *Isaac Newton's Natural Philosophy.* MIT Press.

Buisseret, David J. 1964a. "Ingénieurs du roi sous Henri IV." *Bulletin de la section de Géographie, Comité des travaux historiques et scientifiques* 77: 12–81.

Buisseret, David J. 1964b. "L'Organisation défensive des frontières au temps de Henri IV." *Revue historique des armées* 20: 25–31.

Buisseret, David, ed. 1968. *Sully and the Growth of Centralized Government in France 1598–1610.* London: Eyre and Spottiswoode.

Buisseret, David. 1992. *Monarchs, Ministers, and Maps: The Emergence of Cartography as a Tool of Government in Early Modern Europe.* University of Chicago Press.

Bülow, Dietrich von. 1806. *The Spirit of the Modern System of War.* London: T. Egerton,

Bureaux de Pusy, Jean-Xavier. 1790a. *Considérations sur le Corps du Génie.* Paris: Imprimerie nationale.

Bureaux de Pusy, Jean-Xavier. 1790b. *De la réunion des mineurs au Corps Royal et de celle du génie à l'artillerie.* Paris.

Bury, John. 1985. "Early writings on fortifications and siegecraft: 1502–1554." *Fort* 13: 4–48.

Cadet, Félix. 1970. *Histoire de l'économie politique, les précurseurs: Boisguilbert, Vauban, Quesnay, Turgot.* B. Franklin.

Callahan, William James. 1972. *Honor, Commerce, and Industry in Eighteenth-Century Spain.* Harvard Graduate School of Business Administration.

Cambray, Chevalier de. 1689. *Manière de fortifier de Mr. de Vauban.* Amsterdam: Pierre Mortier.

Cambray, Chevalier de. 1692. *Nouvelle manière de fortifier de Mr. de Vauban.* Paris: Sebastien Mabre Cramoisy.

Camon, Hubert. 1914. *La Fortification dans la Guerre Napoléonienne.* Paris: Berger-Levrault.

Campillo, Antonio. 1986. *La Fuerza de la Razon: Guerra, Estado y ciencia en los tratados militares del Renacimiento, de Maquiavelo a Galileo.* Faculdad de Letras, Universidad de Murcia.

Capel, Horacio, et al. 1983. *Los Ingenieros militares en España Siglo XVIII: Repertorio biográfico e inventario de su labor científica y espacial.* Barcelona: Cátedra de Geografía humana, Universidad de Barcelona.

Capel, Horacio, Joan Eugeni Sánchez, and Omar Moncada. 1988. *De Palas a Minerva: La formacion cientifica y la estructura institucional de los ingenieros militares en el sigolo XVIII.* Barcelona/Madrid: Serbal/CSIC.

Capon, Gaston, and R. Yves-Plessis. 1907. *Vié privée du prince de Conty, Louis-François de Bourbon (1717–1776).* Paris: Schemitt.

Carnot, Lazare. 1784. *Éloge de M. le maréchal de Vauban, Discours qui a remporté le prix de l'Académie des sciences, arts et belles-lettres de Dijon, en 1784.* Dijon/Paris: A. Jombert jeune.

Carnot, Lazare. 1786a. *Observations sur la lettre de M. Choderlos de Laclos à MM. de l'Académie française concernant l'éloge de Vauban.*

Carnot, Lazare. 1786b. *Éloge de Vauban . . . ouvrage enrichi d'observations par un amateur*. La Haye.

Carnot, Lazare. 1789. *Mémoire présenté au Conseil de la guerre au sujet des places fortes qui doivent être démolies ou abandonnées; ou Examen de cette question: Est-il avantageux au roi de France qu'il y ait des places fortes sur les frontières de ses États?*

Carnot, Lazare. 1812. *De la défense des places fortes; ouvrage composé par ordre de S.M. imp. et royale, pour l'instruction des élèves du corps du génie*. Third edition. Paris: Courcier.

Carnot, Lazare. 1823. *Mémoire sur la fortification primitive, pour servir de suite au Traité de la Défense de Places fortes*. Paris: Bachelier.

Carnot, Lazare. 1984. *Lazare Carnot: Révolution et mathématique*, ed. J.-P. Charnay. L'Herne.

Castel de Saint-Pierre, Charles-Irénée. 1713. *Projet pour rendre la paix perpétuelle en Europe*, Utrecht: A. Schouten.

Cavanaugh, Gerald J. 1967. Vauban, D'Argenson, Turgot: From Absolutism to Constitutionalism in 18c France. Ph.D. thesis, Columbia University.

Cervera Vera, Luis. 1978. *El Codice de Vitruvio hasta sus primeras versiones impressas*. Instituto de España.

Chalmin, Pierre. 1951. "La Formation des officiers des armes savantes." In *Actes du 76e congrès des sociétés savantes, Section d'histoire moderne et contemporaine*.

Chalmin, Pierre. 1954. "Les Écoles militaires françaises jusqu'en 1914." *Revue historique de l'armée* 10: 129–166.

Chalmin, Pierre. 1968. "La querelle des bleus et des rouges dans l'artillerie française à la fin du 18ème siècle." *Revue d'histoire économique et sociale* 46: 465–505.

Chandler, David. 1976. *The Art of Warfare in the Age of Marlborough*. Batsford.

Chantereyne, P. de. 1900. *Etudes historiques sur Cherbourg*. Brionne: G. Montfort.

Charbonneau, André. 1990. "The redoubt in New France." *Fort* 18: 43–68.

Charbonneau, André. 1994. *The Fortifications of Ile aux Noix: A Portrait of Defensive Strategy on the Upper Richelieu Border in the 18th and 19th Centuries*. Ottawa: Canadian Heritage, Parks Canada.

Charnay, Jean Paul, ed. 1981. *Guibert ou le Soldat philosophe*. Centre d'Etudes et de Recherches sur les Stratégies et les Conflits de l'Université de Paris-Sorbonne.

Charnay, Jean Paul, ed. 1984. *Lazare Carnot: Révolution et mathématiques*. L'Herne.

Chaussinand-Nogaret, Guy. 1975. *Une Histoire des élites, 1700–1848: recueil de textes*. Mouton.

Chaussinand-Nogaret, Guy. 1976. *La noblesse française au 18e siècle: De la féodalité aux lumières*. Hachette.

Chicoteau, Yves, and Antoine Picon. 1982. "Forme, technique et idéologie: Les ingénieurs des ponts et chaussées à la fin du 18ème siècle." *Culture technique* 7: 188–197.

Chicoteau, Yves, Antoine Picon, and Catherine Rochant. 1984. "Gaspard Riche de Prony ou le génie 'appliqué.'" *Culture technique* 12: 171–183.

Choderlos de Laclos, Pierre-Ambroise-François. 1951. "Lettre à Messieurs de l'Académie française sur l'éloge de M. le Maréchal de Vauban, Proposé pour sujet du Prix d'Éloquence de l'Année 1787." In *Oeuvres complètes*. Gallimard.

Chotard, K. 1890. *Louis XIV, Louvois, Vauban et les fortifications du nord de la France d'après les lettres indédites de Louvois adressées à M. de Chazerat, Gentilhomme d'Auvergne, Directeur de Fortifications à Ypres.* Paris: Plon.

Choumara, P. M. Théodore. 1847. *Mémoires sur la fortification.* Paris: J. Dumaine.

Church, Clive H. 1981. *Revolution and Red Tape: French Ministerial Bureaucracy 1770–1850.* Clarendon.

Clairac, Louis-André de la Mayme de. 1749. *L'Ingénieur de campagne, ou Traité de la Fortification passagère,* second edition. Paris: Jombert.

Clarke, Sir George Sydenham. 1907. *Fortification: Its Past Achievements, Recent Developments, and Future Progress,* second edition. Beaufort.

Clary, David A. 1990. *Fortress America: The Corps of Engineers, Hampton Roads, and United States Coastal Defense.* University Press of Virginia.

Clausewitz, Carl von. 1832. *On War.* Reprint: Princeton University Press, 1976.

Clements, W. H. 1999. *Towers of Strength: The Story of the Martello Towers.* Leo Cooper.

Coehoorn, Menno van. 1706. *Nouvelle Fortification.* Wesel: J. van Wesel.

Cohen, I. Bernard. 1985. *Revolution in Science.* Belknap.

Cohen, Patricia Cline. 1982. *A Calculating People: The Spread of Numeracy in Early America.* University of Chicago Press.

Colin, Jean-Lambert-Alphonse. 1901. *L'Education militaire de Napoléon.* Paris: Chapelot.

Condorcet, Marie-Jean-Antoine-Nicolas-Caritat de. 1847. "Éloge de Fourcroy." In *Oeuvres de Condorcet,* vol. 3. Paris: Firmin Didot.

Constant, Edward W., Jr. 1980. *The Origins of the Turbojet Revolution.* Johns Hopkins University Press.

Contamine, Philippe. 1972. *Guerre, état et société à la fin du Moyen Age: Études sur les armées des rois de rois de France 1337–1494.* Paris.

Conturie, P. M. Jean. 1951. *Histoire de la fonderie nationale de Ruelle (1750–1950) et des Anciennes Fonderies de Canons de Fer de la Marine (Première partie: 1750–1855).* Paris: Imprimerie nationale.

Cormontaigne, Louis de. 1803. *Mémorial de Cormontaingne pour l'attaque des places, ou Recueil fait par ce célèbre ingénieur des préceptes et des méthodes qu'il suivait dans la conduite des sièges, utile à tout militaire employé à l'attaque d'une place, ouvrage posthume publié avec des notes par M. de Bousmard.* Berlin: C. Quieu.

Cormontaigne, Louis. 1835a. *Mémorial pour l'attaque des places,* second edition. Paris: Anselin.

Cormontaigne, Louis. 1835b. *Oeuvres Posthumes de Cormontaigne,* ed. A. Augoyat. Second edition. Paris: Anselin.

Corvisier, André. 1959. "Les généraux de Louis XIV et leur origine sociale." *Dix-Septième Siècle* 42-43: 21–53.

Corvisier, André. 1970. "Hierarchie militaire et hierarchie sociale à la veille de la Révolution." *Révue internationale d'histoire militaire* 30: 77–92.

Corvisier, André. 1976. *Armées et sociétés en Europe de 1494 à 1789*. Presses Universitaires de France.

Corvisier, André. 1978. "La noblesse militaire: Aspects militaires de la noblesse française du 15e et au 18e siècles; état des questions." *Histoire sociale—Social History* 11: 336–355.

Corvisier, André. 1983. *Louvois*. Fayard.

Corvisier, André, ed. 1993. *Actes du colloque international sur les plans-reliefs au passé et au présent: les 23, 24, 25 avril 1990 en l'Hôtel national des Invalides*. Paris: Sedes.

Cosseron de Villenoisy, Louis-Pierre-Jean-Mammès. 1869. *Essai historique sur la fortification*. Paris: Dumaine.

Coste, Anne, Antoine Picon, and Francis Sidot, eds. 1994. *Un Ingénieur des Lumières: Émiland-Marie Gauthey*. Paris: Presses de l'École Nationale des Ponts et Chaussées.

Cresti, Carlo, Amelio Fara, and Daniela Lamberini, eds. 1988. *Atti del Convegno di Studi Achitettura nell' Europa del XVI secolo*. Siena: Edizioni Periccioli.

Crick, Timothy. 1996. "Fortifications from Vauban to Jervois." *Fort* 24: 37–72.

Dard, Emile. 1905. *Le général Choderlos de Laclos, auteur des Liaisons dangereuses, 1741–1803. D'après des documents inédits. Ouvrage orné d'un port. par Carmontelle*. Paris: Perrin.

Darnton, Robert. 1968. *Mesmerism and the End of the Enlightenment in France*. Harvard University Press.

Darnton, Robert. 1982. *The Literary Underground of the Old Regime*. Harvard University Press.

Dartein, Fernand de. 1906. "La vie et les travaux de Jean-Rodolphe Perronet." *Annales des Ponts et Chaussées* (1ème partie), 8e série, 24, October: 5–87

Davidson, F. P., and J. S. Cox. 1983. *Macro: A Clear Vision of How Science and Technology Will Shape Our Future*. Morrow.

Dean, Martin C. 1989. "Changes in the role of fortresses in the French revolutionary wars." *Fort* 17: 37–42.

Decker, Michel. 1984. "Le bassin de la Charente et les fabrications d'artillerie du XVIe au XIXe siècles." *Bulletins et Mémoires de la Société Archéologique et historique de la Charente*, 2d trimestre: 138–152.

Decker, Michel. 1995. "Les matériels de siège et de place dans l'artillerie royale aux XVIe et XVIIe siècles." In *Vauban et ses successeurs en Briançonnais*. Association Vauban.

de Dainville, F. 1954. "L'enseignement des mathématiques dans les collèges des jésuites de France du 17e et 18e siècle." *Revue d'histoire des sciences* 7: 6–21, 109–123.

Deidier, l'abbé. 1734. *Le parfait ingénieur françois ou la Fortification Offensive ou Défensive; contenant la construction, l'Attaque et la Défense des Places Regulières et Irregulières, selon les méthodes des plus habiles Auteurs de l'Europe, qui on écrit sur cette science*. Amsterdam: Compagnie des Libraires.

Dejean, Joan. 1984. *Literary Fortifications: Rousseau, Laclos, Sade*. Princeton University Press.

de la Croix, Horst. 1960. "Military architecture and the radial city plan in sixteenth-century Italy." *Art Bulletin* 42: 263–290.

de la Croix, Horst. 1963. "The literature on fortification in Renaissance Italy." *Technology and Culture* 4: 30–44.

de la Croix, Horst. 1972. *Military Considerations in City Planning: Fortifications*. Braziller.

Delisle de Sales, J. B., and Sulpice Imbert de Laplatière. 1801 [an IX]. "Éloge historique du général Montalembert." Paris.

Demangeon, Alain, and Bruno Fortier. 1978. *Les vaisseaux et les villes: l'arsenal de Cherbourg*. Brussels: P. Mardaga.

Desaudray, Charles. 1800. *Notice sur le Général Montalembert, Lue à la 62e séance publique du Lycée des Arts, le 10 Thermidor an 8 [29 July 1800]*. Paris: Imprimerie du Lycée des Arts.

Deshayes, Philippe. 1985. "Vauban et la norme: de l'intérêt pour l'oeuvre construite à l'importance de la démarche de conception." In *Vauban réformateur*, ed. C. Brisac et al. Association Vauban.

Desportes, Marc, and Antoine Picon. 1997. *De l'espace au territoire: l'aménagement en France XVIe—XXe siècles*. Presses de l'École nationale des ponts et chaussées.

Desquesnes, Rémy, René Faille, Nicolas Faucherre, and Philippe Prost. 1993. *Les Fortifications du Littoral: La Charente-Maritime*. Chauray: Editions Patrimoines et Médias.

De Ville, Antoine. 1640. *Les Fortifications du Chevalier Antoine de Ville*. Lyon: Philippe Borde.

Dhombres, Jean, ed. 1992. *L'École Normale de l'an III—Leçons de Mathématiques (Laplace—Lagrange—Monge)*. Dunod.

Diderot, D., and J. d'Alembert, eds. 1751–1765. *Encyclopédie, ou Dictionnaire raisonné des science, des arts et des métiers*. Paris.

Donkin, S. 1936–37. "The Society of Civil Engineers [Smeatonians]." *Transactions of the Newcomen Society* 17: 51–60.

Dooley, Edwin L., Jr. 1990. "L'instruction militaire à l'École Polytechnique." *Bulletin de la Société des Amis de l'École Polytechnique* 6, June: 14–21.

Dorn, Harold I. 1970. The Art of Building and the Science of Mechanics: A Study of the Union of Theory and Practice in the Early History of Structural Analysis in England. Ph.D. dissertation, Princeton University.

Douglas, Howard. 1859. *Observations on Modern Systems of Fortification, including that proposed by M. Carnot, and a comparison of the Polygonal with Bastion system*. London: John Murray.

Doyle, William. 1972. "Was there an aristocratic reaction in pre-Revolutionary France?" *Past and Present* 57, November: 97–122.

Drake, Stillman, and I. E. Drabkin, eds. 1969. *Mechanics in Sixteenth-Century Italy: Selections from Tartaglia, Benedetti, Guido Ubaldo, & Galileo*. University of Wisconsin Press.

Dubled, H. 1976. "L'artillerie royale française à l'époque de Charles VII et au début du règne de Louis XI (1437-1469): Les Frères Bureau." *Mémorial de l'artillerie française* 50, no. 4: 555–637.

Du Brueil, Jean Pére. 1665. *L'Art Universel des Fortifications, françoises, hollandoises, espagnoles, italiennes, et composées*. Paris: Jacques Dubrueil.

Ducrest, Charles-Louis marquis de. 1777. *Essai sur les machines hydrauliques, contenant des recherches sur la manière de les calculer et de perfectionner en général leur construction; une* méthode nouvelle pour construire les vaisseau; la Description de plusieurs machines nouvelles propres à porter l'hydraulique à un haut point de perfection, et le détail d'un grand nombre d'expériences très-intéressantes. Paris: Esprit, Libraire de S.A.S. Monseigneur le Duc de Chartres, au Palais Royal.

Du Fay, Abbé. 1707. *Manière de fortifier selon la Methode de Monsieur de Vauban, avec un Traité Préliminiaire des Principes de Géométrie.* Paris: J.-B. Coignard.

Duffy, Christopher. 1975. *Fire and Stone: The Science of Fortress Warfare, 1660–1860.* David and Charles.

Duffy, Christopher. 1979. *Siege Warfare,* volume 1: *The Fortress in the Early Modern World.* Routledge and Kegan Paul.

Duffy, Christopher. 1985. *Siege Warfare,* volume 2: *The Fortress in the Age of Vauban and Frederick the Great, 1660–1789.* Routledge and Kegan Paul.

Duffy, Christopher. 1987. *The Military Experience in the Age of Reason.* Routledge and Kegan Paul.

Dulacq, Joseph. 1741. *Théorie nouvelle sur le mécanisme de l'artillerie.* Paris: Charles-Antoine Jombert.

Dull, Jonathan. 1975. *The French Navy and American Independence: A Study of Arms and Diplomacy.* Princeton University Press.

Dupain de, Montesson. 1775. *L'art de lever les plans, de tout ce qui a rapport à la guerre et à l'architecture civile et champêtre,* second edition. Paris: Charles-Antoine Jombert.

Dupuget. 1771. *Essai sur l'usage de l'artillerie, dans la Guerre de Campagne et dans celle des sièges.* Amsterdam: Archistée et Merkus.

Dupuget. 1772. *Procès-verbal des Epreuves faites à Douay, sur les portées des pièces de quatre longues, et de celles de quatre courtes du nouveau modele.* Paris: Jombert père.

Dupuy, Richard Ernest, and Trevor N. Dupuy. 1977. *The Encyclopedia of Military History from 3500 B.C. to the Present,* revised edition. Harper & Row.

Dupuy, Gabriel, et al. 1988. *Réseaux territoriaux, Transports et communication.* Caen: Paradigme

Duranthon, Marc. 1978. *La Carte de France: Son Histoire 1678–1978.* Paris: Solar.

Duthuron, Gaston. 1940. "Un éloge de Vauban." *Annales historiques de la Révolution française* 17: 152–165.

Dutriez, Robert. 1981. *Besançon, ville fortifiée: de Vauban à Séré de Rivières.* Besançon: Cêtre.

Duveen, Denis I., and Roger Hahn. 1957. "Laplace's succession to Bézout's post of Examinateur des Elèves de l'Artillerie: A case history in the 'lobbying' for scientific appointments in France during the period preceding the French Revolution." *Isis* 48: 416–427.

Duvignau, Antoine-Alexandre-Bernard. 1830. *Exercise complet sur le tracé, le relief, la construction, l'attaque et la défense de fortifications.* Paris: Anselin.

Elias, Norbert. 1950. "Studies in the genesis of the naval profession." *British Journal of Sociology* 1: 291–309.

Errard de Bar-le-Duc, Jean. 1584. *Le premier livre des instruments mathematiqes mechaniques.* Facsimile reprint: Berger-Levrault, 1979.

Errard de Bar-le-Duc, I. Jean. 1594. *La geometrie et la practique generalle d'icelle*. Paris: David le Clere.

Errard Bar-le-Duc, Jean. 1600. *La fortification reduicte en art et demonstrée*. Paris.

Ettlinger, Leopold. 1977. "The emergence of the Italian architect during the fifteenth century." In *The Architect*, ed. S. Kostof. Oxford University Press.

Faille, René, and Nelly Lacrocq. 1979. *Les Ingénieurs géographes Claude, François et Calude-Felix Masse*. La Rochelle: Rupella.

Faucherre, Nicolas. 1986. *Places Fortes: Bastion du Pouvoir*. Paris: Rempart.

Faucherre, Nicolas. 1993. Les citadelles du roi de France sous Charles VII et Louis XI. Doctoral thesis, Université de Paris I.

Faucherre, Nicolas. 1995. *Bastions de la Mer: Le Guide des Fortification de la Charente-Maritime*. Cauray-Niort: Éditions du Patrimoine et Médias.

Faucherre, Nicolas. 1996. "Des Villes libres au pré carré: Genèse de l'état monarchique en France." In *La ville et la guerre*, ed. A. Picon. Besançon: Editions de l'Imprimeur.

Faucherre, Nicolas, and Philippe Prost, eds. 1992. *Le Triomphe de la Méthode: Traité de l'attaque des places de Monsieur de Vauban, Ingénieur du Roi*. Gallimard.

Favaro, A., ed. 1932. *Le Opere di Galileo Galilei*. Firenze: G. Barbera.

Favé, Ernest Honoré, ed. 1847. *Mémoires militaires de Vauban et des ingénieurs Hüe de Caligny*. Paris: Corréard.

Favé, Ildefonse. 1846–1871. *Etudes su le passé et l'avenir de l'artillerie*. Paris: J. Dumaine.

Feller, François-Xavier. 1848. *Biographie universelle, ou dictionnaire historique*. Paris/Besançon: J. Leroux Jouby et Cie., Gaumé Frères/ Outh. Chalandre Fils.

Ferguson, Eugene S. 1977. "The mind's eye: Nonverbal thought in technology." *Science* 197: 827–836.

Ferguson, Eugene S. 1992. *Engineering and the Mind's Eye*. MIT Press.

Filion, Maurice. 1967. *Maurepas: Ministre de Louis XV (1715–1749)*. Montréal: Leméac.

Filion, Maurice. 1972. *La Pensée et l'Action Coloniale de Maurepas vis-à-vis du Canada 1723–1749*. Montréal: Leméac.

Fleury de Saint-Charles. 1903. *Un Attaché militaire français à l'armée russe (1759–1760): Le Marquis de Montalembert*. Paris: Plon-Nourrit.

Fontenelle, Bernard le Bovier de. 1709. "Éloge de Vauban." In *Histoire du Renouvellemnt de l'Académie royale des sciences en MDCXCIX et les Éloges historiques de tous les Académiciens morts depuis ce renouvellement: Avec un discours préliminaire sur l'utilité des mathématiques et de la Physique*. Amsterdam. Reprint: Brussels, 1969.

Forbonnais, François-Vernon-Duverger. 1758. *Recherches et considérations sur les finances de France*.

Ford, Franklin Lewis. 1953. *Robe and Sword: The Regrouping of the French Aristocracy after Louis XIV*. Harvard University Press.

Fourcroy, Antoine-François. 1794. *Rapport sur les mésures prises par le comité de Salut Public pour l'établissement de l'école centrale des travaux publics, décrétée par la Convention National, le 21 ventôse dernier; et projet de décret pour l'ouverture de cette école et l'admission des élèves; présentés, au nom des comités de salut public, d'instruction publique et des travaux publics, réunis, par Fourcroy (3 vendémiaire, An 3)*. Paris: Imprimerie du Comité de Salut Public, An 3.

Fourcroy de Ramecourt, Charles-René. 1786. *Mémoires sur la fortification perpendiculaire, par plusieurs officiers du génie*. Paris: Nyon.

Frézier, Amédée-François. 1737–1739. *La Théorie et la pratique de la coupe des pierres et des bois pour la construction des voûtes et autres parties des bâtiments civils et militaires, ou Traité de stéréotomie, à l'usage de l'architecture*. Strasbourg: J.-D. Doulsseker le fils; Paris, L. H. Guérin.

Frézier, Amédée-François. 1760. *Eléments de stéréotomie à l'usage de l'architecture pour la coupe des pierres*. Paris: Charles-Antoine Jombert.

Frost, Robert L. 1991. *Alternating Currents: Nationalized Power in France, 1946–1970*. Cornell University Press.

Fry, Bruce W. 1984. *"An Appearance of Strength": The Fortifications of Louisbourg*. Ottawa: Parks Canada.

Gadol, Joan. 1969. *Leon Battista Alberti: Universal Man of the Early Renaissance*. University of Chicago Press.

Galilei, Galileo. 1932a. "Breve Instruzione all'Architettura Militare." In *Le Opere di Galileo Galilei*, ed. A. Favaro. Firenze: G. Barbera.

Galilei, Galileo. 1932b. "Tratatto di Fortificazione." In *Le Opere di Galileo Galilei*, ed. A. Favaro. Firenze: G. Barbera.

Galilei, Galileo. 1978. *Operations of the Geometric and Military Compass*. Smithsonian Institution Press.

Galilei, Galileo. 1638. *Two New Sciences, Including Centers of Gravity & Force of Percussion*. University of Wisconsin Press, 1974.

Galluzzi, Paolo. 1987. *Leonardo da Vinci: Engineer and Architect*. Montreal Museum of Fine Arts.

Garcin, Nicole. 1972. *De Longwy et Vauban*. Lille.

Gat, Azar. 1989. *The Origins of Military Thought from the Enlightenment to Clausewitz*. Clarendon.

Gaudin, Abbé. 1938. "Marc-René Marquis de Montalembert." *Bulletins et Mémoires de la Société archéologique et historique de la Charente* 51–127.

Gautier, Hubert. 1687. *L'Art de Laver, ou Nouvelle Manière de Peindre sur le papier, suivant le coloris des desseins qu'on envoye à la cour*. Lyon: Thomas Amaulry.

Gay de Vernon, Simon-François. 1805 [an XIII]. *Traité élémentaire d'art militaire et de fortification, à l'usage des élèves de l'École polytechnique, et des élèves des Écoles militaires*. Paris: Allais.

Gayvernon, S. F. 1799. "Discours sur l'enseignement de la géométrie descriptive." *Journal de l'École Polytechnique* 2: 209–269.

Gayvernon, S. F. 1802 [an 10]. *Exposition abrégée du cours de géométrie descriptive appliquée à la fortification; à l'usage des élèves de l'École polytechnique*. Paris: H. Perronneau.

Geiger, Reed G. 1994. *Planning the French canals: Bureaucracy, politics, and enterprise under the Restoration*. University of Delaware Press/Associated University Presses.

Gelfand, Toby. 1984. "A "monarchical profession" of the Old Regime: Surgeons, ordinary practitioners, and medical professionalization in eighteenth-century France." In *Professions and the French State, 1700–1900*, ed. G. Geison. Philadelphia.

Gille, Bertrand. 1960. *Les forges françaises en 1772*, Paris: S.E.V.P.E.N.

Gille, Bertrand. 1964. *Ingénieurs de la Renaissance*. Hermann.

Gille, Bertrand, ed. 1978. *Histoire des techniques*. Gallimard.

Gille, Bertrand. 1980. *Les mécaniciens grecs: La naissance de la technologie*. Paris: Seuil.

Gillispie, Charles Coulston. 1957. "The natural history of industry." *Isis* 48: 398–407.

Gillispie, Charles Coulston, ed. 1959. *A Diderot Pictorial Encyclopedia of Trades and Industry*. Dover.

Gillispie, Charles Coulston. 1971. *Lazare Carnot: Savant*. Princeton University Press.

Gillispie, Charles Coulston. 1980. *Science and Polity in France at the End of the Old Regime*. Princeton University Press.

Gillispie, Charles Coulston, and Ken Alder. 1998. "Exchange: Engineering the Revolution." *Technology and Culture* 39: 733–754.

Gillmor, C. Stewart. 1971. *Coulomb and the Evolution of Physics and Engineering in 18th Century France*. Princeton University Press.

Godlewska, Anne. 1999. *Geography Unbound: French Geographic Science from Cassini to Humboldt*. University of Chicago Press.

Goënaga, Jean-Marie. 1990. "Aspects techniques de la fortification selon Lazare Carnot." In *Lazare Carnot ou Le Savant-Citoyen*, ed. J.-P. Charnay. Presses Universitaires de l'Université de Paris-Sorbonne.

Goldthwaite, Richard A. 1980. *The Building of Renaissance Florence: An economic and social history*. Johns Hopkins University Press.

Gooding, S. James. 1972. *An Introduction to British Artillery in North America*, second revised edition. Museum Restoration Service.

Gordon, J. E. 1978. *Structures, or Why Things Don't Fall Down*. Penguin.

Goulon. 1730. *Mémoires pour l'attaque et la deffense d'une place*. La Haye: Pierre Gosse.

Grandjean de Fouchy, Jean-Paul. 1761. "Éloge de M. de Bélidor." *Histoire de l'Académie royale des Sciences* 167–181.

Grattan-Guinness, Ivor. 1993. "The ingénieur-savant, 1800–1830: A neglected figure in French mathematics and science." *Science in Context* 6: 405–434.

Grelon, André. 1984. "Les ingénieurs, encore." *Culture technique* 12: 11–18.

Grigorian, A. T., et al. 1968. *Russko-frantsuszkii nauchnie sviazi*. Leningrad: Nauka.

Grundy, Mark. 1991. "The Martello Towers of Minorca." *Fort* 19: 23–58.

Guerlac, Henry. 1943. "Vauban: The impact of science on war." In *The Makers of Modern Strategy*, ed. E. Earle. Princeton University Press.

Guibert, Jacques-Antoine-Hippolyte comte de. 1773. *Essai général de tactique*, second edition. Londres: Libraires associés.

Guiffrey, J., ed. 1881–1901. *Comptes des bâtiments du roi sous le règne de Louis XIV*. Paris: Imprimerie nationale.

Guillerm, Alain. 1985. *La Pierre et le Vent: Fortifications et Marine en Occident*. Paris: Arthaud.

Guillerme, André. 1988. "L'Émergence du concept de réseaux 1820–1830." In *Réseaux territoriaux*, ed. G. Dupuy. Caen: Paradigme.

Haber, Samuel. 1971. *Efficiency and Uplift: Scientific Management in the Progressive Era, 1890–1920*. University of Chicago Press.

Hacker, Barton. 1993. "Engineering a new order: Military institutions, technical education, and the rise of the industrial state." *Technology and Culture* 34: 1–28.

Hacker, Barton. 1994. "Military institutions, weapons and social change: Toward a new history of military technology." *Technology and Culture* 35: 768–834.

Hahn, Roger. 1964. "l'Enseignement scientifique aux ecoles militaires et d'artillerie." In *Enseignement et diffusion des sciences en France au XVIIIe siècle*, ed. R. Taton. Reprint: Hermann, 1986

Hahn, Roger. 1971. *The Anatomy of a Scientific Institution: The Paris Academy of Sciences, 1666–1803*. University of California Press.

Hale, John Rigby. 1977. *Renaissance Fortification: Art or Engineering?* London: Thames and Hudson.

Hale, John Rigby. 1983. *Renaissance War Studies*. Hambledon.

Hale, John Rigby. 1985. *War and Society in Renaissance Europe 1450–1620*. Leicester: Leicester University Press.

Hale, John Rigby. 1990. *Artists and Warfare in the Renaissance*. New Haven: Yale University Press.

Halévy, Daniel. 1923. *Vauban, Les Cahiers Verts, pub. sous la direction de Daniel Halévy, 21*. Paris: Grasset.

Hall, A. R. 1952. *Ballistics in the 17th Century: A Study in the Relations of Science and War with Reference Principally to England*. Cambridge University Press.

Hall, A. R. 1983. "Science, technology, and warfare, 1400–1700." In *Science, Technology, and Warfare*, ed. M. Wright and L. Paszek. US Government Printing Office.

Hall, Bert S. 1999. *Weapons and Warfare in Renaissance Europe*. Johns Hopkins University Press.

Harvey, John Hooper. 1971. *The master builders: Architecture in the Middle Ages*. London: Thames and Hudson.

Harvie, C. 1977. "'The sons of Martha': Technology, transport, and Rudyard Kipling." *Victorian Studies* 20, no. 3: 270–282.

Head, Francis Bond. 1869. *The Royal Engineer*. London: John Murray.

Hebbert, F. J. 1990. "Belle-Ile and the siege of 1761." *Fort* 18: 69–82.

Hebbert, F. J. 1997. "Vauban and the building of Phalsbourg in 1679." *Fort* 25: 111–140.

Hebbert, F. J., and G. A. Rothrock. 1989. *Soldier of France: Sebastien Le Prestre de Vauban, 1633–1707*. Peter Lang.

Hecht, Gabrielle. 1998. *The Radiance of France: Nuclear Power and National Identity after World War II*. MIT Press.

Herf, Jeffrey. 1984. "The engineer as ideologue: Reactionary modernists in Weimar and Nazi Germany." *Journal of Contemporary History* 19: 631–648.

Heyman, Jacques. 1972. *Coulomb's Memoir on Statics: An Essay in the History of Civil Engineering*. Cambridge University Press.

Heyman, Jacques. 1998. *Structural Analysis: A Historical Approach*. Cambridge University Press.

Hilaire-Pérez, Liliane. 2000. *L'Invention technique au siècle de Lumières, L'Évolution de l'humanité*. Paris: Albin Michel

Hittle, James Donald. 1994. *The Military Staff: Its History and Development*.

Hoefer, Ferdinand, ed. 1852–1866. *Nouvelle biographie universelle depuis les temps les plus reculés jusqu'à nos jours, avec les renseignements bibliographiques et l'indication des sources à consulter*. Paris: Firmin Didot Frères.

Hogg, Ian V. 1975. *Fortress: A History of Military Defence*. MacDonald and Jane.

Hogg, Ian V. 1981. *The History of Fortification*. Orbis.

Hoog, Simone, et al., eds. 1982. *Manière de montrer les jardins de Versailles, Collection Albums*. Paris: Éditions de la Réunion des Musées nationaux.

Hounshell, David A. 1984. *From the American System to Mass Production 1800–1932: The Development of Manufacturing Technology in the United States*. Johns Hopkins University Press.

Howard, Michael Eliot. 1976. *War in European history*. Oxford ; Toronto: Oxford University Press.

Hublot, Emmanuel. 1987. *Valmy ou la défense de la nation par les armes*. Paris: Centre de Documentation de la Défense nationale.

Hüe de Caligny, Louis-Roland. 1846. *Traité de la Défense des Pacs fortes avec application à la place de Landau, redigé en 1723*. Paris: Corréard.

Hughes, James Quentin. 1974. *Military Architecture*. Evelyn.

Hughes, James Quentin. 1980. A Chronology of Events in Fortification from 1800 to 1914 and An Illustrated English Glossary of Terms used in Military Architecture. Fortress Study Group and Liverpool School of Architecture, University of Liverpool.

Hughes, James Quentin. 1987. "Wellington and fortifications." *Fort* 15: 61–90.

Hughes, James Quentin. 1993. "Kronstadt and the Crimean War." *Fort* 23: 55–81.

Hughes, Thomas P. 1969. "Technological momentum in history: Hydrogenation in Germany 1898–1933." *Past and Present* 44, August: 106–132.

Hughes, Thomas P. 1983. *Networks of Power: Electrification in Western Society 1880–1930*. Johns Hopkins University Press.

Hughes, Thomas P. 1987. "The evolution of large technological systems." In *The Social Construction of Technological Systems*, ed. W. Bijker et al. MIT Press.

Hughes, Thomas P. 1998. *Rescuing Prometheus*. Pantheon.

Humphrey, John H. 1838. *An Essay on the Modern System of Fortification Adopted for the Defence of the Rhine Frontier and Followed in a Greater or Less Degree in All the Principal Works of the Kind Now Constructed on the Continent Exemplified in a Copious Memoir on the Fortress of Coblenz and Illustrated by plans and Sections of the Work at the Place*. London: John Weale.

Huntington, Samuel P. 1972. *The Soldier and the State: The Theory and Politics of Civil-Military Relations*. Belknap.

Hutton, Charles. 1778. *The Force of Fired Gun-Powder, and the Initial Velocities of Cannon Balls, determined by Experiments; From which is also deduced the Relation of the Initial Velocity to the Weight of the Shot and the Quantity of Powder*. London: J. Nichols.

Hutton, Stanley Peerman, and Joel Emery Gerstl. 1966. *Engineers: The Anatomy of a Profession, a Study of Mechanical Engineers in Britain*. Tavistock.

Jackson, Melvin H., and Carel de Beer. 1973. *Eighteenth-Century Gunfounding: The Verbruggens at the Royal Brass Foundry, A Chapter in the History of Technology*. Newton Abbot: David and Charles.

Jähns, Max. 1966. *Geschichte der Kriegswissenschaften vornehmlich in Deutschland: Das 18. Jahrhundert seit dem Auftreten Friedrichs des Grossen. 1740–1800*. Johnson Reprint Corporation.

Janowitz, Morris. 1960. *The professional soldier, a social and political portrait*. Free Press.

Jardin, A. and Tudesqu, A. J. 1973. *La France des notables: Évolution générale 1815–1848*, and *La France des notables: La Vie de la Nation 1815–1848*. Seuil.

Johnson, Stephen B. 1997. "Three approaches to big technology: Operations research, systems engineering, and project management." *Technology and Culture* 38: 891–920.

Joravsky, David. 1970. *The Lysenko Affair*. Harvard University Press.

Kanigel, Robert. 1997. *The One Best Way: Frederick Winslow Taylor and the Enigma of Efficiency*. Viking.

King, James Edward. 1949. *Science and Rationalism in the Government of Louis XIV 1661–1683*. Johns Hopkins University Press.

Kite, Elizabeth S. 1933. *Brigadier-General Louis Lebègue Duportail: Commandant of Engineers in the Continental Army 1777–1783*. Johns Hopkins University Press.

Konvitz, Joseph. 1981. "La cartographie et les travaux publics 1820–1870." *Annales des Ponts et Chaussées*, n.s., no. 19: 96–103

Konvitz, Josef. 1987. *Cartography in France, 1660–1848: Science, Engineering and Statecraft*. University of Chicago Press.

Kostof, Spiro, ed. 1977. *The Architect: Chapters in the History of the Profession*. Oxford University Press.

Krohn, Wolfgang, Edwin Layton Jr., and Peter Weingart, eds. 1978. *The Dynamics of Science and Technology: Social Values, Technical Norms and Scientific Criteria in the Development of Knowledge*. Reidel.

Kruft, Hanno-Walter. 1994. *A History of Architectural Theory from Vitruvius to the Present*. Zwemmer.

Kuhn, Thomas S. 1976. "Mathematical versus experimental traditions in the development of the physical sciences." *Journal of Interdisciplinary History* 7: 1–31.

Lacombe, Henri. 1933. "Un procès contre Marc-René, Marquis de Montalembert." *Bulletin et Mémoires de la Société archéologique et historique de la Charente* xci–xcviii.

Lafayette, M. J. P. Y. R. G. du Motier marquis de. 1837. *Mémoires, Correspondance et Manuscrits du Général Lafayette, publiés par sa famille*. Brussels: Société Belge du Librairie.

Lalande. 1800. "Éloge de Montalembert par Lalande." *Magasin encyclopédique* 1: 123–129.

Lalonde. 1665. *Elémens de Fortification. Première partie, qui contient l'Arithmetique de l'Ingénieur François; Où l'on verra plusieurs nouvelles Methodes qui abbregent et facilitent extrêmement les calculs des Toizes de la Maçonnerie, des Terres, et de la Charpente. Ouvrage utile non-seulement aux Ingénieurs, mais encore aux Architectes, aux Arpenteurs, aux Financiers et aux Marchands*, second edition. Paris: Denys Nion.

Lalonde. 1689. *L'Arithmétique des ingénieurs, contenant le calcul des toises de la maçonnerie, des terres et de la charpente*, second edition. Paris: Denys Nion.

Lamberini, Daniela. 1986. "The military architecture of Giovanni Battista Belluzzi." *Fort* 14: 5–16.

Lamberini, Daniela. 1987. "Practice and theory in sixteenth-century fortifications." *Fort* 15: 5–20.

Langins, Janis. 1987. *La République avait besoin de savants: les débuts de l'École polytechnique: l'École centrale des travaux publics et les cours révolutionnaires de l'an III*. Paris: Belin.

Langins, J. 1990a. "The École Polytechnique and the French Revolution: Merit, militarization, and mathematics." *Llull* 13: 91–105.

Langins, Janis. 1990b. "Carnot, Montalembert, et le Corps royal du Génie dans la polémique sur la fortification perpendiculaire." In *Lazare Carnot ou Le Savant-Citoyen*, ed. J.-P. Charnay. Presses Universitaires de l'Université de Paris-Sorbonne.

Langins, Janis. 1991. "La préhistoire de l'École polytechnique." *Revue d'histoire des sciences* 44: 61–89.

Laudan, Rachel. 1984. "Cognitive change in technology and science." In *The Nature of Technological Knowledge*, ed. R. Laudan. Reidel.

Lauerma, Matti. 1956. *L'artillerie de campagne française pendant les guerres de la Révolution: Évolution de l'organisation de la tactique*. Helsinki: Keskuskirjapaino.

Lavisse, Ernest, ed. 1901–1911. *Histoire de France depuis les origines jusqu'a la Révolution*. Hachette.

Layton, Edwin, Jr. 1971a. *The Revolt of the Engineers: Social Responsibility and the American Engineering Profession*. Case Western Reserve Press.

Layton, Edwin, Jr. 1971b. "Mirror-image twins: The communities of science and technology in nineteenth-century America." *Technology and Culture* 12: 562–580.

Layton, Edwin T., Jr. 1973. "Veblen and the engineers." In *Technology and social change in America*, ed. E. Layton. Harper & Row.

Lazard, P. 1934. *Vauban 1633–1707*. Paris: Alcan.

Leblond, Guillaume. 1742. *Elémens de fortification à l'usage des jeunes officiers*, second edition. Paris: Jombert.

Leblond, Guillaume. 1748. *L'arithmétique et la geometrie de l'officier; contenant la theorie et la pratique de ces deux sciences, appliquées aux differens emplois de l'homme de guerre*. Paris: Jombert.

Le Michaud d'Arçon, Jean-Claude-Éléonor. 1774. *Suite de la correspondance sur l'art de la guerre*. Bouillon: Fantel.

Le Michaud d'Arçon, Jean-Claude-Éléonor. 1783. *Mémoire pour servir à l'histoire du siège de Gibraltar par l'auteur des batteries flottantes*. Cadiz: Hermil frères.

Le Michaud d'Arçon, Jean-Claude-Éléonor. 1786. *Considérations sur l'influence du génie de Vauban dans la balance des forces de l'Etat*. Strasbourg.

Le Michaud d'Arçon, Jean-Claude-Éléonor. 1790. *Réponse aux Mémoires de M. Montalembert, Publiés en 1790, sur la fortification dite perpendiculaire, la composition des casemates inexpugnables, la multiplication illimitée des bouches à feux, le projet d'enceindre le Royaume par des lignes imprenables et d'autres idées d'une apparence très importante; pour servir d'apologie aux principes observés dans le Corps-royal du génie*. Paris.

Le Michaud d'Arçon, Jean-Claude-Éléonor. 1794 [an 2]. *Des Fortifications générales de la guerre de siège, pour servir de Réponse au dernier ouvrage de Marc-René Montalembert*. Paris: Magimel.

Le Michaud d'Arçon, Jean-Claude-Éléonor. 1795 [an III]. *Considérations Militaires et Politiques sur les Fortifications*. Paris: Imprimerie de la République.

Léon, Antoine. 1961. *Histoire de l'Éducation technique*. Presses Universitaires de France.

Léon, Pierre. 1961. *Les techniques métallurgiques dauphinoises au dix-huitième siècle*. Paris: Hermann.

Léonard, Émile G. 1958. *L'armée et ses problèmes au 18ème siècle*. Plon.

Lepelley, Roger. 1990. *Le vieil arsenal de Cherbourg: de 1793 à 1814*. Société nationale académique de Cherbourg.

Lepetit, Bernard. 1988. "L'Impensable réseau: Les routes françaises avant le chemin de fer." In *Réseaux territoriaux*, ed. G. Dupuy. Caen: Paradigme.

Le Pourhiet-Salat, Nicole. 1983. *La défense des îles bretonnes de l'Atlantique, des origines à 1860*. Vincennes: Service historique de la marine.

Le Puillon de Boblaye, Theodore. 1858. *Esquisse Historique sur les Écoles d'Artillerie pour servir à l'histoire de l'École d'application de l'artillerie et du génie*. Metz: Rousseau-Pallez.

Levin, A. 1984. "Venel, Lavoisier, Fourcroy, Cabanis and the idea of scientific revolution: The French political context and the general patterns of conceptualization of scientific change." *History of Science* 22: 303–320.

Livet, Georges. 1985. "Un 'modèle à la Vauban' d'équilibre européen au XVIIème siècle." In *Vauban réformateur*, ed. C. Brisac et al. Association Vauban.

Lloyd, E. M. 1887. *Vauban, Montalembert, Carnot: Engineer Studies*. London.

Lucas, Colin. "Nobles, bourgeois and the origins of the French Revolution." *Past and Present* 60 (August 1973).

Lund, Erik. 1996. Science and the Professionalization of Military Elites: War and Knowledge in the Holy Roman Empire of the German Nation, 1683–1720. Ph.D. thesis, University of Toronto.

Lund, Erik. 1999. *War for the Every Day: Generals, War, and Knowledge in Early Modern Europe, 1680–1740*. Greenwood.

Lundgreen, Peter. 1990. "Engineering education in Europe and the U.S.A., 1750–1930: The rise to dominance of school culture and the engineering professions." *Annals of Science* 47: 33–75.

Lynn, John A. 1980. "The growth of the French Army during the seventeenth century." *Armed Forces and Society* 6: 568–585.

Lynn, John A. 1984. *The Bayonets of the Republic: Motivation and Tactics in the Army of Revolutionary France, 1791–94*. University of Illinois Press.

Lynn, John A., ed. 1990a. *Tools of War: Instruments, Ideas, and Institutions of Warfare, 1445–1971*. University of Illinois Press.

Lynn, John A. 1990b. "The pattern of army growth, 1445–1945." In *Tools of War*, ed. J. Lynn University of Illinois Press.

Lynn, John A. 1990c. "En avant! The origins of the revolutionary attack." In *Tools of War*, ed. J. Lynn. University of Illinois Press.

Lynn, John A. 1991. "The trace italienne and the growth of armies: The French case." *Journal of Military History* 55: 291–330.

Maggi, Girolamo, and Iacomo Castriotto. 1583. *Della Fortificationi delle Citta*. Venice: Camillo Borgominiero.

Mallagh, Christopher. 1981. Science, Warfare and Society in the Renaissance with Particular Reference to Fortification Theory. Ph.D. thesis, Leeds University.

Mandar, C. F. 1801. *L'Architecture des forteresses, ou De l'Art de fortifier les places, et de disposer les Établissemens de tout genre, qui ont rapport à la guerre*. Paris: Magimel.

Marat, Jean-Paul. 1791. *Les Charlatans modernes, ou Lettres sur le charlatanisme académique, publiées par M. Marat*. Paris: Imprimerie de Marat.

Marchi, Francesco de'. 1599. *Della architettura militare libri tre*. Brescia.

Mariage, Thierry. 1990. *L'univers de Le Nostre*. Brussels: P. Mardaga.

Marolois, Samuel. 1627. *Fortification ou Architecture Militaire tant offensive que defensive*. Amsterdam: Jan Jansen.

Marshall, Douglas W. 1976. The British Military Engineers 1741–1783: A Study of Organization, Social Origins, and Cartography. Ph.D. thesis, University of Michigan.

Martin, Georges. 1988. *Histoire et généalogie de la Maison Montalembert*. La Ricamarie: G. Martin.

Massoud, Zaher, and Raoul Piboubès, eds. 1994. *L'Atlas du Littoral de France*. Paris: Edition Jean-Pierre de Monza.

Maury, Claude. 1984. "L'ingénieur, un homme sans image?" *Culture technique* 12: 49–54.

Mayr, Otto. 1986. *Authority, Liberty, and Automatic Machinery in Early Modern Europe*. Johns Hopkins University Press.

McClellan, James E. 1985. *Science Reorganized: Scientific Societies in the Eighteenth Century*. Columbia University Press.

McConnell, David. 1988. *British Smooth-Bore Artillery: A Technological Study to Support Identification, Acquisition, Restoration, and Interpretation of Artillery at National Historic Parks in Canada*. Ottawa: Environment Canada–Parks.

McNeill, William H. 1982. *The Pursuit of Power: Technology, Armed Force, and Society Since A.D. 1000*. University of Chicago Press

Mention, Léon. 1884. *Le Comte de Saint-Germain et ses réformes, 1775–1776*. Paris: A. Clavel.

Mesqui, Jean. 1985. "Vauban et le projet de transport fluvial." In *Vauban réformateur*, ed. C. Brisac et al. Association Vauban.

Michaud, J. F., and L. G. Michaud. 1811–1862. *Biographie universelle, ancienne et moderne*. Paris: Michaud

Michel, Georges, André Liesse, and Léon Say. 1891. *Vauban économiste*. Paris: E. Plon Nourrit.

Migos, Athanasios. 1990. "Rhodes: The knights' battleground." *Fort* 18: 5–28.

Milliet Deschales, Claude François. 1677. *L'Art de Fortifier, de Defendre, et d'Attaquer les Places, suivant les Methodes Françoises, Hollandoises, Italiennes et Espagnoles*. Paris: Estienne Michallet.

Mirabeau, H. G. R. de. 1789. *Tactique prussienne, ou Système militaire de la Prusse*. Paris: Mardan.

Montalembert, Jean-Charles, baron de. n.d *Lettre de M. le Baron de Montalembert à M. de Kéralio, en Réponse au Compte qu'il a rendu dans le Journal des Savans, du Mémoire sur la Fortification perpendiculaire, publiée sous le nom des Plusieurs Officiers du Corps Royal du Génie en 1786*. London: Spilsbury.

Montalembert, Marc-René de. 1755. "Mémoire sur la rotation des boulets dans les pièces de canon." *Mémoires de l'Académie Royale des Sciences* 463–468.

Montalembert, Marc-René de. 1761. *La Fortification perpendiculaire* (prospectus). Paris: Desaint & Saillant.

Montalembert, Marc-René de. 1766. "Cheminée poêle ou poêle françois, mémoire lu à la rentrée publique de l'Académie royale des sciences, 12 novembre 1763." Imprimerie royale.

Montalembert, Marc-René de. 1768. *Extrait de divers mémoires imprimés sur les nouvelles forges établies en Angoumois, par le marquis de Montalembert, et sur les fabrications d'Artillerie qu'il y a fait exécuter pour la Marine, depuis l'année 1750*. Paris.

Montalembert, Marc-René de. 1776–1797. *La Fortification perpendiculaire*. Paris.

Montalembert, Marc-René de. 1777. *Correspondance de Monsieur le Marquis de Montalembert, Étant Employé par le Roi à l'Armée Suédoise, avec Mr. le Marquis d'Havrincour, Ambassadeur en Suède, Monseigneur le Maréchal de Richelieu, Les Ministres du Roi à Versailles, MM. les Généraux Suédois et autres, etc. Pendant les campagnes de 1757, 58, 60, et 61*. Londres.

Montalembert, Marc-René de. 1786. *La Statue, Comédie en deux actes, en prose, mêleée d'ariettes, . . . , Musique de M. de Cambinij Représentée pour la première fois, sur le Théâtre de l'Hôtel de Montalembert, au mois d'Août 1784*. Paris.

Montalembert, Marc-René de. 1787. *Réponse au mémoire sur les fortifications perpendiculaires par plusieurs officiers du corps royal du génie.* Paris.

Montalembert, Marc-René de. 1790a. *Observations sur les nouveaux Forts qui ont été exécutés, & qui doivent l'être pour la défense de la rade de Cherbourg; Où l'on fait mention des travaux faits au Havre, à Dunkerque, & à l'île de France; Où l'on donne enfin les moyens de faire exécuter à l'avenir des Ouvrages moins coûteux & d'une meilleure défense. Avec un Projet de nouvelles lignes frontières permanentes, pour couvrir les Provinces du Royaume*, Paris: Philippe-Denys Pierres.

Montalembert, Marc-René de. 1790b. *Réponse au Colonel d'Arçon, Auteur des Batteries flottantes, sur son apologie des Principes observés dans le Corps du génie.* Paris: Philippe-Denys Pierres et Didot.

Montalembert, Marc-René. 1793. *L'Art Défensif supérieur à l'Offensif, par une nouvelle manière d'employer l'Artillerie et par la suppression totale des bastions, comme étant la principale cause du peu de résistance des Places de guerre. Formant la suite et le dernier volume de la Fortification perpendiculaire*, Paris: Firmin Didot.

Montalembert, Marc-René de. 1795–1797. *L'ami de l'art défensif, ou observations sur le journal de l'école polytechnique.* Paris: Louvet et Maginael.

Montbarey, Alexandre-Marie-Léonor de Saint-Maurice, prince de. 1827. *Mémoires du prince de Montbarey.* Paris: Alexis Eymery.

Montesson, Dupain de. 1757. *Les Amusemens militaires: Ouvrage également agréable et instructif: servant d'introduction aux Sciences qui forment des guerriers.* Paris: Guillaume Desprez.

Morineau, M. 1985. "Tombeau pour un maréchal de France: la Dîme royale." In *Vauban réformateur*, ed. C. Brisac et al. Association Vauban.

Mornet, Daniel. 1910. "Les Enseignements des bibliothèques privées (1750–1780)." *Révue d'histoire littéraire de la France* 17: 449–496.

Mosser, Monique, and Georges Teyssot. 1991. *The Architecture of Western Gardens: A Design History from the Renaissance to the Present Day.* MIT Press.

Mousnier, Roland. 1974–1980. *Les Institutions de la France sous la monarchie absolue.* Presses Universitaires de France.

Muller, John. 1747. *The attack and defence of fortify'd places: in three parts: containing, I. Preparations, and different operations of an attack, from the beginning to the end, digested in a regular and easy manner. II. Preparations, and defence of every particular part of a fortification, with every thing necessary for a good defence. III. A treatise of mines, explaining the manner of making and loading them, with tables of their proper charges, deduced from a new theory: for the use of the Royal Academy of Artillery at Woolwich.* London: J. Millan.

Muller, John. 1780. *A Treatise of Artillery*, third edition. London: John Millan.

Mumford, Lewis. 1964 "Authoritarian and democratic technics." *Technology and Culture* 5: 303–317.

Myatt, Frederick. 1998. *British Sieges of the Pensinsular War.* Spellmount.

Napoleon. 1858. *Correspondance de Napoléon Ier; publiée par ordre de l'empereur Napoléon III.* Paris: H. Plon, J. Dumaine.

Nardin, Pierre. 1982. *Gribeauval: Lieutenant général des armées du roi (1715–1789)*. Paris: Fondation pour les Études de Défense Nationale.

Naudin. 1695. *L'ingénieur françois, contenant la géométrie pratique sur le papier et sur le terrain... la fortification reguliere et irreguliere... Avec la méthode de Monsieur d Vauban, à l'explication de son nouveau système*. Paris: Estienne Michalet.

Nef, John U. 1940. *Industry and Government in France and England 1540–1640*. American Philosophical Society.

Niel, Adolphe. 1855. *Siège de Bomarsund en 1854: Journal des Opérations de l'Artilleirie et du Génie*. Paris: J. Corréard.

Niel, Adolphe. 1858. *Siège de Sebastopol: Journal du Génie, publié avec l'autorisation du Ministère de la Guerre*. Paris: Dumaine.

Noble, David F. 1977. *America by Design: Science, Technology, and the Rise of Corporate Capitalism*. Reprint: Oxford University Press, 1979.

Noble, David F. 1985. "Command performance: A perspective on the social and economic consequences of military enterprise." In *Military Enterprise and Technological Change*, ed. M. Smith. MIT Press.

Noizet, F. 1823. "Mémoire sur la Géométrie appliquée à la fortification." *Mémorial de l'Officier du Génie, ou Recueil de mémoires, expériences, observations et procédés généraux propres à perfectionner la Fortification et les Constructions militaires* 6: 5–224.

Noizet Saint-Paul, Gaspard. 1792. *Traité Complet de fortification; Ouvrage utile aux jeunes militaires, et mis à la portée de tout le monde*. Paris: Barrois.

Norwood, Richard. 1639. *Fortification: or, Architecture military*. Amsterdam: Theatrum Orbis Terrarum.

O'Connell, Charles F., Jr. 1985. "The Corps of Engineers and the rise of modern management." In *Military Enterprise and Technological Change*, ed. M. Smith. MIT Press.

Olivier, Théodore. 1850. "Monge et l'École Polytechnique." *La Revue scientifique et industrielle du Docteur Quesneville*, Février.

Pagan, Blaise-François. 1689. *Les fortifications du comte de Pagan*. Paris: Nicolas Langlois.

Pallière, J. 1979. "Un grand méconnu du XVIIIe siècle: Pierre Bourcet (1700–1780)." *Revue historique des armées* 1: 51–66.

Palmer, Robert R., ed. 1975. *The School of the French Revolution A Documentary History of the College of Louis-le-Grand and its Director, Jean-François Champagne 1762–1814*. Princeton University Press.

Palmer, Robert R. 1986. "Frederick the Great, Guibert, Bülow: From dynastic to national war." In *Makers of Modern Strategy from Machiavelli to the Nuclear Age*, pp. 91–119. Princeton University Press.

Pannabecker, John R. 1998. "Representing mechanical arts in Diderot's *Encyclopédie*." *Technology and Culture* 39: 33–74.

Parent, Michel. 1982. *Vauban: Un encyclopédiste avant la lettre*. Paris: Berger-Levrault.

Parent, M. 1971. *Vauban*. Paris: J. Freal.

Parker, Geoffrey. 1988. *The Military Revolution: Military Innovation and the Rise of the West, 1500–1800*. Cambridge University Press.

Parker, Harold T. 1993. *An Administrative Bureau during the Old Regime: The Bureau of Commerce and Its Relations to French Industry from May 1781 to November 1783*. University of Delaware Press.

Parsons, Talcott. 1968–1991. "Professions." In *International Encyclopaedia of the Social Sciences*. Macmillan.

Pelletier, Monique. 1990. *La carte de Cassini: l'extraordinaire aventure de la carte de France*. Presses de l'École nationale des Ponts et Chaussées.

Pepper, Simon. 1976. "Planning versus fortification: Sangallo's project for the defence of Rome." *Fort* 2: 33–49.

Pepper, Simon, and Nicholas Adams. 1986. *Firearms and Fortifications: Military Architecture and Siege Warfare in Sixteenth-Century Siena*. University of Chicago Press.

Perez-Gomez, Alberto. 1983. *Architecture and the Crisis of Modern Science*. MIT Press.

Pernot, Jean-François. 1987. "La Guerre et l'infrastructure de l'état moderne: Antoine de Ville, Ingénieur du Roi (1596?–1656?), La Pensée d'un technicien au service de la mobilisation totale du royaume." *Revue d'histoire moderne et contemporaine* 34, Juillet-Septembre: 404–426.

Pernot, Jean-François. 1989. "L'ingénieur Pierre d'Argencourt, le 'fidèle' des cardinaux." In *Mélanges André Corvisier*. Paris: Economia.

Perrucci, Robert, and Joel Gerstl. 1969. *Profession without Community: Engineers in American Society*. Random House.

Perrucci, Robert, and Joel Gerstl, eds. 1969. *The Engineers and the Social System*. Wiley.

Peter, Jean. 1994. *Vauban et Toulon: Histoire de la Construction d'un Port-Arsenal sous Louis XIV*. Economica.

Peter, Jean. 1998. *Le port et l'arsenal de Brest sous Louis XIV*, Hautes études maritimes ; 11. Paris: Institut de stratégie comparée: Economica.

Petot, Jean. 1958. *Histoire de l'administration des ponts et chaussées 1595–1815*. Paris: Marcel Rivière.

Picciola, André. 1999. *Le comte de Maurepas: Versailles et l'Europe à la fin de l'Ancien Régime*. Perrin.

Picon, Antoine. 1987. "L'ingénieur et l'idéal analytique à la fin du 18e siècle." *Sciences et techniques en perspective* 13: 70–108.

Picon, Antoine. 1988. *Architectes et Ingénieurs au Siècle des Lumières*. Marseille: Éditions Parenthèse.

Picon, Antoine. 1989. "Les ingénieurs et la mathématisation: L'exemple du génie civil et de la construction." *Revue d'histoire des sciences* 42: 155–172.

Picon, Antoine. 1989. "Naissance du territoire moderne: Génies civil et militaire à la fin du XVIIIe siècle." *Urbi* 11: c–cxiv.

Picon, Antoine. 1992. *L'Invention de l'Ingénieur Moderne: L'École des Ponts et Chaussées 1747–1851*. Presses de l'École Nationale des Ponts et Chaussées.

Picon, Antoine. 1995. "Charles-François Mandar (1757–1844) ou l'architecture dans tous ses détails." *Revue de l'art* 109: 26–39.

Picon, Antoine. 1996. *La ville et la guerre*. Besançon: Editions de l'Imprimeur.

Picon, Antoine, and Yvon Michel. 1989. *L'Ingénieur Artiste: Dessins anciens de l'École des Ponts et Chaussées*. Paris: Presses de l'École Nationale des Ponts et Chaussées.

Pi-Sunyer, O., and Thomas DeGregori. 1966. "Technology, traditionalism and military establishments." *Technology and Culture* 7: 402–7.

Poisson, Georges. 1985. *Choderlos de Laclos, ou l'Obstination*. Paris: B. Grasset.

Pollak, Martha D. 1991a. *Military Architecture, Cartography and the Representation of the Early Modern City: A Checklist of Treatises on Fortification in the Newberry Library*. Chicago: Newberry Library.

Pollak, Martha D. 1991b. *Turin 1564–1680: Urban Design, Military Culture, and the Creation of the Absolutist Capital*. University of Chicago Press.

Pool, Robert. 1997. *Beyond Engineering: How Society Shapes Technology*. Oxford University Press.

Porter, Whitworth. 1889. *History of the Corps of Royal Engineers*. Longmans, Green.

Prévost de Vernois, Le général. 1861. *De la Fortification depuis Vauban ou Examen des principales innovations qui s'y sont introduites depuis la mort de ce grand homme*. Paris: Dumaine.

Pritchard, James. 1987a. *Louis XV's Navy 1748–1762: A Study of Organization and Administration*. McGill–Queen's University Press.

Pritchard, James. 1987b. "From shipwright to naval constructor: The professionalization of eighteenth-century French naval shipbuilders." *Technology and Culture* 28: 1–25.

Pritchard, James. 1995. *Anatomy of a Naval Disaster: The 1746 French Naval Expedition to North America*. McGill–Queen's University Press.

Prost, Philippe. 1991. *Les Forteresses de l'Empire: Fortifications, villes de guerre et arsenaux napoléoniens*. Paris: Moniteur.

Prost, Philippe. 1996. "La ville et la guerre de Vauban à Napoléon 1er." In *La ville et la guerre*, ed. A. Picon. Besançon: Editions de l'Imprimeur.

Pujo, Bernard. 1991. *Vauban*. Paris: Albin Michel.

Puységur, Jacques-François Chastenet de. 1748. *Art de la Guerre, par Principes et par Règles*. Paris: Charles-Antoine Jombert.

Quimby, Robert Sherman. 1957. *The Background of Napoleonic Warfare: The Theory of Military Tactics in Eighteenth-Century France*. Columbia University Press.

Racine, Jean. 1952. *Oeuvres Complètes*. Gallimard,

Rae, John, and Rudi Volti. 1993. *The Engineer in History*. Peter Lang.

Ramelli, Agostino. 1976. *The Various and Ingenious Machines of Agostino Ramelli: A Classic Sixteenth-Century Illustrated Treatise on Technology*. Johns Hopkins University Press.

Rapaille, Roger. 1992. *Le Siège de Mons par Louis XIV en 1691: Étude du Siège d'une Ville des Pays-Bas pendant la Guerre de la Ligue d'Augsbourg*. Mons: Renard Découvert.

Reinhard, Marcel. 1950–1952. *Le grand Carnot*. Hachette.

Reinhard, Marcel. 1956. "Élite et noblesse dans la seconde moitié du 18e siècle." *Revue d'histoire moderne et contemporaine* 3: 5–37.

Reti, Ladislao. 1963. "Francesco di Giorgio Martini's treatise on engineering and its plagiarists." *Technology and Culture* 4: 287–298.

Reti, Ladislao, and Bern Dibner. 1969. *Leonardo da Vinci, technologist: three essays*. Norwalk, Connecticut: Burndy Library.

Reverony Saint-Cyr, Baron Jacques-Antoine de. 1803. *Essai sur le perfectionnement des Beaux arts, Par les Sciences Exactes, ou Calculs et Hypothèses sur la poésie, la peinture et musique*. Paris: Pougens; Heinrichs; et Magimel.

Reverony Saint-Cyr, Baron Jacques-Antoine de. 1808. *Vauban à Charleroi*. Paris: Barba.

Richard, Camille. 1922. *Le Comité de Salut Public et les fabrications de guerre sous la Terreur*. Paris: Rieder.

Richet, Denis. 1973. *La France Moderne: L'Esprit des Institutions*. Flammarion.

Ringrose, Daniel M. 1995. Engineering Modernity: Civil Engineers between National State and Provincial Society in France, 1840–1914. Ph.D. thesis, University of Michigan.

Robinson, Willard B. 1977. *American Forts: Architectural Form and Function*. University of Illinois Press.

Rochas d'Aiglun, A. 1867. *D'Arçon: Ingénieur militaire—Sa vie et son oeuvre*. Paris/Besançon: Dumaine/Marion.

Rochas d'Aiglun, A., ed. 1910. *Vauban: Sa Famille et Ses Ecrits, Ses Oisivetés et sa Correspondance—Analyse et Extraits*. Reprint: Slatkine, 1972.

Rocolle, Pierre. 1973. *2000 ans de fortification française*. Charles-Lavauzelle.

Rocolle, Pierre. "La réalisation du pré carré." 1985. In *Vauban réformateur*, ed. C. Brisac et al. Association Vauban.

Rocolle, Pierre. 1993. "Les Plans en relief et la représentation du pré carré." Paper presented at the Actes du Colloque International sur les Plans-Reliefs au Passé et au Présent (Les 23, 24, 25 avril 1990) en l'Hôtel National des Invalides, Paris.

Roland, Alex. 1992. "Theories and models of technological change: Semantics and substance." *Science, Technology and Human Values* 17: 79–100.

Rolt, Lionel Thomas Caswell. 1973. *From Sea to Sea: The Canal du Midi*. London: Allen Lane.

Romero, Aldemaro. 1999. "The blind cave fish that never was." *National Speleological Society News* 57, no. 6: 180–181.

Rosen, Howard. 1981. The Système Gribeauval: A Study of Technological Development and Institutional Change in Eighteenth-Century France. Ph.D. thesis, University of Chicago.

Rosenberg, Hans. 1958. *Bureaucracy, Aristocracy and Autocracy: The Prussian Experience 1660–1815*. Harvard University Press.

Rosenberg, Nathan. 1982. *Inside the Black Box: Technology and Economics*. Cambridge University Press.

Rosenberg, Nathan, and Walter G. Vincenti. 1978. *The Britannia Bridge: The Generation and Diffusion of Technological Knowledge*. MIT Press.

Rothrock, George A. 1969. "Musée des plans-reliefs." *French Historical Studies* 6 253–56.

Rouse, Hunter, and Ince, Simon. 1957. *History of Hydraulics*. Reprint: Dover, 1963.

Rousset, Camille. 1864–1879. *Histoire de Louvois et de son administration politique et militaire*. Paris.

Rudyerd, Charles William. 1970. *Course of Artillery at the Royal Military Academy*. Ottawa: Museum Restoration Service.

Russell, Jack. 1965. *Gibraltar Besieged 1779–1783*. London: Heinemann.

Saint-Simon, Henri de. 1952. *Henri Comte de Saint-Simon (1750–1825): Selected Writings*. Translated and edited by F. M. H. Markham. Blackwell.

Saint-Simon, Louis de Rouvroy, duc de. 1988. *Mémoires*. Gallimard.

Sakarovitch, Joël. 1989. *Théorisation d'une Pratique, Pratique d'une Théorie: Des Traités de coupe de pierres à la géométrie descriptive*. Paris: École d'Architecture de Paris La Villette.

Sakarovitch, Joël. 1990. "Architecture et représentation: De la projection à la double projection." *In Extenso: Recherches à l'École d'Architecture Paris-Villemin* no. 13 63–102.

Sautai-Dossin, Anne-Véronique, and Nicolas Faucherre. 1994. "L'État du pré carré." In *Vauban et ses successeurs en Hainaut et d'Entre-Sambre-et-Meuse*. Association Vauban.

Saxe, Maurice. 1757. *Reveries, or memoirs upon the art of war by Field-Marshal Count Saxe. Illustrated with copper-plates. To which are added some original letters, . . . Translated from the French*. London: J. Nourse.

Say, Horace. 1796. "Mémoire sur le défilement des fortifications." *Journal de l'École Polytechnique* 2: 588–616.

Scaglia, Gustina. 1993. *Francesco di Giorgio: Checklist and History of Manuscripts and Drawings in Autographs and Copies from ca. 1470 to 1867 and Renewed Copies (1764–1839)*: Lehigh University Press.

Schalk, Ellery. 1986. *From Valor to Pedigree: Ideas of Nobility in France in the Sixteenth and Seventeenth Centuries*. Princeton University Press.

Schäuffelen, Ottmar. 1989. *Bundesfestung Ulm: Ein Führer durch die Festungsanlagen*. Ulm: Armin Vaas.

Scott, Samuel F. 1978. *The Response of the Royal Army to the French Revolution: The Role and the Development of the Line Army 1787–93*. Clarendon.

Seely, Bruce E. 1984. "The scientific mystique in engineering: Highway research at the Bureau of Public Roads, 1918–1940." *Technology and Culture* 25: 798–831.

Seeman, Ute A. 1992. "The Amsterdam Battery, Cape Town, South Africa." *Fort* 20: 45–54.

Sewell, William H., Jr. 1974. "Etat, corps, and ordre: Some notes on the social vocabulary of the French Old Regime." In *Sozialgeschicte Heute: Festschrift für Hans Rosenberg*, ed. H.-U. Wheler. Göttingen: Vandenhoek und Ruprecht.

Shallat, Todd. 1990. "Building waterways, 1802–1861: Science and the United States Army in early public works." *Technology and Culture* 31: 18–50.

Shallat, Todd. 1994. *Structures in the Stream: Water, Science, and the Rise of the U.S. Army Corps of Engineers*. University of Texas Press.

Shelby, Lon R. 1967. *John Rogers: Tudor Military Engineer.* Clarendon.

Shelby, Lon R. 1972. "The geometrical knowledge of the mediaeval master masons." *Speculum* 47: 395–421.

Shinn, Terry. 1984. "Reactionary technologists: The struggle over the École Polytechnique, 1880–1914." *Minerva* 12: 329–345.

Sinclair, Bruce. 1986. "Inventing a genteel tradition: MIT crosses the river." In *New Perspectives on Technology and American Culture*, ed. B. Sinclair. American Philosophical Society.

Six, Georges. 1929. "Fallait-il quatre quartiers de noblesse pour être officier à la fin de l'Ancien Régime?" *Revue d'Histoire Moderne* 4: 47–56.

Smeaton, W. A. 1955. "The early years of the Lycée and the Lycée des Arts. A chapter in the lives of A. L. Lavoisier and A. F. Fourcroy. I. The Lycée of Rue de Valois. II. The Lycée des Arts." *Annals of Science* 11: 257–267, 309–319.

Smiles, Samuel. 1904. *Lives of the Engineers.* London: Murray.

Smith, Cecil O., Jr. 1990. "The longest run: Public engineers and planning in France." *American Historical Review* 95: 657–692.

Smith, Nathaniel B. 1969. "The idea of the French hexagon." *French Historical Studies* 6, Fall: 139–155.

Smith, Merritt Roe, ed. 1985. *Military Enterprise and Technological Change: Perspectives on the American Experience.* MIT Press.

Solon, Paul. 1981. "Frontiers and boundaries: French cartography and the limitation of Bourbon ambitions in 17c France." In Proceedings of the 10th Annual Meeting of the Western Society for French History.

Specklin, Daniel. 1589. *Architectura von Vestungen; wie die zu unsern zeiten mogen erbawen warden.* Strasbourg.

Stapleton, Darwin H., and Roger L. Shumaker. 1986. *The History of Civil Engineering since 1600: An Annotated Bibliography.* Garland.

Steele, Brett D. 1994. "Muskets and pendulums: Benjamin Robins, Leonhard Euler, and the ballistics revolution." *Technology and Culture* 35: 348–382.

Sterne, Laurence. 1759. *The Life and Opinions of Tristram Shandy, Gentleman.* York.

Stoye, John. 1994. *Marsigli's Europe, 1680–1730: The Life and Times of Luigi Ferdinando Marsigli, Soldier and Virtuoso.* Yale University Press.

Straith, Hector. 1849. *Introductory Essay to the Study of Fortification, for Young Officers of the Army.* London: Parker, Furnivall and Parker.

Straub, Hans. 1964. *A History of Civil Engineering: An Outline from Ancient to Modern Times.* MIT Press.

Sturm, Leonhard Christofle. 1708. *Le Véritable Vauban se montrant au lieu du faux Vauban.* The Hague: Nicolas Wilt.

Suleiman, Ezra N. 1978. *Elites in French Society: The Politics of Survival.* Princeton University Press.

Tarbé de Saint Hardouin, F. P. H. 1884. *Notices biographiques sur les ingénieurs des Ponts et Chaussées.* Paris: Baudry.

Tartaglia, Niccolò. 1554. *Quesiti et Inventioni Diversi*, second edition. Published by the author. Reprint: Ateneo di Brescia, 1959.

Taton, René. 1947. "Une lettre inédite de Monge sur la situation en France, en 1791, après la fuite du roi." *Revue d'histoire des sciences* 1: 358–59.

Taton, René. 1951. *L'Oeuvre scientifique de Monge*. Presses Universitaires de France.

Taton, René, ed. 1964a. *Enseignement et diffusion des sciences en France au XVIIIe siècle*. Reprint: Hermann, 1986.

Taton, René. 1964b. "L'École du génie." In *Enseignment et diffusion des sciences en France au XVIIIe siècle*, ed. R. Taton. Hermann, 1986.

Thépot, André, ed. 1985. *L'ingénieur dans la société française*. Paris: Editions ouvrières.

Thoenig, Jean-Claude. 1973. *L'ère des technocrates: Le Cas des Ponts et Chaussées*. Paris.

Tiersonnier, Philippe. 1913. "Regemortes." *Revue du Centre*, 10 février: 220–231.

Timoshenko, Stephen P. 1953. *History of Strength of Materials*. McGraw-Hill.

Trincano, Didier-Grégoire. 1766. *Elémens de fortification, de l'Attaque et de la Défense des Places; Contenant les systêmes des auteurs les plus célèbres, neuf systêmes de l'Auteur, l'Analyse et la Comparaison de tous ces systêmes; la Fortification irregulière et celle de campagne; les changemens qui peuvent contribuer à la perfection des Ouvrages extérieures; le Calcul de mines, la poussée des terres et des voutes; l'art de camper et de retrancher une Armée; deux nouvelles méthodes de conduire les travaux d'un Siège; un Traité abregé du Lavis; un Dictionnaire des Termes de Fortification, de Guerre et d'Artillerie, etc*. Paris: Musier.

Truttmann, Philipe. 1976. *Fortification, architecture et urbanisme aux XVIIe et XVIIIe siècles: Essai sur l'oeuvre artistique et technique des ingénieurs militaires*. Thionville: Service culturel de la ville de Thionville.

Tuetey, Louis. 1908. *Les Officiers sous l'ancien régime: Nobles et roturiers*. Paris: Plon.

Van Creveld, Martin. 1977. *Supplying War: Logistics from Wallenstein to Patton*, Cambridge University Press.

Van Creveld, Martin. 1989. *Technology and war: from 2000 B.C. to the present*. Free Press; Collier Macmillan.

Van Creveld, Martin. 2000. *The Art of War: War and Military Thought*. Cassell.

van Hoof, J. P. C. M. 1984. "Fortifications in the Netherlands." *Révue internationale d'histoire militaire* 58: 97–123.

van Hoof, J. P. C. M. 1992. "Dutch fortifications." *Fort* 20: 87–96.

Vauban, Sébastien le Prestre de. 1685. *Le directeur général des fortifications*. La Haye: Henri van Bulderen.

Vauban, Sébastien le Prestre de. 1686. *Mémoire du Maréchal de Vauban sur les Fortifications de Cherbourg*, ed. J. Menant. Reprint: V. Didron, 1851.

Vauban, Sébastien le Prestre de. 1707. *Projet d'une dixme royale, qui supprimant la taille, les aydes, les doüanes d'une province à l'autre, les décimes du clergé, les affaires extraordinaires, & tous autres impôts onereux & non volontaires, et diminuant le prix du sel de moitié & plus, produiroit au roy un revenu certain et suffisant, sans frais, & sans être à charge à l'un de ses sujets plus qu'à l'autre, qui s'augmenteroit considerablement par la meilleure culture des terres*. Paris.

Vauban, Sébastien le Prestre de. 1968. *A Manual of Siegecraft and Fortification*, ed. G. Rothrock. University of Michigan Press.

Vauban, Sébastien le Prestre de. 1986. *Description géographique de l'élection de Vézelay, contenant ses revenus, sa qualité, les moeurs de ses habitants, leur pauvreté et richesse, la fertilité du pays et ce que l'on pourrait y faire pour en corriger la stérilité et procurer l'augmentation des peuples et l'accroissement des bestiaux*, ed. J.-F. Pernot. Association des amis de la maison Vauban.

Veblen, Thorstein. 1915. *Imperial Germany and the Industrial Revolution*. Macmillan.

Vérin, Hélène. 1984. "Le mot: ingénieur." *Culture technique* 12: 19–28.

Vérin, Hélène. 1991. "Technology in the park: Engineers and gardeners in seventeenth-century France." In *The Architecture of Western Gardens*, ed. M. Mosser and G. Teyssot. MIT Press.

Vérin, Hélène. 1993. *La gloire des ingénieurs: L'Intelligence technique du XVIe au XVIIIe siècle*. Albin Michel.

Vérin, Hélène. 1998. "Ingénieur." In *La science classique, XVIe–XVIIIe siècle: Dictionnaire critique*, ed. M. Blay et al. Flammarion.

Vidal, Laurent, and Émilie d'Orgeix. 1999. *Les villes françaises du Nouveau Monde: Des premiers fondateurs aux ingénieurs du roi (XVIe-XVIIIe siècles)*. Paris: Somogy éditions d'art.

Vignon, E. J. M. 1862. *Etudes historiques sur l'administration des Voies Publiques en France aux Dix-Septième et Dix-Huitième Siècles*. Dunod.

Vigny, Alfred de. 1835. *Servitude et grandeur militaires*. Paris: F. Bonnaire, V. Magen.

Vincenti, Walter G. 1990. *What Engineers Know and How They Know It: Analytical Studies from Aeronautical History*. Johns Hopkins University Press.

Virgin, Jean-Bernard. 1781. *La défense des places mise en equilibre avec les attaques savantes et furieuses d'aujourd'hui*. Stockholm.

Vitruvius. 1914. *The Ten Books on Architecture*. Reprint: Dover, 1960.

Voltaire. 1724. *La Henriade*. Geneva.

Walker, P. K. 1981. *Engineers of Independence: A Documentary History of the Army Engineers in the American Revolution, 1775–1783*. Washington: Historical Division, Office of Administrative Services, Office of the Chief of Engineers.

Weiss, John H. 1982a. "The lost baton: The politics of intraprofessional conflict in nineteenth-century French engineering." *Journal of Social History* 16: 3–20.

Weiss, John H. 1982b. *The Making of Technological Man: The Social Origins of French Engineering Education*. MIT Press.

Weiss, John H. 1984. "Bridges and barriers: Narrowing access and changing structures in the French engineering profession, 1800–1850." In *Professions and the French State 1700–1900*, ed. G. Geison. University of Pennsylvania Press.

Westfall, Richard S. 2001. "The background to the mathematization of nature." In *Isaac Newton's Natural Philosophy*, ed. J. Buchwald and I. B. Cohen. MIT Press.

White, Lynn, Jr. 1967. "Jacopo Aconcio as an engineer." *American Historical Review* 72: 425–444.

White, Richard Wallace. 1999. *Gentlemen Engineers: The Working Lives of Frank and Walter Shanly*. University of Toronto Press.

Wickenden, William Elgin. 1929. *A Comparative Study of Engineering Education in the United States and Europe.* Society for the Promotion of Engineering Education.

Wilkinson, Catherine. 1977. "The new professionalism in the Renaissance." In *The Architect*, ed. S. Kostof. Oxford University Press.

Williams, Rosalind. 1993. "Cultural origins and environmental implications of large technological systems." *Science in Context* 6: 377–403.

Williams, Rosalind. 2000. "'All that is solid melts into air': Historians of technology in the information revolution." *Technology and Culture* 41: 641–668.

Woodbridge, John. 1985. "La Conspiration du Prince de Conti (1755–1757)." *Dix-Huitième Siècle* 17: 97–109.

Woodbridge, John. 1994. *Revolt in Prerevolutionary France: The Prince de Conti's Conspiracy against Louis XV.* Johns Hopkins University Press.

Woronoff, Denis. 1984. *L'industrie sidérurgique en France pendant la Révolution et l'Empire.* Paris: Editions de l'EHESS.

Wright, Monte D., and Lawrence J. Paszek, eds. 1983. *Science, Technology, and Warfare, Proceedings of the Third Miltary Symposium, USAF Academy, 8–9 May 1969.* U.S. Government Printing Office.

Young, Arthur. 1927. *Travels in France during the years 1787, 1788, 1789.* London: Dent.

Zastrow, Heinrich Adolf von. 1848. *Histoire de la fortification permanente.* Paris: J. Corréard.

Zeller, Gaston. 1928. *L'organisation défensive des frontières du Nord et de l'Est au 17e siècle.* Paris: Berger-Levrault.

Zeller, Gaston. 1933. "La monarchie de l'ancien régime et les frontières naturelles." *Revue d'histoire moderne* 8: 305–333.

Index

Abbott, Andrew, 168–171
Académie française, 209, 335, 344, 345
Academy of Dijon, 334
Adams, Nicholas, 37, 123, 129
Aix, Island of, 277, 279, 302–324, 336, 342, 361–366, 384, 389, 408, 424
    Alberti, Leon Batista, 22–24
Alder, Ken, 4, 165, 403, 418, 420–426
Aménagement, 75, 76
American System of Manufacturing, 196
Angoulême, 263, 267
Angoumois, 266, 267
Apollonius of Perga, 251
Arches, 227, 230–231, 369, 370
Archimedes, 35
Architects, 22–25, 37, 145, 155, 390
    Architecture, 22, 25, 31, 234, 408, 430
Arcq, Philippe Auguste de Sainte-Foix Chevalier d', 176, 177, 210
Argenson, Marc-René de Voyer, comte d', 89–94, 97–101, 122
Argenson, René-Louis, marquis de, 263, 272
Arithmetic, 43, 44
Armies, 85, 160
  divisions of, 133, 134, 201
  general staff of, 135, 219, 385, 429
  national character and, 200, 209
  reform of, 7, 154, 178, 184
  size of, 123, 133
Arsenals. *See* Forts
Artillery, 14, 15, 102, 128, 129, 185, 194–197, 214, 287, 362, 372, 384, 387, 388
  Gribeauval system, 194–197, 332, 418
  guncarriages, 63, 332, 361–365, 370, 385
  military engineers and (*see* Military engineers)
  Montalembert and (*see* Montalembert)
  performance of, 46, 283
  schools of, 181, 229
Artois, Comte d', 272
Augoyat, André-Marie, 2, 50, 255
Austrian Empire, 423

Barré de Saint-Venant, Adhémar Jean-Claude, 224
Bastion, 14–16, 21, 283, 284, 287, 291, 342
Bayonet, 109, 110
Belair, A. P. Julienne de, 393
Belhoste, Bruno, 240
Bélidor, Bernard Forest de, 144, 165, 166, 183, 216, 219, 223–226, 256, 257, 265, 408
Belle-Isle (island), 277, 279
Belle-Isle, Charles Louis Auguste Fouquet, duc de, 90, 127, 191, 233
Belluzzi, Giovanni Battista, 27, 37
Bénézech de Saint-Honoré, Raymond-François-Honoré, 379, 385
Bensaude-Vincent, Bernadette, 260, 261
Berlin, 277
Bien, David, 154, 175–183, 403
Blanc, Honoré, 196
Blanchard, Anne, 2, 88, 94, 239
Blondel, François, 42
Borda, Jean-Charles de, 330, 334, 343
Bordeaux, 277

Bory, Gabriel de, 334, 397
Bossut, Charles, 95, 237, 397
Boucher, Hyacinthe, 395
Bouchet, Benoît-Louis de, 394
Boufflers, Marshal, 115
Bouguer, Pierre, 225
Bourcet, Jean de, 305
Bourcet, Pierre, 94, 213
Bousmard, Henri-Jean-Baptiste de, 255, 386
Brassac, 267
Brialmont, Henri-Alexis, 131
Brindley, James, 163
Brunelleschi, Filippo, 22–24, 236
Buache, Philippe, 247
Buchotte, Nicolas, 231–233
Buffon, Georges-Louis-Leclerc, comte de, 332, 334
Bureau brothers, 14, 124
Bureaucracy, 3, 63, 106, 122, 127, 138, 170–172, 411
Bureaux de Puzy, Jean-Xavier, 379, 389, 390

Cadet, Rosalie, 397
Cadet de Vaux, Antoine-Alexis-François, 397
Cambray, Chevalier de, 59, 60
Campredon, Jacques-Martin de, 394, 395
Canals, 68, 149, 168, 192, 431
Candide, 192, 431
Caponier, 26, 289, 291
Caponier. *See* Perpendicular fortification
Carcassonne, 150
Carnot, Claude-Marie, 395
Carnot, Lazare, 1, 57, 131, 202, 239, 255, 329, 334–336, 344, 345, 373–380, 385, 391–393, 399, 403, 407, 411, 419–421, 430, 432
Carnot, Sadi, 165
Carpilhet, Jacques de, 305, 311–315
Cartography, 65–66, 206, 255
Casemates, 56–58, 103, 285, 291, 342, 343, 363–366, 369–370, 384–385. *See also* Perpendicular fortification
Castiglione, Baldassare, 27
Castriotto, Jacomo Fusto, 30, 39
Cataneo, Pietro, 37
Cauchy, Augustin-Louis, 224

Caux de Blacquetot, Jean-Baptiste de, 380
Caux, Pierre-Jean de, 371, 378, 380
Cavaliers, 296
Cavendish, Henry, 239
Cessart, Louis-Alexandre de, 362
Chambre, Jean de, 387
Chamlay, Jules-Louis de Bolay, Marquis de, 127
Charente, 277, 279
Charles V, 123, 129
Charles XII, 192
Chastillon, Nicolas-François-Antoine de, 94, 96, 236, 239–246, 253–256
Chaulnes, Duc de, 266, 272
Chaussinand-Nogaret, Guy, 174
Chemistry, 260
Cherbourg Forts, 289, 292, 378, 381, 385
  Fort du Homet, 366
  Fort Royal (at Cherbourg), 362, 366–370
  Querqueville, 369, 370
Cherbourg, 50, 89, 258, 289, 298, 300–302, 361–370, 380, 388, 402
Chevallier, François, 88
Choderlos de Laclos, Pierre-Ambroise-François, 47, 317, 318, 320, 322, 334–336, 344, 345
Choiseul, Duc de, 120, 195, 276, 281, 282, 353, 362
Chotusitz, battle of, 200
Clairac, Chevalier de, 202
Clausewitz, Carl von, 191, 203, 204, 207, 211, 213
Clerville, Louis Nicolas de, 77, 78
Club of Rome, 166
Coehoorn, Menno, baron van, 63, 117, 118, 212, 389
Colbert, 63, 68, 77, 79
Comité des Fortifications, 188, 255, 379, 380, 394
Commissaire général des fortifications, 78, 80. *See also* Fortifications
Committee of Fortifications. *See* Comité des Fortifications
Committee of Public Safety, 1, 202, 380, 391, 392, 403, 421
Condé, Louis, Prince de, 127

Condillac, Étienne Bonnot de, 409
Condorcet, Jean-Antoine-Nicolas de Caritat, 189, 192, 210, 251, 343, 344, 353–356, 432
Conseil de Guerre, 190, 212, 356, 374
Conservative realism, 404, 431
Constant, Edward, 409, 410
Constituent Assembly. *See* National Assembly
Construction, 44, 155–158, 183, 218, 221, 227, 228, 240, 243, 406, 407, 432
Conti, Louis-François de Bourbon, Prince de, 263–266, 272, 276
Contractors, 99, 154–155, 183, 372, 390
Contrôle générale des finances, 90, 96, 400
Cook, Captain James, 329
Cormontaigne, Louis de, 57, 76, 89, 90, 95, 98–103, 119, 151, 202, 233, 255, 284, 335, 338, 339, 343, 347, 350, 369
Corps des Ponts et Chaussées, 3, 4, 95–97, 170, 220, 357, 362, 385, 396, 401, 402. *See also* Ingénieurs des Ponts et Chaussées
Corps du génie. *See* Corps of Military Engineering
Corps of Military Engineering, 1, 40, 139, 186–188, 217–218, 236, 283, 329–330, 337, 341, 357–360, 371–373, 376, 381, 385, 399–400, 429
  as corps à talents, 101, 138, 181–186, 419
  as service corps, 1, 8, 359
  recruitment and social origins, 82
  regulations for, 81, 91, 92, 150, 175, 241, 401
  unification with artillery, 93, 153–154
Corps of Roads and Bridges. *See* Corps des Ponts et Chaussées
Corps royal du génie. *See* Corps of Military Engineering
Corps savant. *See* Corps of Military Engineering
Corvée, 426
Corvisier, André, 63, 72, 179
Cosimo de' Medici, 27
Cosseron de Villenoisy, Louis-Pierre-Jean-Mammès, 99, 281, 294, 298, 300

Coulomb, Charles-Augustin de, 103, 139, 161, 165, 188, 216, 219, 220, 233, 234, 257, 311–315, 320, 321, 329, 333, 427
Coup d'oeil, 179, 213
Cremilles, Louis-Hyacinthe Boyer de, 91
Crublier, Henry, 395
Cuneo, battle of, 263

D'Alembert, Jean Le Rond, 237, 412
D'Arçon, Jean-Claude-Éléonor Le Michaud de, 123, 128, 153–155, 179, 188, 201, 241, 325, 332, 349, 350, 359, 380–387, 394, 399, 404
D'Asfeld, Claude-François Bidal, Marquis de, 85–93, 100, 121, 146, 148–150, 232, 233
Dahlberg, Erik, 293
Dajot, Louis-Lazare, 188
Danton, Georges Jacques, 420
Davidson, Frank P., 166
De Broglie, Marshal, 133, 200, 201
De Vigny, Alfred, 181
De Ville, Antoine, 45, 46, 56
Dead Ground, 27, 30
Deidier, L'abbé, 61
Desaudray, Gaulhard-Charles-Emile, 393
Descartes, Réné, 45, 127, 213, 420
Deshautchamps, Michel Vandebergue, 386
Design, 23, 30, 158, 240, 241, 243, 254, 406, 424, 432
Directeur général des fortifications des places de terre et de mer, 80, 140–142
Directors of Fortifications, 86, 96, 121, 142, 146–148, 187, 188
Disegno, 22, 23
Dobenheim, Alexandre-Magnus, 368–373, 385, 394–396
Douai, 195
Drawing, 27, 148, 243, 256
Du Breuil, Jean, 40–42
Du Buat, Pierre-Louis-Georges, 239, 254
Du Fay, L'abbé, 60, 61
Du Portail, Louis Lebègue, 2, 155–158, 179, 214–222, 256, 401
Du Teil brothers, 420
Du Vignau, Antoine-Nicolas-Bernard, 237, 246, 247, 251, 254, 386

Dubois, 91, 93
Dubois-Crancé, Edmond-Louis-Alexis, 379
Duc d'Uzès, 150
Duc de Bourgogne, 112
Ducarla, Marc, 247
Duffy, Christopher, 37, 47, 80, 126, 131
Duhamel du Monceau, Henri-Louis, 144, 333, 342
Dürer, Albrecht, 25–27, 34

École centrale des travaux publics. *See* École polytechnique
École des mines, 94, 394, 403
École des Ponts et Chaussées, 94–97, 394, 403, 408
École militaire, 180, 181
École Normale de l'an III, 250
École Polytechnique, 137, 170, 239, 245, 255, 257, 341, 386, 393–402, 405, 421
Edison, Thomas A., 413
Elias, Norbert, 171
Embrasures, 56, 366, 370. *See also* Perpendicular fortification
'Engineer', etymology of, 31, 165, 166, 180, 213
Engineering, 3–6, 37, 163–167, 402–403, 433
Engineering science, 160, 165, 223–261, 407
Engineers
  civil (*see* Ingénieurs des Ponts et Chaussées)
  fortress (*see* Military engineers)
  military (*see* Military engineers)
  naval, 95
  siege (*see* Military engineers)
  topographical, 95, 170
Errard de Bar-le-Duc, Jean, 32–37, 40, 41, 135, 183
Essai Général de Tactique, 209–214, 281
Euclid, 26, 31, 33
Eugene of Savoy, 71, 82, 115, 203
Euler, Leonhard, 224
Experience, 218, 228, 339, 340, 354, 432

Faucherre, Nicolas, 125
Favart, Jean-Baptiste, 120, 157, 394
Ferguson, Eugene, 21
Feuillade, Duc de la, 173, 180
Filarete, Antonio, 16
Filley, Augustin de, 152–158
Filley, Pierre de, 117, 326–328, 336, 362
Flachon de la Jomarrière, Ferdinand, 163
Flanders, 71, 72, 131, 203, 212, 213
Foissac de Latour, François-Philippe de, 395
Folard, Chevalier de, 192, 204, 209
Fontenelle, Bernard le Bovier de, 48, 51
Forrester, Jay W., 166
Fortification(s), 6, 240
  administration of, 77–81, 90, 91, 100, 144, 156, 423
  ancient, 13
  artillery, 9, 15–22,
  classification of, 41
  as conservative art, 13, 189, 399, 404, 405, 429
  costs of, 40, 49, 50, 312, 374
  defilading, 245–256, 341, 397
  Dutch, 32, 39, 42
  features of, 13, 14, 128, 129
  field, 133, 202
  German, 32
  and Guibert (*see* Guibert)
  illustrations and drawings of, 148, 368, 385
  impregnability of, 31, 282, 383, 390, 391, 405
  Italian Renaissance, 9, 16–18, 53, 129, 283
  maxims of, 41, 59, 60, 110, 111
  Medieval, 14
  modern, 57, 101
  national traditions of, 32, 41
  perpendicular (*see* Perpendicular fortification)
  plan views or ground plans of, 283, 284
  principles of, 13, 14, 43, 397
  profile views of, 246
  quality of, 26, 119, 120, 348, 349
  revolutionary courses on, 394
  science of, 6, 26–28, 33, 44, 260, 338, 339, 349–351
  Spanish, 32, 39
  and strategy (*see* War)
  systems of, 57, 101, 254
  treatises on, 25–31, 51, 58–61, 218

Fortress engineers (ingénieurs des places), 78, 79, 86, 144, 145, 220, 377
Forts and fortresses, 292–294, 374
  Acre, 131
  Alessandria, 131
  Antwerp, 16, 39, 131
  Belfort, 55, 60
  Berg(en)-op-Zoom, 89, 117, 118, 205
  Besançon, 48, 55
  Brest, 50, 277, 302
  Candia (Iraklion, Crete), 109
  Charleroi, 60, 130
  Château du Taureau, 57
  Denia, 370
  Dunkerque, 131
  Fontarbia, 370
  Fort Condé, 293
  Fort Conti, 296
  Fort Royal (proposed), 298
  Fort Royal at Cherbourg (*see* Cherbourg)
  Genoa, 131
  Giron, 370
  Island of Aix (*see* Aix)
  Landau, 55, 60, 128, 342
  Longwy, 151–153, 157
  Louisville, 296, 298, 311, 342
  Metz, 89, 98
  Montmédy, 50
  Namur, 64, 113
  Neuf-Brisach, 48, 54–57, 90, 287, 342
  Olmütz, 342
  Palmanova, 16
  San Sebastian, 370
  temporary (du moment), 213
  Toulon, 50
  Tour de Camaret, 57
  Turin, 16
  Valenciennes, 131
Fourcroy de Ramecourt, Charles-René, 62, 91, 101, 102, 119–122, 187–189, 212, 265, 306, 312–318, 333, 334, 343, 356, 373–376, 383, 395, 396, 411, 430, 432
Fournier, Georges, 40
Francesco de' Marchi, 28, 42
Francesco di Giorgio, 21

Frederick the Great (Frederick II), 130, 200, 201, 276, 377, 423
Frézier, Amédée-François, 234–237
Fronde, 124, 127
Fuller, J. F. C., 191, 194

Galileo, 27, 31, 32, 45, 165, 223
Garavague, Jean-François, 370
Garcin, Nicole, 151
Gardens, 68, 69
Gat, Azar, 205
Gaudin, L'abbé, 392
Gaxotte, Pierre, 139
Gayvernon, Simon-François, 245, 246, 255
Genius, 179, 180, 213, 214, 396
Geometry, 9, 24, 26, 33, 35, 207, 235, 255
Gerstl, Joel E., 169
Gibraltar, 279, 312, 334, 349, 387
Gillispie, Charles, 159, 362, 421
Gobert, Jacques-Nicolas, 395
Gourdon de l'Eglizière, Jean, 119
Grenier, Pierre-Dominique de, 227, 334, 337, 347, 350, 383, 408
Gribeauval, Jean-Baptiste Vacquette de, 6, 161, 185, 195, 212, 281–284, 305, 317, 396, 399, 423
Guerlac, Henry, 61
Guibert, Jacques Antoine Hippolyte de, 6, 131, 161, 193, 200–202, 207, 209, 212–214, 374, 399, 404
Guillerme, André, 76
Guncarriages. *See* Artillery
Guntowers, 289–293. *See also* Perpendicular fortification
Gustavus Adolphus, 197

Hale, John Rigby, 15, 17, 25, 146
Hall, A. Rupert, 245
Hassenfratz, Jean-Henri, 239
Havrincour, d', 276
Hebert, 58
Henri IV, 40, 62, 133
Herbort, Jean-Antoine d', 351
Hüe de Caligny, Louis-Roland, 67, 128, 129
Hughes, Thomas Parke, 3–4, 51, 166–167, 412–415, 427, 428

528   Index

Huntington, Samuel, 165, 171, 182–184, 404, 431

Ile d'Aix. *See* Aix
Ile Pelée, 362, 367, 368
Ingénieurs de la marine. *See* Engineers
Ingénieurs des places. *See* Military engineers
Ingénieurs des ponts et chaussées, 75, 76, 97, 135, 138, 139, 258, 369, 390, 401, 402, 426
Ingénieurs des tranchées. *See* Military engineers
Ingenieurs géographes. *See* Engineers
Innovation, 3, 16, 193, 195, 209, 210, 273, 315, 340, 368, 376, 422
Institut National, 393, 397, 419
Interchangeable parts, 4, 425 *See also* American System

Jacobin Club, 388
Jervis, John, 170
Joseph II, 130

Keegan, John, 105
Kepler, 251
Kéralio, Louis Félix Guinement de, 346, 350, 352, 353, 381
Kipling, Rudyard, 429
Kuhn, Thomas S., 258–260, 409–410

L'Ami de l'Art Défensif ou Observations sur le Journal Polytechnique de l'École Centrale des Travaux Publics, 395–397
L'Hôpital, Guillaume-François-Antoine de, 226
La Fère, artillery school at, 229
La Fortification perpendiculaire, 281–302, 324–333, 343–352, 366, 381, 388, 393
La Fortification reduicte en Art et Demonstrée, 32–37
La Hire, Philipppe de, 66, 227, 234
La Hogue, 362
La Rochefoucauld, Duc de, 332
La Rochelle, 40, 277, 302
La Science des Ingénieurs, 224–234

La Tour Dupin-Gouvernet, Jean-Frédéric de, 320, 321, 361, 370
Lafayette, Marquis de, 2, 399
Lafitte-Clavé, André-Joseph de, 188
Lalande, Joseph-Jerome-François de, 397, 398
Lalonde, 43
Lamberini, Daniela, 145, 146
Languedoc, 149, 151
Lanteri, Giacomo, 26
Lapisse, Antoine-Pierre de, 395
Laplace, Pierre-Simon, 181, 343, 354, 419
Lavoisier, Antoine-Laurent, 160, 161
Layton, Edwin, Jr., 170
Le Directeur général des fortifications, 100, 140–142
Le Michaud, *See* D'Arçon
Le Nôtre, André, 68, 126
Le Peletier de Souzy, Michel, 80–85, 140, 146–148
Le Roy, Jean-Baptiste, 330–334, 343
Le Sancquer, Jacques-Julien, 91, 312
Leblond, Guillaume, 42, 54
Légier, Pierre-Joseph, 380
Léonard, Émile, 127
Leonardo da Vinci, 21
Lepetit, Bernard, 75
Lille, 110, 115, 129, 375, 379, 380
Lines of defense, 55
Lines of fire, 30
Lloyd, Henry Humphrey Evans, 205, 206
Lombard, Jean-Louis, 420
Londes, Etienne Poisson des, 370
Louis XIV, 64, 65, 68, 80, 85, 112, 124, 133, 202, 203, 371, 387
Louis XV, 50, 88, 89, 195, 258, 277
Louis XVI, 89, 258, 281, 361
Louvois, 77, 79, 128, 421
Lycée des Arts, 393

Maastricht, 71, 108
Machault d'Arnouville, Jean-Baptiste, 271, 273
Machiavelli, Niccolò, 20, 191
Madone de Ulmo, battle of, 263

Maillebois, Yves-Marie-Desmarets, comte de, 330
Maintenance, 155, 406, 416
Mandar, Charles-François, 322, 408
Marat, Jean-Paul, 341, 354, 355
Maria Theresa, 195
Mariotte, Edmé, 224
Maritz, Jean, 268–276
Marlborough, Duke of, 71, 83, 115, 129
Marolois, Samuel, 42
Martinet, Colonel, 126
Martinique, 427
Masse, Claude, 142
Mathematics, 9, 27, 32, 35, 36, 41, 181, 224–226, 229
Maurepas, Jean-Frédéric Phélypeaux, comte de, 266, 281, 302, 305, 306, 320, 361, 384, 389
Maurice de Saxe, 129–133, 191, 201–205, 209, 219, 293
Maurice of Nassau, 27, 197
Mauritius, 303
Maurois, André, 431
Mazarin, Cardinal de, 124, 128
McNeill, William, 196
Mémoire pour servir d'instruction dans la conduite des sièges et dans la défense des places, 110–113
Mémoires sur la fortification perpendiculaire par plusieurs officiers du Corps royal du génie, 336–343, 347, 368, 375, 395
Merit, 178, 180
Mesmer, Franz Anton, 341, 354–356
Mesnil-Durand, F.-J. de Graindorge d'Orgeville, 200, 201
Metz School of Military Engineering and Artillery, 255, 386, 402
Meulan d'Ablois, Marie-Pierre-Charles, 320
Meusnier, Jean-Baptiste, 239, 247, 249, 250, 254, 255, 333, 362–366, 370, 378, 385
Mezalectre, 336
Mézières (school and fortress), 89, 94–98, 151, 181, 236, 237, 239, 240, 251, 255, 256, 373, 386, 394, 400
Michelangelo, 25

Milet de Mureau, Louis-Marie-Antoine d'Estouf, 386, 395
Military engineers, 31, 32, 111, 215, 257–259. *See also* Fortress engineers; Siege engineers
  artillerists and, 418–428
  attitudes towards and status of, 86, 87, 155, 180, 186, 187, 389, 390
  as cartographers, 66
  casualties among, 89, 115–117
  as conservatives, 4, 161, 223, 399, 403–408, 417, 422, 425–427
  and "details," 134, 153–158, 221, 259, 404, 407, 408
  Dutch, 32, 42
  examinations of, 81, 88
  Italian, 39
  legacy of, 422, 430, 431
  of Colbert, 78, 79, 86
  of Louvois, 78, 79, 86
  numbers of, 186, 187
  origins of, 87, 93
  profession of, 169, 183
  qualities of, 34, 79, 145, 153, 154, 164
  ranks of, 86, 87
  self-image of, 137
  systems of, 412–417
  title of, 31, 183, 185
Military nobility, 175–177
Milliet de Chales, Claude-François, 40–42
Milly, Nicolas-Christian de Thy, comte de, 333
Mines, countermines, and miners, 46, 103, 220, 400
Mirabeau, Honoré-Gabriel-Victor Riquetti de, 322, 361, 377–380, 383, 399
Molière, 174
Moments of fortification, 102, 338, 348
Monge, Gaspard, 95, 235, 237, 239–240, 247, 250–256, 259, 386, 422
Montalembert, Jean-Charles, baron de, 352–355, 381
Montalembert, Marc-René, marquis de, 5, 85, 117, 131, 159, 161, 195, 206, 263ff., 335ff., 343, 360, 388, 398, 408–411
  and artillery, 133, 268–270, 301

Montalembert, Marc-René, marquis de (cont.)
  as cannon founder, 265–276
  French Navy and, 387
  Royal Academy of Sciences and, 265, 274, 322, 351–357, 370–373, 397
  Seven Years' War and, 270, 276–279
  and Vauban, 47, 57, 347, 350, 351, 389
  workers and, 319, 320, 424
Montalembert, Marie-Josephine de Comarieu, marquise de, 352, 384
Montalembert, Paul de, 272
Montbarey, Prince de, 305
Montecuccoli, Raimondo, 191
Montesquieu, Charles de Sécondat, baron de la Brède et de, 211
Morlet de Boisset, Michel-François Beaudet de, 386, 395
Mousnier, Roland, 75, 138
Müller, John, 48

Napoleon I, 130, 131, 196, 201, 202, 206, 213, 322, 379, 393, 411, 419, 420, 422
National Assembly, 371, 372, 375–381, 386, 389
National Convention, 379, 380, 392, 420, 421
Natural frontiers, 73–75, 214
Naudin, 42, 60
Navier, C.-L.-M.-H., 216, 223, 257
Navies, 266–279, 362
Netherlands, 129
Newton, Isaac, 213, 224, 339
Noble, David, 170
Nobles, 173–179, 185, 211
Noizet de Saint-Paul, Jean-François-Gaspard, 255, 392, 395
Noizet, F., 249, 253
Nollet, abbé Jean-Antoine, 95, 259
Novelty. *See* Innovation

Officers, 154, 160, 182–186, 215, 419
  Officier réformé, 86
  Oléron, Island of, 277, 279, 300, 323, 353
Ordre du Tableau, 126, 178
Ordre mince, 197, 200
  Ordre profond, 197, 200

Paciotto, Francesco, 16, 39
Pagan, Blaise-François, 46, 47, 53, 58
Palladio, 25
Parent, Antoine, 224, 229
Parker, Geoffrey, 123
Patronage, 87, 161, 272, 275, 306
Pepper, Simon, 37, 129
Périgord, 266
Perpendicular fortification, 281–302, 373, 384
Perronet, Jean-Rodolphe, 96, 97, 265, 334, 402
Perrucci, Robert, 169
Peter the Great (Peter I), 210
Petty, William, 66
Picard, Jean, 66
Picon, Antoine, 5, 257, 373, 427
Pierron de la Chambre, Elie-Marie, 394
Piètre, 91
Pirch, Baron de, 200, 201
Pitot, Henri, 224
Poisson, Siméon-Denis, 224
Pompadour, Madame de, 266, 272, 306
Poncelet, Jean-Victor, 165
Pondichéry, 303
Portfolio culture, 100, 101
Ports. *See* Forts
Priestley, Joseph, 343
Prieur de la Côte-d'Or, Claude-Antoine, 1, 403, 420
Professions, 22, 168, 169, 183
Prony, Gaspard Riche de, 216, 334, 397, 408
Provveditore di fortezze, 44, 145, 146
Prussia, 377, 423
Puységur, Jacques-François- Chastenet de, 193, 203–205, 209

Quiberon, 277

Racine, 64, 74, 112
Ramelli, Agostino, 21, 33, 35
Ramsault, Charles-Antoine de Rault de, 90, 92, 149
Rankine, W. J. M., 165
Regemorte (or Regemortes) family, 90, 92
Reinhard, Marcel, 345, 373
Reix des Fosses, 266, 267, 270

Réponse au Mémoire sur la Fortification Perpendiculaire par plusieurs Officiers du Corps Royal de Génie, 343–352, 371
Révérony Saint-Cyr, Baron Jacques-Antoine, 137
Revolutions
 French, 177, 188–190, 202, 215, 250, 322, 356–360, 371, 373, 377–380, 388–393, 400, 403, 405, 408, 419–426, 429
 Industrial, 170
 Scientific, 160, 258, 420
 Second Industrial, 171
Richelieu, Cardinal de, 56, 107, 110, 124, 301
Riquet, Pierre-Paul, 62
Robespierre, Maximilien, 392, 420
Rochefort, 266–271, 277, 302, 311, 314, 320
Roffignac, 267
Rogers, G. F. C., 406
Rohault, Jacques, 226
Rosen, Howard, 196
Rosenberg, Nathan, 406
Rossbach, 181
Rouillé, Antoine-Louis, comte de, 266, 271
Rousseau, Jean-Jacques, 179
Royal Academy of Sciences, 3, 96, 144, 196, 224, 225, 229, 232–234, 237, 251, 311, 330–334, 337, 343–346, 361, 363
Royal Corps of Artillery and Military Engineering, 93
Royal Corps of Military Engineering. *See* Corps of Military Engineering
Rozières, Pierre-Louis Antoine de, 380
Ruelle, 267–276

SAGE, 166, 428
Saint Malo, 302, 312
Saint-Auban, Marquis de, 332
Saint-Germain, Claude-Louis, comte de, 91, 99, 122, 133, 149, 153, 154, 178, 184–189, 195, 197, 241, 357, 400
Saint-Pierre, Charles-Irénée Castel de, 192
Saint-Simon, Henri de, 389
Saint-Simon, Louis de Rouvroy, duc de, 47, 48
Sakharovitch, Joël, 243
Sané, Jacques Noël, 397
Sangallo, Antonio da, 22

Sanlot, Auguste-Thomas-Nicolas, 395
Sappers, 400
Sartine, Antoine Jean de, 302, 305
Sauveur, Joseph, 52, 80
Say, Horace, 247, 253, 395, 397
Scamozzi, Vincenzo, 16
Schalk, Ellery, 173
Science, 183
 Baconian, 258, 259
 Newtonian, 191
Ségur, Philippe Henri, marquis de, 175–178, 193, 318–320, 335, 386
Senneton de Chermont, Antoine-François Lemercier de, 114
Senneton de Chermont, Louis-Thomas Lemercier de, 117, 380
Senneton de Chermont, Pierre-Claude Lemercier de, 117
Séré de Rivières, Raymond-Adolphe, 131
Seuillet, Grégoire, 313, 321, 408
Shading, 241, 243
Siegecraft, 34, 56, 105–135, 301
Siege engineers (ingénieurs des tranchées), 78, 79, 220, 377
Siege journals, 106, 113–119, 340
Siege parallels, 71, 108
Sieges. *See also* Forts
 Cartagena, 341
 Château Dauphin, 263
 Demonte, 263
 imaginary, 119, 120, 347, 350
 Kehl, 263
 Mantua, 131
 Maubeuge, 130
 Mons, 114–117, 129
 Philippsburg, 263
 Schweidnitz, 195
 Sevastopol, 131
 Siena, 37
 Tournai, 341
 Villefranche, 263
 Yorktown, 215
 Zaragoza, 131, 202
Siege warfare, 122–133
Smeaton, John, 170
Smiles, Samuel, 163, 429

Soltikov, Prince, 277
Speckle (Specklin), Daniel, 32
Standardization, 100, 101, 142
Stengers, Isabelle, 260, 261
Stereotomy, 234–237, 240, 243
Sterne, Laurence, 113
Strasbourg, 74, 195, 270, 273
Sully, Maximilien de Béthune, duc de, 40, 68
Systems engineering, 45, 428

Tables, 111, 112, 228, 231
Tactics, 197–201, 214, 217
Talent, 179–181
Tarade, Jacques, 146
Tartaglia, Niccolò, 13, 26, 31
Taylor, Frederick W., 427
Technical fantasy, 21, 26, 54
Technological momentum, 4
Terrain, 213, 214, 218, 247, 254, 255
Tholozé, David-Alexis, 381–386
Thomassin, 39, 51, 55
Tinseau, Charles-Marie de, 239, 333
Topography. *See* Terrain
Tower bastions, 54–56
Trace italienne. *See* Fortification
Tressan, Louis-Élisabeth de la Vergne, comte de, 330
Trudaine, Daniel, 96, 402
Truttmann, Philippe, 62, 89
Turenne, Marshal, 127
Turgot, Anne-Robert-Jacques, 159, 189, 329, 426

Vallière, Jean-Florent de, 195, 233, 265
Vallière, Joseph-Florent de, 195
Van Creveld, Martin, 123, 124
Varignon, Pierre, 226, 232, 234
Vasari, Giorgio, 22
Vassieux, maneuvers at, 200
Vauban à Charleroi, 137
Vauban, Sébastien le Prestre de, 46–51, 60, 61, 76–78, 107–113, 126, 212, 220, 233, 334–336
  activities of, 48, 49, 80, 152
  and frontiers, 65, 66, 71
  Louis XIV and, 62, 63, 108, 122–125, 128, 180, 421
  Louvois and, 62, 63, 70
  method of, 55, 58, 62, 72
  and Paris, 73
  politics of, 74, 75
  and pré carré, 70–74
  reputation of, 47–49, 82, 83, 157
  and statistics, 66–68
  systems of, 51–54, 103
Veblen, Thorstein, 128, 433
Vérin, Hélène, 69, 163, 165, 171
Villars des Brosses, 275
Vincenti, Walter, 406
Virgin, Johan Bernhard, 131
Vitruvius, 24, 25
Voltaire, 179, 191–192, 209
Voluntarism, 213
Von Bülow, Adam Heinrich Dietrich, 206, 207
Voyer, Marc-René, marquis de, 319–322

War(s), 105, 160
  American Revolutionary, 95, 215, 311
  Dutch, 71
  fortification and, 202, 404, 405
  of Devolution, 71
  of Austrian Succession, 89, 115, 122, 203, 263, 266, 330
  of League of Augsburg, 80, 82
  of Polish Succession, 85, 263
  of Spanish Succession, 71, 74, 82, 86, 115, 212
  Peninsular, 131
  science of, 27, 60–62, 191–194, 203–211
  Seven Years', 95, 133, 151, 154, 181, 182, 186, 191–193, 241, 275, 323, 362, 374, 418, 423
Washington, George, 215
Westfall, Richard, 65
White, Lynn, 45
Wilkinson, Catherine, 23
William of Orange (William III of England), 114
Williams, Rosalind, 163, 166–168, 415, 428, 429

Zeller, Gaston, 62, 73